541.124.011.2(061.3)
C23914

No. B6759

B6759
23/11

ROCKET PROPULSION ESTABLISHMENT LIBRARY

Please return this publication, or request a renewal, by the date stamped below.

Name	Date
Mr Pickering	26-6-78

(4/64) L23964 442077 Wt29280 D7061 10/64 10M T&Co **G871**.   R.P.E. Form 243

REACTIVITY OF SOLIDS

REACTIVITY OF SOLIDS

Edited by

John Wood
Chalmers University of Technology
Goteborg, Sweden

Oliver Lindqvist
Chalmers University of Technology
Goteborg, Sweden

Claes Helgesson
Arbman Development AB
Stockholm, Sweden

and

Nils-Gösta Vannerberg
Chalmers University of Technology
Goteborg, Sweden

Plenum Press · New York and London

Library of Congress Cataloging in Publication Data

International Symposium on the Reactivity of Solids, 8th, Chalmers University of Technology, 1976.
Reactivity of solids.

Includes indexes.
1. Solids—Congresses. 2. Reactivity (Chemistry)—Congresses. I. Wood, John. II. Lindqvist, Oliver. III. Title.
QD478.I57 1976   541'.042'1   77-785
ISBN 0-306-31021-X

Proceedings of the Eighth International Symposium on the Reactivity of Solids held at Chalmers University of Technology, Goteborg, Sweden, June 14—19, 1976

© 1977 Plenum Press, New York
A Division of Plenum Publishing Corporation
227 West 17th Street, New York, N.Y. 10011

All rights reserved

No part of this book may be reproduced, stored in a retrieval system, or transmitted, in any form or by any means, electronic, mechanical, photocopying, microfilming, recording, or otherwise, without written permission from the Publisher

Printed in the United States of America

ORGANIZING COMMITTEE

President: **PROFESSOR ARNE MAGNÉLI**
University of Stockholm

Chairman, and Chairman of the
International Advisory Committee: **PROFESSOR GEORG LUNDGREN**
University of Gothenburg

Retiring Chairman of the
International Advisory Committee: **PROFESSOR PAUL HAGENMULLER**
University of Bordeaux

Executive Committee:

Chairman: **PROFESSOR NILS-GÖSTA VANNERBERG**
Chalmers University of Technology

Secretary: **DR. CLAES HELGESSON**
Arbman Development AB

Members: **DOCENT GÖRAN GRIMVALL**
Chalmers University of Technology

DOCENT OLIVER LINDQVIST
Chalmers University of Technology

DR. JOHN WOOD
Chalmers University of Technology

Members of the Organizing Committee:

DR. STEN ANDERSSON
University of Lund

DR. BERTIL ARONSSON
Uddeholm AB

PROFESSOR HELMUT FISCHMEISTER
Chalmers University of Technology

PROFESSOR MATS HILLERT
*Royal Institute of Technology
Stockholm*

DR. BO HOLMBERG
*Research Institute for the
Swedish National Defence
Stockholm*

DR. ROLAND KIESSLING
Sandvik AB

PROFESSOR STIG LUNQUIST
Chalmers University of Technology

PROFESSOR STIG RUNDQUIST
University of Uppsala

Preface

This volume contains all the papers presented at the 8th International Symposium on the Reactivity of Solids. The Symposium, the latest of a series started in 1948, was held at Chalmers University of Technology, Goteborg, Sweden, between June 14th and June 19, 1976.

This was the second occasion on which a symposium in this series had been held in Goteborg. On the first occasion, in 1952, the Symposium was arranged by Professor J. Arvid Hedvall, the pioneering scientist in the field of Solid State Chemistry, Professor at Chalmers University since 1929 and one of those responsible for promoting the early symposia of this series. It was a cause of regret to his friends and colleagues that he died in the interval between the venue for the Eighth Symposium being announced and the meeting taking place. In memory of Arvid Hedvall, an exhibition of his books and writings was arranged and put on display for the benefit of participants in the Symposium.

The Symposium comprised 112 papers, about 60% of those originally submitted, and was attended by approximately 250 participants from all over the world. For the first time in the series, papers were presented in two forms, in overlapping but not parallel sessions. Approximately one half of the papers were presented in the form of a poster display. For these, each paper was displayed for longer periods to allow for all the material to be digested.

The Symposium was sponsored and financially supported by the Swedish Natural Science Research Council (NFR), Chalmers University of Technology and Goteborg University, and the Swedish Department of Education. The Organising Committee are also grateful to the following companies who assisted the Symposium both financially and materially, and who enabled an exhibition of current apparatus and techniques to be staged.

Arbman Development AB
Eckersteins Universitetsbokhandel AB
Hugo Tillqvist AB
Kjellbergs Successors AB
Mettler Instruments AG, Switzerland
Scandinavian Airlines Systems, SAS
Spectrospin AB
Springer Verlag, Berlin
Svenska Philips AB
Syntex, Inc.

The Organising Committee thank the large number of people who worked to provide the facilities for the Symposium. The success of any Symposium, however, depends principally on the participants. Therefore to the participants, especially those who performed special functions, Plenary Lecturers, Session Chairmen and the authors of contributed papers, the Organising Committee extend their grateful thanks.

Contents

New Crystallographic Developments Applicable in
       Studies of Reactions in Solids
       (Introductory Lecture) 1
  A. Magnéli

SECTION 1: REACTIONS AT SURFACES AND INTERFACES, ESPECIALLY AT ELEVATED TEMPERATURES

Reactions at Surfaces and Interfaces (Plenary
       Lecture) 15
  P. Kofstad

Reaction Kinetics in the Ca-Mn-O System 43
  J. M. Longo and H. S. Horowitz

Surface Reactivity Towards Olefin Oxidation of
       Cadmium Molybdate Doped with Transition
       Metal Ions 49
  L. Burlamacchi, G. Martini, S. Simonetti,
  and B. Tesi

Chemisorption of Sulfur on Iron and Its Influence
       on Iron-Gas Reactions, Surface Self
       Diffusion and Sintering of Iron 55
  H. J. Grabke, W. Paulitschke, and S. R. Srinivasan

The Influence of Intrinsic Defects on the Mechanism
       of the Solid State Reaction Between
       CdTe and HgSe 63
  V. Leute and W. Stratmann

Characterization and Surface Reactivity of
       Finely-Divided CoO-MgO Solid Solutions 69
  A. P. Hagan, C. O. Arean, and F. S. Stone

The Reactivity of Some Tungsten Oxides 77
   M. Schiavello, S. De Rossi, B. A. De Angelis,
   E. Iguchi, and R. J. D. Tilley

The Growth of Oxide Crystallites During the
       Oxidation of Iron in Carbon Dioxide
       Within the Scanning Electron Microscope 83
   A. M. Brown and P. L. Surman

The Reactivity of Metal Oxides and Sulfides with
       Lithium at 25°C - The Critical Role of
       Topotaxy 89
   M. S. Whittingham and R. R. Chianelli

On the Reaction Between Silver and Copper Iodide
       in Thin Evaporated Films 95
   U. Pittermann and K. G. Weil

Oxygen Diffusion in Strontium Titanate Studied
       by Solid/Gas Isotope Exchange 101
   R. Haul, K. Hübner, and O. Kircher

Oxidation Behaviour of Zirconium Nitride
       in Oxygen 107
   J. G. Desmaison, M. Billy, and W. W. Smeltzer

Oxidation Behaviour of Nickel Tellurides in
       the Composition Region Between $NiTe_{0.8}$
       and $NiTe_{2.0}$ 113
   D. Kolar, V. Marinkovič, and M. Drofenik

Reactions and Equilibrium Wetting in
       Titanium - Molten Glass Systems 119
   A. Passerone, G. Valbusa, and E. Biagini

Oxidation Kinetics Study of Finely-Divided
       Magnetites Substituted for Aluminium
       and Chromium 125
   B. Gillot, A. Rousset, J. Paris, and P. Barret

Crystalline Solid Rearrangement Brought About
       by Gas Removal. Shape and Size of the
       Crystallites Obtained 131
   J. C. Niepce, J. C. Mutin, and G. Watelle

Peroxide and Superoxide as Catalysts. A Model
       Based on Molten Salts Experiments 137
   E. Desimoni, F. Paniccia, and P. G. Zambonin

CONTENTS

Carbon Formation on Ni Foils by Pyrolysis
      of Propene (290° – 800°C) 143
  T. Baird

Oxidation Reaction of Chromium Oxide (Cr_2O_3) in the
      Presence of MM'O_3 (M : Ca, Sr, Ba; M' : Ti, Zr) . . . 149
  T. Nishino, T. Sakurai, and S. Nishiyama

Influence of a Foreign Gas in Heterogeneous
      Reactions 155
  B. Guilhot, R. Lalauze, M. Soustelle,
  and G. Thomas

About the Use of a Solid State Electro-Chemical
      Oxygen Pump for Selective Oxidation of
      Organic Materials 161
  S. Pizzini, C. M. Mari, and D. C. Hadjicostantis

On the High Temperature Reactivity in Air of Some
      Metal Powders and Their Relationship with
      the Self-Sealing Coatings for the
      Protection of High-Temperature Metal Surfaces . . . 167
  G. Perugini

Kinetic Aspects of Some Reactions Between
      Metals and Gaseous Carbon Disulphide 171
  A. Galerie, M. Caillet, and J. Besson

Oxidation of Cobalt at High Temperatures and
      Self-Diffusion in Cobaltous Oxide 177
  S. Mrowec and K. Przybylski

Cation Distribution in Spinel Solid Solutions
      and Correlations Between Structural and
      Catalytic Properties 183
  F. Pepe, P. Porta, and M. Schiavello

Thermal Decomposition of Mg-Al-Formate and Low-
      Temperature Formation of Spinel Thereby 191
  Z. G. Szabó, B. Jóvér, J. Juhász, and K. Falb

Reaction Between Cemented Carbides and
      Molten Copper 197
  T. Yamaguchi and M. Okada

Kinetics of TiB Formation 203
  G. C. Walther and R. E. Loehman

Description of Solid-Gas Reactions. Conclusions
 of Thermoanalytical and Electron Optical
 Studies on Calcium Carbonate Thermal
 Decomposition 209
  Gy. Pokol, S. Gál, J. Sztatisz, L. Domokos,
  and E. Pungor

Dynamic Behaviour of Solid Surfaces 215
  K. H. Lieser

Nucleation Process in the Decomposition of
 Nickel Formate 221
  M. E. Brown, B. Delmon, A. K. Galwey,
  and M. J. McGinn

Kinetic Study on the Phase Transformation of GeO_2 227
  M. Yonemura and Y. Kotera

Kinetic Aspects on the Formation of Solid Solution 233
  M. Yonemura and Y. Kotera

SECTION 2: INFLUENCE OF STRUCTURAL DEFECTS ON THE REACTIVITY OF SOLIDS

Influence of Structural Defects on the
 Reactivity of Solids (Plenary Lecture) 237
  H. Schmalzried

Studies on Non-Stoichiometric Strontium
 Ferrate, $SrFeO_{3-x}$ 253
  B. C. Tofield

Electron Microscopy of Ferroelectric Bismuthate
 Compounds 261
  J. S. Anderson, J. L. Hutchison, and C. N. R. Rao

Some Observations on the Chemical Reactivity of
 Nickel Powder Mechanically Treated by
 Grinding 267
  J. Carrión, J. M. Criado, E. J. Herrera, and C. Torres

Defect Chemistry of La-Doped Barium Titanate 273
  D. Hennings

The Kinetic Processes of Equilibrium Restoration
 in La-Doped $BaTiO_3$ 279
  R. Wernicke

CONTENTS

High Temperature Redox-Reactions of Co:α-Al$_2$O$_3$ 285
   A. Schweiger and H. H. Günthard

Structural Evolution in the FeF$_3$-WO$_3$ System 289
   J. M. Dance, A. Tressaud, J. Portier,
   and P. Hagenmuller

Diffusion and Reactivity in Perovskite Materials of
       the Composition (Pb,La)(Zr,Ti)O$_3$ 297
   K. J. de Vries, A. H. M. Hieltjes, and
   A. J. Burggraaf

Formation of Large Defects in Neutron Irradiated
       Alkali Halide Crystals 305
   F. W. Felix, W. E. Monserrat Benavent

Mechanism and Cationic Rearrangements Involved
       in the Reduction of Co^{2+}[Co$^{3+}_{2-x}$Cr$^{3+}_x$]O$^{2-}_4$
       Spinels 311
   P. Bracconi, L. Berthod, and L. C. Dufour

Correlations Between the Kinetics of Decomposition
       and the Crystal Structure for Some
       Inorganic Azide Single Crystals 319
   H. T. Spath

Influence of P_{O_2} and P_{H_2O} in the Atmosphere
       on the Rate of Changes in the System
       CaO-SiO$_2$ 325
   V. Jesenák and Z. Hrabě

Autocatalytic Kinetics and Mechanism of the
       Reduction of Crystalline Cobalt
       Molybdate in Hydrogen 331
   J. Haber, A. Kozłowska, and J. Słoczynski

An IR and EM Study of the Reactivity of Some
       Divalent Metal Hydroxides with Silica Gel 337
   T. Baird, A. G. Cairns Smith, and D. S. Snell

Unexpected Cases of Reactions Between Solid
       Substances at Room Temperature and Normal
       Pressure. Four Different Examples 343
   M. E. Garcia-Clavel, M. I. Tejedor-Tejedor,
   and A. Martinez-Esparza

Dislocations and Their Role in the Thermal
       Dehydration of Ba(ClO$_3$)$_2$·H$_2$O 349
   G. G. T. Guarini and R. Spinicci

New Insights on the Thermal Decomposition of
    Ammonium Perchlorate from Studies on
    Very Large Single Crystals   355
  P. J. Herley and P. W. Levy

Trapping of Hydrogen by Structural Defects in α-Iron   361
  E. M. Rieke

Cation Diffusion, Point Defects and Reaction in
    the CoO-β-Ga_2O_3 System   367
  W. Laqua, B. Küter, and B. Reuter

The Texture and Heat Treatment of Hydrated Tin
    (IV) Oxides   373
  B. J. Dalgleish, D. Dollimore, and D. V. Nowell

Redox Processes of Manganese Ions Dispersed
    in Oxide Matrices   379
  M. Valigi and D. Cordischi

Studies by Electron Microscopy of the
    Reduction of MoO_3   385
  G. Liljestrand

Influence of Crystallographic, Magnetic
    Transformations and Structural Defects
    on the Reactivity of Simple and Combined
    Iron Oxides   391
  I. Gaballah, C. Gleitzer, and J. Aubry

Non-Quenchability of Some Transition-Metal
    Chalcogenides   397
  M. Nakahira, K. Hayashi, M. Nakano-Onoda,
  and K. Shibata

About the Symmetry Lowering of Some Apatites as
    Evidenced by Their Infrared Spectrometric
    Study   403
  J. C. Trombe and G. Montel

Mechanism and Kinetics of Solid State Reaction
    in the System: $CuCr_2O_4$ - $Cu_2Cr_2O_4$ - CuO   409
  J. Haber and H. Piekarska-Sadowska

Topotaxy and Solid-Gas Reactions   415
  J. Guenot, F. Fievet-Vincent, and M. Figlarz

Microstrain and the Mechanism of the Reversible
    Monoclinic ⇌ Tetragonal Phase Transformation
    in Zirconia   421
  S. T. Buljan, H. A. McKinstry, and V. S. Stubican

CONTENTS

Mechanical Damping of Ionic Crystals with
      Regard to MgO 427
  J. Kriegesmann, G. H. Frischat, and
  H. W. Hennicke

Ferrite Formation Mechanism 433
  P. Y. Eveno and M. P. Paulus

Planar Defects and Reactivity of Perovskite-
      Like Compounds ABO_{3+x}. The Series
      $(Na,Ca)_n Nb_n O_{3n+2} (n>4)$ 439
  J. Galy, R. Portier, A. Carpy, and M. Fayard

Electron Density Fluctuations at the Vicinity
      of the Surface Layers 443
  S. Michalak and L. Wojtczak

Heterogeneous Ion-Exchange Reactions on Non-
      Stoichiometric Tungsten Oxide in
      Aqueous Media 449
  T. Szalay, L. Bartha, T. Nemeth, and J. Lengyel

SECTION 3: SOLID STATE REACTIONS IN ORGANIC MATERIALS

Polymerization and Other Organic Reactions
      in the Crystalline State (Plenary
      Lecture) 457
  H. Morawetz

Laser Induced Reactions in Doped Polymethyl-
      methacrylate and Their Correlation
      With Results Obtained by X-Ray and
      UV Irradiation 475
  D. J. Morantz and C. S. Bilen

Thermal Isomerization of Cisazobenzene in
      the Solid State 481
  H. K. Cammenga, E. Behrens, and E. Wolf

Energy Transfer in the Solid State
      Photopolymerization of Diacetylenes 487
  G. Wegner, G. Arndt, H.-J. Graf, and M. Steinbach

Solid-Gas Reactions. Part V. Bromination of
      Organic Solids 493
  E. Hadjoudis

Mechanisms of Some Room Temperature
        Phosphorescence Phenomena in Doped
        Solid Polymer Matrices 499
   D. J. Morantz, C. S. Bilen, and R. C. Thompson

SECTION 4: REACTIONS IN VITREOUS SOLIDS

Reactions in Vitreous Solids (Plenary Lecture) 505
   W. Vogel

Thermal Decomposition of Rhodium Oxide Gels 519
   E. Morán, M. A. Alario-Franco, J. Soria,
   and M. Gayoso

Polymerization Effects During the Crystallization
        of Silicate Glasses 525
   J. Götz, D. Hoebbel, and W. Wieker

The Use of Laser Raman Spectroscopy in the
        Study of the Formation of Oxide Glasses 529
   H. Verweij and H. van den Boom

Relations Between Properties and Structural
        Evolution of Some Si, Ge, Sn
        Ternary Chalcogenides 535
   E. Philippot, M. Ribes, and M. Maurin

High and Low Temperature Forms of Leucite 541
   L. Hermansson and R. Carlsson

Desorption of Kr (or Kr^{85}) and SF_6 from
        Vitreous and Crystalline SiO_2, and of
        Kr from B_2O_3 and GeO_2 Glasses 547
   W. W. Brandt and H. W. Ko

Surface Crystallization in the $SiO_2 - Al_2O_3 - ZnO$
        System by DTA 553
   Z. Strnad and J. Šesták

SECTION 5: NEW DEVELOPMENTS IN EXPERIMENTAL TECHNIQUES FOR THE STUDY OF REACTIVITY OF BULK SOLIDS AND SURFACES OF SOLIDS

Some Recent Developments in the Investigation
        of Solid-State and Surface Reactions
        (Plenary Lecture) 559
   E. G. Derouane

CONTENTS

The Investigation of Solid State Reactions
      with EPR 587
  W. Gunsser and U. Wolfmeier

Computer Simulation of Crystal Dissolution 593
  A. I. Michaels and M. B. Ives

Study of the Initial Stage of the Oxidation
      Reaction of Pure Titanium and TA6V4 Alloy
      at High Temperature 599
  C. Coddet, G. Béranger, J. Driole, and J. Besson

Study of Oxygen Diffusion in Quartz by
      Activation Analysis 605
  R. Schachtner and H. G. Sockel

Theoretical Study of Simulated Dehydration of
      Crystal Silica Planes and Determination
      of Silica Gel Hydroxyl Distribution
      by NMR 611
  J. Demarquay and J. Fraissard

Thermal Decomposition of Cobalt Oxalate 617
  A. Taskinen, P. Taskinen, and M. H. Tikkanen

A Study of the Different Stages During Nickel
      Oxide Single Crystal Sulfurization
      by Hydrogen Sulfide 625
  A. Steinbrunn, P. Dumas, and J. C. Colson

Oxygen Diffusion in NiO and ZnO 631
  D. Hallwig, H. G. Sockel, and C. Monty

Sulfation Mechanism of CoO and NiO 635
  L. E. K. Holappa and M. H. Tikkanen

High Resolution Electron Microscopy of
      Reacting Solids 641
  G. Schiffmacher, H. Dexpert, and P. Caro

The Electronic Structure of an Adsorbate
      Layer Studied by Angle-Resolved
      Photoemission Spectroscopy 647
  P. Butcher, P. M. Williams, and J. C. Wood

Thermal Decomposition of Alkaline Earth Chromates,
      Oxalates and Chromate-Oxalate Mixtures.
      The Ba, Sr, and Mg Compounds 653
  E. G. Derouane, R. Hubin, and Z. Gabelica

High Resolution Electron Microscopy Studies
of Fluorite-Related Cerium Oxides 659
O. T. Sørensen

The Development of Simultaneous TA-MS for the
Study of Complex Thermal Decomposition
Reactions 663
P. A. Barnes

SECTION 6: SOLID STATE REACTIONS IN TECHNOLOGY

The Status of Understanding Diffusion Controlled
Solid State Sintering, Hot Pressing and
Creep (Plenary Lecture) 669
R. L. Coble

Kinetic Study of the Reduction of Ni^{2+} Ions in
an X-Type Zeolite 689
M. Kermarec, M. F. Guilleux, D. Delafosse,
M. Briend-Faure, and J. F. Tempère

Development of Conductive Chains in RuO_2-Glass
Thick Film Resistors 695
R. W. Vest

Evaluation and Effects of Dispersion and Mixing
Method of Reactant Particles on the
Kinetics and Mechanism of Solid State
Reactions 701
T. Yamaguchi, S. H. Cho, H. Nagai, and H. Kuno

Carbothermal Reduction of Silica 707
J. G. Lee, P. D. Miller, and I. B. Cutler

Electrochemical Cells with Sulphate-Based
Solid Electrolytes 713
B. Heed, A. Lundén, and K. Schroeder

Reversible Topotactic Redox Reactions of
Layered Chalcogenides 719
R. Schöllhorn

Preparation and Properties of FeOCl-Pyridine
Derivative Complexes and Their
Reactivities with Methyl Alcohol 725
S. Kikkawa, F. Kanamaru, and M. Koizumi

CONTENTS

Formation and Properties of Some Subhalides
      of Tellurium 731
  A. Rabenau

Thermal Studies on the Boron-Molybdenum
      Trioxide Pyrotechnic Delay System 737
  E. L. Charsley and M. R. Ottaway

The Kinetics of the Reaction Between Magnesium
      Oxide and Iron (III) Oxide: The Effect
      of Sizes of Particles 743
  J. Beretka, T. Brown, and M. J. Ridge

Co-Precipitation of Metal Ions During Ferrous
      Hydroxide Gel Formation 749
  S. Okamoto and S. I. Okamoto

Dehydration-Reduction Coupling Effects in the
      Transformation of Goethite to
      Magnetite 755
  M. L. Garcia-Gonzalez, P. Grange, and B. Delmon

Steps in Low Temperature Dehydroxylation
      of Clay 761
  M. Gábor, J. Wajand, L. Pöppl, and Z. G. Szabó

Separation of Phases in Some Carbide Systems 767
  V. S. Stubican and M. Brun

Mass Transport by Self-Diffusion and Evaporation-
      Condensation in High Temperature Kinetic
      Processes in $UO_2 \cdot PuO_2$ Nuclear Fuel 773
  Hj. Matzke

The Structural Degradation of Carbon Fibres in
      Nickel at Elevated Temperature 779
  R. Warren and J. C. Wood

Solid State Reaction Studies on La_2O_3-NH_4F and
      $La(OH)_3$-NH_4F 785
  G. Adachi, B. Francis, K. Rajeshwar,
  E. A. Secco, and J. C.-S. Wong

The Corrosion Resistance of Fe-Cr Alloys as
      Studied by ESCA 791
  I. Olefjord and B.-O. Elfström

The Influence of Oxygen on the Current
    Potential Characteristics of Zn/-
        and ZnO/Electrolyte Contacts 797
  O. Fruhwirth, J. Friedmann, and G. W. Herzog

Author Index 803

Subject Index 807

NEW CRYSTALLOGRAPHIC DEVELOPMENTS APPLICABLE IN STUDIES OF
REACTIONS IN SOLIDS

Arne Magnéli

Arrhenius Laboratory, University of Stockholm

S-104 05 Stockholm (Sweden)

INTRODUCTION

In 1938 Professor J. Arvid Hedvall published a book carrying the title "Reaktionsfähigkeit fester Stoffe", i.e. Reactivity of Solids. It is hardly a mere coincidence that this became the heading of this series of symposia which have since developed into a well established scientific tradition. Hedvall undoubtedly played an active and inspiring role in the launching of the early meetings of the series. Actually what is now called the Second Symposium was arranged by him in Göteborg in 1952.

The name Reactivity of Solids for the book, and later, for the Symposia, was a fortunate invention. It was coined at a time when chemical research had essentially been concerned with reactions involving liquids and gases and when the possibility of chemical processes occurring in the solid state was seldom thought of or was even denied. That attitude was very strongly opposed by Hedvall, who had thought and worked a lot on such processes. The title of the book may have sounded rather provocative to many colleagues at that time.

A great advantage of the concept Reactivity of Solids thus created is that it is so broad and even somewhat diffuse in scope. Thus it is not limited to chemistry but open to contributions from many other branches of science, experimental and theoretical as well as pure and applied. The Symposia on Reactivity of Solids thus from the beginning could develop into a forum for cross-disciplinary research.

The scientific contributions which will be presented during this week of the 8th Reactivity of Solids Symposium represent a wide spectrum of research specialities and interests. By exchange of knowledge, experience and ideas at the sessions and at individual discussions we shall be able to identify the present frontier of our science and to find the proper directions for continued work within our different fields of research.

The subject of this talk will be some recent developments in crystallographic techniques, exemplified by studies performed at the University of Stockholm. This is work still in progress - only little of the results have yet been published. These studies are examples of the kind of contributions which structural chemists can give towards the understanding of the chemical behaviour of solids.

PROFILE ANALYSIS OF X-RAY POWDER PHOTOGRAPHS

To most people the X-ray powder photograph is nowadays a finger-print method for identification of crystalline solids. In the 1920's and 30's the situation was different; the data registered on the powder photograph was the basis for most crystal structure analyses and a remarkable mastership was developed in using powder data to derive structures of considerable complexity.

The development of single-crystal instruments of ever-increasing sophistication and the computerization of the structure-solving techniques gradually reduced the application of the powder method for structure analyses. The advance of X-ray single-crystal diffractometry has made possible the use of very small crystal specimens but there are still innumerable materials which have been prepared only in virtually microcrystalline form. The crystals necessary for neutron diffraction investigations are very large compared to those used in X-ray experiments.

Powder-diffraction for structural purposes was literally revived when Rietveld (1) pointed out that a neutron powder diffractogram actually contains much more information than had previously been utilized in structural studies. This information could be exploited by a least-squares refinement, not as previously done by using the integrated intensities of single as well as of overlapping reflections, but rather by employing directly the profile intensities obtained from step-scanning measurements of the diffractogram. This has provided a very powerful technique for structural refinement of substances suited for neutron diffraction analysis. It was pointed out by Rietveld that the profile refinement method is in principle also applicable to X-ray powder diagrams but that this would involve some special difficulties.

Work to improve the techniques for utilization of X-ray powder photographs for qualitative and quantitative analysis and also for structure determination has been conducted for some time by Werner and Malmros. It has been found that the focussing Guinier-type camera designed by Hägg has several advantages compared to diffractometers. It has a very high resolution. It utilizes strictly crystal monochromatized $K\alpha_1$ radiation which makes any computational correction for the $K\alpha_2$ component unnecessary. The thin rotating sample in the transmission position reduces the effects of several sources of error, in particular preferred orientation. All the reflections are registered simultaneously on the film which means perfect counter statistics. The exposure time is short, usually of the order of one hour.

The film intensity data are obtained from measurements with an automatic film scanner connected to a computer. The software system developed by Werner and Malmros (2,3) produces intensity values corrected for the polarization, Lorentz and geometrical factors and corresponding 2θ values at about $0.015°$ steps along the film.

Some results obtained so far may be illustrated by three examples. In order to test the procedures Malmros and Thomas (4) performed a profile analysis of the X-ray powder pattern of α-bismuth oxide. This structure is known from a single-crystal investigation by Malmros (5). In all 1588 measured non-zero intensities comprising 237 contributing reflections were used to refine a total of 25 parameters. These included four parameters corresponding to the asymmetric modified Lorentz function describing the profile of the individual reflection, one scale factor, and for each of the five atomic positions of the structure three positional parameters and one isotropic thermal parameter. Table 1 compares the results obtained from this study with the single-crystal derived data. The positional parameters of the two investigations agree remarkably well. The powder diffraction data was not corrected for absorption which probably affects the thermal parameters.

Another example may serve to illustrate the use of the powder photograph, first to solve and then to refine the crystal structure of a compound only available in microcrystalline form. This is a magnesium niobium oxide of the formula $Mg_3Nb_6O_{11}$ studied by Marinder (6). The crystal structure was solved from the information about interatomic distances given by the radial distribution function, calculated from the film intensity data according to Werner (7) and from geometric discussions. The refinement included 918 non-zero step-scan intensities and 18 parameters and converged to give a reliability index of 5.5 %. Fig. 1 compares observed and calculated intensities and shows also the difference between these.

Table 1. Comparison of structural parameters for α-Bi$_2$O$_3$, (A) single-crystal work (5) and (B) profile refinement (4)

Atom		x	y	z	B (Å^2)
Bi(1)	A	0.5240(1)	0.1831(1)	0.3613(1)	0.60(4)[x]
	B	0.5254(7)	0.1825(5)	0.3608(5)	0.67(9)
Bi(2)	A	0.0409(2)	0.0425(1)	0.7762(1)	0.60(4)[x]
	B	0.0431(6)	0.0416(5)	0.7777(5)	0.64(8)
O(1)	A	0.780(4)	0.300(3)	0.710(3)	0.9(2)
	B	0.767(7)	0.304(5)	0.711(6)	-1.1(8)
O(2)	A	0.242(5)	0.044(4)	0.134(4)	1.2(3)
	B	0.262(8)	0.062(5)	0.144(6)	-1.4(9)
O(3)	A	2.271(4)	0.024(3)	0.513(3)	0.8(2)
	B	0.305(8)	0.026(6)	0.510(7)	-0.8(10)

[x] Values calculated from anisotropic thermal parameters.

Another structural study may serve to illustrate the present state of development of the powder pattern crystal structure analysis. The substance investigated is a rather complicated organometallic, ammonium dimolybdatomalate, studied by Werner et al. (8). The substance has only been prepared in a microcrystalline state.

The powder photograph was indexed by a trial-and-error indexing program designed by Werner (9). The structure was found to be monoclinic with a unit cell of almost 1500 Å^3 and four formula units (NH$_4$)$_2$(MoO$_3$)$_2$C$_4$H$_3$O$_5$ in the cell. A rough Patterson function, calculated from about 120 integrated intensities made it possible to deduce two four-fold Mo positions. By iterative Fourier calculations and profile analyses it was possible to find and to refine positions of the oxygen and nitrogen atoms. The sites of the carbon atoms, which could not be identified from the Fourier map, were deduced from the oxygen atom positions and the known form of the organic molecule. More than 50 parameters were included in the refinement.

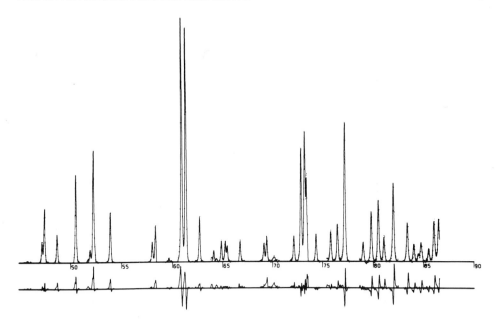

Fig. 1. X-ray powder photograph of $Mg_3Nb_6O_{11}$.
a) calculated (line) and measured (dots) profiles
b) difference

The results of the structural study is a four-centered Mo complex having a twofold symmetry axis. Both carboxy groups and the alcoholic oxygens of the malic radical are coordinated to the molybdenum atoms. It seems quite remarkable that such detailed structural information can be deduced from just one powder photograph.

These examples show that profile analysis applied to X-ray powder photographs will provide a powerful tool for crystal structure analysis and refinement of microcrystalline materials. The progress has been possible by the very accurate intensity data which are obtained by densitometer measurements of Guinier-Hägg type powder photographs. Further developments of the technique are in progress.

The short time required to collect the complete set of data, i.e. the hour or so for taking the X-ray photograph of a minute amount of powder, makes this method for structural studies highly attractive whenever speed is important. One may foresee applications in many fields and not the least in studies of structural changes associated with various types of reactions in solids.

ELECTRON MICROSCOPY AND X-RAY CRYSTALLOGRAPHY - COMPLEMENTARY TECHNIQUES FOR STUDIES OF STRUCTURES AND DEFECTS

In recent years there has been a rapidly growing interest in the application of electron microscopy to materials studies. The possibility of direct lattice imaging has been used to give very visual confirmation of the correctness of crystal structures previously derived by crystallographic techniques. In addition the lattice imaging has made it possible to study various kinds of structural irregularities and extended defects. This means a valuable complement of information to the crystal structure, which always represents an averaged description of the atomic arrangement in a very large number of unit cells.

Recent studies combining electron microscopy and X-ray diffraction carried out by Margareta Sundberg have given some new contributions to the classical problem of reduced tungsten trioxide (10,11). The materials investigated consisted of crystals prepared by prolonged heating of mixtures of tungsten dioxide and trioxide in sealed tubes in a temperature gradient. In one experiment a starting sample of composition $WO_{2.93}$ was placed at the hotter end (1075°C) while the cooler end was kept at 1000°C.

An X-ray crystallographic analysis of crystals found in the residue showed an ReO_3-type based shear structure W_nO_{3n-2} (n = 25) characterized by {103} shear, i.e. groups of six edge-sharing octahedra. This is illustrated in the lower part of Fig. 2 which represents the interpretation of the lattice image of a fragment of a crystal of the same origin in terms of WO_6-octahedra. The image is in good agreement with the crystal structure analysis mainly containing WO_3 slabs of thickness n = 25 with a moderate proportion of slabs of deviating width (Wadsley defects). The formula $W_{25}O_{73}$ corresponds to the composition $WO_{2.920}$ for the residue crystals.

The sample obtained by a transport process at the colder end of the tube differed markedly from the residue at the hotter end. It was richer in oxygen and less homogeneous. An X-ray structure investigation of a fairly good crystal gave a {103} shear structure with the n value 24 corresponding to the member $W_{24}O_{70}$ of the W_nO_{3n-2} series. This structure would thus be slightly lower in oxygen ($WO_{2.917}$) than the structure deduced for the residue crystals.

Several crystal fragments from the transported material were analyzed by electron microscopy. The lattice image of one of these is presented in Fig. 3. The picture shows {103} shear planes which are rather uniformly distributed within broad bands. Here they occur at distances corresponding to n values around 24 which

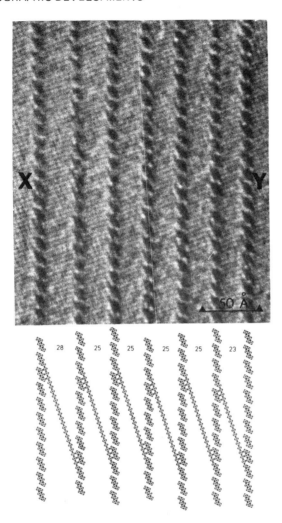

Fig. 2. a) Lattice image of well-ordered crystal flake of reduced tungsten trioxide containing {103} shear planes.

b) Interpretation of the image in terms of WO_6 octahedra. The numbers between the shear planes refer to the n values in the formula W_nO_{3n-2} and correspond to the number of octahedra in the indicated rows. All other octahedra in the WO_3 slabs are omitted for the sake of clarity.

is in accordance with the X-ray structure determination. However, between the sheared areas there are wide bands of unreduced structure. They vary in width within the range 50-160 Å. It seems natural to interpret this picture containing areas of rather well-ordered shear structure intermingled with wide areas of non-sheared structure as a coherent intergrowth of lamellae of $W_{24}O_{70}$ and of WO_3.

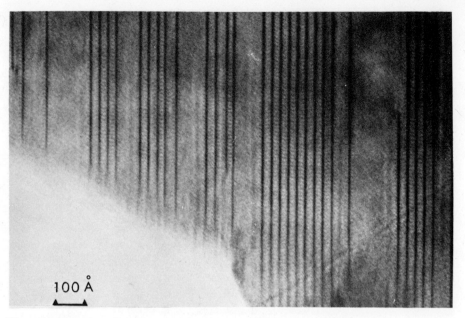

Fig. 3. Lattice image of transport grown reduced tungsten trioxide crystallite showing intergrowth of lamellae of W_nO_{3n-2} ($n \sim 24$) and of WO_3.

An analysis of the micrograph revealed some further interesting details. The distances between the shear planes were carefully measured and the result is shown in the histogram in Fig. 4. The abscissa represents the widths of the slabs in terms of n, i.e. the number of WO_6 octahedra in the characteristic direction. The ordinate gives the number of slabs observed. The most common widths are evidently those with n = 24 and 25 but there is a considerable spread and also very high n values for the WO_3 components. If the result of the measurement is used to calculate the average composition of the entire crystal fragment, this turns out to correspond to an n value of 39 and the formula $WO_{2.949}$.

The parallel investigations by X-ray crystallography and lattice image microscopy thus give entirely divergent results for the compositions of these crystals of reduced tungsten trioxide. The X-ray study gave a shear structure of the formula $W_{24}O_{70}$, i.e. $WO_{2.917}$. The electron microscopy study showed $W_{24}O_{70}$ structure intergrown with slabs of tungsten trioxide to give an overall composition of $WO_{2.949}$. Further studies are in progress on the background of this case of intergrowth.

The reason why the X-ray investigation did not reveal the presence of WO_3 in the crystal is that it represents the substructure of the shear structure. If shear planes were uniformly introduced within the areas of WO_3 the histogram would get the appearance shown in the bottom figure, with a pronounced maximum at n = 24. This relation is further evidenced by the electron diffractogram shown in Fig. 5. The WO_3 lamellae contribute only to the strong substructure diffraction spots. The superstructure reflections due to the shear are the rather sharp, weaker ones lined up in the [103] direction.

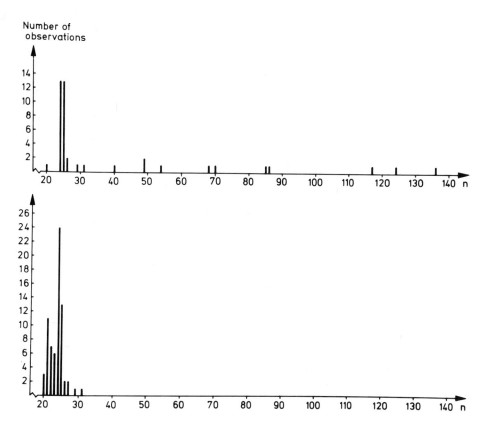

Fig. 4. a) Distribution of n values in the image shown in Fig. 3.
b) Distribution if hypothetic shear planes are uniformly introduced within the WO_3 areas of the image.

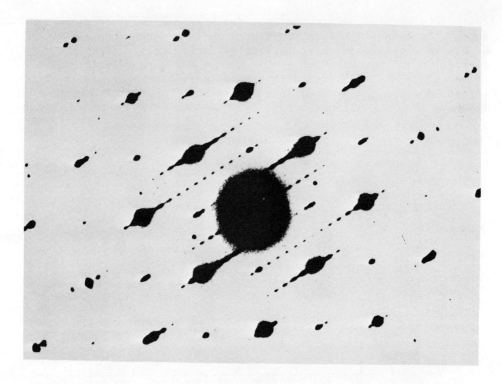

Fig. 5. Electron diffractogram of the reduced tungsten trioxide crystallite referred to in Figs. 3 and 4.

ELECTRON MICROSCOPY FOR CRYSTAL STRUCTURE ANALYSIS

Under favourable conditions the lattice image technique may also be utilized for crystal structure determination of microcrystals. This is illustrated by a recent investigation by Hussain and Kihlborg (12). They have studied the tungsten bronze systems of the heavy alkali atoms, potassium, rubidium, and cesium for low alkali concentrations. Below the range of the hexagonal bronze type ($x < 0.13$ in the formula A_xWO_3) apparently homogeneous products, e.g. $K_{0.10}WO_3$, were obtained consisting of black shining crystals, translucent with a greenish colour in thin sections. The crystals were found to be resistant towards alkaline solutions and to acids, in this respect behaving like the previously known tungsten bronzes.

The X-ray powder pattern of the new material was found to be very complicated. A large number of crystals were investigated and found to be composite and unsuitable for single-crystal work.

NEW CRYSTALLOGRAPHIC DEVELOPMENTS

Electron diffraction patterns from samples obtained by crushing of the crystal aggregates, were registered for a considerable number of crystallites. All the patterns indicated orthorhombic lattices with two rather short axes of about 7.5 Å and a long one of the order of 20 or 30 Å units. Patterns which differed with respect to the long axis were observed but one of about 28 Å was found to occur most frequently.

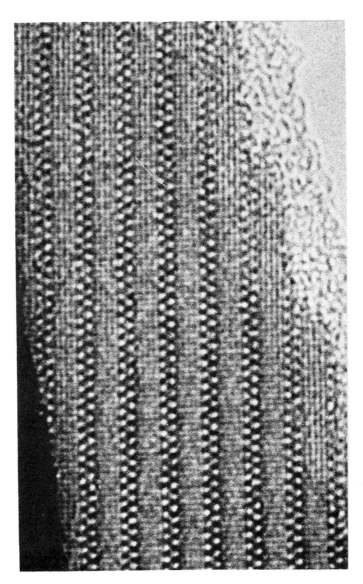

Fig. 6. Lattice image of crystallite from $K_{0.10}WO_3$ sample.

A lattice image of such a crystal is shown in Fig. 6. It contains a regular pattern with zig-zag bands of large light spots surrounded by a dark contour. The areas between the bands possess a regular texture similar to the one found in several tungsten oxides. The image was interpreted in terms of the structure model shown in Fig. 7.

This may be described as an ordered intergrowth of the six-sided tunnels of tungsten-oxygen octahedra, which are present in the hexagonal bronze structure with a somewhat distorted tungsten trioxide structure. This interpretation of the image was verified by comparison with theoretical images obtained by n-beam dynamical calculations using the multislice method. Later on a fairly good single crystal of the material was found and the structure model was also verified by an X-ray crystallographic study.

Lattice images of crystallites showing deviating lengths of the long cell axes were analogously interpreted. The structures were found to fit into a series of phases which differ only in the width of the tungsten trioxide component slabs of the structures. The complex appearance of the powder pattern was evidently due to the composite nature of the original sample, which contained a mixture of several members of the structural series.

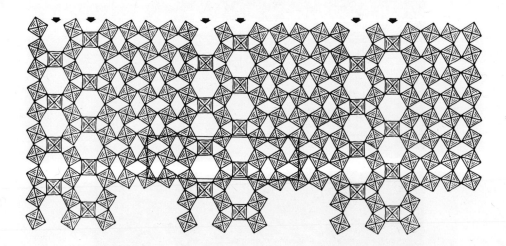

Fig. 7. Structure model derived from the image in Fig. 6. The alkali metal atoms in the hexagonal tunnels are not indicated.

So far crystal structure determination from lattice images may often be looked upon as a somewhat adventurous technique. If the image can be interpreted as composed of recognizable, well established structural elements, it certainly represents an interesting and profitable way to study extremely tiny objects. It is essential, however, that the structural model can be verified by comparisons between the observed and calculated images.

STUDIES OF REACTIONS IN THE ELECTRON MICROSCOPE

Under the influence of the electron beam and of the vacuum in the electron microscope tungsten oxide crystals often undergo further reduction by formation of new shear planes. The progress of the shear formation may be studied from sequences of lattice image pictures taken at appropriate intervals.

Fig. 8. is not a particularly excellent piece of electron microphotograph but is rather a "snapshot" which has registered an interesting observation. The sample studied was a transport grown crystal and the diffractogram revealed fairly disordered {103} shear. Mrs. Sundberg who took the picture describes her observations in the following way:

"I was adjusting the specimen by observing the lattice image on the screen and tried to resolve the groups of six edge-sharing octahedra by slightly changing the defocusing of the crystal fragment. The electron beam was not very strong and the screen accordingly fairly dark but the picture yet clearly visible.

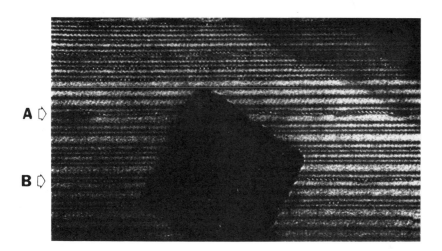

Fig. 8. Reduction process in tungsten oxide (compare text).

Suddenly I observed new shear planes progressing on the screen between those already there. I got very excited. It happened very fast and I tried to concentrate on a recognizable part of the pattern before exposing the photograph. The extra crystal flake on top of the object helped to do so.

The shear plane indicated by A was not originally there. I suddenly saw an elongated very narrow loop, which after a little while, maybe a second, jumped to the right to a new position. Such jumps occurred to new positions at intervals leaving the trace of the new shear plane. I was then able to follow the formation of another shear plane at B. This too proceeded in a discontinuous way as a series of jumps."

So far this picture is our only documentation of this kind of observation. However, it gives good hope that the electron microscope when provided with an image intensifier and equipment for video recording will be very suited for detailed studies of such processes. Work in that direction is in progress.

Professor Gunnar Hägg, the Senior of Swedish crystallographic research, has told how almost fifty years ago he sought the advice of Professor Arne Westgren for a suitable subject for his doctoral thesis. Westgren strongly recommended research in X-ray crystallography and described the thrills of such work, saying: "One nearly has the feeling of touching the atoms." I think that the thrill is not diminished in the new applications of electron microscopy, not only that you nearly see the atoms but also that you may be able to see how they move while taking part in chemical reactions.

REFERENCES

1. Rietveld, H.M. J. Appl. Cryst. 2 (1969) 65.
2. Werner, P.-E. Arkiv Kemi 31 (1970) 505.
3. Malmros, G. and Werner, P.-E. Acta Chem. Scand. 27 (1973) 493.
4. Malmros, G. and Thomas, J.O. J. Appl. Cryst. In print.
5. Malmros, G. Acta Chem. Scand. 24 (1970) 384.
6. Marinder, B.-O. To be published.
7. Werner, P.-E. and Karlsson, J. Univ Stockholm Inorg. & Phys. Chem. DIS No. 26 (1966).
8. Werner, P.-E. and Berg, J.E. To be published.
9. Werner, P.-E. Arkiv Kemi 31 (1970) 513.
10. Sundberg, M. Acta Cryst. B32 (1976) 2144.
11. Sundberg, M. To be published.
12. Hussain, A. and Kihlborg, L. Acta Cryst. A32 (1976) 551.

REACTIONS AT SURFACES AND INTERFACES

Gas-Metal Interactions at High Temperatures

Per Kofstad

Department of Chemistry, University of Oslo

Blindern, Oslo 3, Norway

INTRODUCTION

The title "Reactions at Surfaces and Interfaces" may cover any aspect to be dealt with at this Symposium and will in the widest sense include all reactions with solids, in solids, and on solids. Within a limited space it is both necessary and desirable to limit the scope and to focus on part of this vast field. It is my prerogative to choose, and I shall confine my remarks to gas-metal interactions, and, furthermore, primarily focus on interactions at high temperatures. Even so, this will hopefully bring forth important aspects of reactivity of solids as gas-metal reactions in all their ramifications provide examples of reactions and processes taking place with solids, in solids, and on solids.

THERMODYNAMICS OF GAS-METAL REACTIONS

As a start it may be appropriate briefly to recapitulate the basic thermodynamics of such reactions, i.e. to consider the driving force for such reactions and under what conditions they will take place. At this stage let us for simplicity consider reactions between elemental solids and a gas and as examples use metal-oxygen reactions.

Consider then the formation of an oxide M_aO_b from the metal M and oxygen gas: $aM + b/2O_2 = M_aO_b$. The metal M can only be oxidized to the oxide M_aO_b if the ambient partial pressure of oxygen is larger than the dissociation pressure of the oxide in equilibrium with the metal. The dissociation pressure is given by

$$p(O_2) \gtreqless \exp\left(\frac{2\Delta G^o(M_aO_b)}{b\,RT}\right) \qquad (1)$$

where $\Delta G^o(M_aO_b)$ is the standard free energy of formation of the oxide at temperature T. Such free energy data and the dissociation pressures of oxides as a function of temperature are conveniently summarized through the well known Ellingham diagram. This is illustrated in Fig. 1 with data for a few oxides of importance in high temperature oxidation of metals and alloys. Corresponding diagrams may also be prepared for sulfides, carbides, halides, etc. Thermodynamic data used for preparing Fig. 1 and those used later in the paper are given Refs. 1 and 2.

Constant partial pressures of oxygen are given by straight lines radiating from the common zero point, $\Delta G^o = 0$ and $T = 0\,^{\circ}K$, in the upper left hand corner of the diagram. As examples, the lines corresponding to $p(O_2) = 10^{-20}$ and 10^{-30} atm. are shown on the diagram. Low partial pressures (or activities) of oxygen are commonly achieved by gas mixtures of $H_2O + H_2$ and $CO_2 + CO$.

From this diagram it may for instance be seen that the dissociation pressure of NiO at $1000\,^{\circ}C$ is 10^{-10} atm. while that of CoO is 10^{-12} atm., for Cr_2O_3 10^{-22} atm., and for Al_2O_3 10^{-35} atm. Partial pressures of oxygen higher than these values are necessary in order that the respective metals are to be oxidized to the oxides. Such data are, for instance, of considerable practical consequence in oxidation of alloys.

MAIN ASPECTS OF OXIDATION OF METALS (3-6)

Let us start with a pure metal whose surface is truly clean, i.e. no foreign atoms are adsorbed on the surface. When this surface is exposed to oxygen, almost all of the initially impinging molecules stick to the surface, dissociate, and adsorb on neighbouring metal sites, (7-10). Under clean conditions the adsorbed species will on many metals take on two-dimensional ordered structures, (7-13). An example of this is illustrated in Fig. 2, which shows structures of chemisorbed oxygen on a (100) nickel surface corresponding to 25 and 50% coverage, respectively.

But even before the first layer of chemisorbed oxygen has been completed, oxygen can adsorb on the chemisorbed oxygen, and through place exchange begin to form oxide islands on the surface. These grow laterally at the periphery of the islands until the metal surface becomes completely covered with an oxide film. At this stage the sticking coefficient - or reaction rate - decreases abruptly. The reactants - metal and oxygen - are separated by the

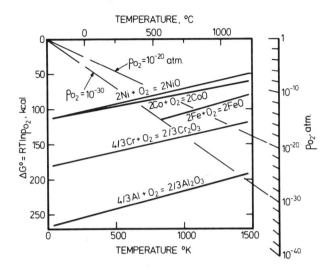

Fig. 1. Ellingham diagram (ΔG° vs. T) illustrating the stability of various binary metal oxides as a function of temperature. Constant partial pressures of oxygen are represented by straight lines originating from the common zero point, $\Delta G^\circ = 0$ and $T = 0\ ^\circ K$, in the upper left hand corner of the diagram. For illustration the lines corresponding to $p(O_2) = 10^{-20}$ and 10^{30} atm. are drawn in the diagram.

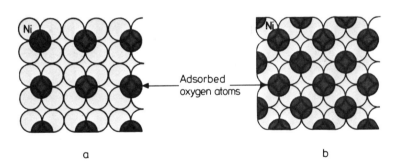

Fig. 2. Structures of oxygen adsorbed (chemisorbed) on a (100) nickel surface, (9,12,13)

a) double-spaced face-centered structure corresponding to 25% coverage

b) single-spaced square array structure at 50% coverage

oxide film; the reaction mechanism changes completely and any
further oxidation can only proceed by a solid state diffusion
through the film, (3-6).

At low temperatures - and when thin films are formed - it is
necessary to postulate a driving force in the form of an electric
field across the oxide in order to explain the transport through
the film. This electric field is created by electron tunnelling
through the oxide film to traps due to the adsorbed oxygen on the
oxide surface, (3-6).

At high temperatures - when thick films or scales are formed -
the oxidation continues through a thermally activated transport
with a driving force in the form of a chemical or electrochemical
potential gradient. It is the reactions under these conditions on
which the following discussion will be focussed.

GAS-METAL REACTIONS AT HIGH TEMPERATURES

The gas-metal reactions at high temperatures may take many
courses and forms depending on the metal, the reacting gas, temperature, elapsed time of reaction, etc., (3-6). If the reaction
products remain on the surface as a continuous, compact scale,
the reaction is governed by the solid state transport through
the scale. At the same time atoms of the reacting gas dissolve
and diffuse into the metal. In many cases, however, the oxide
scale may crack making the scale pervious to the reacting gas in
the form of cracks and porosity. In still other cases liquid
oxide phases may be formed and these will particularly affect
the diffusional transport through the scale; oxides may also
continuously evaporate from the surface.

But let us further consider growth of compact scales at
high temperatures governed by solid state transport of the reactants through the scale. Kinetically the reaction rate decreases
with time. The solid state diffusion will involve lattice (or
volume) diffusion, grain boundary and short circuit diffusion. All
diffusional mechanisms will be operative in a growing scale, but
as a general rule lattice diffusion will predominate at high
temperatures, while grain and short circuit diffusion become of
increasing relative importance with decreasing temperature. When
lattice diffusion predominates, the oxidation kinetics follow the
well-known parabolic rate equation:

$$\frac{dx}{dt} = k_p \frac{1}{x} \quad \text{or} \quad x^2 = 2k_p t + C \tag{2}$$

where x is a measure of the amount of reaction, e.g. oxide thickness, t is time, k_p the parabolic rate constant and C an integration constant.

Our basic understanding of this type of reaction mechanism is due to Wagner who related the parabolic rate constant to the transport properties of the oxide, (14,15). In addition to assuming a rate-determining lattice diffusion of reacting ions or transport of electrons through the scale, Wagner in his theory also assumed that thermodynamic equilibria are established at the outer and inner interfaces of the oxide.

When lattice diffusion predominates, one form of the rate constant k_p for the formation of an oxide M_aO_b may be given through the equation

$$k'_p = \frac{C_O}{2b} \int_{p^i(O_2)}^{p^o(O_2)} (D_O + \frac{z_c}{|z_a|} D_M) \, d\ln p_{O_2} \tag{3}$$

$p^o(O_2)$ and $p^i(O_2)$ denote the partial pressures of oxygen at the oxygen/oxide (outer) and at the metal/oxide (inner) interface, respectively; C_O is the concentration of oxygen in the oxide, z_c and z_a are the valence of the cation and anion, and D_M and D_O the self-diffusion coefficient of metal and oxygen ions.

The equation clearly illustrates the direct relation between the parabolic rate constant and the transport properties of the oxide. In the cases where measured and calculated values of k_p can be compared, good agreement has been obtained, (16). It should be emphasized, however, that the model is an ideal one, and that a number of factors will complicate the picture in a detailed analysis of a real system; these factors include effects of impurities, the development and presence of voids and porosity in scales, etc.

In this connection it may be of interest to make a comparison between oxidation rates of a few metals and measured self-diffusion coefficients for the faster diffusing species in the corresponding oxides. This is illustrated in Figs. 3 and 4. Self-diffusion coefficients of the cations in CoO (17,18), NiO (19,20), Cr_2O_3 (21,22), and Al_2O_3 (23) are shown in Fig. 3, and it is seen that the self-diffusion coefficients decrease from CoO to Al_2O_3 (24). In order to apply these diffusion coefficients in the Wagner equation, their oxygen pressure dependences have to be known and the proper integrations must be made. In oxidation the possibility of grain boundary diffusion, gaseous transport across pores, rate-

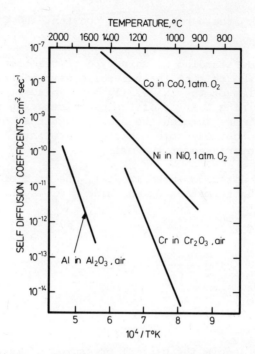

Fig. 3. Literature values of self-diffusion coefficients in CoO (17,18), NiO (19,20), Cr_2O_3 (21,22) and Al_2O_3 (23).

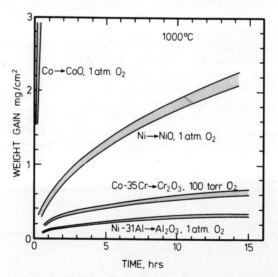

Fig. 4. Oxidation of cobalt (25,26) and nickel (27-30) as a function of time at 1000 °C in 1 atm. O_2. For comparison is also shown oxidation of the alloy Co-35% Cr (26,31), which forms a continuous scale of Cr_2O_3, and of the alloy Ni-31% Al, which forms a continuous scale of Al_2O_3 (32,33).

limiting electron transport, etc., also have to be considered.
Even so these diffusion coefficients may serve as a qualitative
guide as to the relative rates of oxidation of the respective
metals provided these form continuous scales. In this respect
Fig. 4 shows a plot of oxidation of cobolt (25,26) and nickel (27-
30) in 1 atm. O_2; for comparison are also included data for oxidation of an alloy which selectively forms a protective scale of
Cr_2O_3, i.e. Co-35% Cr (26,31), and an alloy selectively forming a
scale of Al_2O_3, i.e. Ni-31% Al (32,33). The same order of reaction
rates are observed as for the selfdiffusion coefficients, and
scales consisting of Cr_2O_3 and Al_2O_3 give low oxidation rates. In
most practical cases involving high-temperature alloys one does in
fact rely on the formation of Cr_2O_3-, Al_2O_3-, and in some cases
SiO_2-scales, to achieve good oxidation resistance at elevated
temperatures.

OXIDATION OF ALLOYS WITH ONE OXIDANT

As a next step it is of interest to consider in some detail
the more complex cases of oxidation of alloys. But let us start
with relatively simple cases and examine the reaction of binary
and ternary alloys with a single oxidant.

A comprehensive treatment of oxidation of alloys comprise a
large number of phenomena (3-6) which can not be covered in a
limited space. Instead of discussing a more general case, important features may be illustrated by specific alloy systems. As
such it is of interest to look at the behaviour of alloys of
nickel and cobalt with additions of chromium and/or aluminium. In
addition to illustrating important fundamental aspects, these
systems are of great practical interest as they constitute the
base alloys for a large number of important commercial super-
alloys.

Let us first consider the oxide phases which can exist in the
Co-Cr-O and Ni-Cr-O systems. These can be summarized as follows:

Co-Cr-O: CoO, Co_3O_4 (stable < 970 °C in 1 atm. O_2), Cr_2O_3, $CoCr_2O_4$

Ni-Cr-O: NiO, Cr_2O_3, $NiCr_2O_4$

Chromium oxide is more stable than both NiO and CoO (Fig. 1), and
on purely thermodynamic grounds one would thus expect Cr_2O_3 to be
the reaction product in oxidation of Co-Cr and Ni-Cr alloys.
However, this would require a continuous supply of chromium from
the inner part of the alloy to the metal/oxide phase boundary,
i.e. it requires that both the concentration and diffusivity of
chromium in the alloy are sufficiently high that chromium is
always in ample supply at the metal/oxide interface. If this

condition is not fulfilled, the other alloy component will be oxidized, and the reaction mechanism becomes a function of several factors which include the composition of the alloy and rates of transport in both the alloy phase and the oxide scale.

Let us consider this in more detail and start with a clean Co-Cr alloy surface. As it is exposed to oxygen gas, both the surface atoms of cobalt and chromium are oxidized and initially form CoO (and Co_3O_4 at $t < 970\,°C$) and Cr_2O_3 on the alloy surface. The oxide grows to form a film consisting of CoO and Cr_2O_3 which thickens by a solid-state diffusion process. As illustrated in Fig. 5 several competing processes take place at or near the interface between CoO and the alloy: i) cobalt continues to dissolve in CoO, diffuses outwards through the oxide and the reaction with oxygen takes place at the outer oxide surface, ii) chromium in the alloy diffuses to the oxide alloy interface, and as CoO is less stable than Cr_2O_3, chromium reduces CoO to cobalt metal while it itself is oxidized to Cr_2O_3, iii) oxygen from CoO dissolves and diffuses into the alloy where it reacts with chromium atoms to form Cr_2O_3-particles in a zone beneath the scale/metal interface. This constitutes internal oxidation.

The internal oxide particles block part of the diffusional cross-section in the alloy, and if sufficient chromium arrives at the internally oxidized zone, a continuous layer of Cr_2O_3 is built up. A protective layer of Cr_2O_3 is thereby established, and the rate of reaction then becomes governed by the transport through this growing layer.

The establishing of such a continuous Cr_2O_3 layer depends on several factors which include i) the concentration and diffusion of chromium in the alloy, ii) the diffusion of oxygen in the alloy, and iii) the growth rate of the CoO-layer. In practice this means that at a particular set of conditions continuous, protective scales of Cr_2O_3 are built up at and above a critical concentration of chromium; this critical concentration is, in turn, a function of temperature and ambient partial pressure of oxygen.

At lower chromium concentrations a continuous layer of Cr_2O_3 fails to develop and the oxidation reaches a steady state and involves formation of CoO and Cr_2O_3; these also undergo a solid state reaction to yield the $CoCr_2O_4$ spinel. A schematic diagram of the composition of the external oxide scale and the internally oxidized zone and of the reaction mechanism is shown in Fig. 5. An outer layer of the scale consists of CoO (under conditions where Cr_3O_4 is not formed), an inner layer consists of CoO + $CoCr_2O_4$, while the internally oxidized zone consists of Cr_2O_3 particles in cobalt metal. Particularly the inner layer of the external scale (consisting of CoO + $CoCr_2O_4$) also develops relative large amounts

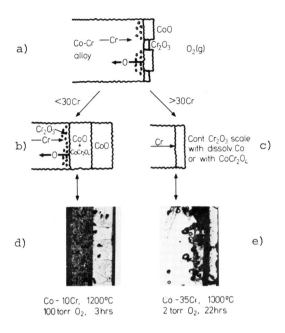

Fig. 5. Schematic illustration of processes taking place in binary Co-Cr alloys (26,31,33).

a) initial, transient oxidation involving formation of both cobalt and chromium oxide, oxygen dissolution in the alloy, diffusion of chromium to the metal/scale interface, internal oxidation of chromium beneath the CoO.

b) insufficient concentration of chromium in the alloy (<∼30% Cr); the reaction products consist of a two-layered surface scale and internal oxidation of chromium; the outer layer of the surface scale consists of CoO (+ Co_3O_4 at t < 970 °C and 1 atm. O_2) and the inner layer of CoO and $CoCr_2O_4$-spinel formed by the reaction between CoO + Cr_2O_3.

c) the development of a continuous layer of Cr_2O_3 on Co-Cr alloy with sufficient concentration of chromium (>∼30% Cr); selective oxidation of chromium.

d)) metallographic cross-section of a Co-Cr alloy oxidized for 3 hrs at 1200 °C and 100 torr O_2.

e) metallographic cross-section of a Co-Cr alloy oxidized for 22 hrs at 1300 °C and 2 torr O_2.

of voids and closed porosity. The rate of reaction is governed by
solid state diffusion of cobalt through the CoO part of the inner
and outer layers; in addition oxygen can move across the closed
pores and this results in an inward growth of the CoO layer.
Finally oxygen dissolves in the alloy substrate forming internal
particles of Cr_2O_3. Because of the inward growth of CoO (caused by
oxygen transport across the pores), the internally formed Cr_2O_3-
particles become embedded in CoO, and these then react to form the
$CoCr_2O_4$-spinel, (26,31,33).

As one increases the concentration of chromium in Co-Cr
alloys, an increasingly larger fraction of the inner layer will
consist of the $CoCr_2O_4$-spinel. Diffusion through this oxide is
slow compared to that in the CoO-phase, and essentially the
$CoCr_2O_4$-particles block part of the area for diffusional transport through the scale, and as a consequence the rate of oxidation of Co-Cr alloys gradually decreases with increasing
chromium concentration until the critical Cr-concentration is
reached where a continuous Cr_2O_3-layer is formed. This dependence
of the parabolic rate constant on Cr-concentration is illustrated
in Fig. 6 for oxidation at 1000 $^\circ$C, (26,31,34,35). At chromium
concentrations lower than 10%, the oxidation rate increases with
increasing Cr-concentrations. This effect is primarily believed
to be associated with effects of voids and porosity developing
in the scale, and that rapid gaseous transport across these pores
overrides the effect of diffusional blockage by the $CoCr_2O_4$-
particles in the inner scale, (26,31,34). Oxidation of many other
alloys is concluded to comprise these same main features, e.g.
oxidation of Ni-Cr, Co-Al, Ni-Al, Fe-alloys, etc., (3-6).

As we increase the complexity of the alloy compositions the
same main principles apply. Let us briefly consider a ternary
alloy such as for instance Ni-Cr-Al. Apart from the good protective properties of possible Al_2O_3-scales which may be formed,
alloy additions of aluminium are important due to the $\gamma'(Ni_3Al)$-
strengthening mechanism.

For such an alloy the oxidation may - depending on the alloy
composition and other experimental conditions - comprise
i) formation of duplex scales of NiO, $NiCr_2O_4$, $NiAl_2O_4$ and
an internal oxidation zone consisting of Cr_2O_3 and Al_2O_3 in a
manner similar to that described for the Co-Cr alloys; ii) selective oxidation of chromium + internal oxidation of Al to Al_2O_3
(as Al_2O_3 is more stable than Cr_2O_3), and iii) selective oxidation of aluminium to yield an external, protective scale of
α-Al_2O_3. It should be noted that chromium may facilitate the
development of a continuous Al_2O_3-layer; this may involve an
initial selective formation of a Cr_2O_3 layer and this in turn
slows down the reaction rate sufficiently that an internal Al_2O_3-

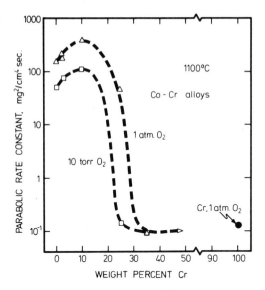

Fig. 6. The parabolic rate constant for oxidation of cobalt-chromium alloys at 1100 °C as a function of chromium content. Results of Kofstad and Hed (26,31,34) and Hagel, (35).

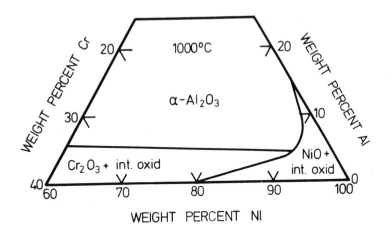

Fig. 7. "Oxide map" for alloys in the Ni-Cr-Al system delineating the composition ranges for formation of different types of oxide scales. In region I the scale consists of NiO, $NiCr_2O_4$ and $NiAl_2O_4$ spinels and in addition the alloy is internally oxidized; in region II the scale consists of Cr_2O_3 and in addition aluminium is internally oxidized; in region III aluminium is selectively oxidized, (36-38).

zone develops and eventually grows into a continuous Al_2O_3 layer. Thus, during an initial transient period a complex interplay of interface reactions and diffusion processes take place until a steady state reaction behaviour is established.

And again certain critical alloy concentration ranges apply where the reaction during steady state conditions changes from one mechanism to another. For some alloy systems sufficient data are available to permit the construction of "oxide maps" which delineate the composition ranges for formation of different types of oxide scales and reaction behaviour. An example of this is illustrated in Fig. 7 for oxidation of Ni-Cr-Al alloys at 1000 °C; the three different types of oxide formation and reaction behaviour summarized above are delineated, (36-38). It should be emphasized that such an "oxide map" today only represents empirical, experimental data. A great challenge for investigators in the field is to be able to derive such maps from more basic data involving the interplay of thermodynamics, diffusion data and interface reactions. A most valuable approach in this respect is through the "diffusion composition path" method, (39).

OXIDATION OF METALS WITH TWO OXIDANTS

Although reactions with oxygen are generally the most important gas-metal reactions - at least as far as high temperature compatibility of metals are concerned - one will in practical situations often encounter environments that consist of two or more oxidants that may react with the metal or alloy. Well-known phenomena in high temperature corrosion are simultaneous oxidation + nitridation, oxidation + carburization, oxidation + sulfidation, etc.

For the sake of illustration and due to their great practical importance, let us focus on reactions involving simultaneous oxidation + sulfidation. And as a start let us consider a relatively simple example: the reaction of an unalloyed metal with oxygen and sulfur. This may be illustrated by the reactions of chromium and nickel in an oxygen- and sulfur-containing gas such as sulfur dioxide.

As a basis it is necessary first to consider the thermodynamics of the Cr-O-S and Ni-O-S systems, i.e. what phases exist and under what conditions are they stable? A convenient way of presenting such data is in the form of Pourbaix-type diagrams. In these diagrams the stability ranges of the different compounds may for instance be plotted as a function of the partial pressures (activities) of oxygen and sulfur or, alternatively, through the partial pressures of other gases through which the partial pressures of oxygen and sulfur are determined, (40-43).

An isothermal section at 1200 °C of the Cr-O-S system is illustrated in Fig. 8, where the stability ranges of Cr-metal, Cr_2O_3 and CrS is shown as a function of $p(O_2)$ and $p(SO_2)$. (Higher sulfides of chromium exist, but these are usually not of importance in oxidation in SO_2 and are not included in the diagram.) In this type of plot the sulfur isobars run diagonally across the diagram as indicated by the broken lines. It may be noted that at this temperature the decomposition pressure of Cr_2O_3 in equilibrium with the metal is 3×10^{-24} atm. O_2, and that of CrS in equilibrium with the metal is 1.7×10^{-12} atm. S_2.

Let us expose chromium metal to pure SO_2 at 1 atm. In this gas $p(O_2) = 1.3 \times 10^{-8}$ and $p(S_2) = 6.3 \times 10^{-9}$ atm. as indicated by the black dot in Fig. 8. Under these conditions - and assuming equilibrium - Cr_2O_3 is the stable phase, and when chromium metal is exposed to 1 atm. SO_2, it would not be unexpected that a layer of Cr_2O_3 is formed on the metal. If the Cr_2O_3 scale is protective, the reaction would be mainly governed by outward chromium diffusion through the scale as for chromium metal oxidation in oxygen gas.

But another process may also take place: sulfur will dissolve in the Cr_2O_3-scale. Furthermore, as the partial pressure of sulfur in the ambient SO_2 gas is 6.3×10^{-9} atm. while the decomposition pressure of CrS in equilibrium with chromium metal is lower and only 1.7×10^{-12} atm. S_2, diffusion of sulfur through the scale may lead to formation of CrS beneath the Cr_2O_3-scale. Apparently no studies have as yet been reported of the reaction of chromium metal with pure SO_2, but Pettit et al. (41) have shown that the reaction of a Ni-20% Cr with pure SO_2 at 1200 °C leads to a Cr_2O_3-scale with CrS-particles in the alloy in a zone beneath the surface scale. One interpretation of this behaviour is that sulfur dissolves and is transported through the Cr_2O_3-scale.

However, this does not appear to be the complete picture: Pettit et al. (41) also made corresponding studies of the Ni-20% Cr alloy in mixtures of $SO_2 + O_2$ with $SO_2/O_2 = 0.75$ in which the partial pressure of sulfur is considerably smaller than the decomposition pressure of CrS as indicated by the cross (X) in Fig. 8. If sulfidation takes place by solid state diffusion of sulfur through a continuous Cr_2O_3 scale, no sulfide formation should take place in this case. However, sulfide was still found to be formed in the alloy beneath the scale, and this shows that partial pressures of sulfur considerably larger than that in the ambient SO_2 gas are built up in the scale and the alloy. A possible explanation for this behaviour is that SO_2 penetrates the scale along grain boundaries and microcracks (42,43) as indicated schematically in Fig. 9. As it penetrates into the scale, the partial pressure of oxygen gradually decreases, and if local equi-

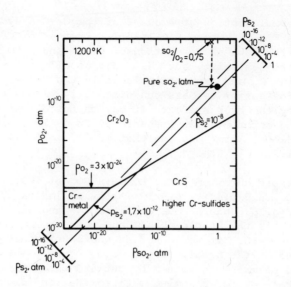

Fig. 8. An isothermal section at 1200°K of Cr-O-S system delineating the stability ranges of the Cr-metal, Cr_2O_3, and CrS as a function of the partial pressure of oxygen and sulfur dioxide. Higher sulfides of chromium exist, e.g. Cr_7S_8, Cr_5S_6, Cr_3S_4, Cr_2S_3, but only the data of CrS are included, as this is the only sulfide for which thermodynamic data are available; it is also the main sulfide in the reaction of Cr with SO_2. Sulfur isobars are shown as broken lines. "o" represents the composition of pure SO_2. Under these conditions the partial pressure of sulfur in SO_2 exceeds the decomposition pressure of CrS. "X" represents the composition of a $SO_2/O_2 = 0.75$ mixture.

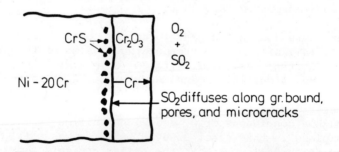

Fig. 9. Schematic representation of processes taking place during reaction of a Cr-alloy in SO_2 leading to a continuous scale of Cr_2O_3. CrS formation takes place even though the partial pressure of sulfur in the ambient gas is smaller than that needed to form CrS in equilibrium with chromium metal. This suggests that SO_2 can diffuse through the Cr_2O_3 layer via grain boundaries, microcracks,

librium with SO_2 can be assumed, eventually a sufficiently small $p(O_2)$ is reached where $p(S_2)$ becomes sufficiently large to permit sulfide formation. In Fig. 8 this reaction path may be illustrated by the vertical stippled line originating from "X".

A similar situation is found for nickel. The Ni-O-S equilibrium diagram is shown in Fig. 10 where the stability ranges of Ni-metal, NiO, and Ni_3S_2 are shown as a function of $p(O_2)$ and $p(SO_2)$ at 1200 °K (923 °C). The partial pressure of oxygen at the phase boundary NiO/Ni is 10^{-11} atm., while the partial pressure of sulfur at Ni_3S_2/Ni is 1.2×10^{-6} atm. In 1 atm. pure SO_2 the stable equilibrium phase is NiO, and the partial pressure of sulfur in the SO_2 is smaller than the decomposition pressure of the sulfide.

If nickel is preoxidized in oxygen to yield a protective NiO-layer with a thickness of at least 2000 Å and is then exposed to pure SO_2 at 1 atm., no sulfide formation is found in the metal or in the scale next to the Ni/NiO phase boundary, (44,45). Thus, for NiO-films preformed in oxygen SO_2 does not penetrate NiO in sufficient amounts to permit formation of the nickel sulfide. However, if unoxidized Ni-metal is exposed to SO_2 gas, considerable sulfide formation is found to take place. Furthermore, the reaction rate is much faster than for oxidation of nickel in oxygen gas, and as illustrated in Fig. 11 the reaction in SO_2 is after more extended reaction approximately linear with time indicating that the NiO layer is non-protective, (44-46). It is reasonable to conclude that this behaviour in SO_2 gas originates with the initial, transient period of reaction. The initial reaction may well involve both oxide and sulfide formation according to the overall chemical reaction:

$$7Ni + 2SO_2 = 4NiO + Ni_3S_2 \tag{4}$$

which with excess Ni metal is thermodynamically favourable.

However, additional SO_2 may oxidize the sulfide to the oxide as illustrated by the overall reaction

$$2Ni_3S_2 + 3SO_2 = 6NiO + 7/2S_2 \tag{5}$$

Through this reaction the local activity or partial pressure of sulfur may then become considerably larger than in the ambient SO_2 gas. An estimate using Eq. 5 and assuming $p(SO_2) = 1$ atm. and equilibrium conditions yields $p(S_2) = 6 \times 10^{-5}$ atm. The sulfur may then have two paths to take: i) reaction with Ni-metal to form additional sulfide, ii) transport outward through the NiO to the NiO/SO_2 phase boundary.

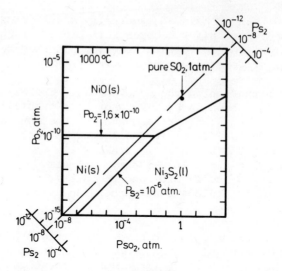

Fig. 10. Isothermal section at 1200 °K of the Ni-O-S system delineating the stability ranges of the Ni-metal, NiO, and Ni_3S_2 as a function of the partial pressure of oxygen and sulfur dioxide.

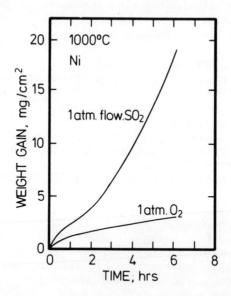

Fig. 11. Comparison between the reaction rates of Ni at 1000 °C in 1 atm. oxygen and flowing sulfur dioxide, respectively. After Goebel and Pettit, (46).

The nickel sulfide is furthermore liquid under these conditions, and the overall result apparently is then formation of a NiO-layer with microcracks and porosity. When steady state reaction is achieved, it is probable that SO_2 fills the porous NiO-layer, and if sulfur can not escape sufficiently rapidly to the ambient atmosphere, a simultaneous oxidation and sulfidation will result.

The details of the reaction behaviour are in need of further clarification. In this respect data regarding the solubility of sulfur in oxides and metals and that of oxygen in sulfides and the rates of transport of the dissolved species in the respective phases are of great interest. Only few such data are as yet available.

The combined oxidation/sulfidation may lead to considerably larger reaction rates than for oxidation alone as, for instance, illustrated above for the reaction of nickel with oxygen and sulfur dioxide, respectively. The same also applies to Ni-Cr alloys when the chromium concentration is not sufficiently large to yield protective layers of Cr_2O_3, i.e. at Cr $<\sim$ 20%. As the sulfides are generally lower melting than the oxides, and may even be liquid at the temperatures of interest, e.g. Ni_3S_2, the sulfidation process may furthermore severely affect the mechanical properties of alloys.

Once sulfidation has been started, the process may prove self-sustaining on subsequent oxidation. As the sulfides are oxidized, the liberated sulfur may penetrate further into the alloy and form fresh sulfide deeper in the alloy, (47). In this way the total amount of sulfur involved in the oxidation remains within the alloy.

HOT CORROSION

In applied systems the level of complexity may increase even further, and in this respect high temperature corrosion is no exception. Combustion gases - particularly from low quality fuels - contain various impurities, notably sodium, sulfur, and vanadium. This composite reaction gas causes an accelerated, and in cases catastrophically rapid attack which involves both oxidation and sulfidation, (48). This accelerated rate of reaction of iron-, nickel-, and cobalt-based heat resistant alloys at high temperatures is usually termed hot corrosion.

An example of this type of attack is given in Fig. 12, which shows a metallographic cross-section of a Ni-9Cr-6Al specimen hot-corroded for 96 hrs at 950 ^{o}C in a combustion gas with excess

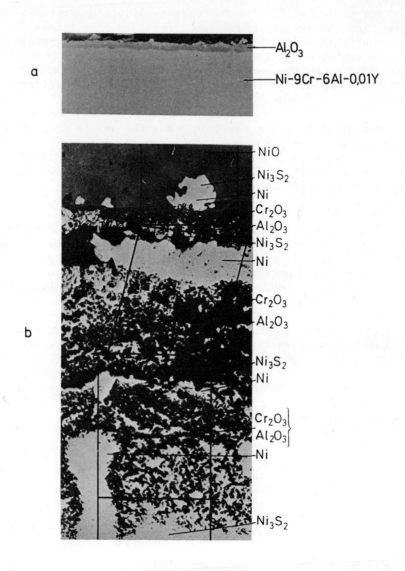

Fig. 12. Metallographic cross-sections of Ni-9% Cr-6% Al-0.01% Y specimens.
a) oxidized for 40 hrs in air at 1000 °C. 1200 x.

b) Hot-corroded for 96 hrs at 950 °C in combustion gas with excess oxygen (air/fuel ratio 50:1) and with 50 ppm sodium and sulfur, respectively. 700 x.

REACTIONS AT SURFACES AND INTERFACES

oxygen and with 50 ppm of sodium and of sulfur. The hot-corrosion leads to formation of porous oxide scales mixed with sulfides. Sulfide formation precedes oxide formation as an advancing front into the alloy. For comparison is shown a cross-section of the same specimen after 40hr in air at 1000 °C; under these conditions a protective Al_2O_3-scale is formed and the alloy exhibits excellent oxidation resistance.

The detailed composition of the reacting, ambient gas will, of course, vary with type of fuel, impurities in the air, etc., but it is generally accepted that this type of corrosion attack is caused by the formation of Na_2SO_4 on the alloy surface. Na_2SO_4 may, for instance, be formed by the reactions, (49)

$$SO_2 + \tfrac{1}{2}O_2 = SO_3 \tag{6a}$$

$$2NaCl + SO_3 + H_2O = Na_2SO_4 + 2HCl \tag{6b}$$

Pure Na_2SO_4 has a melting point of 855 °C, but with dissolved salts, e.g. NaCl, the melting point is lowered. The hot corrosion takes place when a layer of molten salt is present on the metal surface.

The hot corrosion phenomena has been the subject of numerous investigations, (48). Although many aspects are unsettled and is in need of much additional research, the extensive studies of Goebel, Pettit and Goward (33) and others have significantly contributed to the understanding of the main processes which take place.

As a basis for a brief discussion of these reactions it is convenient first to consider the stability diagram for the Na-O-S system. An isothermal section of this at 1000 °C is shown in Fig. 13, where the stability ranges are plotted as a function of $p(O_2)$ and $p(SO_3)$. The system contains four phases which at this temperature are all liquid: Na metal, Na_2O, Na_2S, and Na_2SO_4. The partial pressure of sulfur is fixed by $p(O_2)$ and $p(SO_3)$ through the equilibrium, (49)

$$2SO_3 = S_2 + 3O_2 \tag{7}$$

and the sulfur isobars can be included in the diagram as diagonal lines. The isobars corresponding to p_{S_2} equal to 1, 10^{-10}, and 10^{-20} atm are shown for illustration.

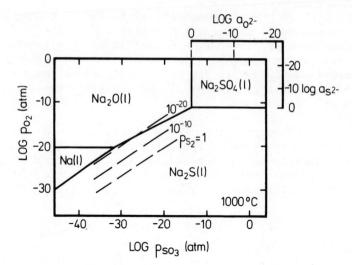

Fig. 13. Isothermal section at 1000 °C of the Na-O-S system delineating the stability ranges of Na(l), Na$_2$O(l), Na$_2$S(l), and Na$_2$SO$_4$(l) as a function of the partial pressure of oxygen and sulfur trioxide. Sulfur isobars run diagonally across the diagram, (33).

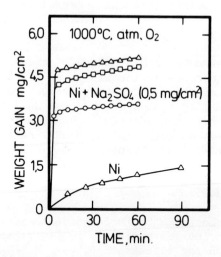

Fig. 14. Comparison of the weight gain (amount of reaction) as a function of time for nickel and Na$_2$SO$_4$-coated nickel oxidized in 1 atm. O$_2$ at 1000 °C.
After Goebel and Pettit, (46).

The region of primary interest is the Na_2SO_4 region of the diagram. In this region the activities of the Na^+ - and SO_4^{2-}-ions can by appropriate choice of standard states be taken to be unity. Other ions such as oxide and sulfide ions are also present through the equilibria

$$SO_4^{2-} = O^{2-} + SO_3 \tag{8a}$$

$$SO_4^{2-} = S^{2-} + 2O_2 \tag{8b}$$

When Na_2SO_4 is in equilibrium with Na_2O it is convenient to take the activity of the oxide ions (or Na_2O) to be unity, while the activity of the sulfide ions is unity at the Na_2SO_4/Na_2S equilibrium.

When making controlled laboratory studies of hot-corrosion it is generally convenient to coat the metal specimens with measured amounts of Na_2SO_4 and then oxidize at elevated temperatures. This method was used by Goebel et al. in their studies of hot-corrosion of nickel and nickel base alloys, (33,46).

Results of weight gain studies of oxidation of uncoated and Na_2SO_4-coated nickel at 1000 °C are shown in Fig. 14, (46). Coated nickel exhibits a greatly enhanced reaction compared to the uncoated nickel, but without continuous supply of Na_2SO_4, the reaction rate eventually subsides. The extent of the rapid oxidation increases with the amount of Na_2SO_4 originally coated on the surface. While the uncoated specimen on oxidation yields a continuous NiO-scale, the reaction products on the coated specimens are a thick porous NiO-layer and Ni_3S_2 at the metal/oxide interface similar to that found for nickel reacted in SO_2. When the reaction subsides, a continuous NiO-scale is formed at the metal/oxide interface.

To explain this reaction behaviour it is appropriate to examine the stability diagram of the nickel phases that are stable in the Na_2SO_4 region. This is shown in Fig. 15; the phases comprise Ni metal, NiO, and Ni_3S_2. Goebel et al. (46) estimate that the starting composition of the Na_2SO_4 corresponds to the cross "X", and under this condition NiO is the stable equilibrium phase. The initial reaction between nickel metal and Na_2SO_4 probably involves formation of NiO through reaction with oxygen supplied by the Na_2SO_4. This, however, establishes an oxygen activity gradient across the Na_2SO_4-layer and the oxygen activity at the Na_2SO_4/NiO interface decreases. This, in turn, increases the sulfur activity and eventually yields conditions favourable for formation of sulfide. But sulfide formation is accompanied by an increase in the oxide ion concentration, and this in turn, dissolves the nickel oxide through formation of nickelate ions:

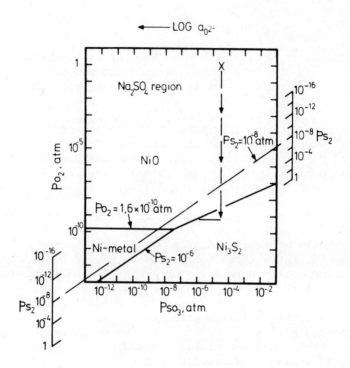

Fig. 15. An isothermal section at 1000 °C showing the stability ranges of Ni(s), NiO(s), and Ni_2S_3(l), in the Na_2SO_4-region as a function of the partial pressure of oxygen and sulfur trioxide. "X" indicates the estimated composition of the Na_2SO_4; as NiO is formed, an oxygen gradient is established in the Na_2SO_4 layer, this increases the partial pressure of sulfur at NiO/Na_2SO_4 interface and results in formation of nickel sulfide. The sulfide formation, in turn, increases the O^{2-}-activity, which leads to a basic fluxing and dissolution of NiO.
After Goebel and Pettit, (46).

REACTIONS AT SURFACES AND INTERFACES

$$NiO + O^{2-} = NiO_2^{2-} \tag{9}$$

Thus the molten sulfate serves as a basic flux.

The nickelate ions migrate through the molten Na_2SO_4 and close to the Na_2SO_4/gas interface the nickelate ions probably decompose and precipitate NiO particles. In this manner NiO is continuously dissolved and redeposited as particles, and this results in a porous non-protective scale. Eventually the Na_2SO_4 becomes saturated with nickel (as nickelate ions) and NiO is no longer dissolved; it forms a continuous layer on the metal and the accelerated attack subsides. The hot corrosion of nickel is thus not a self-sustaining reaction.

A corresponding model applies to alloys which on oxidation in oxygen forms a continuous protective scale of Al_2O_3, e.g. Ni-30% Al alloy, (33). When the alloy coated with Na_2SO_4 is exposed to oxygen, a continuous Al_2O_3 scale is initially formed. As the oxygen gradient is established across the molten Na_2SO_4 layer, the oxygen activity at the Al_2O_3/Na_2SO_4 interface decreases; this correspondingly leads to an increase in the sulfur activity, and sulfide formation may take place. This takes place as formation of aluminium sulfide particles in the alloy beneath the oxide layer probably as a result of sulfur diffusion through the scale. But as sulfur is lost from the Na_2SO_4, the oxide ion activity increases, and this eventually leads to fluxing of the Al_2O_3-scale through formation of aluminate ions:

$$Al_2O_3 + O^{2-} = 2AlO_2^- \tag{10}$$

This in turn leads to a loss of protective properties of the scale and an accelerated rate of attack in a manner similar to that described for nickel. When the Na_2SO_4 becomes saturated with aluminate ions, the Na_2SO_4-caused effects will subside. However, even after this stage the alloy may continue to oxidize at a considerably more rapid rate than that for uncoated specimens. The reason for this is that aluminium is present as sulfide particles beneath the scale, and the oxidation of the aluminium sulfide particles does not yield a continuous protective scale. Thus in this case, the Na_2SO_4-induced attack causes a lasting effect.

While these are examples of basic fluxing, aluminium oxide may as an amphoteric oxide also dissolve under conditions of acidic fluxing following the equation

$$Al_2O_3 = 2Al^{3+} + 3O^{2-} \tag{11}$$

The regions of basic and acidic fluxing and of no reaction are illustrated in the stability diagram shown in Fig. 16.

Acidic fluxing of aluminium oxide scales may take place in hot corrosion of alloys containing molybdenum, tungsten, etc. as alloying components. The oxides of these metals in their highest oxidation state, i.e. MoO_3 and WO_3, decrease the oxide ion concentration in Na_2SO_4 through the reaction of the type

$$MoO_3 + O^{2-} = MoO_4^{2-} \qquad (12)$$

and this may result in acidic fluxing (Eq. 11). In a molybdenum-containing alloy, e.g. Ni-30 Al-Mo, the hot corrosion behaviour becomes as follows: i) initially a protective Al_2O_3 scale begins to develop, ii) due to the oxidation of molybdenum to MoO_3 during the transient oxidation, formation of molybdate ions and acidic fluxing of the Al_2O_3 takes place and it dissolves as Al^{3+}, iii) at the Na_2SO_4/gas interface MoO_3 continually evaporates and this makes the outer boundary of the Na_2SO_4 less acidic and Al_2O_3 is precipitated out as particles yielding a porous non-protective oxide, iv) later the nickel metal also becomes oxidized to NiO.

This overall hot-corrosion process is self-sustaining as MoO_3 is continually formed through oxidation of the alloy and lost at the Na_2SO_4/gas interface through evaporation. This transport of MoO_3 is accompanied by acidic fluxing of Al_2O_3 and precipitation of the oxide as a porous network. This type of process may lead to catastrofically rapid reaction rates.

The alloys which are most resistant to hot-corrosion are the chromium-containing alloys, and in practice the hot corrosion resistance increases with increasing chromium content. Although thermodynamic data are not available for a detailed theoretical consideration, it is believed that the initial step in the reaction mechanism is similar to that described above for nickel and aluminium. As chromium oxide is formed on the surface and chromium sulfide in the alloy close to the alloy surface, basic fluxing dissolves the Cr_2O_3 yielding possibly - in analogy with nickel and aluminium - CrO_2^--ions. But while the transport of NiO_2^{2-} - and AlO_2^- -ions to the Na_2SO_4/gas interface precipitates porous networks of NiO and Al_2O_3, the CrO_2^--ions are further oxidized to CrO_4^{2-}-ions:

$$2CrO_2^- + O^{2-} + \tfrac{3}{2}O_2 = 2CrO_4^{2-}$$

as evidenced by the detection of chromate ions, CrO_4^{2-}-ions, in the Na_2SO_4. This prevents precipitation of Cr_2O_3 as a porous network, and also as the oxide ion activity is reduced, conditions favourable for formation of Cr_2O_3 are established. The result is a formation of a continuous, protective scale of Cr_2O_3 with good hot corrosion resistant properties.

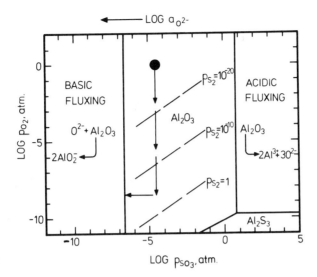

Fig. 16. An isothermal section at 1000 °C showing the phases of aluminium stable in the Na_2SO_4 region as a function of the partial pressure of oxygen and sulfur trioxide. At high O^{2-}-activities basic fluxing of Al_2O_3 takes place while at low O^{2-}-activities acidic fluxing may occur, (33).

CONCLUDING REMARKS

The preceding discussion has attempted to identify some of the main aspects of our current understanding of reactions of metals and alloys with gases at high temperatures. We are gradually achieving a better scientific understanding of the processes and principles involved. But many phenomena, and particularly gas-metal reactions under more complex conditions as for instance those involving two or more oxidants and molten salt layers on the metal surface, require extensive efforts before the detailed mechanisms are elucidated. This offers a considerable challenge both to workers in basic research and to those who are to convert this knowledge to practical deeds.

REFERENCES

1. O. Kubaschewski, E.Ll. Evans, and C.B. Alcock, "Metallurgical Thermochemistry", 4th Ed., Pergamon Press, London, 1967.

2. T. Rosenqvist, "Principles of Extractive Metallurgy", McGraw-Hill, New York, 1974.

3. O. Kubaschewski and B.E. Hopkins, "Oxidation of Metals and Alloys", Butterworths, London, 1962.

4. K. Hauffe, "Oxidation of Metals", Plenum Press, New York, 1965.

5. J. Benard, "Oxydation des Metaux", Gauthier-Villars et Cie., Paris, 1962.

6. P. Kofstad, "High-Temperature Oxidation of Metals", Wiley, New York, 1966.

7. "Conference on Clean Metal Surfaces", Ann. N.Y. Acad. Sci., $\underline{101}$ (1963).

8. "Metal Surfaces: Structure, Energetics and Kinetics", American Society for Metals and Met. Soc. of AIME, 1963.

9. Hayward, D.O. and Trapnell, B.M.W., "Chemisorption", Butterworths, London, 1964.

10. G.A. Somorjai, "Principles of Surface Chemistry", Prentice-Hall Inc., Englewood Cliffs, 1972.

11. R.E. Schlier and H.E. Farnsworth, J. Appl. Phys., $\underline{25}$, 1333 (1954); Advan. Catalysis, $\underline{9}$ (1957) 434.

12. L.H. Germer and A.U. MacRae, J. Chem. Phys., $\underline{36}$ (1964) 319.

13. A.U. MacRae, Science, $\underline{139}$ (1963) 379; Surface Sci., $\underline{1}$ (1964) 319.

14. C. Wagner, Z. physik. Chem., $\underline{B21}$ (1933) 25.

15. C. Wagner, "Atom Movements", Am. Soc. Metals, Cleveland, 1953, p. 153.

16. P. Kofstad, Corrosion, $\underline{24}$ (1968) 379.

17. R.E. Carter and F. Richardson, Trans. AIME, $\underline{200}$ (1954) 1244; $\underline{203}$ (1955) 336.

18. W.K. Chen, N.L. Peterson, and W.T. Reeves, Phys. Rev., 186 (1969) 259.

19. K. Fueki and J.B. Wagner, Jr., Z. Phys. Chem. N.F., 49 (1966) 259.

20. M.L. Volpe and J. Reddy, J. Chem. Phys., 53 (1970) 1117.

21. R. Lindner and Å. Åkerstrøm, Z. Phys. Chem. N.F., 6 (1956) 162.

22. W.C. Hagel and A.U. Seybolt, J. Electrochem. Soc., 108 (1961) 1146.

23. A.E. Paladino and W.D. Kingery, J. Chem. Phys., 37 (1962) 957.

24. P. Kofstad, "Nonstoichiometry, Diffusion, and Electrical Conductivity in Binary Metal Oxides", Wiley, New York, 1972.

25. J.A. Snide, J.R. Myers, and R.K. Saxes, Cobalt, 36 (1967) 157.

26. P. Kofstad and A.Z. Hed in "Proceedings of the Fourth International Congress on Metallic Corrosion", Amsterdam, Holland, 1972, p. 196.

27. R.M. Doerr, U.S. Bureau of Mines, Report of Investigations 6231, 1963.

28. J.A. Sartell and C.H. Li, J. Inst. Metals, 90 (1961-62) 92.

29. K. Fueki and J.B. Wagner, Jr., J. Electrochem. Soc., 112 (1965) 384.

30. G.C. Wood and I.G. Wright, Corrosion Science, 5 (1965) 841.

31. P. Kofstad and A.Z. Hed, Werkstoffe und Korrosion, 21 (1970) 894.

32. F.S. Pettit, Trans. AIME, 739 (1967) 1296.

33. J.A. Goebel, F.S. Pettit, and G.W. Goward, Met. Trans., 4 (1973) 261.

34. P. Kofstad and A.Z. Hed, J. Electrochem. Soc., 116 (1969) 224, 229, 1542.

35. W.C. Hagel, Trans. ASM, 56 (1963) 583.

36. C.S. Giggins and F.S. Pettit, J. Electrochem. Soc., $\underline{118}$ (1971) 1782.

37. G.R. Wallwork and A.Z. Hed, Oxidation of Metals, $\underline{3}$ (1971) 171.

38. I.A. Kvernes and P. Kofstad, Met. Trans., $\underline{3}$ (1972) 1511.

39. A.D. Dalvi and D.E. Coates, Oxid. of Metals, $\underline{5}$ (1972) 113.

40. J.M. Quets and W.H. Dresher, J. Mat., $\underline{4}$ (1969) 583.

41. F.S. Pettit, J.A. Goebel, and G.W. Goward, Corrosion Science, $\underline{9}$ (1969) 903.

42. A. Rahmel, Oxid. of Metals, $\underline{9}$ (1975) 401.

43. A. Rahmel, Corrosion Science, $\underline{13}$ (1973) 125.

44. C.B. Alcock, M.G. Hocking, and S. Zador, Corrosion Science, $\underline{9}$ (1969) 111.

45. M.R. Wotton and N. Birks, Corrosion Science, $\underline{12}$ (1972) 829.

46. J.A. Goebel and F.S. Pettit, Met. Trans., $\underline{1}$ (1970) 1943.

47. C.J. Spengler and R. Viswanathan, Met. Trans., $\underline{3}$ (1972) 161.

48. J.F. Stringer, "High Temperature Corrosion of Aerospace Alloys", AGARDograph No. 200, AGARD, NATO, 1975.

49. J.F.G. Condé, in "High Temperature Corrosion of Aerospace Alloys", AGARD Conference Proceedings No. 120, AGARD, NATO.

REACTION KINETICS IN THE Ca-Mn-O SYSTEM

J. M. Longo and H. S. Horowitz

Corporate Research Laboratories, Exxon Research and Engineering Company, Linden, New Jersey (U.S.A.)

INTRODUCTION

Mixed metal oxides represent a large class of inorganic materials which possess many important properties not found in the more limited binary oxides. Mixed metal oxides can stabilize high oxidation states of transition elements, have extensive ranges of either anion or cation nonstoichiometry and provide new structural arrangements not possible in a simple oxide. This very flexible class of compounds, however, is limited in applications because of the difficulties involved in the preparation of high surface area materials. The usual routes to these compounds generally involve reaction at high temperature for long periods of time. These conditions are required to overcome the slow reaction kinetics that occur when two solids are brought together. Of course these same conditions of high temperature and long times lead to crystallite growth and low surface areas. In addition, the high temperatures prevent the formation of compounds containing high valent cations such as Mn^{4+}.

Historically, the research done on the Ca-Mn-O system has been performed at high temperatures ($\geq 1000°$). The interest has usually been in the electrical and magnetic properties of the well established compositions (Ca_2MnO_4, $Ca_3Mn_2O_7$, $Ca_4Mn_3O_{10}$, $CaMnO_3$, $CaMn_2O_4$). These five compounds are, in fact, the only stable phases in the entire Ca-Mn-O system above 1000°C in air (0.2 atm O_2). Toussaint in 1964 [1] has surveyed the entire phase diagram at lower temperatures and reports two additional phases. We will show that one of his "phases" was a mixture while the other is reported with an incorrect composition. More recently Bochu et al. [2] have reported the preparation of $CaMn_7O_{12}$ at high pressures and temperature (80 kbar/1000°C).

In this paper we will describe our experiences in the manganese rich portion of the Ca-Mn-O phase diagram with particular emphasis on those factors which influence the reaction kinetics of the system. The most important factors that we investigated were temperature, starting materials, heating rate, oxygen partial pressure and residual surface species.

EXPERIMENTAL

The materials preparations that we have carried out have all been at temperatures less than 1000°C in an atmosphere of pure flowing O_2. Unless otherwise specified, the starting materials have been accurately weighed portions of calcium carbonate and manganese carbonate. The $MnCO_3$ we used was freshly precipitated from a manganese nitrate solution with large excesses of ammonium carbonate, dried at 100°C in a vacuum oven and stored in sealed containers until used. Commercially available $MnCO_3$ (brown) was unacceptable because it always contained significant amounts of oxidized manganese products. Starting mixtures of approximately 2 grams were well ground with an agate mortar and pestle and fired in ceramic boats for long periods of time to ensure equilibrium. All reaction products were examined on a Philips X-ray diffractometer to determine which phases were present. Oxygen content of all phases was established using a Fisher Thermogravimetric Analyzer containing a Cahn electrobalance. Samples were reduced in H_2 and weight loss was attributed to manganese with oxidation states higher than 2+.

RESULTS AND DISCUSSION

The results of our phase diagram study of the manganese rich portion of the Ca-Mn-O system are presented in Figure 1. It can immediately be seen that by working at lower temperatures, three compounds appear in the phase diagram that are not stable at 1 atm O_2 above 1000°C. The first compound to appear is $CaMn_7O_{12}$ which is also the most stable mixed valent compound of the system. It contains six Mn^{3+} and one Mn^{4+} and has a structure related to perovskite (2), with three Mn^{3+} and a Ca^{2+} on the A site. Joubert, in a private communication, reports that they also have been able to prepare this phase without high pressure. Above 950°C $CaMn_7O_{12}$ breaks down into $CaMn_2O_4$ and Mn_2O_3.

At lower temperatures (T < 925°C) another mixed valent phase appears with a Ca/Mn ratio of 1/3. Thermogravimetric analysis in H_2 shows that one-third of the manganese are present as Mn^{4+} and therefore indicates a formula of $CaMn_3O_6$. Toussaint (1) reports a "$CaMn_3O_7$" in his study but his published X-ray pattern shows only the strong lines of $CaMn_4O_8$. He states that there were small amounts of the calcium rich phase $CaMnO_3$ present in his preparation of "$CaMn_3O_7$". Below 900°C a compound with a Ca/Mn ratio of 1/4 appears in the phase diagram. In this case thermogravimetric analysis shows that one-half of the manganese are present as Mn^{4+} leading to a formula of $CaMn_4O_8$. Toussaint (1) reports a phase "$CaMn_4O_7$" but his X-ray data shows it to be a mixture containing predominantly $CaMn_7O_{12}$. A thorough analysis of our own X-ray patterns for $CaMn_3O_6$ and $CaMn_4O_8$ is not complete, but it does appear that these two phases are related structurally. They can both be written as Ca_xMnO_2 where $x = 1/3$ for $CaMn_3O_6$ and $x = 1/4$ for $CaMn_4O_8$ suggesting that they are related to the A_xMnO_2 phases described by Professor Hagenmuller's laboratory (3).

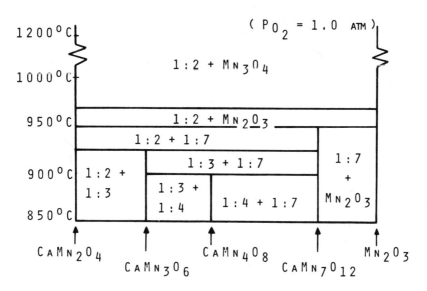

Figure 1 - Phase Diagram of the Ca-Mn-O System

As might be expected from the presence of a number of mixed valent phases, the reaction kinetics of the lower (< 1000°C) temperature phase diagram are very sensitive to a number of experimental variables. We find that a mixture of carbonates is much more reactive than metal oxides and is therefore better suited to a study of equilibrium phase relations. The decomposition of $Mn^{2+}CO_3$ in oxygen leads to the in situ formation of reactive manganese oxides. Care must therefore be taken to use the optimum temperature schedule during reaction. If a mixture of carbonates is brought directly to temperature and fired for a prolonged period of time without grinding, equilibrium is difficult to attain. Generally a mixture of very crystalline phases results, and further solid state reaction is severely retarded. However if the reaction is carried out at increasingly higher temperatures with frequent regrindings, the attainment of equilibrium is much easier. This is particularly important during the early stages of reaction when the carbonates have decomposed to their respective oxides.

Subtle changes in the oxygen partial pressure during reaction have rather dramatic effects on the reaction kinetics. Starting with a mixture of acetates rather than carbonates does not lead to single phase products as easily. Presumably, during decomposition of the acetates, a CO/CO_2 atmosphere is generated at the reaction interfaces. The initial reducing atmosphere will favor formation of phases with Mn^{3+} which must then back react to form the equilibrium phase. In a similar manner, grinding the starting materials or intermediate products under acetone has a marked effect on the ability to attain equilibrium. In fact grinding under acetone and then firing causes a sufficiently reducing atmosphere that $CaMn_4O_8$ (which has the highest Mn^{4+} content) can not be formed. This means that even though the reaction is carried out at the right temperature and oxygen partial pressure, the reaction kinetics for the formation of $CaMn_4O_8$ from $CaMn_3O_6$ and $CaMn_7O_{12}$ are very slow.

The decomposition temperatures for the three phases containing both Mn^{3+} and Mn^{4+} are very sensitive to the equilibrium oxygen pressure. Once the phases have been formed by control of temperature treatment at 1 atm O_2, they can be decomposed by switching to flowing air or more dramatically by firing in stagnant air. For example, $CaMn_4O_8$ whose decomposition temperature in flowing O_2 is 900°C, will decompose if fired at 800°C in stagnant air. The effect on $CaMn_3O_6$ and $CaMn_7O_{12}$ is less pronounced presumably because they contain a lower percentage of Mn^{4+}.

It is interesting to note that for compounds in the manganese rich portion of the phase diagram there is a very clear correlation between the percent Mn^{4+} and the decomposition temperature in 1 atm O_2 and in stagnant air. The difference between these two decomposition temperatures is also directly related to the Mn^{4+} content of the phase. These data are summarized in the table below.

Table 1

Decomposition Temperatures of Ca-Mn-O Phases

	% Mn^{4+}	Decomposition Temperature (°C)	
		1 atm O_2	Stagnant Air
$Ca[Mn_2^{3+}]O_4$	0	>1425	>1425
$Ca[Mn_6^{3+}Mn^{4+}]O_{12}$	14	950	925
$Ca[Mn_2^{3+}Mn^{4+}]O_6$	33	925	850
$Ca[Mn_2^{3+}Mn_2^{4+}]O_8$	50	900	800
$[Mn^{4+}O_{1.991}]$	99	675 (4)	

REFERENCES

(1) H. Toussaint, Revue Chimie Minerale 1, 141 (1964).

(2) B. Bochu, J. Chenavas, J. C. Joubert and M. Marezio, J. Sol. State Chem. 11, 88 (1974).

(3) C. Fouassier, C. Delmas and P. Hagenmuller, Mat. Res. Bull. 10, 443 (1975).

(4) D. S. Freeman, P. F. Pelter, F. L. Tye and L. L. Wood, J. Appl. Electrochem. 1, 127 (1971).

DISCUSSION

R. METSELAAR (University of Technology, Eindhoven, Netherlands). Have you considered the method of spray drying solutions for the preparation of Ca-Mn-O samples? This often leads to powders with high surface area and reactivity.

AUTHORS' REPLY: We have considered but not used this method. It cannot provide mixing on an atomic (10Å) scale, whereas a solid presursor can.

A.K. GALWEY (Queens University of Belfast, Northern Ireland, U.K.) Have you considered the thermal decomposition of calcium permanganate as a preparative route to Ca-Mn mixed oxide phases? While in principle only one Ca/Mn ratio would be used, the phases generated at the interface of a kinetic process could contain high area materials of varied composition. There is still some discussion on the decomposition of $KMnO_4$, a reaction which proceeds at approximately $470°K$.

AUTHORS' REPLY: We have not studied this decomposition as a route to Ca-Mn-O phases. Even though it is subject to the limitation you describe, it should be interesting to examine the decompsition for the preparation of $CaMn_2O_4$.

K.J. de VRIES (University of Technology, Twente, Netherlands) Did you check that in $CaMnO_3$ prepared by thermolysis of the carbonate for one hour no carbon containing species remained? Such residues might be expected to influence the valence state of magnesium and possibly cause problems during densification of the material by ceramic processing.

AUTHORS' REPLY: Based on the methods available to us, we feel there is no carbon containing residue in our materials. The role of carbon is important and showed up most dramatically when we washed our carbonate solid solution presursors with organic solvents such as acetone. The local reducing atmosphere produced during decomposition of the organic residue forced us to work at higher temperatures in order to obtain a single phase product.

D. DOLLIMORE (University of Salford. U.K.) You comment that the preparation of $MnCO_3$ is important to the preparation of ceramic samples. In view of the fact that it is possible for example, to mix two $CaCO_3$ samples together and obtain two DTA peaks due to the decomposition of each separately, do you not consider preparation of $CaCO_3$ to be equally important? Can you give details of the form of $CaCO_3$ and of possible effects of variations in the form of $CaCO_3$ used?

AUTHORS' REPLY: Your point about the importance of variations in particle and agglomerate size on the kinetics of decomposition is an important one. All forms of $CaCO_3$ were first dissolved, mixed with $MnCO_3$ solution and then reprecipitated as a solid solution. Therefore, the initial form of $CaCO_3$ or $MnCO_3$ was not a parameter in our studies.

SURFACE REACTIVITY TOWARDS OLEFIN OXIDATION OF CADMIUM MOLYBDATE DOPED WITH TRANSITION METAL IONS

L.Burlamacchi[¶], G.Martini[¶], S.Simonetti[†], and B.Tesi[†]

Istituto di Chimica Fisica (¶) and Istituto di Chimica Applicata (†); Università di Firenze; Firenze; (Italy)

INTRODUCTION

Molybdate catalysts, when selective in the olefin or alcohol oxidation, are thought to use lattice oxygen for the catalytic reaction (1-4), while gaseous oxygen only restores the original stoichiometry of the catalyst. On the contrary, total oxidation (mainly to CO_2) was suggested to occur at the expense of oxygen adsorbed in an active form on the surface (5,6). Therefore, the non-selective total oxidation given by cadmium molybdate was associated to a strong oxygen adsorption together with the lack of red-ox partner (7).

In this work, the high sensitivity toward structure and localization of paramagnetic centres shown by the Electron Spin Resonance technique is employed for studying the migration and the coordination of transition metal ions (namely Mn^{2+} and Fe^{3+}) added to $CdMoO_4$. The influence of differently localized ions on the catalytic activity is further investigated by flow reactor measurements.

EXPERIMENTAL

$CdMoO_4$ was obtained as described in ref. (7). The appropriate amount of Fe(III) or Mn(II) (nitrate salts) were added to $CdMoO_4$ in two ways: A) impregnation from the salt solutions, and B) coprecipitation in order to obtain atomic ratios Me:Cd:Mo = x:1-x:1 (with x in the range 0.005-0.05).

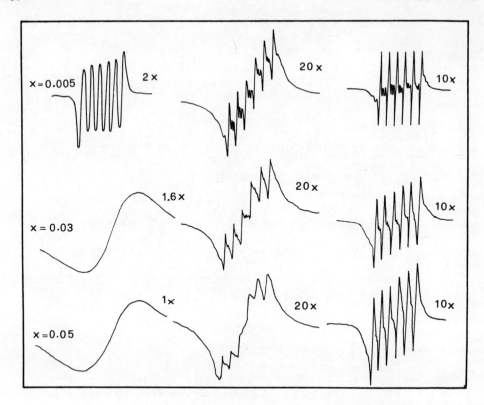

Fig. 1 - ESR spectra of Mn A samples. On the left, simple air drying; on the center, heating at 120 °C for 16 h; on the right, activation at 500 °C for 16 h.

Surface area of all samples, measured by the BET method, are $\sim$ 4-4,5 sqm/g.

Catalytic activity was measured with a flow reactor filled with 7.5 g of catalyst. Flow of the gases was 12 cm^3/min (propene 3%, O_2 35%, N_2 62%); the contact time was 25 sec. The products were analyzed with a Hewlett-Packard gas-cromatograph model 5750 G with Poropak Q columns and thermal conductivity and flame ionization detectors.

ESR spectra were registered with a Varian V4502 spectrometer.

RESULTS AND DISCUSSION

Fig. 1 shows the ESR signals of type A Mn(II) in $CdMoO_4$ at various x values after different heating treatment. The first step

of the preparation of the samples (simple air drying) consists in
covering the surface by a water phase which still contains hexasol-
vated Mn(II), giving rise to liquid-type ESR spectra. Similar results were previously reported for Mn(II) adsorption on γ-alumina
(8). The linewidth of the individual hpf components increases with
the concentration of Mn(II), due to spin-spin intermolecular dipolar interaction. After treatment at 120 °C, desolvation of the Mn(II)
occurs, giving rise to two superimposed solid-type spectra. The
first one is the known sextet with resolved forbidden transitions.
The second is a broader, unresolved spectrum, probably exchange
narrowed to a certain extent. Activation at 500 °C, for 16 h,
strongly reduces the intensity of the broader spectrum, while the
resolved sextet remains practically unchanged. Its intensity is
almost equal in the various samples and substantially less than
expected from the amount of Mn(II) contained in the samples, corresponding roughly to x = 0.0005. It must be attributed to Mn^{2+}
ions migrated to highly symmetrical lattice sites. The broad

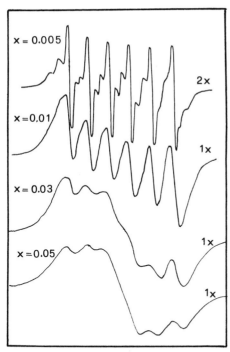

Fig. 2 - ESR spectra of Mn B samples activated at 500 °C.

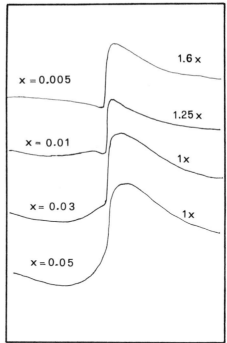

Fig. 3 - ESR spectra of Fe A samples activated at 500 °C.

spectrum can be attributed to the excess Mn(II) which was not accepted in the lattice sites. Its location is most probably on the surface in a dispersed state. Oxidation to higher manganese oxides by heating at 500 °C accounts for the disappearance of the broad signal. We can deduce, therefore, that in samples, above $x \sim 0.0005$, two Mn phase exist, the first being lattice Mn(II), while the second one consists of adsorbed manganese species.

Obviously in samples B the liquid adsorption phase and the desolvation step do not exist, all Mn(II) occurring in lattice positions. Indeed, no dependence of the ESR lineshape is observed upon heating. Fig. 2 shows the ESR spectra from Mn(II) in samples B activated at 500 °C. When the Mn amount is small (x less than 0.005) no differences are observed between samples A and B. This suggests that the same lattice sites are occupied by Mn(II) in both cases. Since the spectral resolution accounts for isolated ions, this would suggest that the lattice ion fraction after impregnation has rapidly migrated into the bulk even at 120 °C. Such a process may involve intergranular defect sites accounting for the limitation of the available sites. The increase of the Mn content leads to a small dipolar broadening of the hpf components. The larger effect, however, is the appearance of an exchange narrowed spectrum particularly evident in the 5% Mn sample. Due to the difference in the crystal structure of $CdMoO_4$ and $MnMoO_4$ (9,10), the latter should form a new phase in which spin exchange dominates the spin Hamiltonian parameters.

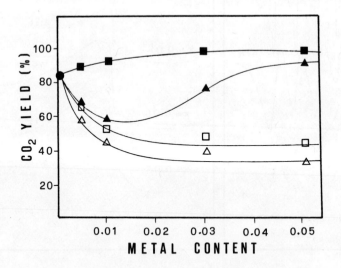

Fig. 4 - Propene conversion (%) to CO_2 at 400 °C as a function of the metal content in cadmium molybdate: ■ Fe(III) A samples; □ Fe(III) B samples; ▲ Mn(II) A samples; △ Mn (II) B samples.

Fig. 3 shows the ESR spectra of Fe(III) in $CdMoO_4$ (type A) after activation at 500 °C. The relatively narrow signal ($\Delta H \sim$ 100 G, g = 2.00), which is predominant at low Fe(III) content and whose intensity seems do not depend on the concentration, can be attributed to Fe(III) ions located in lattice sites with probably tetrahedral environment. In this case too, a limited number of sites is available for the migration. The large unsymmetrical signal superimposed on the above spectrum should be attributed to amorphous oxides adsorbed on the surface. Samples B give always symmetrical ESR signals centered at g = 2.05, whose linewidth decreases from 1200 G at 0.5% Fe up to 500 G at 5% Fe. The signals are attributable to Fe(III) in almost octahedral environment in the lattice sites.

Fig. 4 shows the propene conversion to CO_2 at 400 °C with pure and metal containing $CdMoO_4$ activated at 500 °C. The runs were also extended in the range 300-500 °C. After the catalytic runs, the ESR spectra of the samples did not show variations. In the absence of oxygen, samples A and B give rise to very low propene conversion which is comparable to that observed with $CdMoO_4$. The ESR spectra of Mn A samples after catalytic runs in the absence of oxygen show a significant recovery of the broad signal intensity, while in the Fe samples a decrease of the broader spectrum is observed. No variations occur on B samples after reduction with propene.

The results of fig. 4 show that the total oxidation is influenced by the presence of metal ions. When iron is predominantly located on the surface as amorphous reducible oxides (samples A), the CO_2 formation is enhanced. The same result is given by high manganese content (x > 0.02) while an initial decrease of the catalytic activity is observed for Mn <1%. This decrease appears of the same order as that observed on coprecipitated samples, and stabilizes at $\sim$ 40% conversion for x > 0.02. Since Fe_2O_3 and MnO_2 are known to give high conversion of olefins to CO_2 (11,12), the increase of activity of samples A can clearly attributed to the formation of such oxides supported on the $CdMoO_4$ surface.

At low metal content (x < 0.01), ESR suggests that, in contrast to Fe(III), Mn(II) occupies the same defect sites in samples A and B. This fact further suggests that the sites to be occupied are involved in the catalytic process, possibly influencing the adsorption of molecular oxygen. Further increase of the metal content above x $\sim$ 0.02 and the formation of a new condensed phase seem not to influence the catalytic activity.

Aknowledgments

Thanks are due to the National Council of Research (CNR) for the financial support. The authors are also indebt to Prof. E. Ferroni for the useful discussion during the elaboration of this paper.

REFERENCES

1) Ph.Batist, C.J.Kaptijens, B.C.Lippens, and G.C.A.Schuit; J.Catalysis; $\underline{7}$ (1967), 33.
2) I.Pasquon, F.Trifirò, and P.Centola; J.Catalysis; $\underline{10}$ (1968), 86.
3) J.M.Peacock, A.J.Parker, P.G.Ashmore, and J.A.Hockey; J.Catalysis; $\underline{15}$ (1969), 387.
4) L.Burlamacchi, G.Martini, and E.Ferroni; J.C.S.Faraday I; $\underline{68}$ (1972), 1586.
5) J.Haber, and B.Grzybowska; J.Catalysis; $\underline{28}$ (1973), 489.
6) G.K.Boreskov, V.V.Popovskii, and V.A.Sazonow; Proc. Int. Congr. Catalysis 4th (Russ. ed.), 1970, pg. 343.
7) L.Burlamacchi, G.Martini, F.Trifirò, and G.Caputo; J.C.S. Faraday I; $\underline{71}$ (1975), 209.
8) L.Burlamacchi, P.L.Villa, and F.Trifirò; React.Kin.Catal. Letters; in the press.
9) A.V.Chicagov, L.N.De'myanets, V.V.Ilyukhin, and N.V.Belov; Soviet Physics, Crystallography; $\underline{11}$ (1967), 588.
10) S.C.Abrahams, and J.M.Reddy; J.Chem.Phys.; $\underline{43}$ (1965), 2533.
11) Y.Moro-oka, and A.Ozaki; J.Catalysis; $\underline{5}$ (1966), 116.
12) Y.Moro-oka, Y.Morikawa, and A.Ozaki; J.Catalysis; $\underline{7}$ (1967), 23.

DISCUSSION

B. DELMON (Catholic University of Louvain, Louvain-la-Neuve, Belgium) There are no specific surface area data in your communication. Fig.4 corresponds to overall yield and not to intrinsic (i.e. per unit surface area) activity. We have observed in many systems that surface area changes considerably with composition, even when adding a small fraction of a percent of a foreign ion.

AUTHORS' REPLY: We have measured the surface area of all samples using a Perkin Elmer Sorptmeter (B.E.T. method). The values were in the range 4-4,5 sqm/m, without large variations.

W. GUNSSER (Technical University, Hannover, West Germany) For small dopant concentrations, did you examine variations in relative intensities and in the coupling constant?

AUTHORS' REPLY: In the case of impregnated catalysts, the signal intensity roughly stabilizes above 0.5% Mn content. However, we did not attribute large significance to the measured intensity nor to the forbidden hyperfine transitions.

CHEMISORPTION OF SULFUR ON IRON AND ITS INFLUENCE ON IRON-GAS

REACTIONS, SURFACE SELF DIFFUSION AND SINTERING OF IRON

H.J. Grabke, W. Paulitschke, S.R. Srinivasan

Max-Planck-Institut für Eisenforschung GMBH

D - 4000 Düsseldorf

ABSTRACT

Segregation equilibria of sulfur on iron single crystals are studied at elevated temperatures, 600-900°C, in an UHV apparatus using LEED for the observation of chemisorption structures and AES for the determination of surface concentrations. Even at very low sulfur activities, as established by the concentration of dissolved sulfur (10 and 27 ppm), the surface is covered with an ordered layer of chemisorbed sulfur.

In other experiments, the sulfur activity was established by the H_2S/H_2 ratio in a flowing gas phase at atmospheric pressure. The carburization of iron in CH_4-H_2 and the nitrogenation in N_2 or NH_3-H_2 are strongly retarded by chemisorbed sulfur. The surface self diffusion coefficient of iron is increased by the presence of chemisorbed sulfur.

SURFACE SEGREGATION STUDIES

Iron single crystals with a known sulfur concentration were introduced in the UHV chamber, cleaned by sputtering and heated to the temperature range 600-900°C. In this way equilibria are established

$$S \text{ (dissolved)} = S \text{ (chemisorbed)} \qquad (1)$$

for which the thermodynamic activity of sulfur is determined by the concentration of dissolved sulfur. The desorption of sulfur is negligible. Auger electron spectra were obtained at the equilibrium temperature and also the LEED pattern could be observed at temperatures up to about 750°C. The measurement of the sulfur surface concentration by AES was calibrated using samples doped with S^{35}. After the equilibration in the UHV chamber and the AES measurement the samples were quenched and the radioactivity on the surface was measured in a methane proportional counter.

Samples with 10 ppm S and 27 ppm S and with the orientations (100), (110) and (111) have been studied in a temperature range in which no threedimensional sulfide can be formed. In that range the surface concentration of sulfur is virtually independent of the solute concentration and the temperature, and somewhat dependent on the orientation. The atom ratio is about S/Fe ≃ 0.5 for all orientations. The LEED patterns indicate the c(2 x 2) structure on Fe(100) and a (1 x 1) structure on Fe(111), as indicated by intensity changes on sulfur adsorption. A (2 x 4) structure is formed on Fe(110), but only after quenching at room temperature. The detected structures are chemisorption structures and not "twodimensional sulfides" (1). The sulfur atoms are arranged in an ordered structure on the iron surface - the iron atoms remain in the metal lattice and are not built in the chemisorption layer. Presumably the sulfur atoms are mobile on Fe(110) at high temperatures so that the adsorption structure is formed only on quenched samples.

SURFACE REACTION KINETICS

The carburization of iron in CH_4-H_2 mixtures

$$CH_4(gas) = C(dissolved) + 2 H_2(gas) \qquad (2)$$

and the nitrogenation according to

$$NH_3(gas) = N(dissolved) + \frac{3}{2} H_2 \qquad (3)$$

or

$$N_2(gas) = 2 N(dissolved) \qquad (4)$$

were measured in a flow apparatus at atmospheric pressure by a resistance relaxation technique (2,3,4). On thin iron foils (12.7 μm) these reactions are controlled by the dissociation of the molecules on the iron surface (2,3). The sulfur activity was imposed by the H_2S/H_2 ratio in the gas phase, the chemisorption equilibrium was established:

$$H_2S(gas) = S(chemisorbed) + H_2(gas) \qquad (5)$$

The polycrystalline iron foils show a recrystallization texture, large crystals of 50-200 μm diameter with (100) faces are prevailing.

Therefore the reported kinetics can be attributed to the (100) face. In Fig. 1 the initial rate of reaction (2) is shown in dependence on the sulfur activity. The reaction is strongly retarded by chemisorbed sulfur even at a sulfur activity, which is only 10^{-4} of the sulfur activity at the equilibrium Fe-FeS and corresponds to a concentration of dissolved sulfur of 0.01 ppm S.

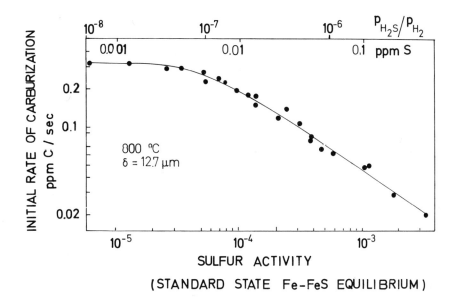

Fig. 1. Initial rate of carburization in a H_2-1%CH_4 mixture at 800°C, measured on a 12.7 μm thick iron foil, plotted double-logarithmically vs. the sulfur activity.

Also for the nitrogenation in NH_3-H_2, which was measured at 700°C and 800°C and for the nitrogenation in N_2 at 850°C, the reaction rate is retarded with increasing pH_2S/pH_2 at sulfur activities $> 10^{-4} \cdot (pH_2S/pH_2)_{Fe-FeS}$. The dependence of the reaction rate v on the sulfur activity can be described by

$$d\log v/d\log(pH_2S/pH_2) = m$$

where the constant is $m = -0.65$ for reaction (2), $m = -0.5$ for reaction (3) and $m = -2$ for reaction (4).

SURFACE SELF DIFFUSION AND SINTERING

The influence of adsorbed sulfur on the surface self-diffusion on iron was studied measuring the grain boundary grooving on large grain iron polycrystals during annealing in pure H_2 and in H_2S-H_2 atmospheres. Controlled H_2S-H_2 mixtures were obtained by flowing hydrogen through a quartz tube with Fe-FeS at the annealing temperature and adding more or less hydrogen. The groove widths were measured by interference microscopy. Experiments were performed on α-iron at the temperatures 790°, 820° and 890°C, the samples were annealed in pure H_2 and H_2S-H_2 mixtures.

The increase of groove width with annealing time is analyzed using the method of Robertson and Srinivasan (5), in which the volume diffusion contribution is taken into consideration. Values are obtained for the product of the self-diffusion coefficient $D_s [cm^2/sec]$ and the surface energy $\gamma [erg/cm^2]$. One typical result for measurements at 820°C is given below

pH_2S/pH_2:	0	$1 \cdot 10^{-3}$	$1.83 \cdot 10^{-3}$
$D_s \cdot \gamma$:	$6.9 \cdot 10^{-5}$	$26 \cdot 10^{-5}$	$21 \cdot 10^{-5}$

It was initially intended to derive the isotherm for the surface concentration in dependence on the sulfur activity, Γ_S vs. a_S, from the surface segregation experiments. Then the surface energy at different sulfur activities may be calculated according to Gibbs' adsorption isotherm

$$\left(\frac{\delta \gamma}{\delta \ln a_S}\right)_T = - RT \cdot \Gamma \quad (5)$$

and D_s can be obtained from the measured values for $D_s \cdot \gamma$. But this was not possible, since even at a concentration of 10 ppm S the surface is already saturated with chemisorbed sulfur and segregation experiments with the low sulfur concentrations down to 0.01 ppm as needed according to Fig. 1 would lead to sulfur depletion in the bulk.

Anyway, at all temperatures the product $D_s \cdot \gamma$ is higher for the experiments in H_2S-H_2 mixtures than for the experiments in pure hydrogen. Since according to Gibbs' adsorption isotherm the surface energy should continuously decrease with increasing sulfur activity, this result indicates a strong increase of the surface self-diffusion coefficient caused by the sulfur chemisorption.

It is already known from other systems, Cu-S (6) and Ag-S (7), that sulfur chemisorption can increase D_s - and several hypotheses have been put forward to explain this phenomenon. Rhead (8) assumed that a two-dimensional compound is formed which melts at a temperature below the melting point of the substrate and forms a quasi two-dimensional liquid in which diffusion is rapid. For the system Fe-S this hypothesis can be excluded, since ordered chemisorption structures were observed in the same temperature range as the increased surface diffusion.

According to the results on $D_s \cdot \gamma$ the sintering of iron powder should be accelerated by the presence of sulfur. Dilatometric experiments on the sintering of porous iron pellets in H_2 and in H_2S-H_2 atmospheres could not confirm this conclusion. The rate of shrinkage at 880°C seems to be independent of the H_2S/H_2 ratio.

CONCLUSIONS

LEED and AES studies at elevated temperatures show that the iron surface is covered with a nearly saturated layer of chemisorbed sulfur atoms even at very low activities of sulfur. According to kinetic studies (Fig. 1) at 800°C, the saturation is approached at about 0.01 ppm S in solid solution, resp. $pH_2S/pH_2 \simeq 2 \cdot 10^{-7}$ or $pS_2 \simeq 10^{-17}$ atm. Reactions of iron with gases, the carburization or the nitrogenation, are strongly retarded by the presence of sulfur since the dissociation of the molecules can take place only in the few vacancies of the chemisorbed layer. The surface self-diffusion of iron is enhanced by the presence of sulfur - presumably the mobility and/or the concentration of iron adatoms is increased on the chemisorbed layer.

REFERENCES

1. J. Bénard, Catalysis Reviews 3 (1), 93 (1967)
2. H.J. Grabke, Ber. Bunsenges. physik. Chem. 69, 409 (1965)
3. H.J. Grabke, Ber. Bunsenges. physik. Chem. 72, 533,541 (1968)
4. H.J. Grabke, Metallurg. Trans. 1, 2972 (1970)
5. W.M. Robertson, S.R. Srinivasan, Metallurg. Trans. 6A, 1653 (1975)
6. H.E. Collins, P.G. Shewmon, Trans. AIME 236, 1354 (1966)
7. J. Perdereau, G.E. Rhead, Surface Sci. 7, 175 (1967)
8. G.E. Rhead, Surface Sci. 15, 353 (1969)

ACKNOWLEDGEMENTS

The authors are grateful to Ms. E. Petersen for assistance with the kinetic experiments. The work was supported in part by the A.I.F. (Bundesministerium für Wirtschaft) and by the DFG (Deutsche Forschungsgemeinschaft).

DISCUSSION

J. HABER (Academy of Sciences, Krakow, Poland) It has been found that the surface composition of a solid solution is sensitive to the composition of the gas phase and differs from that of the bulk. For example, Szymerska recently found (J. Catalysis $\underline{41}$, 197 (1976)) that in the presence of gas phase hydrogen, sulphur migrates from the bulk to the surface of palladium crystals the effect being reversible. Would a similar effect not be expected to operate in the present case, where the surface sulphur activity is determined by H_2S/H_2 mixtures, leading to uncertainties in the surface sulphur concentration.

AUTHORS' REPLY: In our UHV studies the chemisorption equilibrium was established by the segregation

$$S(dissolved) = S(chemisorbed)$$

and in our studies at atmospheric pressure in H_2S-H_2 mixtures the equilibrium was established by the reaction

$$H_2S(gas) = H_2(gas) + S(chemisorbed)$$

Both equilibria are coupled by the well-known equilibrium

$$H_2S(gas) = H_2(gas) + S(dissolved).$$

Thus it cannot be understood why the presence of hydrogen should increase the amount of chemisorbed sulfur. Even in the presence of pure hydrogen desulphurisation should occur, and the amount of dissolved and chemisorbed sulfur decrease continuously.

$$S(dissolved) \quad\quad S(chemisorbed) \quad\quad\quad H_2S(gas)$$
$$\text{Diffusion} \quad\quad\quad\quad\quad + H_2$$

The effect reported by the questioner is very peculiar and cannot apply to our studies.

M. FIGLARZ (University of Picardie, Amiens, France) You say that two-dimensional sulfides observed by other authors in the case of copper-sulfur and silver-sulfur are due to non-equilibrium conditions. Did you observe this type of compound for the iron-sulfur system in non-equilibrium conditions.

AUTHORS' REPLY: Benard, Oudar et al (1-4) reported on the formation of two dimensional sulfides, the symmetry of which is different from that of the substrate, under non-equilibrium conditions, upon imposing pressures of H_2S of the order 10^{-7} - 10^{-1} Torr at room and elevated temperatures in the LEED-system. For the Fe-S system they observed at first formation of an adsorption structure and after that two-dimensional sulfides.

In segregation equilibrium we observed only adsorption structures in the system Fe-S, never two-dimensional sulfides. Only in the case of oversaturation did we observe a reconstruction of the surface When for example a sample with the orientation (110) was held at low temperatures, where it was oversaturated with dissolved sulfur, a reconstruction of the surface and formation of a complicated structure was seen.

We did not study metals other than Fe and therefore conclude from our restricted work, that all two dimensional compounds observed are non-equilibrium metastable structures.

1. J. Bénard, Catalysis Rev. 3 (1), 93 (1969)
2. E. Margot, J. Oudar, J. Bénard, C.R. Acad.Sci.Fr. 270,1261(1970)
3. J. Oudar, E. Margot, Colloque C.N.R.S. n° 187 (1969)
4. J. Oudar, Bull.Soc.fr. Minéral. Cristallogr. 94, 225,(1971)

THE INFLUENCE OF INTRINSIC DEFECTS ON THE MECHANISM OF THE SOLID STATE REACTION BETWEEN CdTe AND HgSe

V. Leute, W. Stratmann

Institute of Physical Chemistry

University Münster, BRD

ABSTRACT

The behaviour of the double conversion between HgSe and CdTe is strongly influenced by intrinsic doping. Because of the miscibility gap in the reacting system the reaction paths with chalcogen saturated and metal saturated samples are totally different. These differences are explained using the diffusion coefficients of the corresponding quasibinary systems.

EXPERIMENTS

The II-VI compounds CdTe and HgSe are prepared from the elements. Crystals are pulled by the Bridgeman technique. Definite densities of intrinsic defects are achieved by annealing the crystals in metal vapour (Cd or Hg) or in chalcogen vapour (Se or Te). CdTe- and HgSe-crystals with polished surfaces are pressed against each other within a matrix consisting of the polycrystalline substance of one of the reacting compounds. The reactions are carried through in sealed and evacuated quartz ampules. The reaction zones between the original crystals are analysed by an electron beam microprobe. By mechanical line scanning X-ray profiles are recorded from which mole fraction profiles are calculated using experimental calibration curves. The interdiffusion coefficients have been calculated from concentration profiles measured with quasibinary diffusion couples. The values of Tab.1 have been ob-

cm^2/s	HgSe	HgTe	CdTe	CdSe	
$\tilde{D}_{ch}^{450°C}$	$2 \cdot 10^{-14}$	10^{-14}	$4 \cdot 10^{-16}$	$2 \cdot 10^{-17}$	
$\tilde{D}_{ch}^{550°C}$	10^{-12}	$3 \cdot 10^{-12}$	$2 \cdot 10^{-14}$	10^{-15}	metal saturated
$\tilde{D}_{me}^{450°C}$	$2 \cdot 10^{-10}$	$5 \cdot 10^{-12}$	$2 \cdot 10^{-14}$	$6 \cdot 10^{-13}$	
$\tilde{D}_{me}^{550°C}$	$2 \cdot 10^{-9}$	$1.5 \cdot 10^{-10}$	$8 \cdot 10^{-13}$	10^{-11}	
$\tilde{D}_{ch}^{450°C}$	$3 \cdot 10^{-11}$	$3 \cdot 10^{-9}$	10^{-13}	10^{-15}	
$\tilde{D}_{ch}^{550°C}$	$3 \cdot 10^{-10}$	10^{-6}	$3 \cdot 10^{-12}$	$3 \cdot 10^{-14}$	chalcogen saturated
$\tilde{D}_{me}^{450°C}$	$7 \cdot 10^{-11}$	$7 \cdot 10^{-12}$	10^{-14}	$2 \cdot 10^{-13}$	
$\tilde{D}_{me}^{550°C}$	$8 \cdot 10^{-10}$	$3 \cdot 10^{-10}$	$6 \cdot 10^{-13}$	$4 \cdot 10^{-12}$	

Tab.1: Interdiffusion coefficients of II-VI-compounds

tained by extrapolation of the concentration dependent interdiffusion coefficients to the pure binary substances. From the dependence of these coefficients on intrinsic doping, i.e. on excess metal or excess chalcogen content, conclusions about the diffusion mechanism can be drawn.

DIFFUSION IN QUASIBINARY SYSTEMS

Measurement of the chalcogen interdiffusion in chalcogen saturated diffusion couples HgSe/HgTe and CdSe/CdTe yields diffusion coefficients which are several orders of magnitude higher than in the case of metal saturated compounds. From this we conclude that in these systems chalcogenes always diffuse via interstitials.

In the systems HgSe/CdSe and HgTe/CdTe the metal exchange is only slightly affected by intrinsic doping. Taking into consideration some other investigations (Kirkendall effect, Hall effect), we can conclude that in these systems Cd and Hg diffuse via interstitials as well as via vacancies. Interstitial diffusion predominates in metal saturated and vacancies in chalcogen saturated samples. Moreover the diffusion experiments show that the diffusion coefficients in the chalcogen sublattice as well as in the metal sublattice

INTRINSIC DEFECTS AND SOLID STATE REACTION

are essentially smaller in Cd-rich regions than in Hg-rich regions. By comparison, the differences between Te-rich and Se-rich regions are much less characteristic.

DOUBLE CONVERSIONS

The double conversion between CdTe and HgSe shows a qualitatively similar behaviour in undoped and in chalcogen saturated samples, whereas in metal saturated samples the behaviour is a distinctly different one. This is demonstrated by the reaction path diagram (Fig.1), obtained from the mole fraction profiles $x_{Hg}(z)$ and $x_{Te}(z)$ by elimination of the space variable z.

Reaction between Te-saturated CdTe and Se-saturated HgSe. This reaction is characterized by the fact that in the original crystals the diffusion coefficients for metal and chalcogen are of the same order of magnitude. Te diffusing into HgSe increases the chalcogen diffusion coefficient $\tilde{D}_{ch}$ and decreases the metal diffusion coefficient $\tilde{D}_{me}$, whereas Se diffusing into CdTe decreases $\tilde{D}_{ch}$ and increases $\tilde{D}_{me}$. Thus in the HgSe-rich zone, the exchange of chalcogens, and in the CdTe-rich zone, the exchange of metals is promoted. As the system (Cd,Hg)(Se,Te) exhibits a miscibility gap the building up of local thermodynamic equilibria between the two original phases causes the precipitation of a CdSe-rich product phase (Fig.3). Because of the fast chalcogen interstitial diffusion in the Hg-chalcogenides, the exchange of the chalcogens also domi-

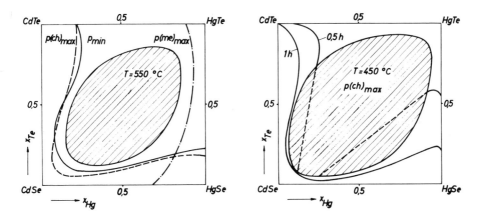

Fig.1   Reaction path diagrams   Fig.2

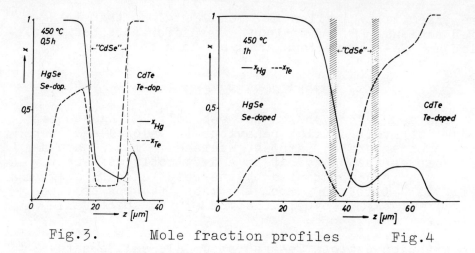

Fig.3.   Mole fraction profiles   Fig.4

nates in the further course of the reaction in the HgSe side of the reaction couple (Fig.4). But as the transport of chalcogens through the CdSe-rich layer is very slow compared to the transport in the Hg-chalcogenides, the original high Te-concentration at the boundary of the Hg-chalcogenide phase is removed in the course of the reaction by the faster transport into the interior of the phase. On the CdTe-side of the reaction couple, however, the superposition of metal and chalcogen transport is maintained during the whole reaction time, because an increase of the Hg-content is always associated with an increase in $\tilde{D}_{ch}$ (Tab.1).

After a certain reaction time the initially sharp phase boundary between the CdSe-rich product phase and the adjacent Hg(Se,Te)- and (Cd,Hg)Te-rich diffusion zones disappear (Figs. 2 and 4). The reaction

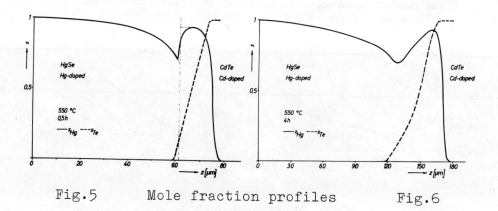

Fig.5   Mole fraction profiles   Fig.6

path now runs around the miscibility gap and in the mole fraction profiles, regions of uphill diffusion substitute the original steps in the mole fractions. The uphill diffusion at the boundary of the miscibility gap is caused by the behaviour of the mean free enthalpy as function of the mole fractions x_{Te} and x_{Hg}.

Reactions between metal saturated samples. In metal saturated samples there is $D_{ch} \ll D_{me}$. Moreover D_{ch} in CdTe is nearly two orders of magnitude smaller than HgSe. As the intermediate region between the HgSe and the CdTe phase is relatively narrow, it works as a smeared phase boundary in which the faster diffusing metals establish local thermodynamic equilibrium (Fig. 5 and 6). This causes a splitting of the metal concentrations on this pseudo phase boundary. With increasing reaction time, the intermediate region - and thus also the region of uphill diffusion of the metals - is becoming broader and broader. As in this region the extremals of the Hg-concentration do not depend on the reaction time, it is concluded, that the initial concentrations at the original phase boundary were the same, so that from this values the affinity of the double conversion can be calculated. One obtains at $550°C$ an affinity of 13.8 kJ/mole. In metal saturated samples the reaction path runs around the miscibility gap on the Hg-rich side, so that no product phase is precipitated.

CONCLUSION

In Fig.1 both the reaction path of the metal saturated system and the reaction paths of the chalcogen saturated system and of the system of minimal vapour pressure are plotted. One can see that metal saturated samples prefer a reaction path totally different from that of samples with smaller metal content. For the behaviour of the reaction path it is critical whether at the beginning of the reaction the exchange of metals or of chalcogens predominates on the HgSe-side of the reaction couple. If the exchange of chalcogens predominates before the reaction path bends into the interior of the diagram, as in the case of chalcogen saturated samples, the formation of a CdSe-product layer results, caused by the miscibility gap. If, however, the exchange of metals predominates the reaction path runs around the miscibility gap on the HgTe-side and forms distinct zones of uphill diffusion.

DISCUSSION

K.J. de VRIES (University of Technology, Twente, Netherlands) You quote diffusion coefficients obtained by extrapolation. It would be useful to know their accuracy particularly as they are concentration dependent.

AUTHORS' REPLY: Interdiffusion coefficients were calculated from mass fraction profiles using the Boltzmann-Matano method. In Table 1, for 10^{-8} cm^2/sec$>\tilde{D}>10^{-13}$ cm^2/sec, where values were obtained by graphical, isothermal extrapolation, the r.m.s. errors in lg $\tilde{D}$ are less than ±0.3. Outside this range for $\tilde{D}$, the values are obtained by extrapolation to lower or higher temperature, thus the errors will be somewhat larger, depending on the accuracy of the values of ΔH and D_0 employed.

A detailed treatment of interdiffusion coefficients will be published in a separate paper.

CHARACTERIZATION AND SURFACE REACTIVITY OF FINELY-DIVIDED CoO-MgO SOLID SOLUTIONS

A. P. Hagan, C. O. Arean and F. S. Stone

School of Chemistry, University of Bath

Bath BA2 7AY, England

ABSTRACT

CoO-MgO (0-10 mol % Co) has been prepared in low, medium and high surface area (LSA, MSA and HSA) forms. a_o for HSA oxides (mean particle size $\overline{D} < 50$ nm) has been determined in vacuo by electron diffraction. The increase of a_o with [Co] is identical to that of sintered LSA oxides prepared at 1300°C. Magnetic data confirm that Co(II) ions are well dispersed in true solid solution in both MSA and HSA oxides.

Reflectance spectra of MSA ($\overline{D}$ = 50-100 nm) and HSA ($\overline{D}$ = 20-50 nm) oxides show that some Co in the surface layers of finely-divided outgassed CoO-MgO is in tetrahedral coordination. Exposure to the atmosphere or oxygen destroys the surface spectrum, though in oxygen the destruction is not complete.

HSA MgO ($\overline{D}$ = 14 nm) has an anomalously low a_o attributed to the surface energy effect predicted by previous authors.

INTRODUCTION

Co(II) is well-known to occupy both octahedral and tetrahedral sites in spinels (1), and in ion-exchanged zeolite it readily adapts its coordination depending on the degree of hydration (2). It is of interest to know how it behaves in a solid surface. As a model system, we decided to examine CoO-MgO dilute in cobalt at three stages of comminution, (i) well-sintered oxide, as studied by Cimino et al. (3), (ii) powdered oxide of mean particle size $\overline{D}$ = 50-100 nm and (iii) a high surface area (HSA) solid with

$\bar{D} < 50$ nm. After this work was started, we became aware of a study by Dyrek and Shvets (4), who showed that CoO-MgO of high surface area contained tetrahedral Co(II), which they considered to be distributed throughout the crystallites. However, their samples were prepared at 500°C, a rather low temperature for solid solution formation. Our HSA samples have been prepared at 1000°C, where cation diffusion is much more rapid, and we have also made precision lattice parameter measurements. We also observe tetrahedral Co(II), but we shall show that the a_o work enables more firm conclusions on its location to be drawn.

EXPERIMENTAL

CoO-MgO was prepared from spectroscopically pure materials. Freshly-precipitated $Mg(OH)_2$, pure or impregnated with Co nitrate solution, was treated as follows: (i) for LSA (low surface area) oxides, 15 h at 400°, 15 h at 1000° and 24 h at 1300°C, always in air; (ii) for MSA (medium surface area) oxides, 15 h at 400° and 15 h at 1000°C, again in air; (iii) for HSA (high surface area) oxides, heating in dynamic vacuum at 350° (15 h), 1000°C (12 h), followed by quenching in vacuo. Surface areas of MSA and HSA oxides were determined by the BET method (N_2, 77 K) after outgassing at 500°C.

Five solid solutions with Co contents from 1 to 10 mol % were prepared, designated MCo 1, MCo 3, MCo 5, MCo 7.5 and MCo 10, respectively. Actual Co contents, determined by (A) EDTA titration and (B) spectrophotometric analysis (nitroso R complex), are shown in Table 1, together with surface areas.

X-ray diffraction (XRD) was used to determine the lattice parameter a_o of LSA and MSA oxides using the Debye-Scherrer method (Cu K_α radiation). Line profiles were determined by diffractometry (MSA oxides) or microdensitometry (HSA oxides). a_o for HSA oxides was determined by selected area electron diffraction at 100 kV. Magnetic susceptibilities were measured between 78 and 373 K using an enclosed Gouy balance (2). UV-vis diffuse reflectance spectra were measured with a Beckman DK 2 spectrometer: the sample cell was attached to a vacuum line and could be heated or dosed with gas without dismantling.

RESULTS

a_o for LSA and MSA oxides was calculated from Nelson-Riley plots of the α_1 components of high angle reflections and was normalized to 21°C (Table 2). a_o for HSA oxides cannot be accurately determined by XRD due to severe line broadening, but this is possible in principle by electron diffraction. However,

TABLE 1

Oxide	[Co]/mole %		Surface area $S_{BET}/m^2 g^{-1}$		
	Method A	Method B	LSA*	MSA	HSA
MgO	0	0	10	43	285
MCo 1	0.94	0.94	10	39	-
MCo 3	3.18	3.10	8	-	230
MCo 5	4.45	4.55	5	29	160
MCo 7.5	7.63	7.63	3	24	122
MCo 10	9.65	9.43	1	16	98

*Estimated values [cf.(5)]

TABLE 2

X-Ray Data for MgO and CoO-MgO

Oxide	LSA	MSA			HSA	
	$a_o/\text{Å}$	$a_o/\text{Å}$	$\bar{D}_{422}/nm$	$S_{XRD}/m^2 g^{-1}$	$\bar{D}_{200}/nm$	$S_{XRD}/m^2 g^{-1}$
MgO	4.2112	-	45.8	38	13.7	122
MCo 1	4.2118	4.2121	55.9	30	-	-
MCo 3	-	4.2133	67.5	24	20.8	78
MCo 5	4.2140	4.2142	82.9	19	26.2	61
MCo 7.5	-	4.2157	98.0	16	35.5	44
MCo 10	4.2165	4.2173	-	-	48.2	32

in practice there are some experimental constraints when the specimen is a highly reactive oxide. First, the sample must be heated in vacuo to remove readily-formed surface hydroxide before analysis. Secondly, the internal standard needed for calibration must be chemically inert on heating in contact with the oxide. The procedure devised employed an evaporated Au film as internal standard and a high temperature attachment which allowed the sample to be heated in situ for ~ 10 sec at 700°C in the vacuum of the electron microscope. a_o was computed from the juxtaposed

TABLE 3

Magnetic moments (μ_{eff}) and Weiss constants (θ)

	LSA		MSA		HSA	
	MCo 3	MCo 7.5	MCo 3	MCo 7.5	MCo 3	MCo 7.5
μ_{eff}/B.M.	*	(5.25 ± 0.05)	5.14	5.28	5.14	5.18
$-\theta$/K	*	(35-40 ± 10)	25	50	50	55

*Data of Cimino et al. (3) for LSA CoO-MgO

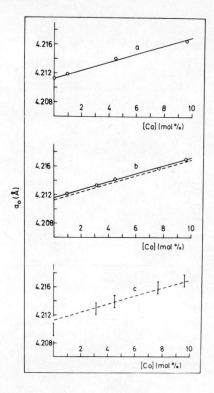
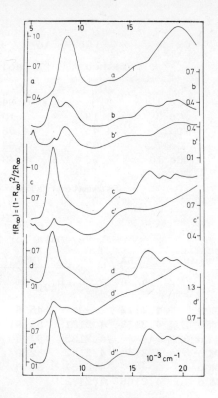

Fig.1. Lattice parameter (a_o) versus cobalt content. a – low surface area (LSA) oxides; b – medium surface area (MSA) oxides; c – high surface area (HSA) oxides. The dashed line in b and c shows the position of the corresponding line for LSA oxides.

Fig.2. Spectra (Kubelka–Munk plots). a-LSA MCo 3, vacuo; b-MSA MCo 3, vacuo; b' – as b, but after air exposure (10 days); c-HSA MCo 3, vacuo; c' – as c, but after air exposure (10 days); d-HSA MCo 7.5, vacuo; d' – after dosing 260 torr O_2; d'' – after outgassing at 1000°C.

(220) and (422) diffraction rings of the oxide and the Au standard [a_o (Au) = 4.0784 Å]. Values are shown in Fig.1, together with a_o vs.[Co] for the LSA and MSA solid solutions.

X-ray crystallite sizes of MSA and HSA oxides were derived from the line breadths of the (422) and (200) reflections, respectively, using the Scherrer equation. From the mean crystallite size $\overline{D}$ an X-ray surface area S_{XRD} was calculated assuming the crystallites were all equal-sized cubes of edge $\overline{D}$ (Table 2). $S_{XRD} < S_{BET}$, presumably because the smallest crystallites escape detection by X-rays but contribute prominently in physical adsorption.

The magnetic moments (μ_{eff}) and Weiss constants (θ) of MSA and HSA MCo 3 and MCo 7.5, determined from the best fit to the equation $\chi = \chi_o + C/(T + \theta)$, are shown in Table 3.

UV-vis reflectance spectra of LSA, MSA and HSA oxides were recorded in vacuo at $20°C$ after outgassing in situ at $1000°C$ for 15 h, and are illustrated for MCo 3 in Fig 2 (a,b,c,). The effect of exposing MSA and HSA oxide to the atmosphere is shown in spectra b',c'. Spectra d,d',d'' show the effect of adsorption and desorption of oxygen on HSA oxide, in this case MCo 7.5

DISCUSSION

Fig.1 shows that the same linear increase of a_o with [Co] is found with LSA, MSA and HSA oxides. The magnitude of Δa_o is consistent with the elastic matrix model (3), and we therefore conclude that Co(II) ions are uniformly dispersed in all three forms. The magnetic results also establish that the cobalt is internally dispersed and coordinated as octahedral Co(II) in all three forms. The most important aspect is the proof that the property of solid solution can be maintained even when crystallite sizes are below 50 nm.

The a_o vs.[Co] line for MSA oxides (determined by XRD in air) lies slightly above the line for LSA oxides. We ascribe this to a thin surface layer of hydroxide which imposes a dilatant volume strain on the crystallite interior, measurable when $\bar{D}$ is in the 50-100 nm range which our MSA oxide possesses (6).

a_o for HSA MgO is significantly lower than a_o for LSA MgO. Such an effect can be expected for pure ionic crystals of very small size on account of their excess surface energy. Our HSA MgO ($\bar{D}$ = 14 nm) lies within the range where a lattice contraction is predicted (7). The effect would not be expected with our HSA CoO-MgO since $\bar{D}$ is much larger and also the ionicity is less.

The spectra of outgassed MSA and HSA oxides (Fig.2) clearly reveal absorption [cf.(4)] which is additional to the well-known octahedral bands of Co(II) in MgO at ~ 8,500 and ~ 20,000 cm^{-1} (8). This absorption, principally at 7,100, 13,900 and 16,800 cm^{-1} is absent in LSA oxide, is well-developed in MSA oxide and is intense in HSA oxide. We therefore ascribe it to cobalt in the surface region of the crystallites. There is no significant difference in μ_{eff} between LSA and other oxides, so the fraction of the total cobalt giving rise to this surface spectrum must be relatively small, even in HSA oxide. It will have a strong spectral response, however, if its coordination is non-centrosymmetric, and the intensities and positions of the above surface bands are consistent with the presence in the surface layers of tetrahedrally-coordinated Co(II) [transitions from $^4A_2$ to $^4T_1(F)$, $^4A_1(P)$ and $^4T_1(P)$ respectively (8)]. 4-fold (near-tetrahedral) coordination of cations is expected (9) on some surface planes, e.g. {110}, but there is no reason why these should be the principal faces present. The predominance of a tetrahedral spectrum therefore suggests that the 4-coordinate Co(II) is not simply present in the outermost layer but that it is distributed over several layers. The experiments of exposure to the atmosphere and to oxygen (Fig.2) are relevant here. Oxygen is chemisorbed at 20°C, but does not completely destroy the surface spectrum. Atmospheric exposure is more effective, due to the more aggressive action of water vapour.

We conclude that the surface layers of finely-divided CoO-MgO solid solutions adapt in vacuo to a structure in which some cations (mostly cobalt) move to tetrahedral holes. This is unlikely to happen unless the anion cubic close packing near the surface is relaxed. A limited occupation of tetrahedral holes could possibly be achieved by introducing stacking faults or anion vacancies, and also by stepping or kinking the surface planes. The rather 'tetraphilic' Co(II) ion may permit the {100} surfaces of MgO to adapt towards a structure in which the normally 5-coordinate surface Mg ions become 6-coordinate in conjunction with the incorporated Co ions becoming 4-coordinate.

ACKNOWLEDGEMENTS

We thank Dr.V.D.Scott for advice on the electron diffraction work and the Spanish C.S.I.C. for a fellowship to one of us (C.O.A.).

REFERENCES

1. G.Blasse, Philips Res.Rep.Suppl.No.3 (1964).
2. T.A.Egerton, A.Hagan, F.S.Stone and J.C.Vickerman, J.C.S. Faraday I, **68**, 723 (1972).
3. A.Cimino, M.Lo Jacono, P.Porta and M.Valigi, Z.phys.Chem.NF, **70**, 166 (1970).
4. K.Dyrek, Bull.Acad.Polon.Sci.Ser.Sci.Chim., **21**, 675 (1973); K.Dyrek and V.A.Shvets, ibid., **22**, 315 (1974).
5. A.Cimino and F.Pepe, J.Catal. **25**, 362 (1972).
6. A.Cimino, P.Porta and M.Valigi, J.Amer.Ceram.Soc., **49**, 152 (1966).
7. P.J.Anderson and A.Scholz, Trans.Faraday Soc., **64**, 2973 (1968).
8. R.Pappalardo, D.Wood and R.C.Linares, J.Chem.Phys., **35**, 2041 (1961).
9. J.Haber and F.S.Stone, Trans.Faraday Soc., **59**, 192 (1963).

DISCUSSION

M. FIGLARZ (University of Picardy, Amiens, France) Is it possible that the difference between surface areas measured by the BET and calculated from X-ray diffraction might be due to porosity? This is common in MgO obtained from $Mg(OH)_2$. Also is there a special reason for using an impregnated hydroxide as starting material rather than solid solutions of the two hydroxides?

AUTHORS' REPLY: I agree that porosity is very probably contributing to the discrepancy between the BET and X-ray surface areas. Four factors can be mentioned as contributing to such a discrepancy: (i) very small crystallites not detectable by X-ray analysis, (ii) porosity (iii) surface roughness and (iv) crystallite aggregation. The first three will cause S_{BET} to be greater than S_{XRD}, and the last will produce the opposite effect. Evidently (i), (ii) and (iii) together outweigh (iv) in the present case, and as to the relative importance of the first three we believe that (i) > (ii) > (iii). There is evidence for porosity in HSA oxide from hysteresis in the adsorption-desorption isotherm, but it is not large enough to account on its own for the magnitude of the discrepancy observed.

As to the second point, there is no special reason for using impregnation. Solid solutions of the two hydroxides could equally well be used.

W. GUNSSER (Technical University, Hannover, West Germany) Can you explain the increase of the Weiss constant with decreasing particle size? One would expect the opposite because of a decrease in exchange coupling with high surface area samples having more surface ions and higher lattice constants.

AUTHORS' REPLY: The lattice constants are the same for LSA, MSA and HSA oxides of equivalent composition, so no change in Weiss constant (Θ) would be expected on that account. However, on other grounds a decrease of Θ with decreasing particle size would indeed be more expected than an increase; one of these grounds is also the distortion of symmetry in the surface layers of HSA oxide. I do not have an explanation for the increase in Θ: The presence of some surface segregation might have such an effect, but if so one would expect it to be greater for MCo 7.5 than Mco 3, which is not observed. There is a substantial error in the determination of Θ, and for the present there must remain some doubt as to whether the trend of an increasing Θ with decreasing particle size has been fully established.

Z.G. SZABÓ (L. Eötvös University, Budapest, Hungary) This research is relevant to catalytic problems. The cobalt ions distributed in the host matrix will create varying electron densities between Co ions and neighbouring O^{2-} ions, implying centres of different catalytic activity. Many reactions, eg. the decompostion of iso-propanol, are sensitive to such electronic factors and might be used to provide information on Co-ion distribution. Have you performed such experiments?

AUTHORS' REPLY: We are indeed interested in the application of these oxides and related systems as catalysts, and LSA oxides have already been studied (1-3). Dr. F. Pepe and I have in fact investigated the decomposition of isopropanol over LSA CoO-MgO^4, which behaves as a dehydrogenation catalyst, but the MSA and HSA oxides have yet to be studied. The effects you mention are very much what we have in mind, but at the present stage it seems safer to infer cobalt distribution information from physical measurements.

1. A. Cimono, M. Schiavello and F.S. Stone, Discuss. Faraday Soc. $\underline{41}$, 350 (1966).
2. A. Cimino and F. Pepe, J. Catalysis, $\underline{25}$, 362 (1972)
3. F. Pepe and F.S. Stone Proc. 5th Internat. Congr. Catalysis (Florida, 1972) ed. J.W. Hightower, p. 137. North Holland/American Elsevier, 1973.
4. F. Pepe and F.S. Stone to be published.

J. HABER (Academy of Sciences, Krakow, Poland) The authors have shown that tetrahedral cobalt ions are present in the samples of CoO-MgO solid solution, these ions being accomodated in the relaxed surface layer. Recently Bielanski et al. (Z.Phys.Chem.NF., $\underline{97}$, 207 (1975)) have found that Schottky defects are present in CoO-MgO samples obtained at low temperatures and they relate the incorporation of Co ions in tetrahedral sites to the presence of these defects. Obviously in the case of samples annealed at high temperatures the Schottky defects are annihilated in the bulk and may remain only at the surface, in agreement with the authors' interpretation.

AUTHORS' REPLY: The work referred to by Professor Haber is a continuation of the study by Dyrek cited in our paper, and we were unaware of this latest publication when we wrote our srticle. Our studies and those of Bielanski et al. are complementary; we have devoted our attention to vacuum-prepared CoO-MgO ex hydroxide, whilst Bielanski et al. have studied air-prepared CoO-MgO ex carbonate. There are many similarities, but some differences. We also have studied CoO-MgO produced at lower temperatures (400-800°C). The role played by hydroxylation and carbonation in the highest area materials has yet to be clarified.

THE REACTIVITY OF SOME TUNGSTEN OXIDES

M. Schiavello[x], S. De Rossi[x], B.A. De Angelis[+], E. Iguchi[*,o] and R.J.D. Tilley[*]

x Centro di Studio sulla Struttura ed Attività Catalitica di Sistemi di Ossidi del CNR, ed Istituto di Chimica dell'Università di Roma, Roma, Italy

+ Laboratori Ricerche di Base, SNAM PROGETTI S.p.A., 00015 Monterotondo, Roma, Italy

* School of Materials Science, University of Bradford, Bradford BD7 1DP, Yorkshire, England

o on leave from Dept. of Metallurgical Engineering, Yokohama National University, Ohka, Minami-ku, Yokohama 233, Japan

ABSTRACT

The catalytic properties of WO_3, WO_2, W_nO_{3n-1}, W_nO_{3n-2}, $W_{18}O_{49}$ for N_2O decomposition and propene oxidation are reported. The results of x-ray photoelectron spectroscopy (XPS) measurements on the same materials are also given. The results are considered in terms of bulk structure and projected surface geometry.

INTRODUCTION

In order to attempt to understand more about the relationship between bulk structure and surface reactivity we have begun a program of work on oxide systems which have a well defined crystal chemistry. That is, the materials chosen have a) a well-defined structure, b) possess features which are expected to be potentially significant for chemical reactivity both in the bulk and at the surface, and c) are structurally flexible, so that a variety of cation and anion types can be incorporated into the parent matrix without destroying the basic structural features of interest.

Such an approach is not completely new, and a number of oxide structures acceptable in this context, the spinels and the rutiles for example, have already been studied in this way. However we have chosen to concentrate on more complex oxides of the sort which rely upon crystallographic shear (CS) as a means of changing anion to cation stoichiometry. Such materials are potentially of great interest and to date have been but little studied from a reactivity point of view (1-4).

The binary oxides which fall in this category are principally TiO_2, VO_2, V_2O_5, Nb_2O_5, MoO_3 and WO_3. This paper summarizes experiments on the catalytic decomposition of N_2O and the catalytic oxidation of propene over binary tungsten oxides. The compositions employed were WO_3, $\sim WO_{2.95}$, $\sim WO_{2.90}$, $WO_{2.72}$ (5) and WO_2 (6). While these are not particularly important materials from a surface chemical viewpoint they fit well into the requirements listed above. The N_2O reaction was chosen because it was fairly simple and well understood and should reflect the overall activity of the materials. The propene reaction, being more complex, should lead to information concerning selectivity of the catalyst. The results are considered from the point of view of the crystal chemistry of the oxides, and the implications of the findings to other structurally related oxides are also discussed.

EXPERIMENTAL

Samples were prepared by sealing appropriate amounts of tungsten metal and WO_3 in evacuated silica tubes which were subsequently heated for several weeks at about 1223 K. The samples were characterized both before and after catalysis by powder x-ray photography, using a Guinier-Hägg focussing camera and strictly monochromatic Cu $K_{\alpha 1}$ radiation, by high resolution electron microscopy using a JEM 100B electron microscope fitted with a goniometer stage and operated at 100 KV, and optically with a Zeiss Ultraphotoptical microscope. Samples were also removed periodically from the catalytic reactor for structural characterization. Some samples were also studied by x-ray photoelectron spectroscopy (XPS) to determine valence states of tungsten as far as possible. XPS spectra were recorded with an AEI ES100 spectrometer, using Al $K\alpha$ excitation. Reflectance spectra were also recorded at room temperature (MgO as reference) by a Beckman DK 1A spectrophotometer on fresh specimens and on specimens used for the propene oxidation.

The N_2O decomposition was carried out in all-glass circulating system, the reactor being made of silica, at initial pressures of N_2O of about 60 Torr and using a mass of 100 to 300 mg according the catalyst. The decomposition was kept low (<1%). The propene oxidation was carried out in a flow system using for the gas analysis a C. Erba Fractovap ATC/f gas-chromatograph equipped with two columns (molecular sieves at r.t. for O_2 and CO, and Porapack R at 419 K for CO_2, propene, water and acrolein).The mass of catalyst was about 1.5 g. Details on the sample preparation and characterization, and on the catalytic measurements can be found elsewhere (7-10).

RESULTS

The N_2O decomposition reaction was carried out in the range of temperature 533-673 K. The following order of activity was found: CS structures ($WO_{2.96}$, $WO_{2.90}$)> tunnel structures ($WO_{2.72}$)> WO_3 ∿ WO_2. The x-ray and electron microscopy analysis carried out on the specimens used for the catalytic reaction showed that the structures were preserved. Moreover in samples which were not reground after catalysis no electron diffraction contrast effects were observed which could be attributed to surface oxidation, reduction or reconstruction.

The propene oxidation was carried out in the temperature range 623-773 K. No great difference was found in the total activity (production of CO, CO_2, H_2O) among all the specimens. However the compounds $WO_{2.96}$, $WO_{2.90}$ and $WO_{2.72}$, at temperatures above 673 K, also perform a partial oxidation to acrolein, although in a limited amount. The selectivity decreased as the number of the runs increased. Reflectance spectra, electron microscopy and x-ray analysis showed that the specimens used in catalytic reaction were fairly oxidized at temperature above 673 K. Reflectance spectra performed on some WO_{3-x} specimens, removed from the reactor when still active and selective, showed the presence of the band characteristic of the fresh specimens (between 500 and 800 nm depending on oxygen deficiency), although of lower intensity and shifted to higher wavelength, and the band due to WO_3 (about 400 nm).

The XPS spectra of the 4f level of W in the sample before reaction showed the presence of tungsten with oxidation states lower than six. $WO_{2.96}$ contains about 10% of tungsten as W^{5+}, and $WO_{2.90}$ contains about 20% of tungsten as W^{5+}. In $WO_{2.72}$ roughly one third of tungsten is present as W^{5+} and one tenth as W^{4+}. Moreover WO_2 is partially oxidized on the surface to W^{5+} and W^{6+}. The spectra in the valence band region showed a peak near the Fermi level which is stronger in $WO_{2.72}$ and WO_2 and absent (or, perhaps, too weak to be revealed in a single scan spectra) in $WO_{2.96}$. This peak is due to 5d electrons which are present in W^{5+} ($5d^1$), and are responsible for the colour and conductivity of the compounds. The spectra of the oxides after the propene oxidation reaction showed that a large part of W^{4+} and W^{5+} are oxidized to W^{6+}.

DISCUSSION

The results show that both WO_2 and WO_3 are low in activity and selectivity, the CS structures are highest in activity and in the propene oxidation also seem to be selective to some extent for the production of acrolein. The $W_{18}O_{49}$ phase also has a significant activity, quite close to the CS phases. Thus there is clearly a correlation between catalytic activity and the degree of the reduction of the tungsten oxide, but it is not a direct one, and we cannot simply relate the chemical behaviour with the amount of W^{5+} or W^{4+} present in the oxide. The observation that catalytic activity becomes high as soon as the reduction becomes appreciable, in $WO_{2.96}$ for example, suggests that two factors may be important: the new structural geometry and the presence of ions in two valence states.

A consideration of the structures of the reacting oxides suggests that two structural groupings may be important. In the CS structures these are the CS planes which consist of blocks of edge shared WO_6 octahedra separated by rectangular tunnels. In $W_{18}O_{49}$ one can distinguish hexagonal tunnels and pentagonal rings of edge shared octahedra which surround a tungsten atom to give it an overall 7-fold pentagonal bipyramidal coordination. Is is also in these regions that the local metal to oxygen ratio is greatly reduced. If, therefore, one can think in terms of localized charges, these geometrical positions are also likely to contain the W^{5+} and W^{4+} ions in greatest proportions.

Concerning the N_2O reaction two main steps can be envisaged: a) adsorption of N_2O and the subsequent rupture of the N-O bond; b) the desorption or the incorporation of the oxygen produced. It seems likely that the first step can occur in one of the sites where the metal to oxygen ratio is reduced. At this stage it is not possible to discriminate which particular position is the most favourable. As for the second step the incorporation process must be considered the most likely. In fact the desorption process requires the injection of electrons to a surface which is already electron-rich, whilst the incorporation process requires electron acceptance from the surface. The sites for the oxygen incorporation process can be the CS planes and/or the rectangular tunnels in the CS phases and in $W_{18}O_{49}$ phase the hexagonal tunnels and/or the pentagonal ring. All these sites are characterized by having incompletely cooordinated W cations, particularly at the surface. It can be noted that, due to the very small percentage (< 1%) of N_2O decomposed, the oxygen produced cannot change significantly the composition of the phase studied and thus structural and compositional changes are not observed.

As for the propene oxidation the following consideration can be made. To obtain the partial oxidation of propene to acrolein it is necessary that the propene is adsorbed on sites which can abstract two hydrogen atoms and can add one oxygen atom. It has been proposed that in the catalysts used for this reaction, based on mixtures of different ratios of Bi_2O_3/MoO_3 (11,12) the various steps of the reaction require different sites. No conclusive proof has been obtained about the nature of these sites and about which step they each perform. Our results suggest that, for WO_{3-x} catalysts, the propene is adsorbed on the sites with high electron density. A close look at the geometry of some of these sites show that the propene molecule can lie flat in such a way to render possible to add one oxygen atom at one end.

Hence, at this stage of the investigation, it would seem that the enhanced catalytic activity of the reduced oxides is due to a combination of a specific surface geometry produced where the CS planes or tunnels intersect the crystal surface plus the mixed valence states present in these regions and which could provide electron transfer sites to facilitate the chemical reaction taking place.

ACKNOWLEDGEMENTS

E. Iguchi and R.J.D. Tilley are indebted to the Science Research Council for an equipment grant and financial support.
M. Schiavello wishes to acknowledge a financial support by the Consiglio Nazionale delle Ricerche (Rome).

REFERENCES

1 R.J.D.Tilley, M.T.P. Int. Rev. Sci., Series 1, Inorg. Chem., Vol. 10, p. 279, Butterworths, London (1972)

2 J.S. Anderson, Surface and Defect Properties of Solids, Vol. 1 Ed. M.W. Roberts and J.M.Thomas, p. 1., The Chemical Society, London, (1972)

3 J.S. Anderson and R.J.D. Tilley, Surface and Defect Properties of Solids, Vol. 3, Ed. M.W. Roberts and J.M. Thomas, p. 1., The Chemical Society, London, (1974)

4 R.J.D. Tilley, M.T.P. Int. Rev. Sci., Series 2, Inorg. Chem., Vol. 10, p. 73, Butterworths, London, (1975)

5 A. Magnéli, Arkiv Kemi, 1, 223 (1949)

6 A. Magnéli, Arkiv Kemi, Mineral Geol., 24A, N° 2 (1946)

7 M. Schiavello, R.J.D. Tilley, S. De Rossi and E. Iguchi, to be published

8 S. De Rossi, E. Iguchi, M. Schiavello and R.J.D. Tilley, to be published

9 B.A. De Angelis, M. Schiavello and R.J.D. Tilley, to be published

10 A. Cimino and V. Indovina, J. Catal., 17, 54 (1970)

11 J. Haber, Zeits. für Chemie, 13, 241 (1973)

12 G.C.A. Schuit, J. Less-Common Metals, 36, 329 (1974)

DISCUSSION

J. HABER (Academy of Sciences, Krakow, Poland) An oxidation catalyst, to be effective, must be able to activate the hydrocarbon molecule in the first step and insert oxygen into the molecule in the second step. Usually the first step is rate determining and hence studies on catalytic activity can give no information on the second step. The second step is more likely to be related to shear structure formation. This then is probably the reason why no correlation has been found, the increased activity of reduced samples being simply due to activating effect of low valent tungsten ions. I think correlation should be looked for not between presence of shear structures and oxygenation ability but between selectivity and the ability to form shear structures in the course of the reaction

THE GROWTH OF OXIDE CRYSTALLITES DURING THE OXIDATION OF IRON IN CARBON DIOXIDE WITHIN THE SCANNING ELECTRON MICROSCOPE

A.M. Brown and P.L. Surman

Surface Chemistry Section, Central Electricity Research Laboratories, Kelvin Avenue, Leatherhead England

ABSTRACT

Pure iron single crystals and polycrystalline iron sheet have been oxidised on a hot stage within the Scanning Electron Microscope (S.E.M.). The oxidant, a carbon dioxide/carbon monoxide gas mixture, was delivered to the specimen surface from a capillary tube which produced a local specimen surface pressure of up to 200 N m^{-2}, without significantly reducing the resolution of the S.E.M.

It is postulated that the rate controlling step for crystallite growth, at low pressure, is the dissociation of carbon dioxide at the surface, whereas at higher pressures the surface diffusion of iron is probably rate controlling. The effect of substrate orientation upon these steps is also discussed.

INTRODUCTION

Oxide films generally grow by reaction on a metal substrate, through a process of nucleation, growth and eventual fusion of individual oxide crystallites. Thus conventional studies of oxidation kinetics, in which the bulk properties of the growing film are observed, tend to blur the detailed mechanistic steps which contribute to the overall process.

In a previous paper (1), we reported a study of the kinetics of growth of the basic crystallite building blocks of a magnetite (Fe_3O_4) film growing on a polycrystalline substrate within the scanning electron microscope. It was shown that, at low pressure, the individual oxide crystallites grew according to a logarithmic rate law, whereas the bulk oxide film thickened according to the parabolic rate law. It was concluded that the dominant kinetic process which determined the overall oxide film growth law, was the interval between successive nucleation steps rather than the actual kinetics of crystallite growth or fusion.

In this paper we present the results of experiments designed to investigate the effects of pressure and crystallographic orientation upon the basic processes which occur during the oxidation of iron.

EXPERIMENTAL

Specimens of 99.999% pure polycrystalline and single crystal iron (Materials Research Limited) were prepared as described previously (1). They were mounted on the hot stage of the modified Stereoscan (Cambridge Instruments Limited, S1) and the oxidant gas was directed onto the surface from a pre-heated capillary tube, so that there was a local elevation of surface pressure. The general experimental arrangements and procedure were as described previously (1). Micrographs of the same area were taken at various time intervals and the structure of the film was subsequently determined by x-ray diffraction.

RESULTS

Effect of Pressure

The effect of total carbon dioxide pressure upon the oxidation growth processes was investigated in an experiment on polycrystalline iron at $460°C$. Oxidation was initially carried out at a calculated surface pressure of 30 N m^{-2} and subsequently the pressure was alternated between this and 3×10^{-3} N m^{-2}.

The changes in surface morphology observed are shown in figure 1a-c. Within ten minutes the metal surface had become covered with a dense layer of oxide crystallites which within 80 minutes had fused together to form a continuous faceted oxide layer. At this stage the pressure was reduced, and the faceted surface structure remained fairly constant, with only slight re-organisation until a total of 25 hours had elapsed when a new generation of oxide crystallites was observed to nucleate apparently at random on the previously formed crystal facets. These nuclei continued to grow (fig. la). Eventually the surface was completely covered with these crystallites, many of which were still independent after 512 hours. Elevation of surface pressure for 2 hours led to the rapid growth and fusion of the crystallites to form a dense crystalline structure. Further exposure to 3×10^{-3} N m^{-2} led to a repeat of the previous sequence of events until at 1180 hours the surface was again covered with an agglomeration of fairly independent oxide crystallites of about 0.2 µm diameter (fig. 1b). Restoration of a surface pressure of 30 N m^{-2} again led to the rapid incorporation of these crystallites into a compact polycrystalline oxide layer (fig. 1c). X-ray diffraction patterns of the final oxide product confirmed that the dominant oxide phase was magnetite.

Effect of Substrate Orientation

The effect of substrate orientation was investigated by oxidising single crystals of iron with the (100) and (111) orientations at 520°C. The (111) crystals were oxidised at a calculated surface pressure of 10 N m^{-2} and the (100) crystals at 100 N m^{-2}.

The original 1 µm diamond polished surface of the (111) crystals developed a network of triangular etch pits (fig. 1d) at the sides of which an apparently oriented network of oxide tetrahedra grew up. After about 20 hours oxidation, a new generation of oxide crystallites nucleated and then grew on the sides of the oxide tetrahedra (fig 1e) until, at ∼200 hours, the whole surface was covered.

X-ray diffraction patterns obtained from the oxidised surface after reaction indicated that the crystal was well within 1° of (111) and suggested that the oxide was thin (<1 µm) and probably epitaxially related to the substrate.

The (100) crystals were oxidised at a higher surface pressure (100 N m^{-2}) and quickly became covered with a mat of fine oxide crystallites. These grew and sintered to form an apparently oriented slab of oxide containing an ordered set of square pits (fig. 1f) which apparently continued to expand as the oxide film thickened.

(a) Growth of second generation crystallites
(3×10^{-3} N m^{-2})

(b) Surface covered by discrete oxide crystallites
(3×10^{-3} N m^{-2})

(c) Compact oxide formed from (b) above
(30 N m^{-2})

(d) Simultaneous formation of etch pits and epitaxial oxide on a (111) single crystal
(10 N m^{-2})

(e) Secondary nucleation of oxide on epitaxial layer developed from above
(10 N m^{-2})

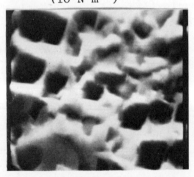

(f) Oriented slab of oxide formed on (100) crystal
(100 N m^{-2})

Figure 1   Micrographs of Iron Surfaces at Various Stages of Oxidation, (a-c) at 460°C and (d-f) at 520°C

⊢ 3 μm ⊣

X-ray diffraction of an oxidised crystal after 71 hours oxidation indicated that the crystal was within $1°$ of (100) and that the oxide tended to be epitaxially related to the surface. The oxide appeared to be stress free, but some smearing of the diffraction lines suggested that there was a degree of disorder in the oxide.

DISCUSSION

The experiment illustrated by figure 1 demonstrates the existence of two distinct oxidation regimes. At high pressure (30 N m^{-2}) crystallite nucleation, growth and fusion were all rapid and difficult to observe separately. At low pressure ($3 \times 10^{-3} \text{ N m}^{-2}$), however, the nucleation and growth were slowed sufficiently to be distinctly observed. The preferential nucleation of a new generation of oxide crystallites, when the pressure was reduced (fig. 1a), implies that the oxide surface had become "supersaturated" with oxidisable iron, whereas the rapid growth and fusion of these separate centres of growth when the pressure was again increased (fig. 1c) suggests that oxidation could then proceed at a rate governed by the supply of iron. We are hence able to define two limiting conditions for crystallite growth. At low pressure it is controlled by the rate of supply of oxygen, presumably by the rate of dissociation of carbon dioxide (cf. ref. 2), whereas at high pressure the rate controlling step is the supply of iron, presumably by surface diffusion (cf. ref. 1).

Oxidation of (111) single crystals at intermediate pressure enabled us to deduce the basic transport processes involved in forming an epitaxial oxide layer. The original polished crystal surface became heavily pitted within two hours (fig. 1d) and at the same time the iron removed by the pitting process was incorporated as oxide on the surfaces adjacent to the pits. The triangular shape and regular orientation of the pits was typical of etch pits formed by evaporation of (111) crystals (3, 4) at high temperature. The driving force for mass transport of iron in this case was presumably the free energy of the oxidation reaction modified by the favourable surface energy changes produced by revealing lower index crystal planes.

When the surface of a (111) crystal was completely covered with epitaxially grown tetrahedra (cf. ref. 5) of magnetite (after about 20 hours oxidation) then the previously observed low pressure reaction to form multiple discrete second generation oxide crystallites occurred (fig. 1e). It is interesting to note that the preferred nucleation sites for these new crystallites was at the base of the previously formed oxide tetrahedra.

A similar process of epitaxial magnetite growth and etch pit formation can also be inferred from figure 1f which refers to reaction on a (100) crystal. In this case the pits were probably square, as would be expected for the (100) orientation (4), but the details of the process were obscured by the faster reaction which occurred at the higher pressure (100 N m^{-2}).

CONCLUSIONS

In-situ oxidation of iron within the S.E.M. has allowed us to directly observe the separate processes which occur simultaneously to produce an oxide film on a clean metal surface. Further work is in progress to quantify these steps and particularly to relate the kinetics of crystallite growth to the overall oxidation kinetics as a function of temperature and crystallographic orientation.

REFERENCES

(1) A.M. Brown and P.L. Surman, Surface Sci., 1975, 52, 85.
(2) F.S. Pettit and J.B. Wagner, Acta Met., 1964, 12, 35.
(3) M.B. Ives, Localised Corrosion, Nat.Assoc.Corr.Eng., Williamsburg Virginia, 1971, p.78
(4) W.H. Hirtle, N.R. Adsit and J.O. Brittain, Direct Observations of Imperfections in Crystals, Interscience 1972, p. 135.
(5) J. Bardolle, J. Chem. Phys. 1956, 53, 639.

ACKNOWLEDGEMENTS

The authors are grateful to Dr E. Metcalfe who carried out the x-ray diffraction investigations.

This work was carried out at the Central Electricity Research Laboratories and is published by permission of the Central Electricity Generating Board.

DISCUSSION

I. GABALLAH (University of Nancy, France) Do you think there is a relationship between recovery or crystallisation in iron and the nucleation in the two regimes observed in your experiments?

AUTHORS' REPLY: The initial density of crystallite nucleii will probably depend on the density of surface defects, which will in turn depend on temperature. However, this does not affect our conclusions that there are two oxidation regimes, one at high and one at low pressure.

THE REACTIVITY OF METAL OXIDES AND SULFIDES WITH LITHIUM AT 25°C -
THE CRITICAL ROLE OF TOPOTAXY

M. Stanley Whittingham and Russell R. Chianelli

Corporate Research Laboratories, Exxon Research and

Engineering Co, P.O. Box 45, Linden, N.J. 07036 U.S.A.

ABSTRACT

It has been known for some time that transition metal oxides, such as those of tungsten and vanadium, can form highly non-stoichiometric ternary compounds with the alkali metals at elevated temperatures, 500-900°C. The sodium tungsten bronzes, Na_xWO_3 where $o < x \leq 1$, are perhaps the best known of these. We have found that many oxides and other chalcogenides of the transition metals, in particular, will also react with lithium at ambient temperatures. The compounds formed are also ternary but usually have crystal structures very closely related to the starting material and different from the known high temperatures phases; that is, a topochemical reaction has taken place.

INTRODUCTION

Although the reactions between lithium metal and various transition and post-transition metal compounds have been used as the basis of electrochemical cells, essentially nothing was known concerning either the mode of reaction or even of the reaction products themselves. It has been assumed that the reaction proceeds by the simple formation of a binary lithium compound, e.g.

$CuS + 2 Li \rightarrow Li_2S + Cu$

$NbSe_3 + 3 Li \rightarrow Li_2Se + NbSe_{1.5}$

$MoO_3 + 2 Li \rightarrow Li_2O + MoO_2$

Although this might be the case for copper sulfide, where a complete destruction of the crystalline lattice takes place, it is not the case for the others. In these reactions the structure of the transition metal compound must be completely rearranged on reaction. Although this might be accomplished at elevated temperatures or if the reaction proceeds via a liquid phase intermediate, it is extremely unlikely that such rearrangements can occur at ambient temperatures. We therefore set out to understand these reactions and to determine what effect, if any, the mechanism of reaction had on its reversibility.

EXPERIMENTAL

The oxides and sulfides studied here were prepared by the direct combination of the elements or were obtained commercially. The lithiation reactions were performed in two ways. In the first the compound studied was pressed in a metallic screen using teflon as a binder and served as the cathode of an electrochemical cell, in which the anode was a sheet of lithium and the electrolyte was usually a lithium salt dissolved in an organic solvent. In the second n-butyl lithium, dissolved in hexane, was allowed to react with a powder sample of the oxide or sulfide for times varying from a few hours to several days.(1) All these reactions were carried out in a Vacuum-Atmosphers glove box filled with helium. The degree of reaction was controlled by the quantity either of current passed or of n-butyl lithium reagent used. The products of reaction were characterized by x-ray and chemical analysis. In addition, the reactions were followed in-situ using optical microscopy on oriented single crystals.

RESULTS AND DISCUSSION

We chose to study titanium disulfide first because of the simplicity of its structure and its well known intercalation compounds. In this case the lithium was found to intercalate into the layered structure with only a minimal expansion of the crystal lattice perpendicular to the basal planes (2,3). This reaction proceeded in a continuous manner indicating the presence of a nonstoichiometric phase, Li_xTiS_2, over the full range of composition, $0 \leq x \leq 1$. This was proven by the changes in lattice parameter, free energy of formation, and nmr Knight shift (3,4). In addition the electrochemical studies indicated complete reversibility of the reaction (4).

On the other hand, titanium trisulfide, which superficially resembles the disulfide, behaved very differently. It reacted with three lithium atoms forming the ternary phase, Li_3TiS_3 (5). Emf and infrared observations indicated that the first two lithium reacted by breaking the polysulfide bond in the structure, TiS(S-S), and the third by a reduction of the titanium:

$$TiS_3 + 2\,Li \rightarrow Li_2TiS_3$$

$$Li_2TiS_3 + x\,Li \rightarrow Li_{2+x}TiS_3, \text{ for } 0 < x \leq 1.$$

The proposed structural changes occurring here are indicated in Fig. 1; the one dimensional chains of this structure are maintained through the reaction, but the x-ray pattern is markedly reduced in sharpness as compared to that of $LiTiS_2$ (5). We found that the reaction in this case is only marginally reversible, at most one lithium can be removed. This is most probably associated with a change in the sulfur environment of the titanium atoms. In TiS_3 itself, the titanium is found at the center of a trigonal prism whereas in TiS_2 it occupies the center of an octahedron, a much preferred environment. We suggest that breakage of the polysulfide bond allows these sulfur atoms to rearrange themselves to the octahedral configuration; this rearrangement is then not reversible. In contrast, for $NbSe_3$ where we also found a ternary compound of formula Li_3NbSe_3, the reaction was completely reversible. This can be associated with the preference of niobium for trigonal prismatic symmetry, so that on breakage of the Se-Se bond there is no tendency for rotation of the chalcogen atoms about the niobium.

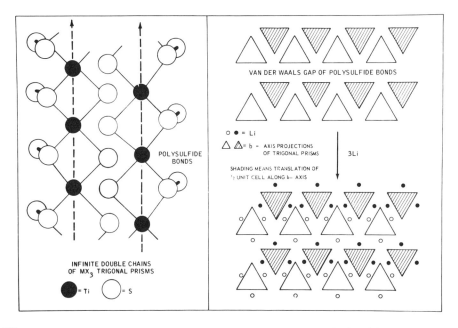

Fig. 1. Structure of TiS_3 and its reaction with lithium.

The formation of these ternary phases was also confirmed by optical microscopy. Crystals still extinguished polarized light along the b axis after reaction even though the larger crystals were heavily striated. In addition, no second phase was observed. We therefore conclude that these reactions proceed topochemically, i.e., without severe structural disruption.

We also found that a number of transition metal oxides on reaction with lithium gave x-ray patterns previously unreported in the literature and were therefore not likely to be those of any simple binary compound. Amongst these were V_2O_5, MoO_3, and TiO_2. The structures of the first two are both layered and might well be expected to react with the alkali metals by a topochemical insertion reaction. Indeed the formation of MoO_3 from its dihydrate is well known to occur by the loss of water in a topochemical or topotactic manner (6). When V_2O_5 was reacted with one mole of lithium, an x-ray pattern intermediate in nature between that of V_2O_5 itself and LiV_2O_5 formed at high temperature was found (3). A possible structure for this ternary phase is indicated in Fig. 2, where it is also compared with the other structures. V_2O_5 itself contains such highly distorted VO_6 octahedra that the vanadium essentially have only five oxygen neighbors, one of which is double bonded, i.e. $\underline{V} = 0$ (7). On reaction with lithium this bond is broken giving V^{IV} species. At high temperatures these V-O polyhedra are arranged so that the bases of the square pyramids alternate up and down singly, in contrast to V_2O_5 in which they alternate in pairs. This rearrangement presumably cannot occur at 25°C so that the LiV_2O_5 formed at low temperatures differs from the high temperature modification in the orientation of these polyhedra as shown in the figure. Such a structure can explain the intermediate x-ray parameters. This oxide has been found to be partially rechargeable (8). The layered MoO_3 was found to react with lithium also by the insertion of lithium giving a ternary compound, Li_xMoO_3. In the case of TiO_2, the lithium is believed to be incorporated into vacant tunnels inside the structure.

CONCLUSIONS

We have found that lithium may be incorporated into the structures of many transition metal compounds by a topochemical reaction, in which the structure of the host material is maintained in some degree. Where the structure is totally unchanged except for a slight expansion in one dimension, such as in Li_xTiS_2, the reaction is completely reversible. Where some distortion of the structure occurs, as for TiS_2, the reaction can only be reversed partially or with difficulty. In those compounds like CuS or CuF_2 where copper metal is formed on reaction with lithium, that is where all the chemical bonds are broken, no reversibility of the reaction is observed in the absence of solu-

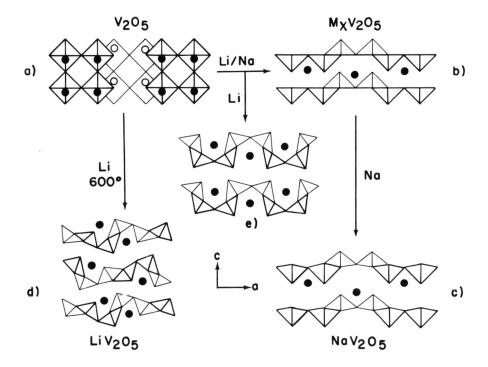

Fig. 2. Structure of vanadium pentoxides looking down the b-axis. (a) V_2O_5; (b) $M_xV_2O_5$ for low x values; (c) NaV_2O_5 formed at high temperature; (d) LiV_2O_5 formed at high temperature; and (e) LiV_2O_5 formed at ambient temperature.

bility of one of the components in the liquid electrolyte phase. Clearly an understanding of the nature of the chemical or electrochemical reaction is critical for the development of a secondary (rechargeable) lithium battery. Such reactions are also important in electrochromic materials that are to be used for optical output on watches, etc. The reaction under most intensive study is that between protons and tungstic oxide forming a hydrogen tungsten bronze, H_xWO_3.

REFERENCES

1. M. B. Dines, Mater. Res. Bull., 10, 287 (1975)
2. M. S. Whittingham and F. R. Gamble, Mater. Res. Bull., 10, 363 (1975)
3. B. G. Silbernagel and M. S. Whittingham, J. Chem. Phys., in press

4. M. S. Whittingham, J. Electrochem. Soc., 123, 315 (1976)
5. R. R. Chianelli and M. B. Dines, Inorg. Chem., 14, 2417 (1975)
6. J. R. Günter, J. Solid State Chem., 5 354 (1972)
7. P. Hagenmuller, J. Galy, M. Pouchard, and A. Casalot, Mat. Res. Bull., 1, 45 (1966)
8. C. R. Walk and J. S. Gore, J. Electrochem. Soc., 122, 68C(1975)

DISCUSSION

J.R. GÜNTER (University of Zürich, Switzerland) In the case of $NbSe_3$ chains of trigonal prisms are observed throughout the reaction. Do these retain their relative orientation, i.e. does the X-ray diffraction diagram show broadened single crystal reflections or does it yield fibre texture patterns?

AUTHORS' REPLY: Only powder patterns have been done to date but single crystal investigations are in progress. However, when $NbSe_3$ is lithiated under mild conditions, the X-ray patterns are almost as sharp in the product as in the starting material. Intensity calculations assuming intact $NbSe_6$ chains surrounded by lithium give a good but not perfect fit to observed intensities. We therefore conclude that chains remain intact and in approximately the same orientation.

ON THE REACTION BETWEEN SILVER AND COPPER IODIDE IN THIN EVAPORATED FILMS

Udo Pitterman and K.G. Weil

Fachgebiet Elektrochemie der Technischen

Hochschule Darmstadt, D 6100 Darmstadt, BRD

INTRODUCTION

In order to be able to understand the processes leading to film growth during evaporation of a substance onto a substrate surface it is necessary to know the surface temperature of the growing film. This quantity cannot be measured by conventional techniques for any probe like a thermocouple or resistance thermometer itself influences the surface temperature via heat conduction.

In this paper we will try to show the possibility of measuring surface temperatures using a temperature dependent chemical equilibrium as a probe for the special case of the evaporated species being a salt. This idea stems from the observation (1), that when CuI is evaporated into thin silver films, mixed crystals $Cu_xAg_{1-x}I$ are formed via the reaction.

$$Ag + CuI = AgI + Cu \qquad (a)$$

EXPERIMENTAL

The substrates used were silver single crystal films grown on freshly cleaved mica sheets. Their surface was a (111)-plane, the film thickness approximately 1000 Å. These specimens were mounted on a steel block, at room temperature. Onto these silver films CuI was evaporated from molydenum boats, the temperature of which could be measured by a Ni/NiCr thermocouple. The evaporation rate could be varied by raising either the size or the temperature of the molybdenum boats. The amount of substance evaporated was measured by weighing the specimens. The composition of the $Cu_xAg_{1-x}I$-films was determined by chemical analysis. The iodides were dissolved in oxygen free KCN-solution and a solution of crystal violet was added. $Ag(CN)_2$ and crystal violet form a compound which can be extracted from the aqueous solution with benzene (2); the extinction of this benzene phase at λ = 605 nm is proportional to the silver concentration.

From the results of these experiments the expression

$$K = \frac{n_{AgI}}{n_{CuI}}$$

was calculated. n_{AgI} is the amount of AgI determined by chemical analysis and $n_{CuI} = n^o_{CuI} - n_{AgI}$ with n^o_{CuI} = evaporated amount of CuI. The conditions under which K represents an equilibrium constant of the reaction (a) will be discussed later.

For some specimens the change of electrical resistivity parallel to the surface during evaporation of the salt was measured. The potential drop along the film was recorded under constant current conditions. Assuming that the increase in resistance of the film during evaporation is caused by the consumption of silver by reaction (a), the resistance measurements also allow the calculation of n_{AgI} and therefore of K.

RESULTS

Figure 1 shows the results for K as a function of the evaporated amount of CuI, n^o_{CuI}, for different evaporation temperatures.

The full lines represent chemical analysis data while the dashed lines indicate values obtained from the resistance measurements. The temperature of the evaporation source is not very well defined because during evaporation it varied within the limits given on the curves.

DISCUSSION

Below 400°C AgI and CuI form mixed crystals with in a wide concentration range (3). These show the μ-structures of the pure components. Electron diffraction investigations of the films formed under the conditions of our experiments show that the salt layer consists of only one phase, the γ-phase (4). From the fact that K-values obtained from the resistance measurements lie above those resulting from chemical analysis, it can be concluded, that the copper formed by reaction (a) does not contribute to the conductivity of the film. We therefore assume, that the copper is precipitated as small aggregates within the matrix of the iodides. Under these conditions, namely CuI and AgI forming a mixed crystal

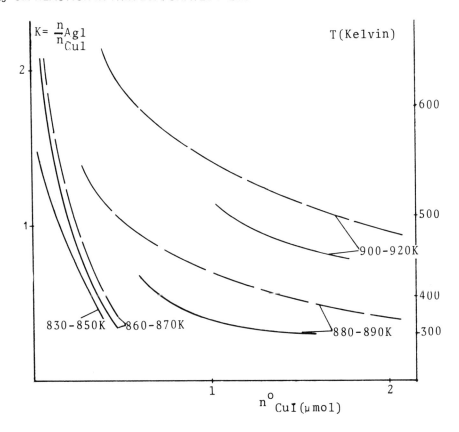

Fig. 1 K values calculated from chemical analysis and electrical resistance measurements

and the metals being present as pure phases, the law of mass action for the reaction (a) gives

$$\Delta G^\circ = -RT \ln \frac{a_{AgI}}{a_{CuI}} \approx -RT \ln \frac{x_{AgI}}{x_{CuI}} = -RT \ln \frac{n_{AgI}}{n_{CuI}}$$

From the dependence of the calculated K-values on the amount of evaporated substance n^o_{CuI}, it is clear that the concentrations of the iodides vary within the mixed crystal film. Only in the immediate neighbourhood of the phase boundary silver/mixed crystal can the reaction (a) be assumed to be in an equilibrium state, whereas at higher film thickness the role of transport processes is no longer negligible. Thus only for small values of n^o_{CuI} can the value of K be interpreted as the true equilibrium constant for reaction (a).

From the standard values of the thermodynamic functions of the reacting species, taken from (5), the equilibrium constant of reaction (a) can be calculated as a function of temperature:
   ln K = $-\Delta G^o/RT$. Table 1 gives some results

T [Kelvin]	300	400	500	600	700
K	0,30	0,54	1,06	1,76	2,52

Table 1 Equilibrium constant for the reaction (a) calculated from thermodynamic functions

It is evident from these data, that equilibrium constants, as deduced from our evaporation experiments, can only be observed when the temperature in the reaction layer is appreciably higher than the temperature of the supporting material. That means that the kinetic energy of the colliding CuI molecules is not completely dissipated by heat conduction during the time of evaporation but is in part stored as internal energy within an extremely local area at the surface.

From the measured equilibrium constants the corresponding mean temperatures in the reaction layer can be calculated. The temperature scale on the right hand ordinate of Fig. 1 is taken from these calculations.

ACKNOWLEDGMENTS

Financial support by the Fonds der Chemischen Industrie is gratefully acknowledged.

REFERENCES

1. R. Ostwald, U. Pittermann, K.G. Weil, Ber. Bunsenges. physik Chem. 78, 1260 (1974)
2. J.J. Markham, Analyt. Chem. 39, 241 (1967)
3. J. Nölting, Ber. Bunsenges. physik Chem. 68, 932 (1964)
4. Dissertation R. Ostwald, Darmstadt 1971
5. I.Barin, O. Knacke, Thermodynamic Properties of Inorganic Substances. Berlin 1973

DISCUSSION

V. LEUTE (University of Münster, West Germany) Considering the small affinity of your reaction, one cannot neglect the real behaviour of the solid solution AgI/CuI. However in calculating activity coefficients assuming a strictly regular behaviour one can see an interaction parameter of only about 1/2 kcal/mole changes the calculated temperatures by up to 30% Do you have any information about the extent of deviation from ideality of the AgI/CuI solid solution?

AUTHORS' REPLY: I agree, that the assumption of ideality of the solid solution needs experimental justification. This was done in the meantime in our laboratory. To this purpose, mixtures of pure CuI and finely powdered silver were annealed under argon atmosphere at different temperatures. After the completion of the reaction the solid solution of CuI and AgI was chemically analyzed and the equilibrium constant

$$K = \frac{n_{AgI}}{n_{CuI}}$$

was calculated. The following table shows the results of these experiments for different temperatures together with values obtained from thermodynamic data of the pure components assuming ideal behaviour of the solution. The agreement between both sets of values confirms this assumption.

T/K	K_{exp}	K_{calc}	(from $K = \exp(-\Sigma \nu_i G_i^o / RT)$)
320	0.32	0.34	
470	0.78	0.80	
510	0.94	1.00	
595	1.55	1.51	

OXYGEN DIFFUSION IN STRONTIUM TITANATE STUDIED BY SOLID/GAS ISOTOPE EXCHANGE

R. Haul, K. Hübner, O. Kircher

Institute of Physical Chemistry and Electrochemistry, Techn. University Hannover, German Federal Republic

1. INTRODUCTION

The method of heterogeneous isotope exchange has been frequently used to study oxygen diffusion in oxide crystals in view of lattice disorder, electrical properties or solid state reactions. In principle, the oxide crystal of known geometry is exposed to a constant amount of gas of a certain oxygen partial pressure at a given temperature. Initially the oxygen -18 concentration is different in both phases. The kinetics of exchange is then followed by measuring mass spectrometically the isotope content of the gas as a function of time without necessarily determining the concentration profile in the solid.

In earlier studies it became apparent that, in general, the influence of the isotope exchange at the solid surface may not be neglected in evaluating diffusion coefficients (1). Owing to the relatively low oxygen diffusion coefficient in most oxides in many experiments reported in the literature exchange was confined to a comparatively thin outer layer of the solid. Frequently fine oxide particles of irregular shape were used. In order to critically judge the potentialities and limits of the method an oxide was chosen with a sufficiently large oxygen diffusion coefficient thus enabling the exchange process to be followed up to isotope equilibrium. $SrTiO_3$ for which oxygen diffusion had already been measured by PALADINO, RUBIN and WAUGH (2) proved suitable for this purpose.

2. METHODS of EVALUATION

On the basis of a reversible first order exchange reaction at the surface, by means of Laplace transforms solutions have been derived from which both rate constants and bulk diffusion coefficients can be obtained (ref. 1, equation 13). This method "A" is, however, only applicable for the model of a thin plate and relatively small degrees of isotope exchange <0.5.

Alternatively, using statistical moments a method "B" has been suggested by KLIER and KUČERA (3) which is valid for the model of a thin plate, sphere and long circular cylinder. For the evaluation of diffusion coefficients the kinetics must, however, be followed up to a relatively high degree of exchange $\phi \approx 0.9$

For comparison some diffusion experiments were also evaluated under the assumption that surface exchange is so rapid that isotope equilibrium is established at the gas/solid phase boundary at any time. Thus only diffusion within the solid is considered to be rate determining (ref. 1, equation 18).

3. EXPERIMENTAL

Parallel sided plates (1 cm^2, 0.2-0.4 mm thickness) were cut from SrTiO$_3$ single crystal boules (National Lead Comp., New Jersey, USA). In addition a narrow sieve fraction was prepared from carefully disintegrated single crystal fragments. The mean particle diameter was 60 μm as measured by low temperature krypton adsorption. Both samples were annealed in air at 1950 K for two days. Oxygen-18 exchange experiments were carried out at oxygen pressures 50-300 Torr in the temperature range 1040-1550 K (4,5).

4. RESULTS and DISCUSSION

<u>Influence of the Phase Boundary Reaction.</u> If the surface exchange reaction is not taken into account diffusion coefficients can be obtained by means of equation 18 in ref. (1). From the measured isotope concentrations in the gas phase the corresponding arguments of e erfc are calculated. A plot of this

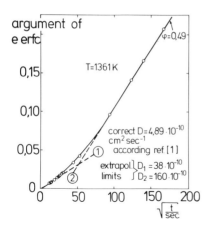

Fig.1. Influence of the surface exchange reaction.

single crystal plate; P_{O_2} = 51.4 Torr; ratio of O-atoms in gas and solid λ = 1.34; surface area per volume S_o = 93.5 cm^{-1}

argument which equals $S_o\sqrt{Dt}/\lambda$ versus $\sqrt{t}$ should result in a straight line through the origin, from the slope of which the diffusion coefficient is to be determined. It is evident from Fig.1 in which a typical experiment is shown that different slopes may be selected resulting in diffusion coefficients at variance (D_1, D_2). The correct value according to method "A" is also given. The results clearly indicate that the effect of the surface exchange reaction has to be realised if diffusion coefficients are evaluated.

Superposition of Surface Exchange and Bulk Diffusion
For experiments with single crystal plates which have been carried out up to isotope equilibrium a comparison is possible between the two methods of evaluation mentioned in section 2.

The results obtained by method "A" are represented by an Arrhenius plot (Fig.2) with a scatter of the diffusion coefficients of ±5%
$D = 3.1 \cdot 10^{-5} \exp(-127 \pm 6 \text{ kJ} \cdot \text{mole}^{-1}/RT) \text{ cm}^2\text{s}^{-1}$.

It should be noted that method "B" - as already mentioned - requires experimental data up to a high degree of exchange ϕ. Under these circumstances the model of a thin plate with negligible exchange through the edges is no longer valid. The diffusion coefficients obtained in this way are therefore, systematically somewhat larger (see Fig.2).

The method "A" evaluation when applied to crystal plate material and experiments $\phi < 0.5$ leads thus to more reliable results, quite apart from the advantage of shorter experimental times.

After oxygen-18 equilibrium had been reached, in some cases exchange experiments were carried out in the reverse direction with the same specimen in an atmosphere of natural oxygen. For instance at T=1363 K the following diffusion coefficients were obtained: 4.34 and $4.49 \cdot 10^{-10}$ cm^{2}s^{-1} respectively. This indicates that the entire transport process can be represented by a uniform diffusion coefficient and that grain boundary diffusion does not play an appreciable rôle.

<u>Exchange experiments with fine Particle Oxide</u>. Since single crystals are not always available, diffusion coefficients have been frequently determined from oxygen isotope exchange experiments with fine particle oxides. The present study offers the possibility to compare such results with those obtained from measurements with single crystal plates.

Although the irregularly shaped particles are certainly better described by a sphere, the experiments were evaluated by means of method "A" using a plate model. This may be justified as long as the depth of penetration is sufficiently small. The total surface area as determined from low temperature krypton adsorption was taken as the diffusion cross section. Fig.3 shows that the results can be well represented by an Arrhenius plot leading to practically the same activation energy as obtained with single crystal

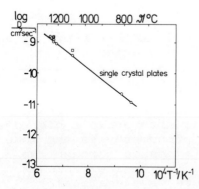
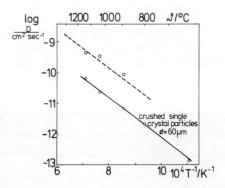

Fig.2   Fig.3
Diffusion coefficients as a function of temperature
  o method "A",   □ method "B",
---- single crystal plate experiments from Fig.2

plates (broken line). The diffusion coefficients are smaller, however, by about one order of magnitude. Apart from the fact that the model is not strictly valid the discrepancy may also be due to surface roughness. The surface area from krypton adsorption is only identical with the diffusion cross section for ideally flat surfaces. Otherwise the true diffusion coefficient is $f^2 \cdot D$, where f is the roughness factor.

Remarkably, the single crystal plate data agree reasonably with diffusion coefficients obtained from experiments with fine particles if the latter are evaluated by method "B" using a sphere model. For this purpose the experiments must, however, be carried out to sufficiently high degrees of exchange *which may be* inconveniently time consuming. On the other hand the surface area cannot be unduly increased since the particles must be reasonably well defined and large in order to measure bulk diffusion.

Diffusion Mechanism. On the whole the oxygen diffusion coefficients determined by PALADINO et al. (2) with $SrTiO_3$ from the same source are confirmed by the present results obtained with higher experimental accuracy and the correct method of evaluation. From measurements of electrical conductivity, Hall constants, thermopower, thermogravimetry and optical absorption reported in the literature (6) it can be concluded that the defect structure of non-stoichiometric $SrTiO_3$ is due to oxygen vacancies. Thus in agreement with PALADINO et al. oxygen diffusion is interpreted in terms of a vacancy mechanism. The question to what extent besides impurities, mainly aluminium, intrinsic thermal defects are responsible for the relatively large vacancy concentration is still open. Diffusion experiments over a wider range of oxygen partial pressures are required to reach definite conclusions.

Oxygen diffusion coefficients in $SrTiO_3$ are larger by orders of magnitude than those for both cations (7) as well as for oxygen in TiO_2 (5,8,9).

Acknowledgement: The authors wish to thank the "Deutsche Forschungsgemeinschaft" for financial support.

1. HAUL, R., DÜMBGEN, G. and JUST, D., Z. Phys. Chem. N.F. **31**, 309 (1962)
2. PALADINO, A.E., RUBIN, L.G. and WAUGH, J.S.,

J. Phys. Chem. Solids 26, 391 (1965)
3. KLIER,K.,KUČERA,E., J. Phys.Chem. Solids 27,1087(1966)
4. KIRCHER, O., Thesis, Techn. Univ. Hannover, 1975
5. see also: HAUL, R. and DÜMBGEN, G. J. Phys. Chem. Solids 26, 1 (1965)
6. for further literature see: YAMADA, H. and MILLER, R.G., J. Sol. State Chem. 6, 169 (1973)
7. RHODES,W.H.,KINGERY,W.D.,J.Am.Ceram.Soc.49(1966)521
8. DERRY, D.H., LEES, D.G. and CALVERT, J.M., Proc. Brit. Ceram. Soc. 19, 77 (1971)
9. GRUENWALD, Th.B. and GORDON, J., Inorg. Nucl. Chem. 33, 1151 (1971)

DISCUSSION

W. W. BRANDT (University of Wisconsin, Milwaukee U.S.A.) Recognising the importance of surface roughness and therefore perhaps some differences between methods A and B, I wonder whether the size distribution of crushed particles (lower line of Figure 3) may also have contributed to a systematic difference between the two sets of points in Figure 3, or may have compensated for part of the existing differences?

AUTHORS' REPLY: Consider uniform spherical particles. In method "A" for low degrees of exchange a plate model is used in which the total surface of all particles is taken as the diffusion cross section. Due to surface roughness the krypton low temperature adsorption area leads to a diffusion cross section which is too large, resulting in too low diffusion coefficients.

The main reason for the discrepancy with respect to single crystal plate experiments is, however, due to the choice of the model. Agreement between plate and sphere model is the better the smaller λ, i.e. the ratio of oxygen atoms in the gas solid phase. But even at $\lambda = 0.1$ good agreement is only obtained up to a degree of exchange of about 0.2. Unfortunately, there exists no convenient solution for a sphere model if method "A" is applied, which makes the necessary allowance for the influence of the phase boundary reaction. On the other hand in method "B" a sphere model can be used which is certainly more realistic for the irregularly shaped particles. In this case good agreement with the single crystal plate experiments were obtained provided that measurements up to sufficiently high degrees of exchange are available.

A sieve fraction of 80-120 µm was used. From the specific surface areas measured by adsorption a mean particle diameter of 60 µm was calculated. The size distribution was not considered.

OXIDATION BEHAVIOUR OF ZIRCONIUM NITRIDE IN OXYGEN

J.G. Desmaison[*], M. Billy[*] and W.W. Smeltzer[**]

[*] Laboratoire de Chimie Minérale et Cinétique Hétérogène, ERA 539 du CNRS, Université de Limoges, 87100 LIMOGES, FRANCE

[**] Department of Metallurgy and Materials Science, McMaster University, HAMILTON, Ontario, CANADA, L8S 4L7

ABSTRACT

The oxidation of 250 µ thick $ZrN_{0.93}$ plates at temperatures in the range 625-800°C over the pressure range 10-730 torr, and of 60 µ diameter $ZrN_{0.84}$ spheres at slightly lower temperatures (600-725°C) over the same pressure range, obey pressure dependent sigmoïdal kinetics. The reaction products consist of monoclinic zirconia accompanied with variable amounts of cubic or tetragonal zirconia. The experimental rate laws and the morphological observations suggest a transformation governed by a phase boundary reaction, the controlling process being the external interfacial reaction.

INTRODUCTION

The oxidation resistance of zirconium nitride is generally considered to be poor but the reaction mechanism is still not well known. The literature data were sparse and qualitative (1, 2) until the first kinetic data were reported recently (3, 4). This paper deals with the oxidation kinetics of Zr-nitride specimens of different symmetries in order to determine the reaction mechanism.

EXPERIMENTAL

Two kinds of samples were prepared : $ZrN_{0.93}$ plates (250 µ thick) by direct nitridation of 99.995 W/O zirconium sheets at 1150°C (5) and $ZrN_{0.84}$ spheres, (60 µ diameter) by dropping 99 W/O substoichiometric powdered zirconium nitride ($ZrN_{0.57}$) through an electric arc (6) and renitriding the spheroidized material for ten days at 1050°C and two days at 1200°C. The main characteristics of these samples : atom ratio (N/Zr), measured density (d), lattice parameter (a) and computed density (d') are given in Table I.

TABLE I

N/Zr	d (g/cm^3)	a (Å)	d' (g/cm^3)
0.93±0.01	7.180±0.008	4.5755±0.0007	7.228±0.013
0.84±0.01	6.77 ±0.02	4.5774±0.0015	7.132±0.017

Kinetics were measured as a function of temperature and oxygen pressure using thermogravimetric techniques. While the furnace was brought to temperature the sample ($\sim$ 40 mg) was kept under vacuum in the cold zone. It was plunged in the hot zone just as the reaction temperature was reached and dry oxygen introduced at the chosen pressure. The kinetic curves were obtained by plotting the fractional weight change $\alpha(=\Delta m/\Delta m_\infty)$ vs.t (time in hr)

The reaction products were studied by scanning electron microscopy, optical microscopy and x-ray diffraction techniques.

RESULTS AND DISCUSSION

<u>Effect of Temperature</u>. For both samples I (plates) and II (spheres) the curves $\alpha = f(t)$ have a sigmoïdal shape with a maximum rate for $\alpha_i \sim 0.45$ (fig. 1) and can be superimposed. A master run roughly in the middle of the series was chosen and a factor A found for each run such that multiplication of the time scale of the run by A would superimpose it onto the master run curve. Log A was found to be a linear function of 1/T. This implies that the activation energies are unique over the whole temperature range (7). Such a result was checked by plotting the logarithm of the instantaneous rate ($v_i = d\alpha/dt$) vs.1/T at different constant values of α. The activation energies obtained by both methods are E_I = 32 ± 6 Kcal/mole for type I samples and E_{II} = 48 ± 4 Kcal/mole for type II.

<u>Effect of Pressure</u>. The sigmoïdal shape is conserved (fig. 2) and the curves can be superimposed. By plotting log A or log v_i vs. log P, families of parallel straight lines are obtained whose slopes are n_I = 0.33 ± 0.03 and n_{II} = 0.19 ± 0.02. For the samples of type I Delmon's "Constant Interface Method" (8) has been used.

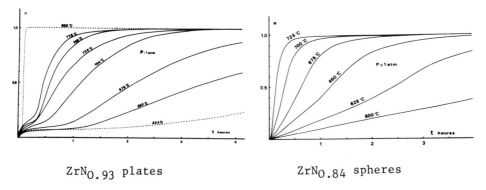

ZrN$_{0.93}$ plates   ZrN$_{0.84}$ spheres

Fig. 1. - α = f(t) curves. Temperature law.

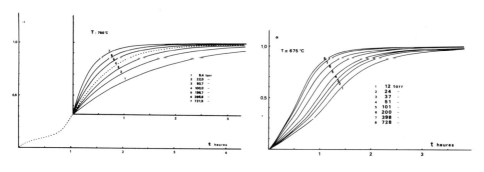

ZrN$_{0.93}$ plates (Delmon's method)   ZrN$_{0.84}$ spheres

Fig. 2. - α = f(t) curves. Pressure law.

Rate Law. These results indicate that the controlling process of the reaction is unique and identical to itself over the whole reaction. The rate laws can be expressed under the form of a separated variables expression (1) where f(α) is a morphological term

$$v = \frac{d\alpha}{dt} = f(\alpha) \cdot g(P) \cdot h(T) = K_P K_T f(\alpha) P^n e^{-E/RT} \quad (1)$$

characteristic of the reaction area. At constant temperature and pressure the shape of the curves is determined only by geometrical factors. The expression (2) derived from the PROUT and TOMPKINS (9) equation (3) gives a good representation of f(α) as shown by the

$$d\alpha/dt = k_1 \alpha (1 - \alpha/2\alpha_i) \quad (2)$$

$$F(\alpha) = \log \left(\alpha/(1 - \alpha/2\alpha_i)\right) = k_1 t + k_2 \quad (3)$$

F(α) vs. t plots which are linear. Using the slopes of these lines the following values have been found $E_I = 32 \pm 4$ Kcal/mole, $E_{II} = 49 \pm 4$ Kcal/mole, $n_I = 0.36 \pm 0.05$ and $n_{II} = 0.17 \pm 0.03$ which are very close to the preceding ones.

Structural and Morphological Observations. The reaction products consist of monoclinic zirconia accompanied by variable amounts of cubic or tetragonal zirconia. These two last forms which are predominate at lower temperatures almost disappear as T increases up to 800°C. In the case of the plates a uniform thin grey film appears at the beginning of the reaction and then oxide protuberances grow at preferential points. The propagation of the reaction towards the interior begins with the first cracks for $0.2 < \alpha < 0.4$ (fig. 3). For $\alpha > 0.5$ the surface is completely covered of bulges and the interior is badly cracked and consists of yellow "islands" of nitride surrounded by grey oxide. In the case of the spheres the surface, very smooth before oxidation, becomes rippled and faceted ($\alpha < 0.2$) then cracks appear around the facets ($0.2 < \alpha < 0.4$) (fig. 3) and finally the sphere breaks up ($\alpha > 0.7$). In both cases samples completely oxidized are transformed to powder at the slightest shock.

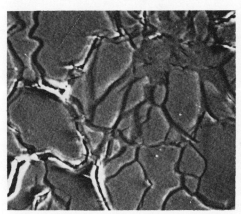
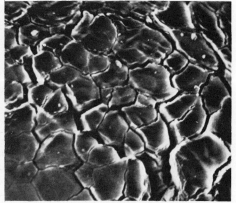

Cross section of a $ZrN_{0.93}$ plate ($\alpha = 0.31$, ×1000)   Surface of a $ZrN_{0.84}$ sphere ($\alpha = 0.39$, ×3000)

Fig. 3. - Scanning electron micrographs of reacted samples.

Mechanism. From these observations it seems that the reaction proceeds towards the interior according to an auto-catalytic type phenomenon rather than by a branched chain nucleation i.e. by a PROUT and TOMPKINS type model (9). It must be recalled that any autocatalytic process can be represented by the equation (4) which

$$d\alpha/dt = k' \alpha (1-\alpha) \qquad (4)$$

leads to (3). The term $\alpha(1-\alpha)$ is proportional to the effective area of the reaction front (7, 8). On the other hand the pressure laws correspond to the equation of the Freundlich's isotherm (heterogeneity of the adsorption sites), so it is probable that the controlling process is the external interfacial reaction. Such a reaction would proceed at a constant speed and the global rate would be modulated by the morphological factors according to (1).

Now, by analogy with the sulfidation of tungsten (10) the $\alpha = f(t)$ curves can be interpreted for $\alpha < \alpha_i$ by an increase of the area of the solid-gas interface owing to the expansion coefficient Δ (here ~ 1.5). In that case the oxidation of the spheres must obey to the well known kinetic law (5) relative to the external interfa-

$$F_A(\alpha) = \left(1 + (\Delta-1)\,\alpha\right)^{1/3} - 1 = k_A t \tag{5}$$

cial process (11). $F_A(\alpha)$ vs. time plots at various temperatures and pressures are effectively represented by straight lines. When $\alpha > \alpha_i$ the "active" interface decreases because of the diminution of the volume of the unreacted nitride "islands", it is then possible to interpret the kinetics curves by a law of type (6) which

$$F_B(\alpha) = \left(1 - (1-\alpha)^{1/3}\right) = k_B t \tag{6}$$

takes in account this geometrical factor (11). This can be checked by the linear plots of F_B vs. time. The activation energies and the n values computed from the slopes of these plots are very close to the experimental ones : for $\alpha < \alpha_i$, $E_{II} = 53 \pm 7$ Kcal/mole, $n_{II} = 0.21 \pm 0.02$ and for $\alpha > \alpha_i$, $E_{II} = 48 \pm 4$ Kcal/mole, $n_{II} = 0.19 \pm 0.02$.

CONCLUSION

Once again this study shows the importance of morphological data in the kinetic study of a gas-solid reaction. Whilst a PROUT and TOMPKINS type model gives a good representation of the kinetic curves, the morphological observations suggests a transformation governed by an external phase boundary reaction, this being confirmed by the oxygen pressure dependence of the rate law.

REFERENCES

(1) D.R. GLASSON and S.A.A. JAYAWEERA. - *J. Appl. chem. Lond.*, 1969, 19, 182.
(2) M.D. LYUTAYA, D.P. KULIK and E.T. KACHKOVSKAYA. - *Porosh. Met.*, 1970, 10, 72.
(3) H.F. AYEDI, M. CAILLET and J. BESSON. - *J. Chim. Phys.*, 1975, 72, 417 and *J. Less Common Met.* (in press).

(4) J. DESMAISON and M. BILLY. - *J. Chim. Phys.*, 1975, 72, 417 and *J. Chim. Phys.*, 1976, 73 (in press).
(5) J.G. DESMAISON and W.W. SMELTZER. - *J. Electrochem. Soc.*, 1975, 122, 354.
(6) J.G. DESMAISON and W.W. SMELTZER. - (To be published).
(7) P. BARRET. - Cinétique hétérogène. Gauthier-Villars, Paris, 1973.
(8) B. DELMON. - Introduction à la cinétique hétérogène. Technip., Paris, 1969.
(9) E.G. PROUT and F.C. TOMPKINS. - *Trans. Farad. Soc.*, 1944, 40, 488.
(10) M. LAMBERTIN. - Thèse, Université de Dijon, 1975.
(11) M. BILLY and G. VALENSI. - *J. Chim. Phys.*, 1956, 53, 832.

OXIDATION BEHAVIOUR OF NICKEL TELLURIDES IN THE COMPOSITION REGION BETWEEN $NiTe_{0.8}$ AND $NiTe_{2.0}$

D. Kolar, V. Marinkovič and M. Drofenik

Faculty of Natural Sciences and Technology and Institute J. Stefan, University of Ljubliana, Yugoslavia

ABSTRACT

Influence of chemical composition of nickel tellurides $Ni_{1+p}Te_2$ (0 p 1.5) on oxidation during heating up to 900°C in air was investigated by thermal, X-ray and electron diffraction analysis.

As a first oxidation step $Ni_{1+p}Te_2$ compounds with p 0 give rise to NiO and $NiTe_2$ phases. At higher temperatures, $NiTe_2O_5$ and $Ni_2Te_3O_8$ are formed, which oxidizes above 600°C to Ni_3TeO_6. This decomposes above 900°C leaving NiO as a final solid product. Formation of oxidized layers significantly slows down the kinetics of the oxidation processes.

INTRODUCTION

In the system Ni-Te, several phases are known to exist. Published data on this subject were reviewed and latest results added in reference (1). The phase richest in tellurium is $NiTe_2$ (66 at % Te), which has a layer structure of the $Cd(OH)_2$ type. By accomodation of additional metal atoms, a continuous solid solution of composition $Ni_{1+p}Te_2$ with gradual change towards NiAs type structure results. The limiting composition of solid solution (δ - phase) is at 52 at % Te. In compositions still richer in metal an orthorombic phase (γ_-) with approximate composition $NiTe_{0.8}$ (or $NiTe_{0.775}$ with 43.7 at % Te), a tetragonal γ_2 ($Ni_{3+q}Te$) phase and several high temperature phases were found.

Present work reports on the oxidation of nickel tellurides in the composition range $NiTe_{0.8}$-$NiTe_{2.0}$ on heating in air.

EXPERIMENTAL

$Ni_{1+p}Te_2$ samples ($1.5 \geq p \geq 0$) were prepared by melting appropriate mixtures of 99.999% pure nickel and tellurium (Johnson & Matthey and VEB Halbleiterwerk, DDR) in evacuated quartz ampules.

Powdered samples were subjected to X-ray analysis. The sample with the nominal composition $NiTe_{1.0}$ was found to be a mixture of the orthorombic $NiTe_{0.8}$ phase and the hexagonal $NiTe_{1.1}$ phase, in accordance with published data. Other samples in the composition region $NiTe_{1.1}$-$NiTe_{2.0}$ corresponded to the δ phase with cell parameters varying according to composition. In the sample with nominal composition $NiTe_{2.0}$ some free Te was also detected. The oxidation process was followed by thermogravimetric analysis (Stanton Instruments), differential thermal analysis (Linscis Apparatus) and X-ray diffraction, using a Guinier type focusing camera and nickel filtered CuKα radiation. Selected area electron diffraction patterns of oxide films stripped from oxidized specimens were made in a Siemens Elmiskop 1A microscope at 100 Kv.

RESULTS AND DISCUSSION

Depending on the composition of compounds, different oxidation sequences have been observed. The hexagonal $Cd(OH)_2$ type phase with composition close to $NiTe_2$ is fairly stable up to $400°C$. A 0.5% weight increase could be detected after heating for 15 hours at $400°C$. X-ray analysis of the product after heating showed the presence of $NiTe_2$ only, but with a small decrease in cell parameters.

By comparison samples with higher Ni content ($NiTe_{1.5}$ — $NiTe_{0.8}$) are far more prone to oxidation. Increase in weight could be detected already at $100°C$. X-ray analysis of the oxidized products after heating at $400°C$ showed the presence of NiO and $NiTe_2$ phases. The former was also confirmed by electron diffraction in all oxidized samples with compositions $NiTe_{0.8}$-$NiTe_{1.8}$. One the basis of this evidence, the first stage of oxidation of nickel tellurides in the composition range between $NiTe_{0.8}$-$NiTe_{2.0}$ give rise to NiO and $NiTe_2$:

$$Ni_{1+p}Te_2 + \frac{p}{2} O_2 \rightarrow NiTe_2 + pNiO \qquad \{1\}$$

The proposed first oxidation step was also confirmed by low temperature oxidation ($450°C$) to constant weight (Table 1)

Above $450°C$ $NiTe_2$ slowly oxidizes in air to $NiTe_2O_5$. This product could be readily identified by X-ray analysis. The presence of both these phases together after prolonged heating up to $550°C$ (Table II) indicates a slow oxidation process, probably governed by the diffusion of oxygen through the oxidized layer. Crushing of the sample after several hours of oxidation produced an immediate accelerated oxidation on renewed heating.

In $NiTe_2$ samples oxidized above 600°C, another phase appeared which could be identified at Ni_3TeO_6. This observation is in accordance with the already reported observation (2) that $NiTe_2O_5$ oxides to Ni_3TeO_6.

Table I

Weight gain after isothermal heating of $Ni_{1+p}Te_2$ samples at 400–450°C (20^h) in air.

Initial composition	Weight gain (%)	Calculated weight gain (%), eq. /1/
$NiTe_{0.8}$	6.6	6.3
$NiTe_{1.0}$	4.6	4.3
$NiTe_{1.2}$	3.3	2.9
$NiTe_{1.5}$	2.1	1.8
$NiTe_{1.8}$	1.1	1.2

Ni_3TeO_6 remains as the only oxidation product of $NiTe_2$ after heating at 800°C. It decomposes above 900°C (3), leaving NiO as the final solid product.

Oxidation of $Ni_{1+p}Te_2$ samples with higher Ni content (0 <p<1.5) above 450°C results in $NiTe_2O_5$ as well. However, in oxidation products obtained after heating at 500-750°C another product $Ni_2Te_3O_8$ could be detected as well as $NiTe_2O_5$. Its presence may be explained by the reaction

$$NiO + 3NiTe_2O_5 \rightarrow 2Ni_2Te_3O_8 \quad \{2\}$$

NiO necessary for the above reaction is present as a result of the first oxidation step, described by the reaction {1}.

The validity of the above explanation is further supported by the following arguments: (a) intensities of $Ni_2Te_3O_8$ reflections increased with increased Ni content in $Ni_{1+p}Te_2$ samples. (b) absence of $Ni_2Te_3O_8$ in oxidation products of $NiTe_2$ and (c) $Ni_2Te_3O_8$ may be prepared by heating NiO + $3TeO_2$ mixtures at 700°C, (2).

Above 800°C, $Ni_2Te_3O_8$ oxidizes to Ni_3TeO_6 as already reported (3). The final residue after heating above 900°C is NiO.

Table II

Phases identified by X-ray analysis after heating $Ni_{1+p}Te_2$ compositions at different temperatures in air.

Composition	400 ± 20°C	540 ± 20°C	600 ± 20°C	670 ± 20°C	860 ± 20°C
$NiTe_{0.8}$	$NiTe_2$ NiO	$NiTe_2$ $NiTe_2O_5$	$NiTe_2O_5$ $Ni_2Te_3O_8$ $Ni_3Te_3O_6$ (traces)	$NiTe_2O_5$ $Ni_2Te_3O_8$ Ni_3TeO_6	Ni_3TeO_6
$NiTe_{1.0}$	$NiTe_2$ NiO	$NiTe_2$ $NiTe_2O_5$	$NiTe_2$ $NiTe_2O_5$ $Ni_2Te_3O_8$ Ni_3TeO_6 (traces)	$Ni_2Te_3O_8$ Ni_3TeO_6	Ni_3TeO_6
$NiTe_{1.5}$	$NiTe_2$	$NiTe_2$ $NiTe_2O_5$	$NiTe_2O_5$ $Ni_2Te_3O_8$ Ni_3TeO_6 (traces)	$NiTe_2O_5$ $Ni_2Te_3O_8$ Ni_3TeO_6	Ni_3TeO_6
$NiTe_{1.8}$	$NiTe_2$	$NiTe_2$ $NiTe_2O_5$	$NiTe_2O_5$ Ni_3TeO_6 (traces)	$NiTe_2O_5$ Ni_3TeO_6	Ni_3TeO_6
$NiTe_{2.0}$	$NiTe_2$ TeO_2	$NiTe_2$ TeO_2 $NiTe_2O_5$	$NiTe_2O_5$ Ni_3TeO_6 (traces)	Ni_3TeO_6	Ni_3TeO_6

REFERENCES

1. K.O. Klepp and K.L. Komarek, Mh. Chem 103, 943, (1972)
2. G. Peser, F. Lasserre, J. Moret and B. Frit, C.R. Acad. Sci. Paris 272, 77, (1971).
3. D. Kolar, V. Urbanc, L. Golic, V. Ravnik and B. Volavsek, J. Inorg. Nucl. Chem. 33, 3693, (1971).

REACTIONS AND EQUILIBRIUM WETTING IN TITANIUM - MOLTEN GLASS SYSTEMS

A. Passerone, G. Valbusa and E. Biagini

Centro Studi di Chimica e Chimica Fisica Applicata CNR

Piazzale J.F. Kennedy - Genova (Italy)

INTRODUCTION

Studying interactions and wetting between molten glasses and metals gives information useful to technology, as in glass-to-metal sealing and in coating metals with glassy enamels.

The compositions of the glasses studied are shown in Table 1. We chose sodium disilicate (glass N° 1) as a reference glass as its physico-chemical characteristics are well-known. Glasses containing lanthanum and titanium are very well suited to join with titanium due to their thermal expansion coefficient (Table 1).

Molten glass interacts with metallic titanium through weak image forces induced in the metallic structure by ions in the molten mass, and by chemical reactions which may produce large amounts of energy.

A small molten drop on a solid flat surface is at equilibrium when the total free energy of the system is at minimum (4). This condition, under all the hypotheses reported in (4), may be expressed by the relationship

$$\gamma_s = \gamma_i = \gamma_1 + \cos \theta \qquad 1)$$

between the surface and interface tensions and contact angle θ.

For low energy surfaces, where solid-liquid interaction occurs only through dispersion forces, eqn. 1) forecasts a decrease in contact angle (increase in wetting) as the solid surface tension increases and liquid surface tension decreases. Such behaviour is not generally valid for high energy surfaces and for reactive systems like molten glass-metal at high temperature. It has, in fact, been observed that the addition of oxides, which increase the surface tension of molten glasses, sometimes improves glass to metal adhesion. This is due to the decrease in interfacial tension caused by chemical reaction (5). The sessile drop method provides an easy way to follow the interaction between metal and molten glass as a function of time, without the need of a crucible or different polluting agents (6).

EXPERIMENTAL

We prepared the glasses from silica, sodium carbonate and titanium dioxide pure reagents. Weighed quantities, ball milled for a long time, were melted in platinum crucibles at 1500°C for 12 hour with intermediate stirring. As a support for the sessile drops pure titanium metal 99,7% was used in the form of slides 12 x 12 x 1 mm, mirror finished.

Tests have been carried out in a horizontal furnace where a graphite susceptor is heated by high-frequency. The system may work under inert atmosphere or under a vacuum of 10^{-6} Torr (7). All tests have been performed under He at 400 Torr.

glass	Composition (mol %)	$\alpha_{20-400} \cdot 10^7$	ref.
1	Na_2O 33.33, SiO_2 66.67	142-176	(1)
2	Na_2O 17.5, SiO_2 62.5, TiO_2 20	89	(2)
3	Na_2O 17, SiO_2 78, La_2O_3 5	90.9	(3)

TABLE 1

At the solid-liquid interface, these reactions will be coupled with titanium oxidation so that

$$Ti^° \longrightarrow Ti^{3+} + 3 e^- \quad : \qquad \qquad 7)$$

in fact the equilibrium $Ti^{4+} + e^- \rightleftharpoons Ti^{3+}$ is strongly shifted towards the reduced state because the oxygen partial pressure, which governs the reaction

$$Ti_2O_3 + \tfrac{1}{2} O_2 \longrightarrow 2\ TiO_2 \qquad 8)$$

is kept at a very low value by the buffer action of the graphite susceptor. After reactions 5) and 6), sodium and silicon leave the molten glass either in a gaseous form or by entering the solid phase, as it was shown by electron probe microanalysis.

E.P.R. Analysis

Titanium Ti^{3+} is paramagnetic while Ti^{4+} is not: it is then possible to detect it by E.P.R. spectrometry. Previous work on glasses containing TiO_2, reduced by carbon or gamma radiation showed the possibility of detecting the presence of Ti^{3+} at ambient temperature (15,16,17). We made EPR spectra from drops of molten glass No. 1 on Ti detached from the metal and then powdered.

We found a signal fairly symmetric at ambient temperature and asymmetric at 77K. g values are 1.930 at 300 K and 1.915 at 77 K. The signal is wide, about 80 Gauss at 300 K and 130 Gauss at 77 K. As the temperature at which the specimen was held in contact with Ti increases, the area of the adsorption peak increases, suggesting an increasing content of Ti^{3+} ions in glass.

Wetting

Equilibrium contact angles between different glasses and titanium are found to follow the relationship:

$$\cos \theta = a + b\,T + c\,T^2$$

Constants a, b, c and standard deviations are given in Tab. 2.

glass	a	b	c	std.dev. σ
base	-9.3276	$1.344 \cdot 10^{-3}$	$-4.401 \cdot 10^{-6}$	$7.699 \cdot 10^{-2}$
La 5%	-2.910	$3.838 \cdot 10^{-3}$	$-9.538 \cdot 10^{-7}$	$3.546 \cdot 10^{-2}$
Ti 20%	-10.346	$1.352 \cdot 10^{-2}$	$-4.067 \cdot 10^{-6}$	$1.738 \cdot 10^{-1}$

TABLE 2

RESULTS AND DISCUSSION

Chemical Equilibria

The system consists of metallic titanium, molten glass and a gaseous phase around the glass-metal couple. In the absence of glass, the composition of the gaseous phase is determined by the presence of graphite and by impurities contained in the helium.

The most important equilibrium is represented by the reactions

$$2C + O_2 \rightleftarrows 2CO \qquad (2)$$

$$2CO + O_2 \rightleftarrows 2CO_2 \qquad (3)$$

For reaction 3) $K_p = P_{CO_2}^2 / (P_{CO}^2 \cdot P_{O_2})$: hence the oxygen partial pressure is determined by the ratio P_{CO}/P_{CO_2}.

To avoid titanium oxidation P_{O_2} must be kept below a critical value which may be calculated from the free energy of formation of TiO_2, equation 4), assuming Ti and TiO_2 to be at unit activity.

$$Ti + O_2 \longrightarrow TiO_2 \qquad (4)$$

Thus $P_{O_2} = 3,31.10^{-26}$ atm and $P_{CO}/P_{CO_2} = 3,48.10^6$ at $1100°C$. The P_{CO}/P_{CO_2} ratio, determined by the graphite, may be dramatically increased by condensation of CO_2. This shifts equilibrium 3) to the right, thereby lowering P_{O_2}. eg at $77°K$, $P_{CO_2} = 1.3.10^{-11}$ atm, and $P_{CO} = 0.7$ atm. Titanium slides appear shiny with no visible trace of oxidised layers (interference colours, opalescences, etc) after treatment at temperatures above $1000°C$. Molten glass, entering such a highly reducing system, induces new equilibria which fix final conditions for wetting and adhesion. Redox equilibria in molten glasses, whether in the presence or absence of solid metals, have been widely studied (8-14). At high temperature, under reducing conditions, the following reactions take place in sodium silicate:

$$Na_2O \longrightarrow 2Na + \tfrac{1}{2} O_2 \qquad (5)$$

$$SiO_2 \longrightarrow Si + O_2 \qquad (6)$$

whose equilibrium constants will increase with increasing temperature.

Our values are reproducable except around 900°C where some scattering has been observed. A value of Θ = 26° at 1260°C has been reported for sodium disilicate on titanium in argon (19); this value compares favourably with our findings.

If we compare the magnitude of contact angles at each temperature as a function of silica content in the glass we always find a minimum corresponding to the composition of base glass. Addition of titania or lanthana increases contact angle at low temperatures; at high temperatures contact angles become nearly the same for the three glasses, touching a value within 15° and 25°. Specimens of glasses 2 and 3 always show crystalline precipitates at the interface, which, mainly in titanium glass, growth as whiskers from titanium metal into glass phase. Solidified drops of glasses 2 and 3 are without cracks, and the titanium slides remain perfectly flat. In the case of glass No. 1 the presence of high stresses at the interface causes either the detachment of the drops after solidification or a considerable bending of metal.

REFERENCES

1) W.A.Weyl and E.C.Marboe. "The Consitution of Glasses" Interscience, New York, 1964. Vol. 2, P757
2) M.H. Manghnani, J. Phys. Chem. Soc. 55 (7), 360, 1972.
3) S.K.Dubrovo and A.D.Shnypikov. Inorganic Mat. 2, (9), 1417, 1966.
4) R.E.Johnson Jr. J.PhysChem. 63, (10), 1955, 1959.
5) See ref 1) Ch. 13
6) J.F.Padday. "Surface and Colloid Science" (E.Matijevic ed.) Wiley Interscience, New York. Vol. 2, P100
7) A.Passerone, E.Biagini and V.Lorenzelli. Ceramurgia 5,(2),81,1975
8) W.D.Johnston. J.Am.Ceramic Soc. 47, (4), 198, 1964.
9) W.D.Johnston, J.Am.Ceramic Soc. 48, (4), 184, 1965.
10) S.B.Holmquist. J.Am.Ceramic Soc. 49, (4), 228, 1966.
11) W.D.Johnston. J.Am.Ceramic Soc. 49, (9), 513, 1966.
12) W.D.Johnston and A.Chelk. J.Am.Ceramic Soc. 49, (10), 562, 1966.
13) M.P.Boron and J.A.Pask. J.Am.Ceramic Soc. 49, (1), 1966.
14) C.E.Hoge, J.J.Brennan and J.A.Pask. J.Am.Ceramic Soc. 56, 51, 1973
15) N.R. Yafaev and Yu.V. Yadokov. Sov. Phys. Solid State 4, 1123, 1962
16) S. Arafa and A.Bishay. Phys. and Chem. of Glasses 11, 75, 1970
17) C.R.Kurkjian and C.E.Peterson. Phys. Chem. of Glasses 15, 12, 1974
18) T.P.Shvaiko-Shvaikovskya, O.V.Mazurin and Z.S.Bashun. Inorganic Mat. 7, (1), 128, 1971.
19) K.K.Visotskis. Zh. Prkl. Khim. 39, (7), 1645, 1966.

OXIDATION KINETICS STUDY OF FINELY DIVIDED MAGNETITES

SUBSTITUTED FOR ALUMINIUM AND CHROMIUM

>Bernard Gillot, Abel Rousset, Jacques Paris
>and Pierre Barret
>Laboratoire de Recherches sur la Réactivité des
>Solides associé au C.N.R.S. - Faculté des Scien-
>ces Mirande - 21000 - Dijon - (FRANCE)
>Laboratoire de Chimie Minérale, U.E.R. de Chimie
>Biochimie, 48, Boulevard du 11 Novembre 1918 -
>69621 - Villeurbanne - (FRANCE)

INTRODUCTION

It has long been known (1-3) that the cubic iron sesquioxide γFe_2O_3 only forms during low-temperature oxidation of finely-divided magnetites. However, "cubic" iron sesquioxides substituted for trivalent ions such as aluminium or chromium have been described only recently by Rousset (4-5). This difference may be explained by the fact that substituted magnetites prepared at temperatures close to 1000°C, by usual procedures are only slightly reactive towards oxygen.

Preparing such magnetites made up of submicronic particles, at temperatures of about 500°C, Rousset succeeded in oxidizing these compounds into "cubic" sesquioxide according to :

$$2(Fe^{2+}Fe^{3+}_{2-x}M^{3+}_x)O_4 + 1/2\ O_2 \rightarrow 3\ \gamma(Fe^{3+}_{1-y}M^{3+}_y)_2O_3$$

$$M^{3+} = Cr^{3+},\ Al^{3+} \quad 0 \leqslant x \leqslant 2 \quad x = 3y$$

SAMPLES

Spinel preparation conditions of type $(Fe^{2+}Fe^{3+}_{2-x}M^{3+}_x)O_4^{2-}$ have been reported in the references (4-6). They mainly consist in treating, under an oxydo-reducing atmosphere, a sesquioxide resulting from decomposition in the air of an oxalic compound of type $(NH_4)_3 Fe_{1-y}M_y(C_2O_4)_3 \cdot 3H_2O$ ($0 < y < 2/3$). The morphology of the spinels and, particularly, the mean crystallite size are determined both by the air-treatment temperature following compound decomposition as well as by the water vapour content of the hydrogen.

The samples prepared are shown in Table 1. Pure magnetites were used for comparison with known results for Fe_3O_4, the influence of size was studies using the $(Fe^{2+}Fe^{3+}_{1.73}Al^{3+}_{0.27})O^{2-}_4$ and the effect of substitution on oxidation kinetics, since these show irregular variations with crystalline parameter using the ferrichromites (7,8).

Sample composition $Fe^{2+}Fe^{3+}_{2-x}M^{3+}_xO^{2-}_4$		Crystallite size in Å	Surface area in m²/g	Crystalline parameter in Å	D.T.A. temperature in °C	Temperature range in °C	Activation energy in kcal mole⁻¹	Crystalline parameter after oxidation in Å
x = 0	(A)	600	16,30		140	150-255	23,70	
	(B)	950	13,12	8,396	170	175-230	26,85	8,346
	(C)	1400	8		230	190-240	27,90	
$M^{3+} = Al^{3+}$								
	(11)	200	52,94		120	170-220	22,80	
	(1)	400	40,15		120	170-230	23,80	
x = 0,27	(2)	600	31,09	8,373	140	180-250	28,72	8,325
	(3)	600	14,25	8,371	200	200-270	31,50	
	(4)	3200	3,14	8,367	290	255-340	32,85	
x = 1,20		350	30	8,261	220	240-310	37	8,182
x = 2		300	40,20	8,149	360	350-410	47,42	8,058
$M^{3+} = Cr^{3+}$								
x = 0,27		480	24,40	8,389	160	179-227	27	8,335
x = 0,80		300	30	8,386	140	192-250	28,40	8,319
x = 1,20		780	15,80	8,405	200	250-308	32,80	8,298
x = 1,59		1300	10	8,390	260	280-350	26,54	8,281
x = 1,87		3000	3,60	8,383	350	350-430	27	8,264
x = 2		800	20	8,378		330-450	25	8,262

Table 1
Sample characteristics - Experimental kinetic results

X-ray diffraction measurements performed under vacuum using monochromatised Cr Kα radiation showed that the samples contained only the spinel phase. In the case of ferrialuminates the slight variation in the crystalline parameter depending on crystallite size may be attributed to a distribution variation of the cations on the crystallographic sites (9-11).

Morphology examination was effected by means of radio-Xray diffraction and electron microscopy. The mean size of the crystallites obtained by means of both methods is in good accordance with the mean diameter of the particles deduced from specific surface area measurements (Table I).

D.T.A. of the samples was effected in presence of air and with a temperature rise of 600°C per hour. The exothermic phenomena are attributed as follows : the former to the temperature at which Fe^{2+} ions oxidation occurs at maximum rate (Table I) and the latter to the conversion of the cubic phase γ into the rhombohedral phase α.

RESULTS

Kinetic studies were effected in a Mac-Bain-type thermoadjustable thermobalance, described previously (12). A 17 mg sample is uniformly spread at the bottom of the skoop so that it oxidizes in the same way as N independent particles of same mean radius. The gas, i.e. oxygen, is introduced only after the sample has undergone prior degassing at 430°C, under a vacuum of 10^{-6} torr.

The curves plotted showing the conversion rate α versus time for a similar pressure, 6 torr of oxygen and at various temperatures, show that the reaction starts at once with a maximal rate. These curves coincide when we look for their affinity versus time until $\alpha = 0.6$ or so. The affinity rate variation versus temperature allows the experimental activation energy to be calculated (Table I). For $\alpha > 0.6$ and especially for intermediate compositions the affinity is less and activation energy calculation from it is less accurate. Under similar temperature and pressure conditions the oxidation rate increase is related to grain size decrease as shown by the activation energy values (case of $Fe^{2+}Fe^{3+}_{1.73}Al^{3+}_{0.27})O^{2-}_4$.

Moreover, the temperature range where the oxidation occurs increases with the substitution rate (Table I). The activation energy compared to that of magnetite accounts fairly well for this phenomenon in the case of ferrialuminates, but in the case of ferrichromites this energy variation relative to the substitution rate is irregular.

DISCUSSION

Although mixed working conditions (diffusion + surface reaction) are established for the reaction onset in the case of pure and slightly substituted magnetites (13), in all the other cases the reaction is only limited by bulk ionic diffusion, which must be considered under variable working conditions. This question has already been dealt with for the oxidation kinetics of iron chromite considering a diffusing substance initially contained in uniform concentration within a sphere of radius a (14).

However, to interpret the kinetic curves we were led to consider that the diffusion coefficient D was no longer constant as was the case for iron chromite but that it decreased as the reaction proceeded. The comparison of our experimental curves with the various theoretical diffusion ones $\alpha = f(t/t_{1/2})$ for $\alpha > 0.5$

(for α < 0.5 all the theoretical curves practically coincide) where $t_{1/2}$ is the half-reaction time (Fig. 1 and 2) leads us to the following conclusions.

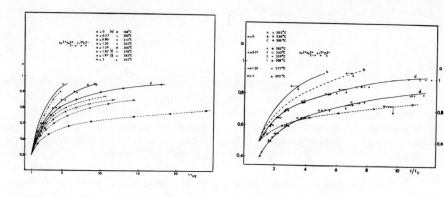

Fig. 1
Case of ferrichromites

Fig. 2
Case of ferrialuminates

Curve c : case of a plane
Curve a : case of the sphere with D constant
Curve d : case of the sphere with D variable.

For pure magnetites, the diffusion coefficient decreases as the reaction proceeds, as previously found by Feitknecht (15). As substitution of M^{3+} ions increases, this decrease is even more marked. However for compounds having compositions of, or close to, FeM_2O_4, the diffusion coefficient tends to become constant

Its decrease during reaction is explained by the increase in the number of vacancies (16-17) and the replacement of Fe^{3+} ions by Cr^{3+} and Al^{3+} ions whose stabilisation energy is higher for octahedral sites (18). The different behaviour of FeM_2O_4 - type compounds might be attributed to a less important rearrangement due to a single type of cation on the B sites.

For the chromium substituted magnetites, the activation energy (calculated at α=0.5) passes through a maximum and mirrors the crystalline parameter variation. The latter shows minor variations during substitution, the values equalling those of magnetite for x = 1 and 1.6

In the case of ferrialuminates the situation is different in that the small size of the aluminium ion causes the crystalline parameter to decrease sharply (4). The result is that activation energy calculated for α = 0.5, i.e. over the range where the diffusion coefficient still increases slightly, may express the high stresses related to volume variation.

CONCLUSION

The oxidation kinetics of $S_1 + G \rightarrow [S_1 + d]$ type reactions is well interpreted by diffusion under variable working conditions of the cations through the compact lattice of oxygen ions. In the present study, we showed the influence of crystallite size, substitution rate, **vacancy proportion and substituent nature** upon the oxidation kinetics. Other factors, such as ions and vacancy distribution occur without any possibility, for the time being, to define exactly the part they play. It appears that the nature of the substituted cation should be the prevailing factor in the oxidation kinetics, at least in the case of high subsitution.

REFERENCES

1 - F. Malaguti, C.R. Acad. Sci., Paris, 56, 467, (1863).
2 - O. Baudisch and I. Welo, Nature 13, 749, (1925).
3 - J. Huggett, Ann. Chim. 5, 1, 627, (1919).
4.- A. Rousset, Thèse Lyon, (1969).
5 - A. Rousset, J. Paris et P. Mollard, Ann. Chim. 7, 119, (1972).
6 - A. Rousset and J. Paris, Bull. Soc. Chim. Fr. 10, 3729, (1972).
7 - M. Robbins, G.K. Wertheim, R.C. Sherwood and D.N.E. Buchanan J. Phys. Chem. Solids, Vol. 32, 717, (1971).
8 - P. Poix, F. Basile and C. Djega-Mariadassou, Annales de Chimie, Vol. 10, n° 3, 159, (1975).
9 - W.L. Roth, Le Journal de Physique, 25, 507, (1964).
10 - C.M. Yagnik and H.B. Mathur, J. Phys. C. Proc. Phys. Soc. 2, 469, (1968).
11 - G. Fagherazzi and Garbassi, J. Appl. Cryst. 5, 18, (1971).
12 - B. Gillot, Thèse Dijon, (1972).
13 - B. Gillot, J.F. Ferriot and A. Rousset (to be published).
14 - B. Gillot, D. Delafosse and P. Barret, Mat. Res. Bull. 8, 1431, (1973).
15 - W. Feitknecht and K.J. Gallagher, Nature 228, 548, (1970).
16 - R.L. Levin and J.B. Wagner, Transaction of Metallurgical Society of AIME, Vol. 233, (1965).
17 - P.E. Childs, L.W. Laub and J.B. Wagner, Proc. Brit. Ceram. Soc. G.B., n° 19, 29, (1971).
18 - M.C. Cox, B. Mc Enaney and U.D. Scott, Philos. Mag. G.B. 26, 839, (1972).

CRYSTALLINE SOLID REARRANGEMENT BROUGHT ABOUT BY GAS REMOVAL.
SHAPE AND SIZE OF THE CRYSTALLITES OBTAINED

J.C. NIEPCE, J.C. MUTIN and G. WATELLE

Laboratoire de Recherches sur la Réactivité des Solides
Facultés des Sciences Mirande, Dijon, FRANCE

A number of studies have shown the existence of orientation relationships in Solid 1 → Solid 2 + gas - type transformations, concerning mineral solids. Little research has been done to know the way in which the structural edifice rearranges, as has been done for metals and metal alloys. This problem has been the aim of the present investigation considering both the following reactions

$$Cd(OH)_2 \rightarrow CdO + H_2O \qquad (I)$$
$$H_2C_2O_4, BaC_2O_4, 2H_2O \rightarrow \beta(H_2C_2O_4), BaC_2O_4 + 2H_2O \qquad (II)$$

For the latter, both sequences : gas removal (IIA) and structure change (IIB) may be made distinct. They will be expressed as follows

$$1/1/2 \rightarrow \alpha(1/1/0) + 2H_2O \text{ (IIA)} \quad \text{and} \quad \alpha(1/1/0) \rightarrow \beta(1/1/0) \text{ (IIB)}$$

EXPERIMENTAL RESULTS

Threedimensional orientation relationships appear in these three transformations. Pattern 1 analysis leads to the following relationships between the hydroxide (H) and the oxide (O) lattices.

$$[001] (H) \rightarrow [1\bar{1}1] (O) \quad \text{and} \quad (11\bar{2}0)(H) \rightarrow (220)(O)$$

Weissenberg patterns of a hydrated barium oxalate crystal (2a) and of the same crystal after dehydration in $\alpha(1/1/0)$ (2b) show conservation of the elementary lattice translations. Pattern 3

 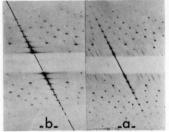 

Pattern 1   Pattern 2   Pattern 3

reveals orientation relationships between the lattices of both anhydrous α and β.

Identical morphological features are observed in these transformations. In the case of barium oxalate, water removal brings about, from the onset of the reaction, a steady breaking of the crystal bulk, following cleavage planes that are accurately defined relatively to its habit. A dehydrated crystal looks like an orderly piling up of small "blocks" (fig 1) whose shape and size (a few μ) are independent of the size and the habit of dihydrated crystals (table 1). The second splitting up seen on the pattern 3, which accompanies the structure change (IIB) has not been studied yet.

In the case of the hydroxide, X-ray diffraction analysis shows that the apparent mean dimensions D_{hkl} of the oxide crystallites produced are independent of those of the hydroxide crystallites, provided however that their thickness (e) and diameter (φ) are higher than what we called "critical values" (samples 1-2-3 table 2). Otherwise, the transformation brings about little or no splitting up (samples 4 and 5 table 2). As for the shape of the oxide crystallites, only a study of the hydroxide texture ob-

 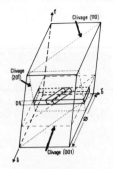 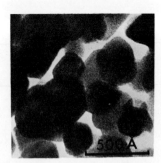

Figure 1   Figure 2   Figure 3

1/1/2		α(1/1/0)
φ (μ)		D (μ)
80		4,3
150		4,9
300		4,7
1000		4,4
2000		4,3

Table 1

Sample	Cd(OH)$_2$		CdO		
	e (Å)	φ (Å)	D_{111} (Å)	D_{200} (Å)	D_{220} (Å)
1	680	3500	134	122	104
2	346	590	138	122	106
3	257	590	135	122	109
4	54	268	75	66	72 - 75
5	21	103	22	26	40 - 43

Table 2

tained through rehydration allowed them to be likened to thin slabs generated according to (111) (0). Electron microscopy shows a homogeneous breaking up of the dehydrated hydroxide crystals with habit preservation and each oxide grain appears as a thin hexagonal plate parallel to (111) (0) (fig 3).

EXPERIMENTAL RESULTS INTERPRETATION BY MEANS OF CRYSTALLINE STRUCTURES

The comparison of oxalate 1/1/2 and α(1/1/0) structures shows that any slight variation in the atomic position causes the crystalline hydrate edifice to become α-anhydrous by water molecule removal. Splitting in dihydrate crystals follows three least resistance planes in the structure, (001), (201), (110) (fig 4). These coincide with three planes containing the water molecules and Ba^{2+} ions. Their representation in the basic, monoclinic prism allows the shape of the α anhydrous oxalate "blocks" to be found (fig 2) (1).

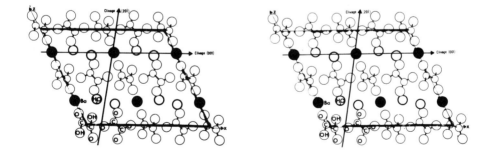

Figure 4

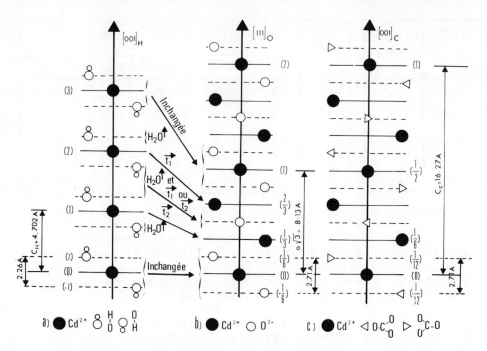

Figure 5

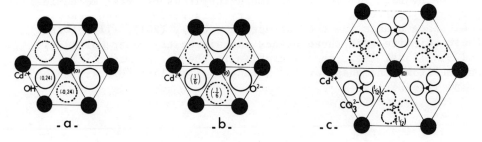

Figure 6

Comparison of Cd(OH)$_2$ and CdO structures. The hydroxide consists of parallel two-layer sets of OH$^-$ ions surrounding a layer of Cd^{2+} ions (fig 5-a and 6-a) and three such sets are required to form the oxide structure (fig 5-b and 6-b) ; one of them is wholly preserved, save small changes in distance, whereas the other two, besides the oxygens lost, undergo plane translations ($\vec{t}_1$ or $\vec{t}_2$ fig 5-a and b). The preservation of a major part of the atomic arrangement suffices to justify the existence of o-rientation relationships. Examination of the reorganisation of the piling up allows the nature of the morphological changes observed to be understood. Indeed, during dehydroxylation water molecules form from two contigous OH⁻ ionic layers; thus, statistically one anion layer out of two is removed. This results in a large structure packing following axis [001] and a simultaneous rearrangement of Cd^{2+} and O^{2-} ions located between the layers whose atomic distribution is not changed. The strains field resulting from these changes must be anisotropic ; it is more intense along axis [001] than in plane (001) where it must have ternary symmetry if account is taken of the atomic translations that have occured. This accounts for the hydroxide crystals splitting up into hexagonal oxide plates.

BASIC FEATURE OF THE STRUCTURAL REARRANGEMENT MODE

From those structural considerations it may be deduced that the crystallites formed result from directional breaks which appear when the strains followings the structural rearrangement reach a tolerance limit. Their magnitude and thus the crystallite size formed, is a function of the "extent of relationship" connecting both crystalline edifices. The tolerance limit depends on the initial structure which is somewhat the "canvas" of the transformation. Therefore, the shape and size of the crystallites produced depend on the parent edifice. These considerations imply a reaction mechanism according to which the final phase formation results from an "on the spot" rearrangement of the initial phase crystalline edifice. The atoms and ions remaining in the solid would keep interacting strongly from the transformation onset. This process, of course, excludes their diffusion over distances longer than the cell dimensions and therefore opposes any notion of growth of the final phase at the expense of the initial one. Finally, let us note that this concept is still supported by analysis of several bibliographical results (2), as well as by our own observations. According to these the dimensions and the shape of the crystallites formed depend neither on the percentage decomposition nor on the rate (3)(4) but, on the contrary, on the crystalline edifice they come from. The latter fact is explained by the schemes in figures 5-c and 6-c concerning CdO oxide formation either from the carbonate or the hydroxide.

REFERENCES

(1) J.C. MUTIN, Thesis Dijon (1975)
(2) J.F. GUILLIATT and N.H. BRETT, Phil. Mag. (1971) 183 617
(3) J.C. NIEPCE, P. DUMAS and G. WATELLE
    "Fine particles 2ⁿᵈ international conference"
    The Electrochemical Society. (1973) Boston (U.S.A.) 256
(4) J.C. NIEPCE Thesis Dijon (1976).

DISCUSSION

D. DOLLIMORE (University of Salford, U.K.) Is the appearance of the α form of $(H_2C_2O_4)BaC_2O_4$ dependant on water vapour partial pressure during dehydration? In other words, does the possible operation of the Smith-Topley effect involve a transition of

$$1/1/2 \longrightarrow \text{amorphous } (1/1/0) + 2H_2O$$

at low water vapour pressure and

$$1/1/2 \longrightarrow \alpha(1/1/0) + 2H_2O$$

at higher water vapour pressure or is the β phase involved?

AUTHORS' REPLY: I do not think it necessary to consider the Smith-Topley effect as arising during this transformation. We have shown that there is a coupling between the solid and the vapour during the endothermic transformation. The effect thus resembles the evaporation of a liquid except that in this case there is a solid present and a chemical reaction occurring.

PEROXIDE AND SUPEROXIDE AS CATALYSTS. A MODEL BASED

ON MOLTEN SALTS EXPERIMENTS

E. Desimoni, F. Paniccia, and P.G. Zambonin

Istituto di Chimica, Università di Bari

Bari, Italy

INTRODUCTION

Peroxide and superoxide species are often postulated as reactive intermediates in many chemical processes both in homogeneous and heterogeneous phases. In our laboratories, particularly exhaustive studies were performed on the processes:

$$2\ O_2^- + 2\ NO_2^- \longrightarrow O_2^{2-} + 2\ NO_3^- \qquad (1)$$

$$2\ O_2^- + H_2O \longrightarrow 1.5\ O_2 + 2\ OH^- \qquad (2)$$

$$2\ NO_2^- + O_2 \underset{}{\overset{O_2^{2-}/O_2^-}{\rightleftarrows}} 2\ NO_3^- \qquad (3)$$

in the molten $(Na,K)NO_3$ equimolar mixture.

Reaction 1 was studied at 500 K under nitrogen atmosphere and in the presence of an excess of nitrite (1). The relevant rate could be expressed by the relation

$$-\frac{d[O_2^-]}{dt} = \frac{k_a k_b [NO_2^-][O_2^-][O_2^{2-}]}{k_{-a}[NO_3^-] + k_b [O_2^-]} + k_i [NO^-][O_2^-] \qquad (4)$$

which is consistent with the following mechanism:

$$NO_2^- + O_2^- \xrightarrow{i} NO_3^- + O^-$$

$$\begin{array}{l} NO_2^- + O_2^{2-} \underset{-a}{\overset{a}{\rightleftarrows}} NO_3^- + O^{2-} \\ O^{2-} + O_2^- \xrightarrow{b} O^- + O_2^{2-} \\ \tfrac{1}{2}(O^- + O^- \xrightarrow[fast]{c} O_2^{2-}) \end{array} \qquad (5)$$

$$k_i < k_a \simeq k_{-a} < k_b < k_c$$

<u>a</u> is the autocatalytic step and <u>c</u> summarizes the fast steps following the formation of O^- free radicals. For these conditions the kinetic parameters were estimated as $k_i = (5 \pm 1)10^{-4}$ mol^{-1} kg sec^{-1}; $k_a = (1.7 \pm 0.1)10^{-2}$ mol^{-1} kg sec^{-1}; $k_b = (4 \pm 1)10^3$ mol^{-1} kg sec^{-1}.

The kinetics relevant to process 2 were followed mainly at constant concentrations of H_2O and O_2 (2). Analysis of the results leads to the following rate equation:

$$-\frac{d[O_2^-]}{dt} = k_2 \frac{[O_2^-]^2[H_2O]}{[O_2]} \qquad (6)$$

which is consistent with the mechanism

$$2\, O_2^- \rightleftarrows O_2 + O_2^{2-} \qquad (7)$$

$$O_2^{2-} + H_2O \xrightarrow{slow} Products \qquad (8)$$

$$Products \xrightarrow{fast} 2\, OH^- + \tfrac{1}{2}\, O_2 \qquad (9)$$

A value of (0.52 ± 0.03) mol^{-1} kg sec^{-1} was calculated for k_2.

It is interesting to note that the breaking of the O-O bond occurs at the level of the peroxide molecule instead of superoxide. Generally speaking the higher reactivity of peroxide relative to superoxide was a common factor in the course of our studies.

An obvious investigation suggested itself, viz. to collect information on the reactivity of the uncharged oxygen molecule whose bond is stronger than that of both peroxide and superoxide. We studied first the oxidation of nitrite by dissolved oxygen (the equivalent of the process <u>a</u> and <u>i</u> in mechanism 5).

$$2\, NO_2^- + O_{2(s)} = 2\, NO_3^- \qquad (10)$$

Even in the presence of a large NO_2^- excess and at quite high temperature (up to 700 K) the reaction was found to be very slow (3). The rate of reaction 10 was found to be

$$-\frac{d[NO_2^-]}{dt} = k_{10}[NO_2^-][O_2] \tag{11}$$

which is consistent with the simple mechanism

$$NO_2^- + O_2 \xrightarrow{slow} NO_3^- + O \tag{12}$$

$$NO_2^- + O \xrightarrow{fast} NO_3^- \tag{13}$$

where $k_{10} = 5 \times 10^{-4}$ mol^{-1} kg sec^{-1} at 500 K.

On the basis of these results, the hypothesis was made that the presence of peroxide and/or superoxide might catalyse reaction 10.

EXPERIMENTAL

The experiments were performed in the following way (4). After vacuum-degassing of the melt containing a certain excess of nitrite (0.1 - 0.5 m) peroxide ions were produced *in situ* by electroreduction of the melt itself.

$$NO_3^- + 2e = NO_2^- + O^{2-} \tag{14}$$

$$O^{2-} + NO_3^- = NO_2^- + O_2^{2-} \tag{15}$$

The gases produced at the anode by oxidation of the solvent

$$NO_3^- = NO_{2(s)} + \tfrac{1}{2} O_{2(s)} + e \tag{16}$$

were immediately pumped off.

Experiments performed under the following conditions have shown that the accelerating power of peroxide and superoxide ions on reaction 10 is very high: $0.5 < P_{O_2} < 1$ atm; $0.1 < [NO_2^-] < 0.5$ m; $5 \cdot 10^{-4} < [O_2^{2-}] < 4 \cdot 10^{-3}$ m; $10^{-3} < [O_2^-] < 8 \cdot 10^{-3}$ m.

RESULTS

When only the O_2^{2-} species was present the experimental kinetic data were consistent with the following mechanism:

$$\text{Mechanism I} \begin{cases} O_2(g) \rightleftharpoons O_2(s) & (17) \\ O_2(g) + O_2^{2-} \rightleftharpoons 2\,O_2^- & (18) \\ 2(NO_2^- + O_2^{2-} \xrightleftharpoons{k_a} NO_3^- + O^{2-}) & (19) \\ 2(O_2^- + O^{2-} \rightleftharpoons O^- + O_2^{2-}) & (20) \\ O^- + O^- \rightleftharpoons O_2^{2-} & (21) \end{cases}$$

$$2\,NO_2^- + O_2 = 2\,NO_3^- \qquad (10)$$

while the disappearance of nitrite was expressed by the relation

$$-\frac{d[NO_2^-]}{dt} = k_a\,[NO_2^-][O_2^{2-}] \qquad (22)$$

It is to be noted that the k_a value obtained in the course of this experiment was the same as obtained in the course of the study for elucidating mechanism 5.

As reaction 18 proceeded the O_2^{2-} species disappeared, giving rise to superoxide ions. The overall process was controlled by the relation

$$-\frac{d[NO_2^-]}{dt} = k_i\,[NO_2^-][O_2^-] \qquad (23)$$

The picture of the experimental kinetic findings obtained under these conditions could be summarized with the mechanism

$$\text{Mechanism II} \begin{cases} NO_2^- + O_2^- \xrightleftharpoons{k_i} NO_3^- + O^- & (24) \\ O_2(g) \rightleftharpoons O_2(s) & (17) \\ O_2(s) + O^- \rightleftharpoons O + O_2^- & (25) \\ O + NO_2^- \rightleftharpoons NO_3^- & (26) \end{cases}$$

$$2\,NO_2^- + O_2 = 2\,NO_3^- \qquad (10)$$

Again the k_i value obtained in the course of the present experiment was the same that the one obtained for mechanism 5 was studied.

A shift from mechanism I to mechanism II was explained on considering that when mechanism I was prevailing, other mechanisms had to parallel it. In effect, in the course of the catalyzed reaction the consumption of peroxide and the formation of a corresponding amount of superoxide was observed with the oxidation of nitrite. This fact was tentatively expressed by reaction schemes such as

Mechanism III
$$\begin{cases} 3\ (O_{2(g)} \rightleftharpoons O_{2(s)}) \\ O_{2(s)} + O_2^{2-} \rightleftharpoons 2\ O_2^- \\ NO_2^- + O_2^{2-} \rightleftharpoons NO_3^- + O^{2-} \\ O^{2-} + O_{2(s)} \rightleftharpoons O_2^- + O^- \\ O^- + O_{2(s)} \rightleftharpoons O_2^- + O \\ NO_2^- + O \rightleftharpoons NO_3^- \\ \hline 3\ O_{2(g)} + 2\ NO_2^- + 2\ O_2^{2-} = 2\ NO_3^- + 4\ O_2^- \end{cases}$$

and/or

Mechanism IV
$$\begin{cases} 2\ (O_{2(g)} \rightleftharpoons O_{2(s)}) \\ O_{2(s)} + O_2^{2-} \rightleftharpoons 2\ O_2^- \\ NO_2^- + O_2^- \rightleftharpoons NO^- + O_2^{2-} \\ O^{2-} + O_2^- \rightleftharpoons O^- + O_2^{2-} \\ O^- + O_{2(s)} \rightleftharpoons O_2^- + O \\ NO_2^- + O \rightleftharpoons NO_3^- \\ \hline 2\ O_{2(s)} + 2\ NO_2^- + O_2^{2-} = 2\ NO_3^- + 2\ O_2^- \end{cases}$$

The value found for the kinetic constants k_a and k_i are 1.9×10^{-2} mol^{-1} kg sec^{-1} and 1.9×10^{-4} mol^{-1} kg sec^{-1} respectively. Both values are, as mentioned, in good agreement with the ones found on studying mechanism 5.

The following conclusion can be drawn from the present work: the presence of a 10^{-3} m concentration of O_2^- or O_2^{2-} could raise the oxidation rate of nitrite by molecular oxygen by 10^3 to 10^5 times respectively or, more suggestively, that at the pressure of 1 atm after the addition of 10^{-3} m of catalyst, the rate of process 10 was approximately the same as that under a pressure of $10^3 - 10^5$ atm, but in the absence of catalyst.

REFERENCES

(1) P.G. Zambonin and A. Cavaggioni, J. Amer. Chem. Soc. 93, 2854 (1971).
(2) P.G. Zambonin, F. Paniccia and A. Bufo, J. Phys. Chem. 76, 422 (1972).
(3) F. Paniccia and P.G. Zambonin, J. Phys. Chem. 77, 1810 (1973).
(4) F. Paniccia and P.G. Zambonin, J. Phys. Chem. 78, 1693 (1974).
(5) E. Desimoni, F. Paniccia and P.G. Zambonin, J. Electroanal. Chem. 38, 373 (1972).
(6) F. Paniccia and P.G. Zambonin, J. C.S. Faraday I, 68, 2083 (1972).

CARBON FORMATION ON Ni FOILS BY PYROLYSIS OF PROPENE ($290°-800°C$)

T. Baird

Department of Chemistry, University of Glasgow

Glasgow, G12 8QQ (Scotland)

ABSTRACT

The formation of carbon deposits on Ni foils by pyrolysis of propene ($290-800°C$) has been studied. Structural and kinetic features of the deposition process are discussed.

INTRODUCTION

The deposition of carbon on to metal surfaces, whether desirable or otherwise, is of importance in many systems. Resistance-heated metal foils have been frequently used as catalysts in deposition studies (1-5). In this work it was found that different types of deposit involving different growth mechanisms were produced within rather well-defined temperature regions. The various deposits could be formed simultaneously on a foil as a consequence of the temperature gradient across the foil surface.

EXPERIMENTAL

Pre-reduced Ni foils were heated in propene (50 Torr) for ½ hr. in the range $290-800°C$. Temperatures quoted are those measured at the centre of the foils. The deposits were examined by electron optical techniques and others. Details are described elsewhere (1).

RESULTS AND DISCUSSION

<u>General deposition characteristics</u>. Deposit weight v. temperature data are given in Fig.1 and Fig.2 (Arrhenius plots).

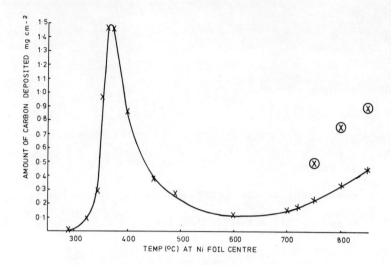

Fig.1. Weight of C deposited v. temperature curve; ⊗'s represent total carbon (filaments and turbostratic) above 700°C; X's above 700°C represent turbostratic graphite only.

Maximum deposition occurred at 375°C (± 5°C). The different types of deposit found within certain temperature regions are described below. Owing to the temperature gradient along the length of the foil, foils heated to >380°C exhibited heavier deposition at the cooler edges (Fig.3). Temperature profile measurements showed that the boundary between the heavy deposition and the thinner carbon layer always occurred at 380°C. Very similar results were

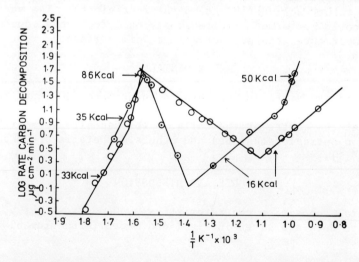

Fig.2. Arrhenius plots for the deposition of carbon on Ni; O = propene; ⊖ = 1:3 butadiene (50 Torr).

Fig.3. Ni foils (50 x 7 mm) after heating to 450°C (a) and 600°C (b) in propene.

obtained with 1:3 butadiene.

Deposition at 290°C - ~320°C. Both the acid treatment required to remove the deposit from the foil and the subsequent washing procedure influenced the morphology of the deposit as observed by EM (Fig.4). Careful preparation revealed continuous films and rectangular arrays of Ni_3C crystals. All the Ni_3C films were highly textured giving single crystal or almost single crystal spot patterns, the most common zone axes being [331] and [411]

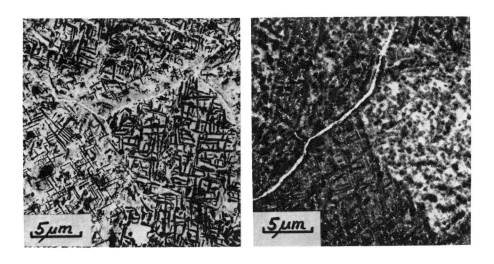

Fig.4 (a) Ni_3C crystals; short acid treatment; ethanol wash.
(b) Carbon framework remaining after prolonged acid treatment; washed with water.

Fig.5 (a) Continuous film of Ni_3C. (b) [331] reciprocal lattice network from an area within (a).

(Fig.5). The patterns could be indexed only on the basis of formation of a superlattice structure (6).

<u>Deposition at</u> $\sim 320°C - 380°C$. Progressively thicker carbon ($d_{(0002)} = 3.40 Å$) deposits were obtained with increase in temperature up to 380°C. Visual growth of deposits was observed as reaction proceeded. SEM and EM revealed the columnar growth features of these deposits (Fig.6). An activation energy of 33 kcal/mole is suggestive of a growth mechanism involving migration

 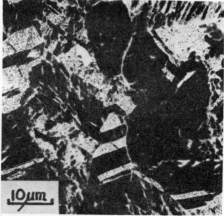

Fig.6 (a) Scanning electron micrograph of a fractured foil with deposit. (b) Preferred carbon growth features as revealed by EM.

of carbon atoms through the Ni to active growth centres (2). The structural features of this deposit and an observed constant rate of deposition tend to support such a mechanism. The value of 86 kcal/mole (Fig.2) is probably an artefact since deposit weights were very sensitive to small temperature changes in this region (350 - 380°C). Experiments with butadiene gave only the 35 kcal/mole value.

<u>Deposition at >380°C</u>. The main product above 380°C was turbostratic graphite ($d_{(0002)}$ = 3.36 Å). Some carbon filaments were also observed. The rate of carbon deposition here dropped sharply to a fairly steady low value. An activation energy of 16 kcal/mole (also found for butadiene) may be more consistent with a surface diffusion mechanism (5). At the highest temperatures studied masses of carbon filaments were formed at the cooler edges of the foil (see Fig.1). These could be easily removed from the foil. Migration of mobile metal particles from the foil centre appears to have occurred here.

References

1. T. Baird, J.R. Fryer and B. Grant, Carbon, 1974, <u>12</u>, 591.

2. L.S. Lobo, D.L. Trimm and J.L. Figueiredo, Proc. Int. Congr. Catal. 5th, 1972, 1125 (1973).

3. A.E.B. Presland and P.L. Walker, Jnr., Carbon, 1969, <u>7</u>, 1.

4. S.D. Robertson, Carbon, 1970, <u>8</u>, 365.

5. F.J. Derbyshire and D.L. Trimm, Carbon, 1975, <u>13</u>, 189.

6. S. Nagakura, J. Phys. Soc. Japan, 1957, <u>12</u>, 482.

OXIDATION REACTION OF CHROMIUM OXIDE (Cr_2O_3) IN THE PRESENCE OF MM'O_3 (M : Ca, Sr, Ba ; M' : Ti, Zr)

T.Nishino, T.Sakurai and S.Nishiyama

Musashi Institute of Technology

Tamazutsumi, Setagaya, Tokyo (JAPAN)

INTRODUCTION

When Cr_2O_3 is heated alone in an oxidizing atmosphere, the oxidation reaction takes place at the surface of Cr_2O_3 particles, but the amount of oxidized product is as much as 0.4 mg of Cr^{6+} ion per gram of Cr_2O_3 assuming CrO_3 as a product.
As shown in the equations (a) and (b), however, Cr_2O_3 is oxidized readily to the corresponding acid-soluble chromate in the presence of an alkali or an alkaline earth compound of basic nature; the reaction of which, for example, has been applied to the production of chrome compounds from chrome ore and to the estimation of the solid basicity in glass technology.

$$2CaCO_3 + Cr_2O_3 + 3/2\ O_2 \longrightarrow 2CaCrO_4 + 2CO_2 \qquad (a)$$

$$2Na_2CO_3 + FeCr_2O_4 + 3.5/2\ O_2 \longrightarrow 2Na_2CrO_4 + 1/2\ Fe_2O_3 + 2CO_2 \quad (b)$$

The study had been carried out to explore the oxidation reaction of Cr_2O_3 in oxidizing atmosphere in the presence of an alkaline earth carbonate with or without other oxides such as SiO_2, TiO_2, ZrO_2, Al_2O_3 (1).
The present paper deals with, at first, the thermal oxidation of Cr_2O_3 in the presence of a double oxide MM'O_3 where M and M' represent Ca, Sr, Ba and Ti, Zr respectively. Successively, a concept of solid basicity will be proposed by taking into account the results which are concerned with the reactivity of a series of solid solution between two kinds of the double oxide with Cr_2O_3.

EXPERIMENTAL

The raw materials used were Cr_2O_3 and alkaline earth carbonates of analytical reagent grade. The double oxides of 99.5% purity were supplied by Kyoritsu Yogyo Co., and the several kinds of solid solution between the double oxides were prepared by heating the pellet of proper composition to a high temperature. The formation of the solid solution was confirmed by the linear relationship of the lattice parameter(s) between the initial double oxides. Either the initial double oxide or the solid solution was ground in an agate mortar until grain size was reduced to <10μm, then, was mixed with a slight excess of Cr_2O_3 (MO / Cr_2O_3 = 2.0/1.25 in molar ratio) and was allowed to react at proper temperature in O_2 stream of 200 ml/min.

Thermal analyses (DTA and TG) were carried out not only to predict the reaction temperature but also to analyze the reaction process quantitatively; X-ray phase analysis with Cu Kα radiation was used to identify the phase present before and after acid-treatment of the heated samples.

In order to monitor the fractional reaction completed as a function of heating temperature and time, quantitative determinations of the formed Cr^{6+} ion and the dissolved alkaline earth ion(s) in dilute HCl solution were made by the conventional techniques of chemical analysis using Mohr's salt and $KMnO_4$ solutions and by an atomic absorption spectrophotometry, respectively.

The fractional conversion to chromate (α) was calculated as the ratio of the amounts of the chromate formed and of the theoretical value.

RESULTS and DISCUSSION

Whether a certain compound exerts an influence on the oxidation of Cr_2O_3 in positive or negative sense will be of great interest in view of the solid basicity of the compound. The results so far obtained in our laboratory are grouped into the following three items.

[A] Compound, foreign to the oxidation of Cr_2O_3, ie., the α-value is zero such as $CaAl_2O_4$, $CaSiO_3$, $CaTiO_3$, $SrSiO_3$, $SrTiO_3$ etc.

[B] Compound, effective in chromate conversion but insufficient to establish completion because of the formation of intermediate phases ie., the α-value is below 100% such as $CaZrO_3$, $SrAl_2O_4$, $BaTiO_3$ etc.

[C] Compound, effective for complete chromate conversion, ie., the α-value is 100% such as $CaFe_2O_4$, $Ca_2Fe_2O_5$, $BaSiO_3$, $SrZrO_3$, $BaZrO_3$ etc.

What is obvious on comparing the homologous oxide series such as silicate, titanate and zirconate with their maximum α-values, the contributive effect of M^{2+} ions on the basicity is assumed to be in the order of Ba >Sr >Ca.

At first, according to the order of the classification, the results obtained regarding to the thermal behavior of some titanates and zirconates in the presence of Cr_2O_3 are summarized as follows.

[A] Both $CaTiO_3$ and $SrTiO_3$ remained unaltered up to 1200°C which was confirmed by DTA and TG curves of no response. This supports that these oxides eventually will be classified as acidic oxide groups.

[B] DTA and TG curves of the mixed powder of $CaZrO_3$ and Cr_2O_3 indicated the occurence of $CaCrO_4$ formation at about 600°C as an abrupt exothermal with weight gain. However, the reaction did not proceed completely independently of heating condition and the molar ratio $CaZrO_3/Cr_2O_3$; the maximum α-value was close to 80% with grey coloured cubic ZrO_2 remaining as residue. It must be stressed that the result suggests that the cubic phase in the binary system $CaO-ZrO_2$ does not react with Cr_2O_3. Therefore, the amount of $CaZrO_3$ coexisted with cubic and/or monoclinic ZrO_2 could be analyzed independently for the selective reaction which may provide some information to establish phase boundary of the cubic field in the $CaO-ZrO_2$ system, and may be applied for the study of solid state reaction such as $CaCO_3-ZrO_2$ system (2).

The reactions of $BaTiO_3-Cr_2O_3$ and $Ba_2TiO_4-Cr_2O_3$ systems were examined isothermally up to 1200°C in O_2 stream. It is worth noting that the reactions attain equilibrium at temperature around 950°C and the maximum α-values are 80 and 90% respectively, with an unknown product of dark colour remaining. None of TiO_2 to be liberated in the course of reaction was detected. These facts suggest that an intermediate will be formed. On the basis of the α-values obtained at equilibrium, it may be supposed the phase consists of BaO and TiO_2 in the molar ratio 1 : 5. As a result of examination of the heat-treated samples of the mixtures of $BaCO_3$+ $5TiO_2$+ xCr_2O_3 (x = 0.1 - 3.0) by the disappearing-phase method, the intermediate phase could be represented as $BaO \cdot 5TiO_2 \cdot Cr_2O_3$. Hence, the formation of the new compound as an intermediate, by the reaction between $BaCrO_4$ formed and TiO_2 liberated will bring about the lowering of the maximum value of α.

[C] Both of $SrZrO_3$ and $BaZrO_3$ react readily with Cr_2O_3 to yield the corresponding chromates and monoclinic ZrO_2 without forming any intermediate.

Many papers have appeared concerned with the estimation of the strength of solid basicity (3), but the results are not necessarily in agreement. Furthermore, representation of the acid-base boundary, pH 7 in aqueous solution, remains obscure. In order to examine this boundary, solid solution between the double oxides with different α-values was prepared and the dependence of α-value on the composition of the solid solution was investigated.

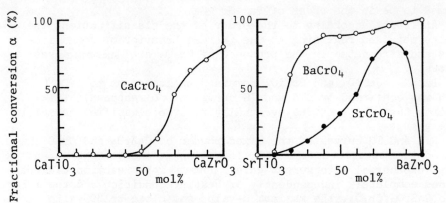

Fig.1 Relations between α and the composition of solid solution

When two kinds of double oxide, whose α-values are 0 and 100% respectively, are blended mechanically in all proportions, a linear relationship would be anticipated between the fractional conversion to chromate and the composition of the mixture. In a series of solid solutions between them, one might expect the α-value to drop to zero at a boundary region - "reaction limit" - where the basicity of the solid solution disappears. By taking into account the crystallographic symmetry, the following two groups of solid solutions were examined at several temperatures. Part of the experimental results are indicated in Fig. 1.

[Orthorhombic system]

$CaTiO_3$-$CaZrO_3$ system : α-value decreased, as shown in Fig.1, as the increase of $CaTiO_3$ content and approached to zero at the composition of 60 mol% $CaTiO_3$.

$CaTiO_3$-$SrZrO_3$ system : the solid solution richer in $SrZrO_3$ reacted with Cr_2O_3 to yield $(Ca,Sr)CrO_4$ and monoclinic and/or cubic ZrO_2. The "reaction limit" was observed at 70 mol% $CaTiO_3$.

[Cubic system]

$SrTiO_3$-$BaZrO_3$ system : fairly large amounts of $SrCrO_4$ and $BaCrO_4$ were produced over most of the composition range as shown in Fig.1, the result of which indicated the system might fall under the category of basic oxide except narrow range near $CaTiO_3$.

$SrTiO_3$-$BaTiO_3$ system : owing to the formation of $BaO \cdot 5TiO_2 \cdot Cr_2O_3$ as an intermediate, the tendency of conversion to chromate was not so remarkable and α-value decreased to zero abruptly at around equimolar composition of the solid solution.

A proposal was presented to estimate the strength of basicity of a double oxide by adopting the molar volume V/Z (Å^3) as a

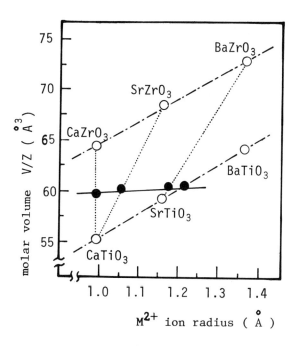

Fig.2
Changes of molar volume of double oxide series and "reaction limits" (filled circles) observed for the solid solutions

parameter where Z was represented the number of molecules in a unit cell. Fig.2 shows the relation of the molar volume of two groups of titanate and zirconate to the ionic radius of M^{2+}, indicating the molar volume is proportional to the size of the ion radius. As is seen in Fig.2, the ordinate may be associated closely with the strength of basicity, that is, double oxides of larger molar volume such as $BaZrO_3$ and $SrZrO_3$ are more basic oxides with higher α-values. It is also found that the α-value obtained for such as $CaZrO_3$ and $BaTiO_3$ was located, by chance, on the same level of V/Z. The "reaction limits" so far obtained in a series of solid solutions were marked on the respective line between the double oxides. The line connecting the filled circles in the diagram may be regarded as a boundary line between acidic and basic properties for the double oxides used.

REFERENCES

(1) T.Nishino et al., Kogyo Kagaku Zasshi, 66, 34, 151 (1963); 68, 31 (1965); 69, 18, 30 (1966)

(2) T.Nishino et al., ibid., 70, 36 (1967)

(3) A.Adachi and K.Hagino, Tetsu to Hagane, 51, 1857 (1965), R.Didtschenko and D.G.Rochow, J. Am. Chem. Soc., 76, 3291 (1954)

INFLUENCE OF A FOREIGN GAS IN HETEROGENEOUS REACTIONS

B. Guilhot, R. Lalauze, M. Soustelle and G. Thomas

Laboratoire de Chimie Physique du Solide, Ecole

Nationale Supérieure des Mines de Saint-Etienne, France

INTRODUCTION

Let us consider an heterogeneous reaction in which a gas G is present with one or more solids. The mechanism of such a reaction is generally very complex, but substantial progress has been realized in this domain by studying the gas pressure dependence of the reaction rate (1) (2) (3).

In the same way and as a concrete example, we have studied the pressure dependence of a gas G' different from the gas G and called the "foreign" gas of the reaction. This "foreign" gas changes the kinetics of the reaction, and the rates of the process can be increased or decreased. Such changes may be induced by different phenomena :

1 - Evolution of the thermal conductivity of the system, which can modify the effects of "subtemperature" or "ultratemperature", owing to the endo or exothermicity of the reaction.

2 - Adsorption of the foreign gas which makes the number of vacant sites decrease. This number intervenes in the determination of the reaction rate.

3 - Variations of the stoichiometry of the reacting solids and the creation or annihilation of point defects which very often are the intermediary species necessary to the reaction.

4 - A catalytic or inhibiting effect produced by a slight combination of the reacting gas in the sorbed state with the foreign gas. Such a combination can enhance or inhibit the desorption of the reacting gas.

5 - Different coefficients or diffusion in the gas phase.

Experimentally we have studied the kinetics of different heterogeneous reactions, each of them illustrating one or more of these phenomena.

REACTION : $SOLID_1 + GAS \rightarrow SOLID_2$

Example (4) (5) : oxygen + niobium, "foreign" gas : chlorine

The reaction between oxygen and niobium results in formation of the metallic oxide phases : NbO_x and NbO_z under 400°C, and the formation of a Nb_2O_5 scale above this temperature.

In an oxygen and chlorine gas mixture where P_{O_2} = 12 torr and P_{Cl_2} = 20 torr, we observe several changes :

1 - The Nb_2O_5 oxide phase appears in the low temperature range, that is to say under 400°C.

2 - The formed compounds $NbCl_5$ and Nb_2O_5 are initially localized at the grain boundaries (Fig. 1).

3 - The rate of the oxide formation increased with the chlorine pressure (Fig. 2).

This result can be explained by a large dissolution of chlorine in the metal.

REACTION: $SOLID_1 \rightarrow SOLID_2 + GAS$

The decomposition of silver carbonate, at a controlled pressure of carbon dioxide leads to the formation of silver oxide (6).

$<Ag_2CO_3> \rightleftarrows <Ag_2O> + [CO_2]$

The reaction rate increases with the water vapour pressure (Fig. 3).

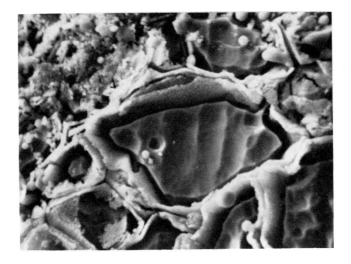

Fig. (1) : Surface state of the sample during the oxides formation in the grain boundaries.

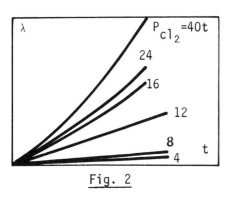

Fig. 2

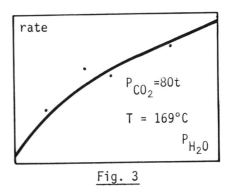

Fig. 3

Fig. 2 : Chlorine pressure dependence of the corrosion of niobium at 240°C and a P_{O_2} = 12 torrs.

Fig. 3 : Influence of water vapour on the reaction
$Ag_2CO_3 \rightarrow Ag_2O + CO_2$

The oxygen pressure dependence is more complicated. We note a minimum value for the reaction rate at 50 torr (Fig. 4).

This catalytic effect may be explained as follows : the water vapour and the carbon dioxide react to form an adsorbed complex, which enhances the carbon dioxide desorption.

The influence of oxygen pressure is more difficult to interpret :
  the inhibitor effect is due to the trapping of carbon dioxide adsorption sites by oxygen molecules,
  the evolution of the reaction rate observed for the high oxygen pressures can be attributed to the creation of defects in the crystal lattice of silver carbonate.

In the same manner we have noticed an inhibitor effect of foreign gas on the dehydration rates of some solids e.g.
  gypsum ($CaSO_4$, $2H_2O$) submitted to methanol pressure (7),
  borax ($Na_2B_4O_7$, $10H_2O$) submitted to helium, nitrogen, argon or carbon dioxide pressures (8).

REACTION: $SOLID_1 + SOLID_2 \rightarrow SOLID_3 +$ GAS

Example : $BaCO_3 + TiO_2 \rightarrow BaTiO_3 + CO_2$

The following experiments show the influence of oxygen and nitrogen on this reaction rate (9).

At 697°C and a partial pressure of carbon dioxide P_{CO_2} of 250 torr in the binary gas mixture $CO_2 - N_2$, it is noticed that the reaction rate is hyperbolic function of the nitrogen partial pressure P_{N_2} (Fig. 5).

With a $CO_2 - O_2$ gas mixture (Fig. 5), the rate exhibits a minimum at P_{O_2} = 150 torr, as in the case of the silver carbonate decomposition.

Such phenomena can be explained by the trapping of adsorption sites at the gas-solid interface, by the gas phase diffusion (inhibiting effect), or by the influence of stoichiometry defects (catalytic or inhibiting effect).

The modifications due to the diffusion in the gas phase are lowered (the gas mixture is agitated). The temperature is high enough to eliminate the conductivity effects.

So, it seems that in this particular example, the most important phenomena are the trapping of sites and the stoichiometry defects of the reacting solids.

In conclusion our examples show that the presence of a "foreign" gas is very important for the evolution of heterogeneous reaction. Such a gas induces a complete change in the reaction process and the energetic path is fundamentally different from the initial one.

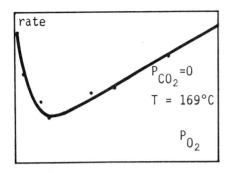
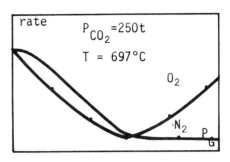

Fig. 4   Fig. 5

Fig. 4 : Influence of oxygen pressure on the reaction
$Ag_2CO_3 \to Ag_2O + CO_2$

Fig. 5 : Influence of nitrogen and oxygen on the reaction
$BaCO_3 + TiO_2 \to BaTiO_3 + CO_2$

REFERENCES

(1) J. BESSON and al., J. Chim. Phys., 63, 1049 (1966)
(2) M. SOUSTELLE and al., J. Chim. Phys., 3, 375 (1972)
(3) R. BARDEL and M. SOUSTELLE, J. Chim. Phys., 1, 21 (1974)
(4) A. SOUCHON, M. SOUSTELLE, R. LALAUZE, Bull. Soc. Chim. sous presse
(5) A. SOUCHON and R. LALAUZE, communication privée
(6) R. BARDEL, thèse, Grenoble, 1973
(7) J.J. GARDET, thèse, Grenoble, 1974
(8) G. THOMAS, thèse, Grenoble, 1972
(9) G. THOMAS and M. COURNIL. Journée de la Société de Chimie-Physique, Limoges, 20 Mars 1975.

DISCUSSION

P.A. BARNES (Leeds Polytechnic, Leeds, U.K.) I would like to comment on the interpretation of the results obtained for decomposition of silver carbonate. Work published with Professor Stone supports your observations on the effect of water vapour (1,2). However, our experiments suggested that water vapour did not act merely at the surface of silver carbonate. The high reactivity of some samples of this material was attributed to the presence of linked defect centres involving the incorporation of water, in the form of OH^- and HCO_3^- ions into the lattice. These linked centres provide diffusion pathways for carbon dioxide thus enhancing the decomposition reaction. The presence of water vapour reduces the rate of thermal destruction of the defects thus giving rise to increased reactivity.

1. P.A. Barnes and F.S. Stone Proc. 6th Int. Symp. on Reactivity of Solids ed J.W. Mitchell, Wiley, New York, 1969
2. P.A. Barnes, M.F. O'Connor and F.S. Stone J.Chem.Soc. Section A pp 3395-3398 1971.

ABOUT THE USE OF A SOLID STATE ELECTROCHEMICAL OXYGEN
PUMP FOR SELECTIVE OXIDATION OF ORGANIC MATERIALS

S.Pizzini[*], C.M.Mari, D.C.Hadjicostantis

Istituto Elettrochimica e Metallurgia
dell'Università di Milano (Italy)

[*]Now head of Material Science Dept.
Istituto Donegani, Novara (Italy)

ABSTRACT

The use of an high temperature electrochemical oxygen pump fitted with catalytically active anode material is described.

It has been proved to be useful for the investigation of heterogeneous oxidation reaction and for the quantitative determination of the concentration of trace amount of organic substances in inert gases.

INTRODUCTION

Solid electrolytes having purely oxygen ion conductivity found widespread use in devices designed for the monitoring of the oxygen partial pressure in inert gases[1,2] or of the chemical potential of oxygen in CO/CO_2 and H_2/H_2O mixtures (oxygen meters), for the electrochemical generation of oxygen (oxygen pump)[3,4] and for the high temperature electrolysis of steam[5].

In their most conventional version, such devices consist of a tube of doped ZrO_2 or ThO_2 fitted with a couple of porous Pt electrodes.

It is known from literature that the lowest detection limit of the oxygen meter (O.M.) in inert gases is around 0.1 ppm with an accuracy better than 5% at 1 ppm[6].

As concerns the oxygen pump (O.P.), oxygen could be electrochemically generated, in inert gases, at 100% faradaic efficiency in a wide range of current densities at temperatures higher than 700-750°C.

At any temperature however, the current density should not exceed a critical value, in excess of which the efficiency drops to lower values. At 550° the critical current density is around 10 mA/cm^2.

As the coupling of an O.P. and of an O.M. provides the means for the generation and the determination of very small amounts (less than 0,1 ppm) of oxygen in a gas stream, it seems worthwhile to extend the use this device for the investigation of oxidation reaction kinetics and for the titration of trace amounts of oxidisable substances in a gas phase (7,8).

EXPERIMENTAL

The O.P.-O.M. device is operated by putting the pump at temperatures between 550°C and 600°C and the oxygen sensor at 750°C (this temperature has been found to meet the requirement of a very short time lag and good sensitivity).

He is used as the carrier gas for benzene and the reaction mixture is fed to the O.P. at a constant rate throught silica tubulations.

From the known value of the constant current passing in the O.P., supplied by a suitable d.c.supply, the oxygen content could be calculated by means of the following equation:

$$ppm_{O_2} = \frac{208,912 \; i \; (mA)}{\phi \; (Nl/h)}$$

when the gas flow rate is known and constant.

From this equation the E.M.F. which should be read at the O.M. could be calculated by using the Nernst equation

$$E = \frac{RT}{4F}\left(\ln 210.000 - \ln \frac{208,912 \; i \; (mA)}{\phi \; (Nl/h)}\right)$$

Poisoning of the electrode of the O.M. by unreacted benzene or reaction products could be avoided by using a trap filled with Linde molecular sieves type 13X between the P.O. and the O.M..

It has infact been experimentally shown that benzene is quantitatively removed from a He/benzene mixture in the range of concentration used (40 ppm max), in good agreement with the Union Carbide specifications. In the same conditions oxygen adsorption takes place in undetectable amounts.

Before carrying out the oxidation experiments on benzene, we tested the actual performances of the O.P. fitted with a Pt/V_2O_5 anode and a Pt cathode at temperatures ranging between 550°-600°C, which represent a top operation limit with benzene. Fig. 1 reports the experimental results at 575°C: the oxygen in the mixture was determined by the O.M. working at 750°C. It is shown that the performances of the pump are remarkably good as it provides for a controlled oxygen flow in He with the theoretical efficiency.

Oxidation experiments have been carried out at different flow rates (36,48,60 Nl/h) when using an He/benzene mixture containing 39 ppm of benzene which first flows throught the O.P. and then though the molecular sieves trap before entering the O.M. to determine the un reacted oxygen.

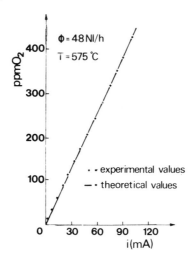

Fig.1 - Faradaic efficiency of the pump using V_2O_5 as the anode material.

Results are reported in figures 2a, 2b and 2c as the amount of unreacted oxygen vs. the applied current.

It appears that the amount of reacted oxygen corresponds to the amount of electrolytically generated (solid line) oxygen until a a critical value (which appears to depend only on the flow rate and on the temperature) is reached.

When calculating the stoichiometric amount of oxygen needed for the oxidation of benzene to maleic anhydride

$$C_6H_6 + 4,5O_2 \rightarrow C_4H_2O_3 + 2CO_2 + 2H_2O$$

or to carbon monoxide

$$C_6H_6 + 4,5O_2 \rightarrow 6CO + 3H_2O$$

for the three experimental flow rates, one observes that at the lowest flow rate the experimental amount of reacted oxygen corresponds to the stoichiometric one at any temperature.

At higher flow rates a deviation occurs, which increases when increasing the flow rate (and decreasing the contact time).

Table I which reports the deviations observed at the different temperatures and fluxes, show that even in the worse condition the deviations observed do not exceed the 30%.

Table I - Benzene determination accuracies.

T(°C)	(Nl/h)	Δ%
550	36	- 2.1
	48	-10.37
	60	-30.21
575	36	+ 0.71
	48	-14.36
	60	-22.34
600	36	+ 3.19
	48	-18.88
	60	-16.60

Is worthwhile to remark that when using as the anode porous Pt instead of Pt/V_2O_5 the amount of unreacted oxygen is very much lower with respect to the stoichiometry.

DISCUSSION AND CONCLUSIONS

Albeit we could not determine quantitatively the reaction products except the unreacted oxygen, it is apparent that in the proper flow rate conditions, and in the presence of V_2O_5, benzene reacts quantitatively with the electrochemically generated oxygen at temperatures between 550°-600°C.

While, on the basis of known catalytic properties of V_2O_5 a selective oxidation of benzene is expected, it is peculiar of the electrochemical method to obtain conversion efficiencies of 100% without any need of excess oxygen while, in the same temperature conditions

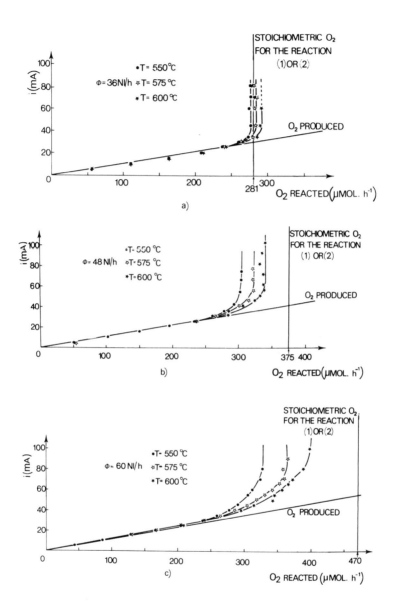

Fig. 2 - Amount of reacted oxygen vs. the pump current at different flow rates for a benzene (39 ppm)/ He mixture.

the conversion efficiency depends on the oxygen/benzene ratio. With a Carberry type of reactor, a conversion efficiency of 90% was infact obtained only when a 10:1 oxygen to benzene ratio was employed.

This peculiarity of the high temperature electrochemical oxidation method enables in principle, both the determination of the kinetics of an oxidation reaction and the quantitative elimination of oxidable material from an inert gas.

The most interesting feature of this electrochemical method, however, is its capability to determine quantitatively, with an accurancy better then 5%, trace amount of organic substances in a inert carrier gas.

REFERENCES

1. T.H.Etsell, S.N.Flengas - J.E.S. $\underline{118}$ (12), 1891 (1971)
2. J.Fouletier, H.Seinera, M.Kleitz - J.Appl.Electroch. $\underline{4}$, 305 (1975).
3. J.Fouletier, G.Vitter, M.Kleitz - J.Appl.Electroch. $\underline{5}$, 111 (1975).
4. C.B.Alcock, S.Zadoz - J.Appl.Electroch., $\underline{2}$, 289 (1972).
5. S.Pizzini, G.Bianchi - La Chimica e l'Industria, $\underline{55}$ (12), 966 (1973).
6. C.Mari, S.Pizzini - J.Appl.Electroch. pres.for publication.
7. S.Pizzini, A.Corradi, C.Mari - Proc.24 I.S.E. Meeting, 1973.
8. S.Pizzini, C.Mari, L.Zanderighi - Ann.Chim.,in press.

ON THE HIGH TEMPERATURE REACTIVITY IN AIR OF SOME METAL POWDERS AND THEIR RELATIONSHIP WITH THE SELF SEALING COATINGS FOR THE PROTECTION OF HIGH-TEMPERATURE METAL SURFACES

Giancarlo Perugini

Montedison - Instituto Richerche G. Donegani

Dipartimento di Corrosione ed Elettrochimica, Novara, Italia

ABSTRACT

The paper deals with the reactivity of titanium (chosen as a test metal) with air at high temperature, the protection given by self-sealing coatings, and the properties of metal and alloy powders useful for the preparation of the mentioned coatings.

RESULTS AND DISCUSSION

<u>Reactivity of titanium in air at high temperature and its protection.</u> Titanium powder-compacts (pressed at 2000 kg/cm^2) when exposed to air at 850°C for several hours suffer both oxidation and nitridation, while the metal sheet and the loose powder suffer oxidation only. From cross-sections of the attacked region three different layers can be identified: X-ray analysis shows the outer layer to be titanium dioxide, the inner one to be titanium nitride, the intermediate layer being a mixed phase formed by oxide and nitride.

Self-sealing coatings (1) protect the powder-compacts; as a matter of fact neither oxidation nor nitridation were observed after a 100-hour long exposure in air at 850°C. The 0,6 mm-thick coating has a three-layered, polyfunctional structure giving protection against oxidation, good bonding characteristics, and superior resistence to thermal shocks.

The first layer (0,1 mm thick) is obtained by plasma-spraying a 50%$_w$ Cr and 50%$_w$ Ni-brazing alloy (type AWS/BNi-1) powder mixture:

the same technique is used for the application of the second layer (0,3 mm thick, 60%$_w$ Cr, 25%$_w$ calcia stabilized ZrO$_2$, and 15%$_w$ Ni-brazing alloy) and of the third one (calcia stabilized ZrO$_2$).

Oxidation rates of some metal powders. This part is concerned with the oxidability of Co, Ti, Ni and Cr powders of fine and medium grain size (respectively 29-53 and 53-75μm).

The experiments were carried out with a thermobalance (Stanton model ME-H5) under flowing air (7 l/hr) and a heating speed of 6°C/min. up to 900°C. The results are reported in figs.1 and 2. Cobalt powders were most readily oxidised. No metal was left well before the temperature reached the preset value; the same behaviour was exhibited by titanium powders.

Nickel powders also oxidised completely but at lower rate: no metal was left after about 20 hours at 900°C; in addition from the curves of fig. 1 it appears that the fine powder is less easily oxidised. This fact is due to a faster sintering process which lowers the gas permeability and the specific surface area. Chromium powders are resistant to oxidation. The film of oxide formed on the surface of the particles is strongly adherent and protective and it hinders effectively further oxidation.

The mean diameter of the metallic core was 61.4 μm and 61,2 μm respectively after 9 hrs and 18,5 hrs at 900°C. The initial mean diameter was 64 μm (see fig. 2). This property of chromium powders is advantageously utilized in the formulation of multilayered polifunctional self-sealing coatings for metal protection at high temperature.

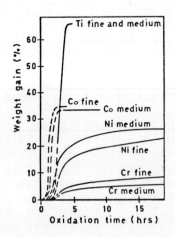

Fig. 1. Oxidation rates of metal powders

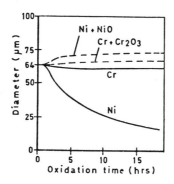

Fig. 2. Residue diameter of metal core (solid lines) and final diameter of oxidised metal particles (dashed lines) versus oxidation time.

Oxidation rates of metal-alloy powders useful for self sealing coatings. This part deals with the oxidation of metal-alloy powders and powder mixtures utilized for the formation of self-sealing coatings. The following types were taken into consideration:

1) Yellow brass 67%$_w$ Cu, 33%$_w$ Zn
2) 80%$_w$ Ni, 20%$_w$ Cr Alloy (Nichrome 8020)
3) AWS^w alloy B Ni-1 (a Ni alloy containing Cr, Fe, Si, B, C)
4) 50%$_w$ Cr, 25%$_w$ AWS, 25%$_w$ Al$_2$O$_3$ (powder mixture)
5) 50%$_w$ Cr, 25%$_w$ Ni, 25%$_w$ Al$_2$O$_3$ (powder mixture)

and tested under the experimental conditions described above (part 2). The weight gain is plotted versus exposure time in fig. 3.

For the sake of comparison the oxidation curves of nickel and chromium powders are also reported. It is interesting to point

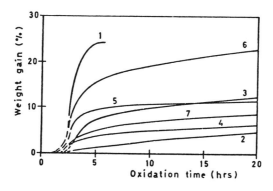

Fig. 3. Oxidation rates of metal-alloy powders and powder mixtures.

out that pure nickel is not resistant to oxidation, but its $20\%_w$ Cr alloy is the most resistant of all tested materials. In the formulation of the self-sealing multilayered coatings, nickel alloys or nickel metal (as a first layer) have a bonding function with respect to the metal substrate (diffusion and alloying processes). Chromium (as the second layer) fulfils several important functions : Protection against oxidation, by means of formation of a protective oxide film; sealing of the coating, as a consequence of substantial volume expansion due to the oxide (the original open porosity is changed into a closed porosity); cermet formation, which assures a gradual change of the properties along the thickness of the coating; finally the protection against thermal shock. The proper balance of porosity, heat conductivity, heat capacity and elastic characteristics is the basic condition for adequate resistance and protective properties.

When nickel is also present in the second layer, a higher ceramizing effect because of the formation of mixed oxides is obtained. A ceramic refractory oxide (the outermost layer of the coating) strongly bonded to the second layer, gives to the coating the required additional thermal and corrosion protection.

REFERENCES

1. G. Perugini: High Temperatures - High Pressures (1974) Vol.6 p 565-575

KINETIC ASPECTS OF SOME REACTIONS BETWEEN METALS AND GASEOUS

CARBON DISULPHIDE

A. Galerie, M. Caillet and J. Besson

Laboratoire d'Adsorption et Reaction de Gaz sur Solides

ERA CNRS n° 368

Ecole Nationale Superieure d'Electrochimie et d'Electrometallurgie - Domaine Universitaire - BP 44 - 38401 - Saint Martin d'Heres - France

INTRODUCTION

The problem of corrosion of metals by sulphur-containing gases has led for several years researchers to study on a theoretical basis the mechanisms of the phenomena which occur during this kind of reaction. But, if a lot of research has been carried out with hydrogen sulphide or sulphur vapour, there is no systematic study of corrosion by gaseous carbon disulphide, CS_2. This compound, a product of big industry which possesses significant vapour pressure at room temperature, seems however very suitable for a theoretical study of the sulphidation of metals. That is the major reason why we have undertaken the study of its behaviour with metals, firstly from a qualitative point of view (thermodynamic diagrams and resulting products), secondly from a quantitative one (kinetics of sulphuration, influence of temperature and pressure).

The study was carried out in apparatus made of Pyrex glass and quartz. In the first one the metallic sample is heated at a fixed temperature and carbon disulphide is introduced at a constant pressure. The second one is a quartz spiral balance connected with a liquid carbon disulphide reserve, thermostated to within $\pm 2°C$ between $-50°C$ and $-105°C$ which fixes low pressures of gas.

The metallic samples, in form of foils, are of different sizes, depending on the metal used. The thick ones are polished with abrasive SiC paper and then with metallographic alumina. The thin ones are only cleaned with ethanol. These two sorts of samples are always rinsed with doubly distilled water before introduction into the reaction tube.

THERMODYNAMIC STUDY

In order to forsee the occurence of every possible product we drew the diagrams $P(CS_2) = f(t)$ for all univariant equilibria. We included also in these diagrams the divariant equilibria of carburation by CS_2, under the assumption that the ratio $P(S_2)/P(CS_2)$ is fixed by the decomposition of CS_2. These equilibria are therefore represented by vertical straight lines (independent of $P(CS_2)$.

The general aspect of a diagram is determined by the thermodynamic properties of the possible carbide. Three examples of these diagrams are presented in the following cases:

- no carbide
- one carbide stable at high temperatures (endothermic formation).
- one carbide stable at low temperatures (exothermic formation):

RESULTING PRODUCTS

The products obtained under a pressure of 150 Torr of carbon disulphide are registered on table 1. It will be noted that most of these products are sulphides and that only two carbides ($NbC + Nb_2C$) and two sulphocarbides (Nb_2SC and $Ta_3C_2S_4$) occur. The first of these sulphocarbides was already known. For the second one which has never before been mentioned, we proposed an orthohombic unit cell with the following constants : a = 6,655Å b = 15,584 Å, c = 3,223 Å. Among these reactions, we selected three particularly interesting cases, which are developed here:

- The reaction of tantalum which forms a new sulphocarbide
- The reaction of iron which forms porous and stratified iron (11) sulphide.
- The reaction of copper which forms compact copper (1) sulphide.

SULPHURATION OF TANTALUM

$$3\,Ta + 2\,CS_2 \rightarrow Ta_3C_2S_4$$

<u>Kinetic results</u>. The kinetics of the reaction between Ta and CS_2 have been studied over the temperature range 800-1085°C at pressures between 10 and 250 Torr. The thermogravimetric curves obtained show an initial parabolic stage during 3-5 hours. This stage is

followed by a linear one, continuing until the end of the experiment (15-20h). Consequently the form of the rate law is $\Delta m^2 = k_p t$ during the first stage and $\Delta m = \Delta m_o + k_1 t$ during the second one. But this law cannot be compared with a classical paralinear one because the linear rate is slower than the parabolic one at the transition point.

TABLE I

800-1000°C Ti_3S_4	800-1000°C V_3S_4	800-1000°C Cr_2S_3	600-800°C MnS	300-1000°C FeS	800°C Co_4S_3 + Co_6S_5	300-700°C Ni_3S_2 + Ni_7S_6	230-750°C Cu_2S	300°C ZnS
800-1000°C ZrS_2 + Zr_3S_4 at 1000°C	800-1000°C NbC Nb_2C Nb_2SC	800-1000°C Mo_2S_3 +MoS_2 at 1000°C					800°C Ag_2S	200°C CdS
800-1000°C HfS_2	800-1085°C $Ta_3C_2S_4$	800-1000°C WS_2				800-1000°C PtS		

At elevated temperatures, we often noted some periods of increased rate. The frequency of these periods grows with increasing temperature.

The parabolic and linear reaction rate constants k_p and k_1 obey an Arrhenius' law between 870 and 1070°C. The activation energies are respectively 26,5 K.Cal.mol^{-1} and 15,5 K.Cal.mol^{-1}.

The influence of the pressure of CS_2 has been studied at 965°C. During the parabolic stage the constant k_1 depends on pressure according to $k_p = Cp^n$ where C is a constant. The values of n lie between 0,56 and 0,68.

The linear constant k_1 depends also on pressure according to the relation $k_1 = \frac{AP}{1 + BP}$. A and B are constants:

<u>Interpretation</u>. These results and those of micrographic studies are tentatively explained by a diffusion mechanism, involving

neutral cation vacancies in the reaction product, during the parabolic stage.

The linear stage corresponds to an external half - reaction process.

SULPHURATION OF IRON

$$2\ Fe + CS_2 \rightarrow 2\ FeS + C$$

<u>Kinetic results</u>. In the temperature range 300-600°C the kinetic curves seem to have a similar aspect as those observed for the tantalum (parabolic then linear). But at the parabolic linear transition point, the rate of the reaction increases considerably. The scale of iron sulphide is non-porous and adherent before the transition point, porous and stratified after.

The linear constant k_l agrees with an Arrhenius' law (activation energy 8,2 KCal.mol^{-1}). The values of k_p are more dispersed At 580°C, the linear constant is an homographic function of pressure, and the parabolic one depends on pressure according to $k_p = AP^{1/4}$ (A is a constant).

<u>Interpretation</u>. The first stage of the kinetic curves corresponds to the growth of a compact layer. During this stage the influence of pressure can be explained by the diffusion of partly ionized cation vacancies. The observed accelerations are the results of the breakdown of the layer. The following linear period corresponds to the growth of a porous scale.

SULPHURATION OF COPPER

$$4\ Cu + CS_2 \rightarrow 2\ Cu_2S + C$$

<u>Kinetic results</u>. The form of the gravimetric curves depends on temperature. Between 230 and 400°C, the law is linear after an initial period during which the reaction rate decreases. At higher temperatures up to 730°C, the law is parabolic. The layer of cuprous sulphide Cu_2S is, in both cases, adherent and compact.

Both linear and parabolic constants agree with an Arrhenius' law. The measured activation energies are respectively 17,0 and 8,2 KCal.mol^{-1}. The study of the influence of pressure was carried out at 305°C and 360°C for k_l and at 655°C for k_p. Both constants are linear functions of pressure.

Interpretation. At lower temperatures, the results can be explained by an interfacial process : sorption or decomposition of CS_2. At higher temperatures, the parabolic law is tentatively explained by a diffusion mechanism.

DISCUSSION

P.L.SURMAN (CERL, Leatherhead, U.K.) The kinetics of the reaction between Fe and CS_2 are similar to those of

$$3Fe + 2CO_2 = Fe_3O_4 + 2C$$

both showing initial parabolic form followed by formation of porous oxide at a linear rate. In the latter, the change in morphology and kinetics is thought to occur when the metal becomes saturated with carbon which can initially diffuse into it. The carbon content of protective oxide is only 1/10th of that in the non-protective oxide. Could the authors' results be explained in similar terms?

AUTHORS' REPLY: It is obvious that the presence of carbon in the reacting gas (CS_2) is of great importance for the rate law of sulphuration of iron. It is well known that the reaction of this metal with carbon-free gases eg. H_2S or sulphur vapour reveals a parabolic rate law, even after a great number of hours.

As you suggest, the parabolic-linear transition which we observe is certainly connected to carbon dissolution in iron at the sulphide-metal boundary. Our microhardness experiments show this dissolution. However, the problem is more complex : the aspect of the rate law is connected to sulphide morphology which depends particularly on the rates of the following processes:

- appearance of carbon at the external interface,
- its diffusion in the increasing sulphide scale,
- its dissolution in the metal at the internal interface.

OXIDATION OF COBALT AT HIGH TEMPERATURES AND SELF-DIFFUSION IN COBALTOUS OXIDE

S.Mrowec and K.Przybylski

Institute of Materials Science

Academy of Mining and Metallurgy, Kraków, Poland

INTRODUCTION

The cobalt-cobaltous oxide-oxygen system has been for many years a subject of much interest owing to its model character in studies on kinetics and mechanism of metal oxidation, as well as on semiconducting and transport properties of transition metal oxides. In the present work an attempt has been made to determine precise values of parabolic rate constants of cobalt oxidation in the wide range of temperatures /950-1300°C/ and oxygen pressures /$6.58 \cdot 10^{-4}$-0.658 atm/. From these data self-diffusion coefficients of cobalt in cobaltous oxide /$Co_{1-y}O$/ and activation energy and the pressure dependence of this process have been calculated. The total concentration of defects in cobaltous oxide as dependent on two parameters mentioned above have been also derived.

EXPERIMENTAL

Kinetics of oxidation of rectangular plates /19x15x0.5 mm/ of spectrally pure cobalt was studied in the original volumetric apparatus, enabling the continous recording of the mass of bound oxygen with the accuracy of 10^{-6} g (1). In calculation of the unit mass gains of the oxidized metal samples, the account was taken of the change in the surface area of the metallic core resulting from its consumption in the process of oxidation and from thermal expansion (2). The shape and dimensions of the metal samples were selected in such a way as to ensure - in the entire range of temperatures and oxygen pressures under study - the compactness of the formed scale and its good adherence to the metallic base during the full course of determining the kinetic curve. The experiments were conducted in such conditions in which cobaltous oxide was the only thermodynamically stable oxide phase. In keeping with the results of the other authors (3) it was found that irrespective of the temperature and the oxygen pressure the oxidation of cobalt follows a parabolic rate law Fig. 1 , the reaction rate increasing exponentially with the increase in temperature /Fig. 2/ and to a certain power with the increase in oxygen pressure. These relationship could be described by the following equation:

$$k_p'' = \text{const.} \; P_{O_2}^{1/n} \exp\left(-\frac{E_k}{RT}\right) \qquad /1/$$

where k_p'' is a parabolic rate constant of the oxidation $/g^2 \cdot cm^{-4} \cdot s^{-1}/$.

The exponent, $1/n$, decreases with the increase in temperature from $1/3.40$ at $950°C$ to $1/3.96$ at $1300°C$, whereas the activation energy, E_k decreases with the increase in the oxygen pressure from 41.7 to 38.1 kcal/mole. The studies carried out with use of platinum markers /wires, 20 μm in dia./ showed that irrespective of temperature and the oxygen pressure the scale is formed by outward diffusion of metal /the marker located at the scale , metal phase boundary/.

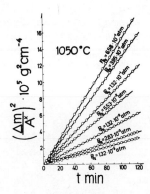

Fig. 1 Kinetics of cobalt oxidation at various oxygen pressures for $1050°C$ /parabolic plot/.

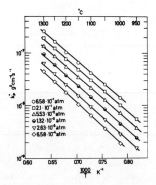

Fig. 2 Temperature dependence of the rate of cobalt oxidation for several oxygen pressures.

DISCUSSION

On the basis of the results described above it can be stated that at appropriately selected conditions of the oxidation process, there is formed on the surface of cobalt a monolayer compact scale built of the $Co_{1-y}O$ phase, closely adherent to the metallic core. The growth of the scale takes place under these conditions due to outward diffusion of the metal, the overall reaction rate being determined by diffusion of cation vacancies in the scale according to the Wagner theory of metal oxidation (4). Full contact of the scale with the metallic core during the whole kinetic run curve ensures that at the metal/scale phase boundary no perturbations of local thermodynamic equilibrium occur. Hence there exist grounds to employ the kinetic data for calculations of self-diffusion coefficients of cobalt in cobaltous oxide, as dependent on pressure and temperature, making use of the Fueki-Wagner equation (5). For the case under consideration this equation assumes the form:

$$D_{Co} = 0.229 \left(\frac{d\, k_p''}{d \log p_{O_2}} \right) \qquad /2/$$

It follows from equation /2/ that for calculation of coefficients D_{Co} from the determined values of k_p'' it is necessary to differentiate the function $k_p'' = f /\log \bar{p}O_2/$. These functions were calculated with the numerical methods for a number of temperatures in the studied range of the oxygen pressures.

The coefficients D_{Co} in $Co_{1-y}O$ calculated as dependent on the oxygen pressure with the above described method are shown in Fig. 3. For comparison in the same figure we plotted also the direct results of radioisotopic studies of Carter and Richardson (6) taking into account the correlation effect (7). As seen from the figure the calculated and directly determined values of D_{Co} are in very good agreement, Fig. 4 presents the collective plot of D_{Co} obtained so far (6, 8, 9) plotted in the Arrhenius system of coordinates. The accordance of the results of direct determinations with the data calculated in the present work constitutes a proof of applicability of the Fueki-Wagner method (5) to determination of self-diffusion coefficients in metal oxides from the measurements of the oxidation kinetics. The advantage of this method lies on one hand in its simplicity, and on the other hand in considerably higher precision of the results obtained.

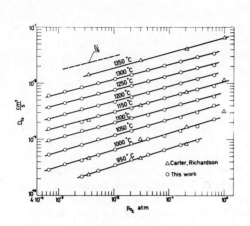 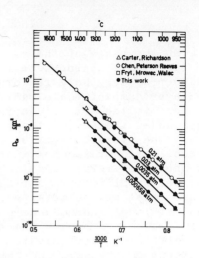

Fig. 3 Dependence of self-diffusion coefficient of cobalt in cobaltous oxide on oxygen pressure for several temperatures.

Fig. 4 Collective plot of temperature dependence of self-diffusion coefficient of cobalt in $Co_{1-y}O$.

The dependence of the calculated values of D_{Co} on oxygen pressure and temperature can be described by the following empirical equation:

$$D_{Co} = \text{const.} \ p_{O_2}^{1/n} \exp\left(-\frac{E_D}{RT}\right) \qquad /3/$$

where exponent $1/n$ decreases with the increase in temperature from $1/3.3$ at $950°C$ to $1/3.95$ at $1300°C$. The activation energy of diffusion, E_D on the other hand increases with the decrease in the oxygen pressure from 37 kcal/mole at $p_{O_2} = 0.331$ atm to 39.6 kcal/mole at $p_{O_2} = 6.58 \cdot 10^{-4}$ atm.

From theory of diffusion in solids it follows that total concentration of defects in the oxide under discussion [def], is related to the self-diffusion coefficients, D_{Co}, chemical diffusion coefficient $\widetilde{D}_{CoO}$ and to the degree of ionization of defects, p, by the following equation:

$$[\text{def}] = \frac{D_{Co}(1+p)}{\widetilde{D}_{CoO}} \qquad /4/$$

It is generally assumed that the total concentration of defects in $Co_{1-y}O$ in the studied range of temperature and oxygen pressure is equal to deviation from stoichiometry, y, resulting from the presence of singly ionized cation vacancies /p=1/. These latter defects are formed in a quasi-chemical surface reaction:

$$\frac{1}{2} O_2 \rightleftharpoons V'_{Co} + h^\bullet + O_O \qquad /5/$$

The concentration of these defects has so far only been determined from measurements of deviation from stoichiometry.

The precise values of D_{Co} determined in the present work, together with equally accurate values of $\tilde{D}_{CoO}$ (10)(11) and p (12)(13) reported by other authors made it possible to calculate from eq. /4/ the total concentration of defects in $Co_{1-y}O$ and thus to verify the assumption mentioned above that y = [def]. The results of these calculations are shown graphically in Fig. 5. It follows from these data that in contrast to the current viewpoints the total concentration of defects in the cation sublattice of $Co_{1-y}O$ is, above the temperature of 1050°C, higher than the concentration of cation vacancies resulting from the deviations from stoichiometry, y, these differences increasing with the increase in temperature. It can be then assumed that in these conditions besides the defects formed in reaction /5/ there are also formed intrinsic ionic defects of Frenkel or Schottky type:

$$Co_{Co} \rightleftharpoons Co''_i + V^{\bullet\bullet}_{Co} \qquad /6/$$

or

$$zero \rightleftharpoons V''_{Co} + V^{\bullet\bullet}_O \qquad /7/$$

Since it has been found in the present work that irrespective of temperature and oxygen pressure the process of growth of $Co_{1-y}O$ layer on the surface of cobalt takes place solely by outward diffusion of metal it can be then assumed that intrinsic defects are mainly due to reaction /6/.

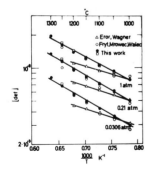

Fig. 5 Temperature dependence of defect concentration in cobaltous oxide for several oxygen pressures.

The total enthalpy of defect formation, ΔH_{def}, in $Co_{1-y}O$ calculated from Fig. 5 is a function of the oxygen pressure and varies from 23 to 30 kcal/mole in the pressure range from $1 \cdot 10^{-3}$ to 0.658 atm. This value is almost twice as high as the enthalpy of formation of cation vacancies in reaction /5/ (13).

REFERENCES

1. S.Mrowec, M.Lasoń, E.Fryt, K.Przybylski and A.Ciembroniewicz, J.Therm.Anal. 6, 193 /1974/.
2. S.Mrowec, A.Stokłosa, Werkstoffe u.Korrosion 21, 934 /1970/.
3. P.Kofstad, "High Temperature Oxidation of Metals", J.Wiley, 1966, p. 122, 127.
4. C.Wagner, "Diffusion and High Temperature Oxidation of Metals" in Atom Movements A.S.M. Cleveland, 1951, p.153.
5. K.Fueki, J.B.Wagner, J.Electrochem.Soc., 112, 384 /1965/.
6. R.E.Carter, F.D.Richardson, Trans.AIME., 200, 1244 /1954/.
7. J.Manning, "Diffusion kinetics for Atoms in Crystals", Van Nostrand, New Jersey 1968, p. 95.
8. W.K.Chen, N.L.Peterson, W.T.Reeves, Phys.Rev., 186, 887 /1969/.
9. E.Fryt, S.Mrowec, T.Walec, Oxidation of Metals, 7, 117 /1973/.
10. G.J.Koel, P.J.Gellings, Oxidation of Metals, 5, 3 /1972/.
11. J.M.Wimmer, R.N.Blumenthal, I.Bransky, J.Phys.Chem.Solids 36, 269 /1975/.
12. B.Fisher, D.S.Tannhauser, J.Chem.Phys., 44, 1663 /1966/.
13. N.G.Eror, J.B.Wagner, Jr.J.Phys.Chem.Solids., 29, 1597 /1968/.

CATION DISTRIBUTION IN SPINEL SOLID SOLUTIONS AND CORRELATIONS
BETWEEN STRUCTURAL AND CATALYTIC PROPERTIES

F.Pepe, P.Porta and M.Schiavello

Centro di Studio sulla Struttura ed Attività Catalitica

di Sistemi di Ossidi del CNR, ed Istituto di Chimica

Inorganica dell'Università di Roma, Roma, Italy

ABSTRACT

The cation distribution in several spinel solid solutions such as $Ni_xMg_{1-x}Al_2O_4$, $Ni_xZn_{1-x}Al_2O_4$, $Cu_xMg_{1-x}Al_2O_4$, $Co_xMg_{1-x}Al_2O_4$, $CoGa_yAl_{2-y}O_4$ and $CoGa_yRh_{2-y}O_4$ has been studied by reflectance spectroscopy, magnetic susceptibility and X-ray analysis.

The results show that there is small but definite change towards a random cation distribution with increase in temperature and that the cation distribution varies within each system with composition $\underline{x}$ and $\underline{y}$. This variation, except for the $CoGa_yRh_{2-y}O_4$ system, is not linear and this interesting behaviour may be explained in terms of anion polarisation and/or crystal field stabilisation energy effects. The catalytic activity towards N_2O decomposition has also been determined for some compounds and it has been established that the site symmetry of the transition metal ions, their dilution in the host matrix and the different chemical properties of the matrix all effect the activity of the solids.

INTRODUCTION

The distribution of cations among the available octahedral, B, and tetrahedral, $\underline{A}$, sites in spinels, whose structure forms the basis for a large class of compounds having a wide range of chemical and physical properties of considerable technological importance, has been experimentally proved to be an equilibrium function of temperature and pressure (1). A fine balance of the respective octahedral and tetrahedral preferences of the ions concerned, and other factors such as the ionic charge and the ionic radius, crystal and ligand field effects, anion polarisation, etc.; each contribute to determine the cation distribution in spinels (2). Moreover, by making use of the property of solid solution formation, gradual changes in the solid state chemistry can be effected, other than with temperature and pressure, by varying the composition of the solution. Studies of cation distribution in the spinel structure are thus of considerable interest in solid state chemistry because they may allow investigation of the relative stabilities of ions in octahedral and tetrahedral coordination and elucidation of some aspects of solid state reactivity. Knowledge of the cation distribution is, moreover, useful for a better understanding of the correlations between structure, and physical and chemical properties.

An extensive research programme was started many years ago in our laboratory on solid solutions of transition metal ions dispersed in oxide spinel matrices with the aim of a detailed study of their solid state chemistry and catalytic properties. We here report on the following systems:

$Ni_xMg_{1-x}Al_2O_4$, $Ni_xZn_{1-x}Al_2O_4$, $Cu_xMg_{1-x}Al_2O_4$, $Co_xMg_{1-x}Al_2O_4$
$CoGa_yAl_{2-y}O_4$, $CoGa_yRh_{2-y}O_4$.

EXPERIMENTAL, RESULTS and DISCUSSION

The compounds were prepared by solid state reaction of appropriate amounts of the component oxides calcined in air at different temperatures and quenched in water. The solid solutions were prepared in a wide range of x and y composition; for brevity only some of the spinels prepared and characterized are presented here. Precise lattice parameters, magnetic susceptibilities between 100 to 295 K, and the electronic d-d transitions by reflectance spectroscopy have been measured. The ratios of the X-ray diffraction intensities 400/220 and 400/422 expected for various distributions of the cations have been calculated for each compound and compared with experimental measurements (taken by careful pulse-counting methods) to determine the fraction of cations in octahedral coordination.

The Table lists for each system, the temperatures of calcination, the x and y compositions, the values of the lattice parameters a (at 294 K), the magnetic moments μ, the fractions of the divalent transition metal ions (Ni^{2+}, Cu^{2+}, Co^{2+}) in octahedral sites (γ) as derived from comparison between observed and calculated X-ray intensity ratios, and the catalytic data deduced from the test reaction of N_2O decomposition.

The structural results show first, that in all series of compounds there is a small but definite change towards a random cation distribution with increase in temperature. It may be recalled that the completely random cation distribution in spinels is that which contains 67% and 33% of total divalent ions in the B and A sites, respectively. The entropy effect caused by the increase in temperature is thus clearly observed in all the systems. In addition, both X-ray and magnetic results show that the transition metal ion distribution varies, within each series of solid solutions treated at the same temperature, with $\underline{x}$ and $\underline{y}$ compositions. This variation except probably for the $CoGa_yRh_{2-y}O_4$ system, is not linear. The interesting results found for the influence of composition on cation distribution may be explained in terms of anion polarisation and/or crystal field stabilisation energy effects. For example, for the $CoGa_yAl_{2-y}O_4$ spinels the trend of cobalt octahedral occupation with $\underline{y}$ (minimum value at $\underline{y}$ = 0.25 - 0.50) may be explained as follows; on going from pure $CoAl_2O_4$ to gallium spinels the Ga^{3+} ions, due to their strong preference for tetrahedral coordination, tend to substitute Al^{3+} ions in the A sites. A lower positive charge density thus results in the A sites, since the ionic radii of tetrahedral Ga^{3+} and Al^{3+} ions are respectively, 0.47 and 0.39 Å (3). As a consequence, oxygens are more polarised towards the B sites favouring a major tendency for trivalent ions (namely Al^{3+}) to occupy the B sites and for divalent ions (Co^{2+}) to occupy the A sites. This occurs for relatively small values of gallium content. With increase of gallium, no more Al^{3+} ions are available to change their coordination from tetrahedral to octahedral and the Ga^{3+} ions are forced to occupy the A sites and the spinels become more and more inverse.

Several spinel systems were tested for the catalytic decomposition of N_2O studied in the 623 - 753 K temperature range. This reaction has been chosen on the basis of the elementary steps believed to occur: a) adsorption on a surface cation which depends on the nature of the transition metal ion and of its available orbitals; b) cation-oxygen bond formation and nitrogen desorption; c) rupture of the cation-oxygen bond depending on the cation coordination; d) oxygen migration and/or desorption related to the ability of the matrix to accept oxygen atoms and to allow an easy migration (4). Since spinels have structures containing the same cations in both octahedral and tetrahedral sites, it is possible to investigate the role played by the cations in different site symmetries (symmetry effect), and by comparing spinel systems of a given ion dispersed in different matrices it is possible to gain

TABLE*

	1673 K					$Ni_xMg_{1-x}Al_2O_4$		1273 K				
x	a	μ	γ	SA	E_a	logk	a	μ	γ	SA	E_a	logk
0.00	8.0835						8.0832		0.9	31		-5.7
0.10	8.0809	3.27	0.78	1.7	26	-5.6	8.0797	3.23	0.81	1.4	26	-5.4
0.25	8.0758	3.21	0.86	1.1	25	-5.7	8.0747	3.18	0.89	1.0	26	-5.3
0.50	8.0665	3.22	0.85	2.0	26	-5.7	8.0655	3.21	0.88	1.3	26	-5.2
1.00	8.0514	3.27	0.79	2.0	30	-5.2	8.0478	3.24	0.78	2.2	32	-5.1

	1473 K					$Co_xMg_{1-x}Al_2O_4$		1073 K				
0.00	8.0831			1.9	28	-5.9	8.0831			1.9	29	-5.7
0.05	8.0842	4.60	0.40	1.7	21	-4.3	8.0844	4.42	0.31	2.2	29	-4.7
0.10	8.0849	4.67	0.25	1.9	20	-4.5	8.0865	4.58	0.20	1.4	29	-5.1
0.25	8.0869	4.80	0.23				8.0885	4.71	0.18	0.5	28	-5.2
0.50	8.0941	4.76	0.23	1.0	20	-4.4	8.0950	4.71	0.18			
1.00	8.1051	4.82	0.23	0.9	20	-4.6	8.1058	4.76	0.18	0.8	29	-5.7

	1473 K		$Ni_xZn_{1-x}Al_2O_4$				1073 K		$Cu_xMg_{1-x}Al_2O_4$ - 1173 K		
	a	μ	γ	SA	E_a	logk	γ	x	a	μ	γ
0.00	8.0874						0.88	0.00	8.0828		
0.05	8.0862	3.31	0.80	1.0	33	-5.6	0.80	0.05	8.0846	2.30	0.68
0.10	8.0846	3.36	0.70	0.9	18	-5.6	0.69	0.10	8.0844	2.29	0.57
0.20	8.0813	3.45	0.64	0.8	19	-4.8	0.72	0.20	8.0842	2.34	0.51
0.40	8.0762	3.46	0.68				0.77	0.40	8.0825	2.35	0.42
0.60	8.0684	3.42	0.70				0.83	0.60	8.0814	2.38	0.38
0.90	8.0545	3.34	0.75	1.0	22	-4.4	0.84	1.00	8.0794	2.31	0.35
1.00	8.0509	3.32	0.77								

	1073 K					$CoGa_yAl_{2-y}O_4$		1473 K	$CoGa_yRh_{2-y}O_4$ - 1273 K		
y	a	μ	γ	SA	E_a	logk	γ	y	a	μ	γ
0.00	8.1058	4.76	0.18	0.8	29	-5.7	0.23	0.00	8.5008	4.55	0.00
0.25	8.1264	4.64	0.15	1.2	29	-5.7	0.20	0.50	8.4741	4.75	0.28
0.50	8.1444	4.69	0.22				0.26	1.00	8.4531	4.79	0.44
0.75	8.1764	4.78	0.36				0.38	1.50	8.3910	4.82	0.50
1.00	8.1985	4.87	0.49				0.50	2.00	8.3213	4.97	0.71
1.50	8.2583	4.97	0.67				0.65				
2.00	8.3243	4.97	0.74	0.6	33	-6.0	0.72				

$CoGa_yRh_{2-y}O_4$ - 1073 K

y	a	μ	γ
0.00	8.5013	4.53	0.00
0.50	8.4742	4.68	0.22
1.00	8.4571	4.79	0.28
1.50	8.3913	4.82	0.38
2.00	8.3243	4.97	0.74

*For brevity, only the most significant data for some compounds are reported. Symbols: x and y = atomic composition ; a = lattice parameters (Å); μ = experimental magnetic moments (BM),(theoretical magnetic moments : Ni_{oct}=3.20, Ni_{tet}=4.10, Co_{oct}=5.20, Co_{tet}=4.5 BM) γ= fraction of divalent transition metal ions in octahedral sites; SA = surface area (m^2/g); E_a = apparent activation energy (Kcal/mol) log k_{abs} at 1/T=1.4 x 10^{-3}, from Arrhenius plots.

also information on point d) of the reaction scheme (matrix effect).

Symmetry effect – As reported in the Table the Co_{oct}/Co_{tet} ratio increases as the quenching temperature is increased; the opposite holds for the nickel-containing spinels. Comparison of activity between pairs of catalysts of equal transition metal ion content and with different cation distribution, will therefore give information on the relative activity of octahedral and tetrahedral transition metal ions. It is immediately apparent from the Table that the specimens with a higher oct/tet ratio are more active. The transition metal ions in octahedral symmetry are thus more active than those in tetrahedral. This behaviour may be due to different oxygen-cation bond strengths for the two site symmetries available in the spinel structure. The $O-Me_{oct}$ bond length is higher than the $O-Me_{tet}$ one in the spinel structure, reflecting a lower bond strength for the octahedrally coordinated cation. Since in the catalytic process the metal-oxygen bond must be repeatedly formed and broken, the oxygen release and migration would be easier for an octahedral coordinated cation. A proof of the major difficulty in the mobilization of the oxygen bonded to a tetrahedral cation is found in the values of the apparent activation energy, constantly higher when more tetrahedral cations are involved.

Matrix effect – The influence of the chemical properties of the matrix itself on their activity for some reactions is well known and must adequately be considered in catalysis (4). By considering the influence of the site symmetry effect alone (higher activity for transition metal ions in octahedral sites) one would expect the specimen $CoGa_2O_4$ to be more active than $CoAl_2O_4$ the former having a higher octahedral cobalt content, whereas the reverse was found. A similar apparent discrepancy is presented by the systems $Ni_xMg_{1-x}Al_2O_4$ and $Ni_xZn_{1-x}Al_2O_4$. The conclusion which can be drawn is that for a given matrix the level of activity is governed by the oct/tet ratio, whilst for a given ion in different matrices the level of activity is determined by the above ratio and by the chemical properties of the matrix.

Another problem to be taken into account is that the cation distribution on the surface layers involved in the catalytic process may not correspond to that measured by X-ray diffraction methods in the bulk. Such an effect is likely to be present in all the solid solutions for thermodynamic reasons, but in some cases it can be revealed by catalysis; thus in our spinels the nickel octahedral occupation for $NiAl_2O_4$ does not vary appreciably with temperature as shown in the Table, but the corresponding spinels at 1673 and 1273 K have substantially different catalytic levels. This behaviour cannot be accounted for on the basis of different nickel distribution nor by difference in matrix properties, but the explanation should rather be found in a probable difference of surface cation distribution not detectable by X-ray methods.

REFERENCES

1 R.K.Datta and R.Roy, Amer.Min., 53, 1456 (1968)

2 P.Porta, F.S.Stone and R.G.Turner, J.Solid State Chem., 11, 135 (1974), and other references therein

3 R.D.Shannon and C.T.Prewitt, Acta Cryst., B 25, 926 (1969)

4 A.Cimino, F.Pepe and M.Schiavello, V Int.Congress on Catalysis, Miami Beach 1972, North Holland Publ. Co., 1, 125 (1973).

DISCUSSION

F.S. STONE and C. OTERO AREAN (University of Bath, U.K.) As pointed out by the authors, random cation distribution in a 2-3 spinel corresponds to two-thirds of the divalent cations on octahedral sites, i.e. $\gamma = 0.67$. However, for a solid solution in which one of the three cations present has a strong site preference, one would expect two stages in the randomization process. In the first stage the limit will be randomization of only two of the cations, the third remaining anchored to its preferred site. In the second stage all three cations randomize.

One of the systems studied by Pepe et al., namely $CoGa_yRh_{2-y}O_4$, affords a good example. Rh^{3+} has a strong octahedral preference; indeed, this is why $CoRh_2O_4$ shows $\gamma = 0$ both for 1073 K and 1273 K. First-stage randomization in $CoGa_yRh_{2-y}O_4$ will thus correspond to randomization of Co^{2+} and Ga^{3+}, with Rh^{3+} anchored. γ for first-stage randomization will be zero for $y = 0$, and will increase with y according to $\gamma = y/(y+1)$, as shown by the solid line in our figure. One would not expect this stage to be passed for $CoGa_yRh_{2-y}O_4$ at 1273 K, and the data confirm this.

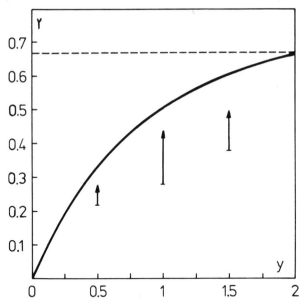

First and second stage randomization for cations in a spinel solid solution

The arrows show the change in γ on raising temperature from 1073 to 1273 K for the three solid solutions reported in the authors' table. The effects explains the relatively small increase in γ for the $y = 0.5$ solution.

J. HABER (Academy of Sciences, Krakow, Poland) The authors attribute differences in catlytic activity to variations in the Me-tetra/Me-octa ratio. The comparison is also being made between samples annealed at temperatures differing by 400° and thus the sintering of samples should also be considered as affecting catalytic activity.

THERMAL DECOMPOSITION OF Mg-Al-FORMATE AND LOW TEMPERATURE FORMATION OF SPINEL THEREBY

Z.G. Szabó, B. Jóvér, J. Juhász, K. Falb

L. Eötvös University, Budapest

INTRODUCTION

In spite of its simplicity, we know little about the mixed Mg-Al-formate. In the literature there is only one uncertain reference to its existence, (1). Its properties, including its thermal behaviour have not been studied.

We prepared this compound by the vacuum-evaporation of the aqueous solution of the component-formates. This process yields a white puffed-up substance. The compound obtained was analysed by various methods and a composition of $MgAl_{2.3}HCOO_{7.9}O_{0.5} \cdot 8H_2O$ was determined. It is to be noted that the deviation from the formula $MgAl_2(HCOO)_8 \cdot 8H_2O$ is significant from a mathematical-statistical point of view but it is unimportant from a chemical point of view.

Analysing the IR spectra of the component-formation together with that of the mixed-formate, we could establish that our sample was indeed a double-formate and not simply a mixture of the Mg-and Al-formates.

The primary aim of the preparation of the Mg-Al-formate was the preparation of a well-crystallized $MgAl_2O_4$- spinel with a large surface area. (That is why we intended to prepare $MgAl_2(HCOO)_8$ and not $MgAl(HCOO)_5$.) This attempt has been successful. The specific area of the $MgAl_2O_4$- spinel obtained by ignition of the double-formate for 15 hrs at $1000°C$ is 59 m^2/g, (2).

During these experiments we observed some peculiarities of the thermal decomposition of the mixed-formate, and made up our mind to perform a detailed analysis of this process.

The thermal decomposition of the various formates has been much studied, (3), (4). The thermal behaviour of the component-formates of the $MgAl_2(HCOO)_8$ has also been investigated, (5), (6). The decomposition of the formates generally yields either the carbonates of the metal (with elimination of H_2 and CO), or the oxide of the metal (with elimination of H_2O and CO). Besides these substances, many other products of decomposition have also been detected during these processes. For example Zn-formate produces methanol, formaldehyde and methylformate, (3). Ca-formate produces formaldehyde, acetone, methanol, carbon and oxalate, (3). Mg-formate produces CH_4, H_2, methanol, acetone, formaldehyde and carbon, (5).

This shows the complexity of this process and indicates the insufficiency of the usual thermoanalytical methods as a tool for determination of the exact decomposition-scheme of these compounds.

EXPERIMENTAL

The Mettler-vacuum thermoanalyser, fitted with a quadrupole mass-spectrometer, used in our experiments, enabled us to study the thermal decomposition of the $MgAl_2(HCOO)_8$ in vacuo ($p < 10^{-4}$ Hgmm) with simultaneous detection of the substances evolved.

The results are shown in Fig. 1. Under the applied conditions the decomposition took place with production of carbon. The sample obtained at the end of the process was black in colour. Ignition in a stream of O_2 at $1000^\circ C$ yields the white $MgAl_2O_4$ spinel.

RESULTS AND DISCUSSION

Fig. 1. shows four characteristic changes which can be identified as follows:

<u>Steps 1 and 2.</u> Two endothermic steps at $86^\circ C$, and $190^\circ C$. Loss of 1.5 and 2.0 molecules of water. No carbon-containing substances evolved.

<u>Step 3.</u> A complex endothermic step beginning at 270°-$275^\circ C$ reaching its maximum rate at $292^\circ C$. Evolution of CO, CO_2, HCOOH and HCHO is detected. The analysis by mass-spectrometry leaves some elements of uncertainty. In this case HCHO could have been partially or totally produced from HCOOH as a parent-material. Analogously CO could have been partially or totally produced by CO_2.

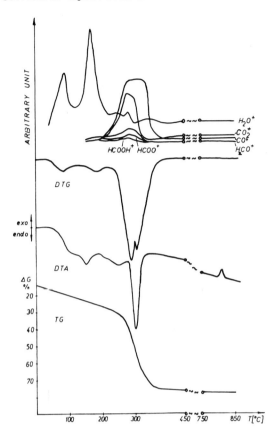

Fig. 1. Thermoanalytical data and mass-spectrum of the decomposition of Mg-Al-formate.

To get more information a complementary experiment was performed. A sample of the double-formate was heated in a stream of purified He and the products of the thermal decomposition were analysed by classical methods. It was shown that below 275°C only water is evolved and at 275°C CO, CO_2 and HCOOH are simultaneously evolved while no HCHO or H_2O can be detected.

Step 4. An exothermic step at 810°C, with no loss of weight.

This step is the transformation of amorphous $MgO \cdot Al_2O_3$ into well-crystallized $MgAl_2O_4$ spinel. This was evidenced by X-ray analysis of samples ignited at 650°C and 1000°C, respectively.

Fig. 2. shows the formulation of these processes and the loss of weight calculated on this basis, as compared with the observed values.

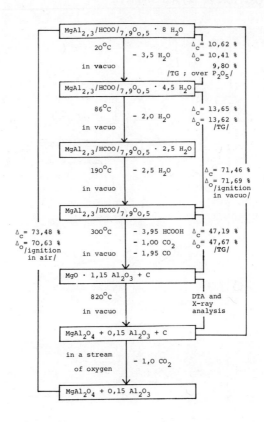

Fig. 2. Decomposition of Mg-Al-formate Δ_c = loss of weight calculated. Δ_o = loss of weight observed.

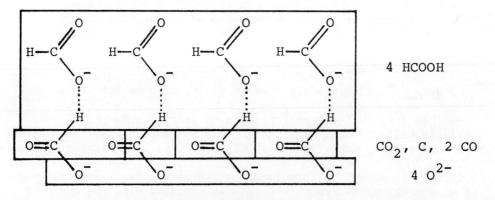

Fig. 3. Decomposition of formate ions

Fig. 3. suggests a probable decomposition-mechanism of the formate ions, yielding CO, CO_2, HCOOH, elementary carbon and oxide ions.

REFERENCES

1. J.A. Wülfing, O.P. 339091
2. Proceedings of Symposium on Scientific Bases for the Preparation of Heterogenous Catalysts, Brussels 1975 (in press)
3. C. Duval, Inorganic Thermogravimetric Analysis, Elsevier, 1963
4. R.C. Mackenzie, Differential Thermal Analysis, Academic Press London, 1970
5. Gmelins Handbuch der Anorg. Chem. Magnesuim Teil B.
6. Joan Taek Kwon Dissertation Abstr. 24, 64-5 (1963)

DISCUSSION

M.E. BROWN (Rhodes University, Grahamstown, South Africa) Were IR spectra of the partially decomposed solid recorded in order to confirm or deny the existence of the suggested intermediates?

AUTHORS' REPLY: The partially decomposed solid (190°C) was studied both by IR and by gas chromatographic analysis of the pyrolysis products. These showed the absence of any polycarboxylic acid and permitted us to revise the previous mechanism as included in the preprint.

REACTION BETWEEN CEMENTED CARBIDES AND MOLTEN COPPER

T. Yamaguchi and M. Okada

Department of Engineering, Keio University

Yokohama, Japan

ABSTRACT

Interdiffusion in the system cobalt-bonded WC-molten copper has been studied with special emphasis on the effects of tool parameters and of Ni-, Co- and Ag-addition to molten copper. A tentative mechanism of interdiffusion has been proposed on the basis of experimental results.

INTRODUCTION

The present work was undertaken in an attempt to better understand the basic reaction between cobalt-bonded carbide tools and molten copper and also to find some useful information for adequate control of the welding performance.

EXPERIMENTAL

Tools employed are basically WC-Co alloys of various compositions. The finished specimens were heated in contact with a sheet of copper in vacuum of approximately 10^{-3} mmHg under the desired time-temperature conditions. The time required to reach a desired temperature was approximately 4h. The couples obtained were sectioned, polished and subjected to optical microscopy, EPMA and microhardness measurement.

EXPERIMENTAL RESULTS

Structures of Diffusion Layers. Photo. 1 shows typical microstructures. A Cu-bonded carbide layer C was formed at copper/tool interface. On further heating, layer C penetrated into the bulk tool with simultaneous development of Co-rich layer A as shown in Photo. 2. Summarizing with the results of EPMA shown in Fig. 1, there are three layers of importance, of which thickness depends on the temperature, time, the amount of copper used, tool parameters such as cobalt content, WC grain size and so forth. No substantial interaction between liquid metal and the WC grains were observed irrespective of heating conditions and tool parameters, and the data for tungsten are not given for the sake of clarity in Figs. 1 and 2.

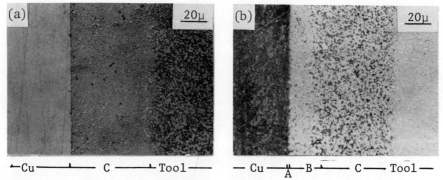

Photo. 1 Typical microstructures of diffusion layers of WC20%Co-Cu couples, heated at 1120°C for, a) 0h, etched in 30%H_2O_2; b) 3h, etched in Murakami's reagent.

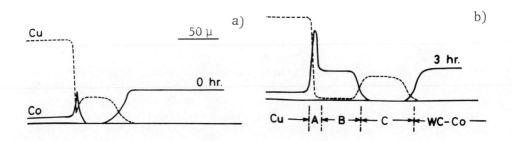

Fig. 1. Effect of heating time on the elemental distribution across the diffusion layers. 1120°C, WC20%Co-Cu, WC grain size, 4.5μ

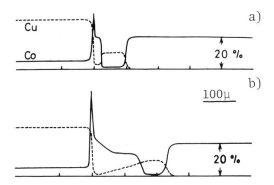

Fig. 2. Effect of WC grain size on the elemental distribution 1120°C, 1h, WC20%Co-Cu, WC grain size, a) 4-5μ, b) 1-2μ

Effects of Time and Temperature. Layers A and B were observed above 1112°C, the peritectic temperature in the system Cu-Co as shown in Fig. 3. Between the melting point of copper and 1112°C, the layer structures were similar to that shown in Photo. 1a irrespective of the heating conditions. Above 1112°C, the average copper content in layer B decreased with increasing heating time. After the departure from the copper/tool interface, the layer C virtually did not grow. Obviously, the formation of Co-rich layer A

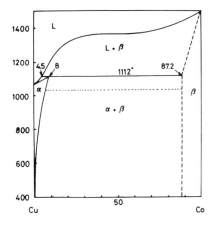

Fig. 3. Phase diagram of the system cu-co(1)

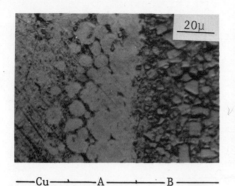
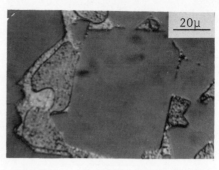

―Cu―•―A―•―B――   ――C――――Tool――

Photo. 2 Structure of layer A in Photo. 1b, etched in Murakami's reagent.

Photo. 3 Structure of Cu-front of 40-50μ-WC20%Co-Cu couple, heated at 1120°C for 3h, etched in 10%HNO₃.

checked the diffusion of copper into tool.

<u>Effects of Tool Parameters.</u> Fig. 2 shows the effect of WC grain size on the layer thicknesses. Fine-grained WC promotes the penetration of copper into tool. Thus, WC grain surfaces play an important role in the penetration of copper as shown in Photo. 3. Increase in cobalt content in tool favored the formation of layers A and B. Increase in carbon content resulted in the decrease of cobalt concentration in layer C and in rather sharp interlayer boundaries. This is possibly explained by the reduced solubility of cobalt in copper as caused by carbon. Addition of TaC and TiC to tool increased the thickness of layer C with more pronounced effects on the latter. For TiC-modified tools, the formation of layers A and B required long-term heating.

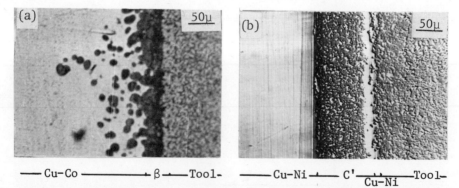

――Cu-Co――――→β―Tool-   ――Cu-Ni―•―C'―•―Tool-
                                        Cu-Ni

Photo. 4 Effect of additions on layer structure of couples WC20%Co-, a) Cu-10%Co, etched in 10%HNO₃; b) Cu-5%Ni, as polished.

Effects of Additions to Copper. While cobalt addition inhibited the copper-to-tool diffusion remarkably, no significant effect was observed with silver, which has no solid solubility to cobalt binder phase. A typical layer structure of WC20%Co-10%Co·Cu couples is illustrated in Photo. 4a, in which the dark areas are the Co-rich solid solution(cf. Fig. 3). Addition of nickel to copper developed a WC-free layer in tool(Photo. 4b), whose composition was the same as that in bulk liquid metal as confirmed by EPMA. Penetration of nickel into tool is more pronounced than that of copper, reflecting the preferential solid solubility of nickel over that of copper to cobalt binder phases. Increase in nickel content in liquid copper increased the thickness of the WC-free layer, whereas heating time had little effect on the layer thickness.

MECHANISM OF DIFFUSION

The interaction between tool and molten copper proceeds in three steps: initial, intermediate and steady. In the initial stage, cobalt dissolves in liquid copper until the latter becomes saturated with cobalt. Copper migrates into tool filling up the cavities produced by the displacement of cobalt, resulting in the formation of layer C. Penetration of copper into tool takes place along the WC-grain surfaces, as is demonstrated in Photo. 3. Growth of layer C continues until liquid copper becomes saturated with cobalt. In the intermediate stage, layer A acts as a barrier in copper migration. Migration of copper into tool at C/tool interface would cause the deficiency of copper in layer C, which enhances the deposition of β-solid solution(cf. Fig. 3), since the binder phase in layer C is saturated with copper at this stage. The deposition of β-solid solution would occur preferentially on the WC grain surfaces. Thus, the precipitation of β-solid solution at the Cu/C interface would develop continuous layers A and B while layer C travels into tool with very little change in thickness. The time required for the formation of layer A depends on how soon liquid copper becomes saturated with cobalt. Increasing amounts of copper or reduction in cobalt content in tool, for instance, increased the time for developing layer A. The effect of Co- and Ag-addition to copper is explained along this line. The main process in the steady stage is the penetration of layer C into tool. Addition of TaC and TiC to WC-Co alloys hindered wetting of copper and impeded the formation of layer A with more pronounced effects on the latter. Phase relationships in the systems WC-TaC and WC-TiC are responsible for layer structures, and the effect of carbide addition is possibly interpreted in terms of copper/carbide interfacial properties.

ACKNOWLEDGMENT

The authors wish to express their gratitude to H. Kuno for his encouraging discussion. They are also grateful to Mitsubishi Kinzoku Co. for financial support of this work.

REFERENCE

1. M. Hansen, Constitution of Binary Alloys, 2nd ed., p469-71, McGraw-Hill Book Co., New York, 1958.

KINETICS OF TiB FORMATION

G. C. Walther and R. E. Loehman

Department of Materials Science and Engineering

University of Florida, Gainesville, Florida

INTRODUCTION

During an X-ray examination of the reaction between Ti and $B_{12}C_3$ powders, it was observed that TiB formed rapidly as one of the major initial products rather than the expected TiB_2. A study of the kinetics of forming TiB by the simpler reaction $Ti+TiB_2 \rightarrow 2TiB$ was begun to determine the mechanism for its rapid formation. Strashinskaya and Stepanchuk have reported growth of diffusional TiB films on TiB_2(1) in which TiB was formed after three hours at 900°C while the reaction was complete after one hour at 1400°C. This study did not take account of the powder compact structure nor could it explain the rapid formation cited above. Therefore it was decided to follow the evolution of the reaction using Quantitative Microscopy to learn what influence the compact structure has on the kinetics.

Quantitative Microscopy provides a means of examining the volume fraction of the solid and porosity phases, their associated surface areas, and total surface curvature from measurements made on a two dimensional polished section of the powder compact(2). These relations are simple and require only that the sections are a suitably random sampling of a uniform distribution of particles or structural features. In addition, enough placements of the grid or fields of view must be made to provide adequate statistics of the distribution of measured quantities.

EXPERIMENTAL PROCEDURE

The starting powders obtained commercially were 99.7% purity

Ti produced by grinding under neutral atmosphere (Atomergic Chemetals Co.) and 99.8% purity TiB_2 (Ventron) produced by reducing the oxides of B and Ti with C. The Ti particle size distribution was primarily 20-50μm with some particles as large as 120μm and having a platey morphology characteristic of grinding swarf. The TiB_2 powders were predominately 20-30μm with some finer fraction and some larger 30-80μm conglomerates; the particles were rough but primarily equiaxed. These were mixed in equimolar amounts in a V-blender for 24 hours and then compacted in a uniaxial die at 3500kg/cm^2 pressure. The resulting discs were heated in a tungsten resistance element furnace under flowing high purity gettered Ar gas for 0, 5, 15, 30, 60, or 120 minutes at 1200, 1300, 1400, or 1500°C. A heating rate of 100°C/min. was used until the desired temperature was achieved. Reaction time was considered as holding time plus heating time above 1100°C.

These sinter-reacted discs were broken into nuggets for metallographic preparation or pulverized for X-ray diffraction analysis. The nuggets were mounted in epoxy and the porosity filled by vacuum impregnation to improve their strength during polishing. These were polished using metal bonded diamond discs and 0.3μm alumina-slurry on nylon cloths. The solid phases were decorated by electrostaining in 28% NH_4OH; with an optical microscope the Ti appeared blue-purple, the TiB was tan, and the TiB_2 was white.

The Quantitative Microscopy measurements were made using a combination of manual and electronic counts on an Imanco Quantimet 720 image analyzing computer. The number of picture points of a given intensity (or voltage pulse) range may be detected and separated by setting suitable threshold windows electronically. For this study each grid placement analyzed a 200μm square area with a grid of 400x400 or 160,000 picture points, and 50 such placements were made for each sample.

The results of these 50 grid placements were averaged, and each parameter for each of the four phases was determined from the following relations:

$$V_V^i = \bar{P}_P^i/160,000 \qquad S_V^i = (0.50)\bar{N}_L^i \qquad M_V^i = 15,708\bar{T}_{Anet}^i$$

where $T_{Anet}^i = T_{A+}^i - T_{A-}^i$, and V_V^i, S_V^i, and M_V^i are the volume fraction, surface area per volume, and total curvature per volume, respectively. The N_L counts for the (Ti+TiB) and (TiB+TiB_2) phases combined were also measured to make solution of the following equations for specific interface area possible:

$$S_V^\alpha = S_V^{\alpha\beta} + S_V^{\alpha\gamma} + S_V^{\alpha\delta}$$
$$S_V^\beta = S_V^{\alpha\beta} + S_V^{\beta\gamma} + S_V^{\beta\delta}$$

$$S_V^\gamma = S_V^{\alpha\gamma} + S_V^{\beta\gamma} + S_V^{\gamma\delta}$$

$$S_V^\delta = S_V^{\alpha\delta} + S_V^{\beta\delta} + S_V^{\gamma\delta}$$

$$S_V^{\alpha+\gamma} = S_V^{\alpha\beta} + S_V^{\alpha\delta} + S_V^{\beta\gamma} + S_V^{\gamma\delta}$$

$$S_V^{\beta+\gamma} = S_V^{\alpha\beta} + S_V^{\alpha\gamma} + S_V^{\beta\delta} + S_V^{\gamma\delta}$$

The symbols α, β, γ, and δ are for the Ti, TiB_2, TiB and porosity phases, respectively.

RESULTS AND DISCUSSION

A qualitative examination of the series of metallographic samples for each temperature shows the quantities of Ti and TiB_2 decreasing and TiB increasing. The porosity was seen to increase and at higher temperatures and longer times was estimated to be about 50%. Some pullout due to polishing was observed, especially at the lower times or temperatures, where less sinter bonding would have occurred. Edge rounding of the TiB and TiB_2 and relief polishing of the epoxy was seen, as would be expected from samples of this particle size and range of hardness. During the grinding stage with 15μm metal bonded diamond discs, the brittle TiB_2 often fractured and left many small pits on the surface. These were often sites of pitting corrosion during the electrostaining.

The TiB product first appeared at the contact regions between Ti and TiB_2 and began growing into the TiB_2. As the reaction proceeded the TiB began extending to the sides of the contact region and coating the TiB_2 particles. It took about 30 minutes to coat the larger particles at 1200°C, about 15 minutes at 1300°C, while at 1400 and 1500°C most particles had a TiB product layer shell surrounding a TiB_2 kernel after only 5 minutes above 1100°C. Many smaller grains were fully converted to TiB by the time coating of the larger particles was complete.

The Ti phase evolution showed these larger platey particles decreasing in both length and breadth but maintaining a nonequiaxed aspect until the reaction was near completion. At this stage the particles were usually observed bonded to the TiB network with large voids surrounding them; these were presumably the regions formerly filled with larger Ti grains. In some well reacted specimens a few Ti particles much larger than the rest were seen. These suggested that the largest Ti grains were not completely consumed, that thick TiB layers hindered Ti diffusion to the unreacted TiB_2, and thus the rate of converting the interior portions of TiB_2 particles was greatly reduced.

The Quantitative Microscopy measurements further confirmed the qualitative observations: a decrease in Ti and TiB_2 volume fraction, surface area, and total curvature; an increase in TiB parameters; and a general increase in volume fraction porosity. These results were accelerated with increasing temperature. However, the porosity measured 45-60%, which was considered high although in qualitative agreement with the observed microstructures. Measurements by an independent immersion method showed a porosity in the range 30-45%. This discrepancy was explained in the early stage of reaction by pullout and in all stages by the rounding of grains during polishing. This effect was not attributed to instrumental errors because the sum of individual volume fractions was equal to unity within 0.5% for almost every case. A plot of the mole fractions of solid phases also showed Ti and TiB_2 present in equimolar amounts for 1200 and 1300°C; the rounding affected the solid phases in approximately the same proportions.

The formation of two moles of TiB for each mole of TiB_2 consumed creates a volume increase of 67%. If this expansion does not totally replace the volume of Ti consumed, the powder compact will expand unless greater densification due to sintering occurs. The influence of sintering of TiB or TiB_2 on the density of the powder compact was determined by weighing and measuring the volume of pressed discs of these powders before and after hearing to 1500°C for two hours; densification of less than one percent was realized. Thus the 10-15% porosity increase that occurs during this reaction is due primarily to the volume increase upon transformation of TiB_2 to TiB.

The appearance of TiB on TiB_2 showed that Ti is the transported species rather than B. The fact that it does not completely and uniformly cover the TiB_2 suggests that vapor transport is not active; the weight change of a pressed Ti powder powder disc heated for two hours at 1500°C was less than 0.5%. The coating motion away from the Ti-TiB contact area also suggests that surface diffusion of Ti over the TiB_2 occurs. Placing separate particles of TiB_2 on a Ti disc and heating produced a definite ring of TiB on the TiB_2 particles, as seen by polishing and staining; since the Ti can diffuse from only one direction, surface diffusion was seen to be faster than volume diffusion.

The specific interface areas calculated form S_V^i data also follow qualitative observations. The TiB-TiB_2 interface, $S_V^{\beta\gamma}$, was of primary interest and was seen to increase with increasing TiB and then to decrease as the amount of TiB_2 became smaller. The velocity of this interface can be determined from the relation:

$$\bar{v}_S = \frac{dV_V^\beta/dt}{S_V^{\beta\gamma}}$$

Initial values of approximately 25μm/min. were determined. Plotting ln $\bar{v}_S$ with ln time showed that the interface velocity follows a $t^{5/2}$ dependence up to 30 minutes at 1200°C and up to 15 minutes at 1300°C while at later times this changes to a $t^{1/2}$ relationship. The reactions at 1400 and 1500°C are so fast that early stage data could not be obtained. A model to explain the 5/2 power has not yet been deduced. Once the initial TiB layer is established the geometry appears to satisfy very closely the assumptions of a series of spherical particle models reviewed by Hulbert(3). However, there is poor agreement between these models and the V_V data collected here. The $\bar{v}_S$ parameter is independent of particle size or shape and may explain the satisfactory results obtained using it.

CONCLUSIONS

The evolution of the reaction between mixed powders of Ti and TiB_2 was followed using Quantitative Microscopy. It was found that a rapid coating of TiB_2 by surface diffusion of Ti is followed by diffusion controlled growth of TiB into the TiB_2. Mechanisms of vapor transport or B diffusion were excluded. There is a 10-15% increase in porosity of the powder compact, due primarily to the volume increase upon transformation of TiB_2. This study demonstrates the importance of microstructural information for understanding the kinetics of mixed powder reactions.

REFERENCES

1. L. V. Strashinskaya and A. N. Stepanchuk, "Contract Interaction of titanium diboride with titanium, zirconium, and vanadium in vacuo," Fiz.-Khim. Mekh. Mater., 6(6)76-9(1970).

2. Quantitative Microscopy, Ed. by R. T. DeHoff and F. N. Rhines, McGraw-Hill Book Co., 1968.

3. S. F. Hulbert, "Models for Solid-State Reactions in Powdered Compacts: A Review," J. Brit. Ceram. Soc., 6(1)11-20(1969).

DESCRIPTION OF SOLID-GAS REACTIONS. CONCLUSIONS OF THERMOANALYTICAL AND ELECTRON OPTICAL STUDIES ON CALCIUM CARBONATE THERMAL DECOMPOSITION

Gy.Pokol, S.Gál, J.Sztatisz, L.Domokos, E.Pungor

Institute for General and Analytical Chemistry, Technical University
Budapest, Hungary

ABSTRACT

Thermal decomposition of calcium carbonate was studied with thermoanalytical and electron optical methods. Experimental results detailed in previous papers are discussed from the point of view of kinetic description. Imperfectness of the common kinetic equation using the reacted fraction, α, as the co-ordinate of the reaction and its effect on kinetic parameters are demonstrated. A new type of general equation analogous with physical kinetic laws is proposed for simple, reversible heterogeneous reactions. It is applied for the determination of some characteristics of $CaCO_3$ powder thermal decomposition.

INTRODUCTION

The authors - who have dealt with the thermal behaviour of alkaline-earth /Mg,Ca,Sr,Ba/ carbonates for some years - intended on the one hand to investigate special parameters of the decomposition and re-formation in various precisely controlled conditions. On the other hand these reactions can give information on general characteristics of solid-gas processes. The results on calcium carbonate thermal decomposition are discussed in the present work with the latter aim.

In previous papers the results of thermoanalytical investigations, carried out on a Sartorius Thermo- Gravimat, are reported on three types of sample: thick /some mm/, powder and thin /10-80 nm/. Real equilibrium was not reached even in isothermal runs under controlled carbon dioxide pressures, though the well-known reversibility of the reaction is not in doubt (1). In repeated isothermal decomposition - re-formation cycles the carbon dioxide uptake of calcium oxide could be divided into a fast and a slow part, and its conversion decreased owing to changes in the morphology. These were studied in an electron microprobe analyser (2).

The experiences of kinetic calculations carried out with both isothermal and non-isothermal data (1,2) led to some fundamental problems of the description of solid-gas reactions.

KINETIC DESCRIPTION OF SOLID-GAS REACTIONS

The reacted fraction, α as the co-ordinate of the reaction

Kinetic calculations for heterogeneous reactions are commonly based on the

$$\frac{d\alpha}{dt} = k \cdot f(\alpha) = Z \cdot \exp\left(-\frac{E}{RT}\right) \cdot f(\alpha) \qquad 1$$

rate equation, where k, Z, E, R, T and t denote the rate constant, pre-exponential factor, enthalpy of activation, gas constant, absolute temperature and the time, respectively. Eq. 1 was obtained from the homogeneous rate law by changing the activity of the reactant for the reacted fraction. The discussion of this analogy is missing or incomplete even in general monographs (3,4,5,6).

The reacted fraction is really not a local quantity like the activity. After strict discussion it was found, that the great differences of kinetic parameters /E,Z/ in the literature might be assigned partly to the imperfectness of eq.1. E.g. for isothermal phase boundary decomposition of a sphere one gets the well known "contracting envelope" equation from eq.1.

$$\frac{d\alpha}{dt} = k\left(1-\alpha\right)^{2/3} \quad \text{and} \quad 1-\left(1-\alpha\right)^{1/3} = kt, \qquad 2$$

where k should be constant. However, applying eqs.2 to spheres of different diameter k appears to depend on the size otherwise the time needed for complete decomposition of spheres of different diameter would be equal. In this case, when the dependence appears in the value of Z, the problem can be solved using the mass of sample and the area of the reacting surface of course - some types of solid-solid reactions are described this way

DESCRIPTION OF SOLID–GAS REACTIONS

(6), which is not equivalent to eq.1. It is worth mentioning that neither is eq. 1 valid in the homogeneous case, except first order reactions.

Rate of reversible reactions

For reversible processes eq.1 was completed by Bradley with a driving force factor, $1-\exp(\Delta G/RT)$, where ΔG is the Gibbs free energy difference of the reaction (7). In a solid solid-gas reaction it equals $(1-p/p_e)$ used by Jüntgen and van Heek, where p and p_e represent the actual and equilibrium pressure of the gas, respectively (8).

An attempt to describe the rate of simple, reversible heterogeneous reactions in general

Starting from the considerations outlined above a new form is proposed to describe the rate of reactions in question. The gross rate of the reaction (W) expresses the transformation rate of the chosen reactant related to unit stoichiometric coefficient $/\nu/$ for the whole system:

$$W = \frac{1}{\nu} \cdot \frac{dN}{dt} = \int_Q kF \, \partial Q , \qquad 3$$

where N denotes the amount of the reactant. The right-hand side of eq.3 is analogous with some laws describing physical processes relating to time e.g., heat conduction etc. It contains the rate constant analogous with thermal conductivity, a driving force, F,- a function of the chemical potentials accordingly to Bradley's proposal (7) - which corresponds to the temperature gradient and the reaction cross-section, Q, which is defined to be proportional to the number of situations geometrically suitable for the reaction. F and k are local quantities of course, i.e. functions in space. The rate of reaction, which is a local quantity can also be obtained from eq.3:

$$w = \frac{\partial W}{\partial V} = kF \frac{\partial Q}{\partial V} = k \cdot (1-\exp \frac{\Delta G}{RT}) \, q \qquad 4$$

where V and q denotes the volume and the density of the reaction cross-section, respectively, e.g. in a simple solid-gas reaction Q and q correspond to the free surface area and the free surface area in a unit volume. Deduction of the reaction cross-section for simple, one step, homogeneous reactions from the general rate law is to be published elsewhere. In this case Q and q are equal to Va^n and a^n, where a and n represent the activity of the reactant and the order of reaction, respectively.
Taking into account that these considerations concern only

one step of the process the authors propose to study this reaction type further with the aid of all the balances and rate equations of the system - similarly to reactor modelling. However, for nearly stationary systems the proper estimation of one step can give a good approximation.

CHARACTERISTICS OF CALCIUM CARBONATE POWDER DECOMPOSITION

In the decomposition of powder samples the reaction goes on either in all particles at the same time (case A) or there is a moving reacting zone in the sample (case B). In order to decide non-isothermal decomposition runs were carried out with an inert gas flow on a Du Pont 951 Thermogravimetric Analyzer. The samples were of different initial mass(m_o) but of the same grain size and similar arrangement. Under these conditions the rate constant is the same in the whole sample at any moment, the driving force does not change significantly, and the reaction cross-section is equal to the reacting surface:

$$Q = b \cdot m_o^\beta \cdot f(\alpha) \quad , \qquad 5$$

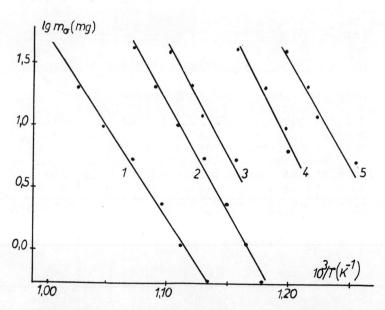

Fig.1. Relation between the initial mass of $CaCO_3$ powder and reciprocal temperature at fixed values

1 - 10°C/min heating rate, $\alpha = 0.5$, $d\alpha/dt = 9.1 \cdot 10^{-2} min^{-1}$
2 - 10°C/min " " , $\alpha = 0.07$, $d\alpha/dt = 2.2 \cdot 10^{-2} min^{-1}$
3 - 2°C/min " " , $\alpha = 0.26$, $d\alpha/dt = 1.1 \cdot 10^{-2} min^{-1}$
4 - 10°C/min " " , $\alpha = 0.005$, $d\alpha/dt = 2.2 \cdot 10^{-3} min^{-1}$
5 - 2°C/min " " , $\alpha = 0.03$, $d\alpha/dt = 1.1 \cdot 10^{-3} min^{-1}$

where b is a constant. In case A $\beta = 1$, in case B $\beta < 1$ - for spherical sample $\beta = 2/3$. Thus, using the molecular weight /M/ of the reactant:

$$W = \frac{m_o}{M} \cdot \frac{d\alpha}{dt} = Z \cdot \exp/-\frac{E}{RT}/ \cdot b \cdot m_o^\beta \cdot f(\alpha) \cdot F, \quad 6$$

which after rearrangement, using a new constant, K, gives

$$f(\alpha) = K \cdot \frac{d\alpha}{dt} \cdot \exp \frac{E}{RT} \cdot m_o^{1-\beta} \quad 7$$

Taking the points of the same α from the different measurements and considering that at these points $d\alpha/dt$ values were nearly the same in the first part of the reaction, $\lg m_o$ can be plotted against $1/T$, accordingly to the following obtained from eq. 7 (C and D are constants)

$$\lg M_o = C - \frac{D}{1-\beta} \frac{1}{T} \quad 8$$

Slopes of straight lines in Fig.1 show that β differs from 1, so case B is valid, i.e. the rate depends mainly on transport processes (probably heat transport) among the particles. The authors intend to repeat this series of measurements with spherical samples, since this way provides an opportunity to estimate kinetic parameters of the reaction.

REFERENCES

1. Gy.Pokol, S.Gál, K.Tomor, L.Domokos, in Proc. 4 Int. Conf. on Thermal Analysis, Budapest, 1974 ed. I.Buzás, Akadémiai Kiadó, Budapest, 1975, vol 1. 479.
2. S.Gál, Gy.Pokol, E.Pungor, Thermochim.Acta, in press
3. D.A.Young: Decomposition of solids, Pergamon Press, London, 1966.
4. W.W.Wendlandt: Thermal methods of analysis, Wiley-Interscience, New York, 1973.
5. J.H.Sharp, in Differential thermal analysis (ed.R.C. Mackenzie) vol 2., Academic Press, London, 1972, 61
6. P.P.Budnikov, A.M.Ginstling: Reactions in mixtures of solids (in Russian), Izd.Lit. Stroitel., Moscow, 1966
7. R.S.Bradley, J.Phys.Chem. 6P (1956) 1347
8. H.Jüntgen, K.H.van Heek, in Proc. 3. Int.Conf. on Thermal Analysis, Davos, 1971 (ed. H.G.Wiedemann), Birkhäuser Verlag, Basel, 1972. vol.2, 423.

DYNAMIC BEHAVIOUR OF SOLID SURFACES

K.H.Lieser

Professor Dr., Technische Hochschule Darmstadt

D-61 Darmstadt, Germany (Fed.Rep.)

ABSTRACT

The investigation of the kinetics of heterogeneous isotopic exchange in systems like NaI/CH_3^*I provides information about the dynamic behaviour of the surface of ionic crystals. The number of surface layers taking part in fast exchange as well as their mobility increases sharply with temperature.

INTRODUCTION

Heterogeneous isotopic exchange reactions between a labelled gas and a solid may be assumed to follow an exponential rate law.

(1) If only the first surface layer of the solid is taking part in the exchange reaction the following equation is applicable:

$$\frac{I}{I_o} = \frac{n_g}{n_g + n_s} + \left(1 - \frac{n_g}{n_g + n_s}\right) e^{-kt} \qquad (1)$$

where I is the concentration or activity of the labelled molecules in the gaseous phase at the time t, I_o the initial value, n_g the mole number of the atoms in the gas which are taking part in the exchange, n_s the corresponding number of the first layer of the solid and k the coefficient of exchange. The corresponding curve is plotted in Fig. 1.

(2) Taking into account that the exchange proceeds via volume diffusion into the interior of the solid, the following equation is obtained:

$$\frac{I}{I_o} = \frac{n_g}{n_g+n_t} + \left(1- \frac{n_g}{n_g+n_s}\right) e^{-kt} + \left(\frac{n_g}{n_g+n_s} - \frac{n_g}{n_g+n_t}\right) e^{-k_D t} \qquad (2)$$

where n_t is the total mol number of the ions in the solid taking part in the exchange reaction ($n_t \gg n_s$) and k_D the coefficient of exchange due to diffusion in the solid ($k_D \ll k$). The corresponding curve is plotted in Fig. 2. By extrapolation of the branch caused by diffusion within the solid the value for $n_g/(n_g + n_s)$ is found.

(3) If it is assumed that several surface layers have the same mobility and are taking part in the exchange simultaneously like a liquid film, the following equation holds:

$$\frac{I}{I_o} = \frac{n_g}{n_g+n_t} + \left(1 - \frac{n_g}{n_g+n \cdot n_s}\right) e^{-kt} + \left(\frac{n_g}{n_g+n \cdot n_s} - \frac{n_g}{n_g+n_t}\right) e^{-k_D t} \quad (3)$$

where n is the number of those surface layers. The corresponding curve for the exchange reaction is plotted in Fig. 3 for n = 9. By extrapolation the value for $n_g/(n_g+n \cdot n_s)$ and hence the value of n is obtained.

(4) Finally, if several surface layers are contributing to the exchange reaction with different rates (exchange coefficients) the following equation is derived:

$$\frac{I}{I_o} = \frac{n_g}{n_g+n_t} + \frac{n_g}{n_g+n \cdot n_s} \sum_{i=1}^{i=n} e^{-k_i t} + \left(\frac{n_g}{n_g+n \cdot n_s} - \frac{n_g}{n_g+n_t}\right) e^{-k_D t} \quad (4)$$

The corresponding curve is plotted in Fig. 4, again for n=9.

EXPERIMENTS

In order to study the fast heterogeneous exchange reaction on the surface of solids in systems like NaI/CH_3I it was necessary to build up a device which allowed to follow the reaction continuously. The apparatus used consisted of a electromagnetic piston-driven pump made from glass, a reaction cell, containing the solid and heated by an electrical furnace to constant temperature, a cell for continuous measurement of the activity in the gaseous phase by means of a NaI scintillation counter, combined with a plotter, and a high vacuum pump with manometer.

Sodium iodide of small grain size was prepared by precipitation using organic solvents as well as by grinding in a mill. The fraction with a grain size between 5 and 20 μm was used. The specific surface area as determined by the BET method was 0,5 m^2/g.

Labelled methyl iodide was obtained by isotopic exchange from CH_3I and carrier free $Na^{131}I$. It was frozen in liquid nitrogen, weighed and introduced in the vacuum system. The pressure of CH_3I was varied between 1 and 500 torr. At methyl iodide pressures below 100 torr argon was added in order that the glass pump was working properly. The ratio of the mol numbers of CH_3I and of NaI in the first surface layer was varied between 0.06 and 53. The temperature range was restricted up to 255°C with respect to the decomposition of methyl iodide at higher temperatures.

RESULTS

The exchange reaction between NaI and CH_3^*I starts at about 150°C. At CH_3I pressures above 100 torr the rate determining step is the desorption of methyl iodide and the reaction follows the rate law

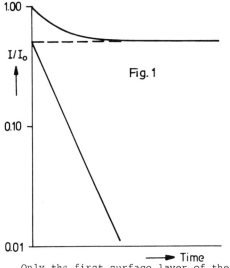

Fig. 1
Only the first surface layer of the solid is taking part in the heterogeneous exchange ($n_s : n_g = 1$)

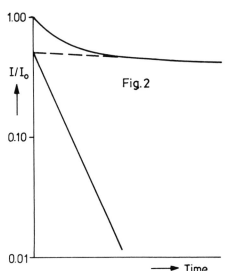

Fig. 2
The first surface layer of the solid is taking part in the fast heterogeneous exchange, followed by volume diffusion ($n_s : n_g = 1$)

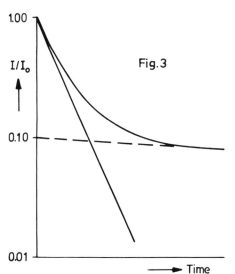

Fig. 3
Several surface layers of the solid are taking part in the fast heterogeneous exchange simultaneously, followed by volume diffusion ($n_s : n_g = 1$, $n = 9$)

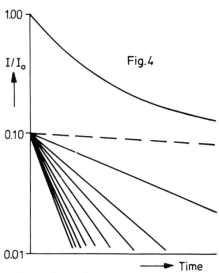

Fig. 4
Several surface layers of the solid are taking part in the fast heterogeneous exchange with different rates, followed by volume diffusion ($n_s : n_g = 1$, $n = 9$)

$$\ln(1-\lambda) = 2{,}45 \cdot 10^5 \exp(-9700/RT) \frac{n_g + n_s}{n_g} t \qquad (5)$$

where t is measured in seconds. At CH_3I pressures below 10 torr the second order reaction becomes rate determining

$$\ln(1-\lambda) = 2{,}78 \cdot 10^6 \exp(-5400/RT) \frac{n_g + n_s}{V} t \qquad (6)$$

where V is the volume of the gas in ml and t the time in seconds.

The analysis of the experimental curves leads to the conclusion that they follow equation (2) at low temperatures near 150°C whereas equation (4) is valid at higher temperatures. Equation (3) is not applicable. With other words, near 150°C where the reaction starts only one surface layer contributes to the fast exchange reaction. The number of surface layers taking part in the fast reaction increases sharply with temperature as shown in Table 1. On the other hand the mobility decreases with the depth. By formal analysis of the exchange curves the activation energies may be calculated for each layer. Results for the temperature range from 236 to 255°C are given in Table 2. They increase from the value found for the first layer (about 10 kcal/mol) to those calculated from the slow step of the exchange curves for bulk diffusion in the solid.

Temp. (°C)	Mean number of surface layers taking part in the fast exchange reaction
177	1,1 ± 0,07
208	2,3 ± 1,5
236	14,7 ± 1,9
255	31,0 ± 7,7

Table 1: Number of surface layers taking part in the fast exchange reaction between NaI and CH_3I

Number of the layer	Activation energy [kcal/mol]
1 (surface)	9,7 ± 0,2
2	11,0 ± 0,7
3	11,8 ± 0,9
4	12,7 ± 1,1
5	15,5 ± 1,3
6	18,6 ± 1,5
7	22,3 ± 1,8
8	25,6 ± 2,2
9	29,2 ± 2,6
10	31,3 ± 3,2
Bulk diffusion	47,7 ± 3,0

Table 2: Formal activation energies for the exchange reaction between NaI and CH_3I for different layers.

Thus the investigation of heterogeneous exchange reactions supply an instructive picture of the mobility of the ions in the different surface layers of a crystal as a function of temperature. At a distinct temperature (in the case of the system NaI/CH_3^*I near $150°C$) the first surface layer becomes mobile and the exchange reaction is measurable. At higher temperatures the second, third and the following layers also become mobile gradually and contribute to the exchange reaction. It should be pointed out, however, that these layers do not form a quasi-fluid film on the surface, but the mobility decreases with the depth.

It should be well understood that the picture of the different layers is a rather formal approximation because the surface will become more and more rough with rising temperature so that an increasing number of layers will be intersected by furrows whereas hills and ridges will be formed at the same time. Therefore the surface area and the number of ions exposed to the gas phase will increase considerably with temperature. A qualitative picture of this roughening of the surface is presented in Fig. 5. Computer calculations of surface roughening of metals as a function of temperature are reported in the literature. The results of the investigations of the dynamic behaviour of solid surfaces by means of heterogeneous reactions correspond rather well with the picture coming out from these calculations.

Applications of the results may be seen in the field of catalysis and for the special case of systems like NaI/CH_3^*I in the possibility of selective separations of radioactive iodine from the off-gas of nuclear plants, where the iodine is to a great extent present in the form of methyl iodine.

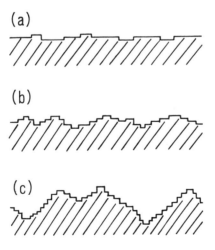

Fig. 5: Roughening of the surface as a function of temperature, a) low temp., b) medium temp., c) high temp.

NUCLEATION PROCESS IN THE DECOMPOSITION OF NICKEL FORMATE

M.E. Brown,[a] B. Delmon,[b] A.K. Galwey[c] and M.J. McGinn[c]

[a] Chemistry Department, Rhodes University
Grahamstown, 6140, South Africa

[b] Laboratoire de Chimie du Solide et de Catalyse
Université Catholique de Louvain
1348, Louvain-la-Neuve, Belgium

[c] Chemistry Department, Queen's University
Belfast BT9 5AG, Northern Ireland

ABSTRACT

Information concerning the nucleation process in the thermal decomposition of nickel formate has been obtained from kinetic and electron microscopic observations. The rate of the initial acceleratory stage of reaction and shape of the yield-time curve for isothermal reaction are sensitive to reactant configuration. This is attributed to variations in the local availability of water vapour, believed to inhibit the nucleation step. The formation of nuclei within the reactant mass is also inhomogeneous, again attributable to changes in the prevailing water vapour pressure. Some kinetic and mechanistic implications of these observations are discussed.

INTRODUCTION

Following a previous communication (1) and previous discussions (2), we have investigated further the nucleation processes in the thermal decomposition of nickel formate.

EXPERIMENTAL

The same preparation of nickel formate dihydrate was used as in earlier studies (1,3). Kinetic measurements were again based on pressure of gas evolved in the constant volume apparatus used previously (1). The weighed sample of salt was outgassed (2-3 h) and then introduced into the heated zone ($\pm$ 0.5 K) with a cold trap (193 K) maintained between the heated reactant and the McLeod pressure gauge.

Partially decomposed samples for electron microscopic study, using an EM6B instrument, were replicated (as previously (1)) by a two-stage technique, preshadowed with gold and palladium at $\cot^{-1} 2$. Any residual reactant adhering to this material decomposed in the electron beam, a problem which was eradicated by water washing of the replica. These preparations showed variations in stability so that it was necessary to examine many areas to obtain examples of surface texture representative of the crystallites as a whole and avoid basing conclusions on the preferred selection of specific, perhaps unusual, features.

RESULTS AND DISCUSSION
Kinetic Measurements

The influence of reactant disposition upon the kinetic characteristics of the acceleratory period of salt decomposition was investigated by measuring the rate of product evolution under accumulatory conditions. Isothermal experiments (450 $\pm$ 1 K) were made in triplicate using similar samples of equal mass (9.0 $\pm$ 1.0 mg) for three different degrees of crystal aggregation. Samples specified by letter C were compacted with a thin glass rod into a 1 mm glass tube and retained by a small glass wool plug. Crystallites loosely aggregated together, L, were retained (1 mm length) at the sealed end of a 5 mm reaction vessel by a glass wool plug. Dispersed salt, D, was mixed with excess glass wool over a length of $\sim$ 7 mm and similarly retained in a 5 mm tube. As can be seen from Figure 1, the fractional decomposition (α)-time curves and kinetics of reaction, $\alpha < 0.3$ (identified by the reference letters) show differences of behaviour for the three crystallite dispositions. The significant variations apparent within each triad of nominally identical reactant assemblages is evidence of considerable sensitivity of onset of reaction to the relative positions of salt particles. Visual inspection of several samples of partially decomposed reactant (0.04 < α < 0.2) revealed the inhomogeneous onset of decomposition within the reactant mass. The distinctive black product was readily identified and at values α ∿ 0.05 darkened, isolated and localised regions of appreciable metal production were easily recognised within the pale green reactant.

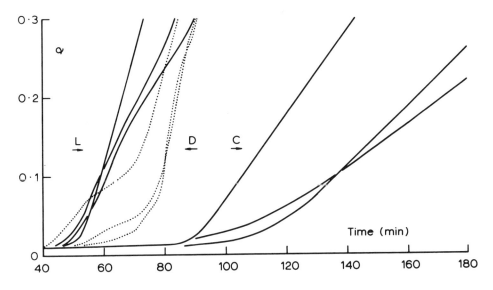

Figure 1. α-time plots for decomposition of nickel formate.

Kinetic analyses of data for the complete reaction (4) confirmed and extended the available evidence supporting the previous conclusion (2) that the maximum on the normalized $(d\alpha/dt)$ against α plot showed a systematic increase with temperature in the value of α at which the maximum rate $\{(d\alpha/dt)_{max} = 1\}$ was attained. There is, therefore, a relative increase in the rate of nucleation at higher temperatures.

Electron Microscopy

Although only a small number of representative examples of surface textures observed in electron micrographs can be illustrated here (Figure 2A-D), the conclusions which we report were reached after examination of many areas, including over 200 photographs of surface features and textures. The materials studied included salt decomposed to various values of α < 0.1 (notably α = 0.022, 0.054 and 0.088) at temperatures extending across the interval for which kinetic data had been obtained (∿ 440-490 K). After the onset of reaction, characteristic surface textural features, of which representative examples are shown in Figure 2, became apparent. The opaque water-insoluble particles are identified as product metallic nickel adhering to the replica, though we cannot be certain that all such particles of product were so retained. Regions where these features were apparent were invariably roughened in comparison with other surfaces: the latter are therefore identified as undecomposed since these were indistinguishable from unreacted material. Sometimes product particles were retained in surface pits. All such areas are identified as decomposition nuclei, resembling those described previously (1), but greater detail can be discerned at the higher magnifications employed here. Nuclei were apparently randomly dispersed on surfaces, though, on occasion, there was evidence of extensive decomposition of one crystal face while other areas nearby were little changed. The diversity of sizes, shapes and disposition of nuclei, which sometimes overlap, even at very low α, make it impossible to determine meaningfully the number of points of independent initiation of reaction per unit area. The differences in appearance, attributable to decomposition, of individual particles within one sample were generally greater than those apparent between particles of samples of salt which had been decomposed at different temperatures. Increased reaction was not accompanied by the immediate appearance of nuclei on all crystallites of reactant. The onset of decomposition of a proportion of the particles present was markedly deferred.

Kinetic and Mechanistic Considerations

It was suggested (1) that water is an efficient inhibitor of the nucleation process in the decomposition of nickel formate and the present work supports this conclusion. Determination of the precise role of this participant in delaying the onset of reaction is not straightforward, since it is available in two forms within the reactant mass: the last traces of water of crystallization may be tenaciously retained and water is also a product of salt breakdown. The visual detection of very localized areas of product formation at low α is attributable to the occurrence of reaction near channels for water escape within the reactant mass while in the more tightly packed regions particles may remain without nuclei long after the commencement of decomposition elsewhere. The observed increase in relative rate of nucleation at higher temperature may be attributed to increased desorption and ease of escape of the volatile participant. The variations in shapes of α-time curves show the sensitivity of behaviour to prevailing conditions and this irreproducibility of data reduces the reliability of kinetic interpretations. It is, therefore, concluded that all crystallites within the reactant assemblage are not equivalent, local inhomogeneities in the distribution of par-

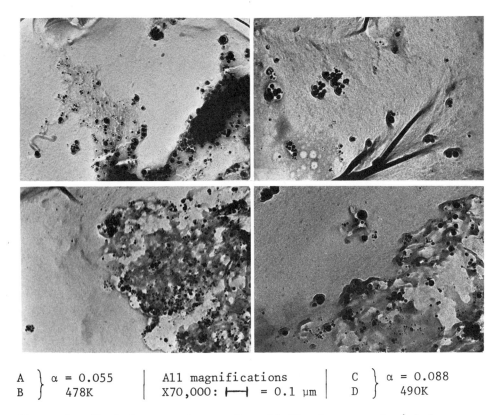

| A | α = 0.055 | All magnifications | C | α = 0.088 |
| B | 478K | X70,000: ⊢―⊣ = 0.1 μm | D | 490K |

Figure 2. Nuclei on surfaces of partially decomposed nickel formate. Opaque particles identified as metallic product nickel situated in areas of pitted and eroded boundary faces.

ticles introduces an additional factor controlling reaction rate, through variations in the availability of water. The effect of such inhomogeneities must be removed before it is possible to make meaningful kinetic analyses. Since the volume of the metallic product is relatively small, the development of nuclei seems to require the mobility of either the reactant or the nickel particles, such mobility of nuclei is not inconsistent with the appearance of eroded surfaces in Figure 2. The role of water in the inhibition of nucleation may be through stabilization of a susceptible surface site, by co-ordination with a cation. There is also the possibility of surface interactions or perhaps equilibration which can be schematically represented:

$$(HCOO)_2Ni \leftarrow H_2O \rightleftharpoons HCOONiOH + HCOOH$$

REFERENCES

1. A.K. Galwey, M.J. McGinn and M.E. Brown, Reactivity of Solids, 7th I.S.R.S. (Chapman & Hall, London, 1972), p.431.
2. P. Grange and B. Delmon, discussion to ref. 1, p. 442.
3. B.R. Wheeler and A.K. Galwey, J. Chem. Soc., Faraday Trans. I, 70 (1974) 661.
4. B. Delmon, Introduction à la Cinetique Hétérogène (Technip Publ., Paris, 1969).

DISCUSSION

P.J. HERLEY (State University of New York, U.S.A.) From the figures in your text (A-D) it would appear that some degree of clustering of the nuclei has occured. Have you quantitatively examined this effect on the initial distribution of the nuclei? Also, have you made any estimates of the size distribution within the clusters? To what do you attribute this phenomena?

AUTHORS' REPLY: The primary objective of this research was to make a quantitative study of the kinetics of nucleus formation from measurements of the changes in numbers and distributions of product particles on unit area of reactant surface with extent of reaction (α) and with temperature. Such information would provide a complete answer to Dr. Herley's question. This objective could not, however, be achieved satisfactorily due to the marked variations between different reactant areas in the structures of those regions identified as sites of establishment of salt breakdown.

The metallic product does not occupy the volume of solid from which it was derived and within the pits and areas of textural roughening attributed to the occurrence of reaction (i.e. the developed nuclei) the finely divided nickel appears as clusters of particles.

The appearance of differing surfaces of crystallites, from a particular sample of partially decomposed salt, showed very considerable variations: some areas were extensively reacted while others were apparently unchanged. This is believed to be a consequence of differences in case of dehydration at various locations within the reactant mass and is consistent with the observed sensitivity of kinetic behaviour to sample disposition. Clustering of nuclei is, therefore, attributable to preferred nucleation at those surfaces which have been most effectively dehydrated. Moreover, the appearance of individual nuclei varies considerably: in some regions distinct deparate nuclei are found, while elsewhere there was evidence of either highly anisotropic growth and/or coalesence of several nucei.

KINETIC STUDY ON THE PHASE TRANSFORMATION OF GeO$_2$

M. Yonemura and Y. Kotera

National Chemical Laboratory for Industry

Mita, Meguro, Tokyo 153, (JAPAN)

INTRODUCTION

The stability relation of the three predominant modifications of germanium oxide are considered by Laubengayer and Morton as follows:

hexagonal $\rightleftharpoons$ tetragonal $\longrightarrow$ glass structure

No systematic rate studies have yet been made on germanium oxide to date, in the same way as on silica and titanium oxide. The rate of transformation in germanium oxide may be complicated by variation in particle size, temperature, surrounding atmosphere, and by the presence of impurities, especially when the reaction takes place entirely in the solid state.

The roles of the alkali germanates and the surrounding atmosphere during phase transformation were studied in detail under standardised conditions. It was found that lithium and sodium germanate catalyse the transformation in the solid state, while potassium germanate does so in the liquid state, and that the phase change of alkali germanates affect the rate considerably. It was also found that the effect of surrounding atmosphere is not so large as that of the catalyst.

EXPERIMENTAL

Spectroscopically pure germanium oxide was used as a starting material. Alkali germanates were prepared by a solid state reaction between Li_2CO_3, K_2CO_3 or Na_2CO_3 and GeO_2. The measurement of the rate was carried out by means of high temperature X-ray diffraction.

RESULTS

Some typical plots of the logarithm of the hexagonal fraction, α, against time are shown in Fig. 1, where α was calculated by H/H+T, H or T being the peak height of hexagonal or tetragonal modification. The catalyst and atmosphere used are given in the figures.

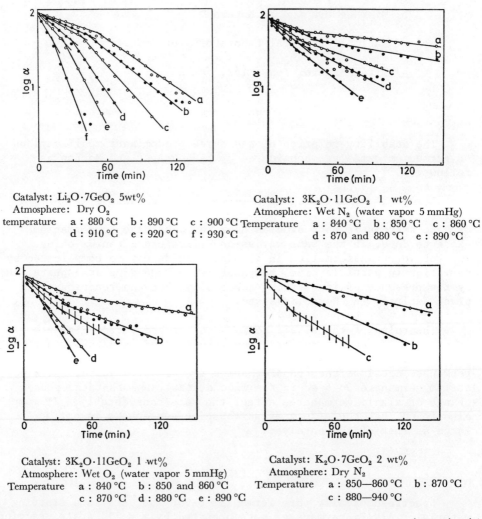

Catalyst: $Li_2O \cdot 7GeO_2$ 5wt%
Atmosphere: Dry O_2
temperature   a : 880 °C   b : 890 °C   c : 900 °C
  d : 910 °C   e : 920 °C   f : 930 °C

Catalyst: $3K_2O \cdot 11GeO_2$ 1 wt%
Atmosphere: Wet N_2 (water vapor 5 mmHg)
Temperature   a : 840 °C   b : 850 °C   c : 860 °C
  d : 870 and 880 °C   e : 890 °C

Catalyst: $3K_2O \cdot 11GeO_2$ 1 wt%
Atmosphere: Wet O_2 (water vapor 5 mmHg)
Temperature   a : 840 °C   b : 850 and 860 °C
  c : 870 °C   d : 880 °C   e : 890 °C

Catalyst: $K_2O \cdot 7GeO_2$ 2 wt%
Atmosphere: Dry N_2
Temperature   a : 850—860 °C   b : 870 °C
  c : 880—940 °C

Fig. 1. First-order plot hexagonal   tetragonal germanium dioxide

KINETIC STUDY OF PHASE TRANSFORMATION OF GeO_2 229

Activation energies were calculated from the Arrhenius plots (Figs 2), where k was determined from the slope of straight lines. The results are given in Table 1.

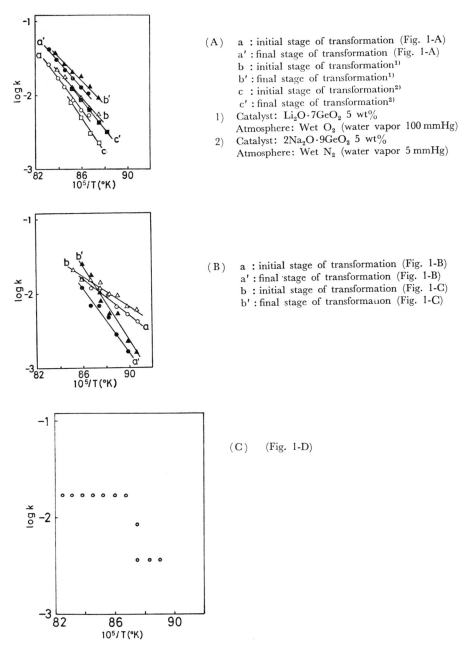

(A) a : initial stage of transformation (Fig. 1-A)
     a' : final stage of transformation (Fig. 1-A)
     b : initial stage of transformation[1]
     b' : final stage of transformation[1]
     c : initial stage of transformation[2]
     c' : final stage of transformation[2]
  1) Catalyst: $Li_2O \cdot 7GeO_2$ 5 wt%
     Atmosphere: Wet O_2 (water vapor 100 mmHg)
  2) Catalyst: $2Na_2O \cdot 9GeO_2$ 5 wt%
     Atmosphere: Wet N_2 (water vapor 5 mmHg)

(B) a : initial stage of transformation (Fig. 1-B)
     a' : final stage of transformation (Fig. 1-B)
     b : initial stage of transformation (Fig. 1-C)
     b' : final stage of transformation (Fig. 1-C)

(C) (Fig. 1-D)

Fig. 2. Arrhenius plot of log k against reciprocal temperature.

TABLE 1. ACTIVATION ENERGY OF TRANSFORMATION

Atmosphere		Activation energy (kcal/mol)				
		$Li_2O \cdot 7GeO_2$ 5 wt%	$3K_2O \cdot 11GeO_2$ 1 wt%	$K_2O \cdot 7GeO_2$ 2 wt%	$2Na_2O \cdot 9GeO_2$ 5 wt%
N_2	Dry	83—91 (122—123)	70—98 (72—102)	50±5	157—149
	Wet { 5 mmHg	96—91	55—98	52—90	101— 90
	{ 100 mmHg	92—78			
O_2	Dry	93—80 (92—82)	47—130 (52—128)	40±5	142—106
	Wet { 5 mmHg	83—90	46—114	115—119	107—83
	{ 100 mmHg	81—74		60—65	

$\ln k = \ln A - E/RT$ k: rate const. E: activation energy (kcal/mol). A: pre-exponential factor

Note: 1) The first number denotes the value for the initial stage of the transformation, and the second number that for the final stage. 2) Data given in parentheses were calculated for $Li_2O \cdot 7GeO_2$ catalyst prepared at low temperature and $3K_2O \cdot 11GeO_2$ catalyst prepared at high temperature,

KINETIC STUDY OF PHASE TRANSFORMATION OF GeO$_2$ 231

In the case of the transformation from tetragonal to hexagonal modification at higher temperature a typical plot of the logarithms of a tetragonal fraction, α, against time is shown in Fig 3, where α was calculated by T/H+T. Activation energies were calculated from the Arrhenius plot and are shown in Table 2.

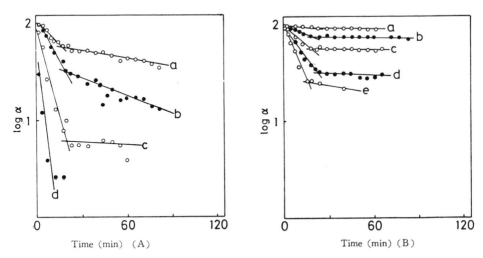

Fig. 3 First-order plot conversion of tetragonal to hexagonal germanium dioxide
(A) Catalyst: Li$_2$O·7GeO$_2$ 5 wt%, Atmosphere: Dry N$_2$
Temperature a: 1000°C, b: 1010°C, c: 1020°C, d: 1030°C
(B) Catalyst: 3K$_2$O·11GeO$_2$ 1 wt%, Atmosphere: Dry O$_2$
Temperature a: 1010°C, b: 1020°C, c: 1030°C, d: 1040°C, e: 1050°C

Table 2. Activation energy of transformation from hexagonal form to tetragonal form (Kcal/mol)

Catalyst wt% Atmosphere	Li$_2$O·7GeO$_2$ 5	3K$_2$O·11GeO$_2$ 1
N$_2$	220- ---	200- ---
O$_2$	220- ---	175- ---

DISCUSSION

It was found that added lithium and potassium react with germanium dioxide to form germanate compounds. It was determined for the lithium salt that the formation of lithium germanate, LiO$_2$·7GeO$_2$, occurs at a lower temperature than that of the transformation. Lithium germanate thus formed seems to catalyse the transformation in the solid state. Potassium germanate in contrast, seems to act as a catalyst in the liquid state.

It was found that the surrounding atmosphere affects the rate and activation energy of the transformation significantly. From the values of activation energy it confirmed that the effect of the surrounding atmosphere upon the phase transformation is smaller than that of the catalyst.

In the case of $K_2O.7GeO_2$ catalyst, the mean value of the activation energy between 850°C and 900°C was found to be 50 ± 5 Kcal/mol (Tab. 1). The temperature where the Arrhenius plot shows a discontinuity coincides with the eutectic point (895°C) of $K_2O.7GeO_2$ and GeO_2. It is concluded that the rate of transformation is much affected by the state of catalyst and its phase change.

Comparing results from the transformation at low and high temperatures, there exists a large difference between the two values of activation energy. These results may be explained by the fact that the former transformation takes place to a stable tetragonal modification from metastable hexagonal one at low temperature while the latter occurs to a stable hexagonal modification from a stable tetragonal one.

REFERENCES

1. Y. Kotera, M. Yonemura: Trans. Faraday Soc. <u>59</u>, 147 (1963)
2. Y. Kotera, M. Yonemura: J. Amer. Ceram. Soc. <u>52</u>, 210 (1969)
3. M. Yonemura, Y. Kotera: Bull. Chem. Soc. Japan <u>47</u>, 789, 793 (1974)
4. M. Yonemura, Y. Kotera: Denki Kagaku <u>43</u>, 660 (1975)

KINETIC ASPECTS ON THE FORMATION OF SOLID SOLUTION

M. Yonemura and Y. Kotera

National Chemical Laboratory for Industry

Mita, Meguro, Tokyo 153, (JAPAN)

INTRODUCTION

Although the study of solid-state reactions has received considerable attention recently, because of their complexity no general explanation has yet been obtained. The simplest reactions involve only a single phase, such as sintering and phase transformation. Germanium oxide is an example of the latter (1). In this research these basic ideas have been extended to two phases in the study of the kinetics of formation of a solid solution. The NaCl-KCl and CaO-SrO systems have been studied by high temperature Xray diffraction.

Previous work on these systems studied the decomposition kinetics of NaCl-KCl, when an activation energy of 22 Kcal/mole was obtained (2). For CaO-SrO, the temperature of formation of a solid solution was determined from changes in the Xray pattern of calcined mixtures of the two oxides (3)

EXPERIMENTAL

The starting materials used in this study were analytical grade reagents. NaCl and KCl were mixed while molten, and pulverized after cooling to room temperature. The starting materials of CaO-SrO system were prepared as follows: mixture of $CaCO_3$ and $SrCO_3$ or $Sr(OH)_2 \cdot 8H_2O$, and $CaCo_3 - SrCO_3$ solid solution coprecipitated by the reaction between $(NH4)_2CO_3$ solution and mixed solution of $Ca(NO_3)_2$ and $Sr(NO_3)_2$.

The kinetics of the formation of a solid solution was determined by means of high-temperature X-ray apparatus. The percentage of formation of a solid solution (100 - α) was calculated as the change of the distance of Xray diffraction lines of the starting materials relative to those of the solid solution.

RESULTS AND DISCUSSION

The change of (200) diffraction patterns of a NaCl(50mol%)-KCl(50mol%) mixture heated at different temperatures is shown in Fig. 1. It is clear that the solid solution forms at temperatures above 480°C. The formation rate is so high that it was difficult to obtain the exact kinetic data. Typical plots of residual amount of unreacted salt against time are shown in Fig. 2.

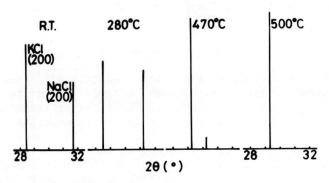

Figure 1 Xray diffraction pattern of NaCl-KCl system

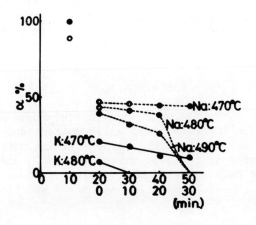

Figure 2. Reaction of NaCl-KCl system.

KINETIC ASPECTS OF SOLID SOLUTION FORMATION

The activation energy calculated from Arrhenius plot is a little larger than 10 Kcal/mol. which may be the activation energy for the mutual diffusion of sodium and potassium ions in the solid solution. It would be possible to analyse the overall kinetics by this research, but measurement of the formation rate is necessary to elucidate the formation mechanism.

As for the CaO-SrO system, the change of Xray diffraction patterns is shown in Fig. 3.

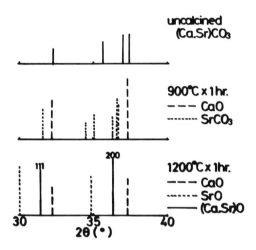

Figure 3. Xray diffraction pattern of CaO-SrO system.

It was found that $(Ca,Sr)CO_3$ changes by heating to the mixture of CaO and $SrCO_3$, and $SrCO_3$ changes to SrO by heating at higher temperature, then CaO and SrO form the solid solution. It was difficult to obtain the kinetic data because of the complicated manner of patterns obtained.

REFERENCES

1. M. Yonemura and Y. Kotera, Bull, Chem. Soc. Japan 47, 789, 793 (1974)
2. E. Schell and H. Stadelmaier, Z. Metalik 43, 227 (1972)
3. K.H. Obst and W. Münchberg Tonind. Zeitg., 92, 201 (1968)

INFLUENCE OF STRUCTURAL DEFECTS ON THE REACTIVITY OF SOLIDS

H. Schmalzried

Institute for Physical Chemistry, Technical

University, Hannover, West-Germany

INTRODUCTION

The broader a topic, the more careful one must handle the concepts. Therefore, this lecture begins with a discussion of the two main concepts involved: (1) Reactivity and (2) Structural defects. Every chemist feels what reactivity is, but he usually hesitates to give a straight answer if he is asked for a quantitative formulation. However, a scientific concept in chemistry must be quantifiable. A tentative formulation can be: "Reactivity is the reaction rate of a system under normalised driving forces."

There is no ambiguity about the term reaction rate. It can be extremely difficult to measure a reaction rate in solid systems, but this does not invalidate the definition. The concept of driving forces is less obvious and is therefore discussed in more detail. One has to distinguish between two situations: (a) Local equilibrium and thermodynamic state functions can be defined. (b) Cannot be defined.

If (a) is true, and consequently local densities of energy and entropy do exist, then one can use in the case of homogeneous solid state reactions the chemical affinity

$$-\Delta G = - \Sigma \mu_i \nu_i \quad (\Sigma \nu_i A_i = \sigma)$$

as the driving force. In the case of inhomogeneous or transport controlled heterogeneous systems, the local equilibrium concept, in the frame of irreversible thermodynamics, allows one to deduce potential gradients as driving forces for solid state reactions.

Furthermore, the introduction of the concept of local equilibrium immediately has consequences with regard to the existence of structural defects: Only a limited selection of all known crystal defects exists in strict local equilibrium; their concentration dependence on the independent variables and on the locus in the reacting crystal is theoretically assessable. Therefore one concludes: Reactivity can unambiguously and quantitatively be defined in solid nonequilibrium systems, provided that local thermodynamic equilibrium is known to be established.

The next point is a discussion of "Reactivity", if local state functions cannot be defined. No doubt, solids not being in local thermodynamic equilibrium do exist and do react. As an example, Fig. 1 shows a surface of a single crystal of NiO, briefly brought into contact with solid Cr_2O_3. Dislocation lines, which are nonequilibrium defects ending at the surface of NiO, are decorated with $NiCr_2O_4$. The reaction thus occurs at nonequilibrium sites. The reaction rate can for example be measured by the weight gain of the NiO-crystal per second. In contrast to the above situation however, one cannot formulate a driving force with the limited number of state functions necessary to describe the macroscopic thermodynamic state. One must know and control an extremely large number of parameters in order to achieve a quantitative formulation of the reaction. Here it is - beside other variables - the number and sort of dislocations, their distribution, their individual orientations etc.

In addition, difficulties in assessing the driving forces are obvious. By and large, in non-local-equilibrium systems, the local energies involved in a reaction are not well known, and entropies are not defined. With regard to Fig. 1 this means that it is difficult or impossible to estimate the Gibbs energy that a dislocation penetrating through the surface of NiO, adds at a given time and locus to the free enthalpy change of the reaction and how this then may influence the reaction rate.

Generalising one may say that if local thermodynamic equilibrium in a solid nonequilibrium system is not established, the term "Reactivity" is very difficult to quantify. Instead of applying phenomenolgical theories, one has to explain an infinite number of situations, all for their own. In a certain sense, those reactions are topochemical, since in nature and rate the reactivity depends on the locus in the system.

EQUILIBRIUM DEFECTS

From the introduction it is clear that fundamental studies are not able to explain the manyfold topochemical reaction problems individually. Those studies can only shape the proper concept and

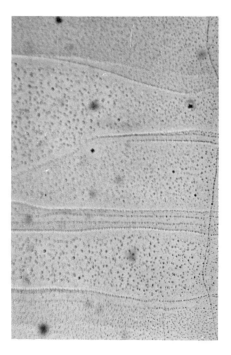

Figure 1: Surface of NiO-single crystal, briefly reacted at $T \simeq 1300°C$ with solid Cr_2O_3 in contact. The adherent reaction product is $NiCr_2O_4$.

guide-lines with regard to the influence of structural defects on reactivity. The first question in this context must be: Which defects do occur in local equilibrium in a bulk crystal? Since the days of Frenkel, Wagner, Schottky and Jost — amongst others — it is known that (only) point defects are equilibrium defects. Defect thermodynamics — which treat point defects essentially as dilute solutes in the solvent crystal — gives the answer to the question on the relation of defect concentrations and thermodynamic variables. Although well known in principle, many details are not at all understood, as will be shown in the specific section.

Since mobilities of the crystal components often depend directly on the number of point defects responsible for the atomistic movement, in principle the dependence of the transport coefficients of the crystal components on the full set of thermodynamic variables is known. Therefore, since in the case of local equilibrium and given the correct number of boundary conditions, driving forces and transport coefficients are locally known, here the problem of the reactivity of solids can in principle be regarded as being solved. Some current problems will be mentioned later. Are other than point defects to be included in the local equilibrium treatment of reactions? Disregarding small defect clusters for a moment, defects of the next higher dimensionality are line defects, i.e. dislocations. To calculate their equilibrium concentration, one may estimate the configurational entropy of a dislocation line in a solid cube according to Fig. 2. Since dislocation possess a line energy, at sufficiently high temperature in equilibrium the (at least partially) straight line is the most probable one. This line can be arranged in the cube in approximately $n \cdot N_o^{2/3}$ ways. The corresponding configurational entropy is

$$S_1 = k \ln W_1 = k \ln n \cdot N_o^{2/3} \cong \frac{2}{3} R \cdot \frac{\ln N_o}{N_o} \qquad (1)$$

n is a small number. Comparing the configurational entropy S_2, of $N_o^{1/3}$ defect atoms lined up along the dislocation line, if distributed at random as defects in the crystal, with S_1, one obtains

$$S_1/S_2 \cong N_o^{-1/3} \quad (\sim 10^{-8})$$

This low configurational entropy ratio along with a considerable line energy (comparable to point defect energies per atom on the dislocation line, assuming similar vibrational entropies) shows that the equilibrium number of dislocation is negligible.

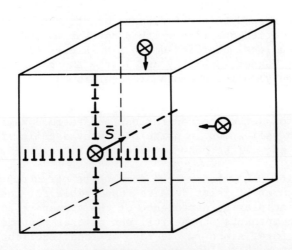

Figure 2: Dislocation arrangement in a solid cube.

The same considerations apply in principle to extended defects of still higher dimensionality. Unless the energy per atom of those defects is vanishingly small, they will not exist in local thermodynamic equilibrium. Therefore one can assume that crystal defects of higher than zero dimensionality do not play a significant role in equilibrium crystals, although they may be decisive for the kinetics of point defect equilibration {1} as sites of repeatable growth inside the crystal. Before discussing the role of nonequilibrium defects on the reactivity - a very special discussion by necessity - let us first inquire into some current problems regarding point defects.

SOME CURRENT PROBLEMS IN POINT DEFFECT CHEMISTRY

The selection of problems in this section is somewhat arbitary. Let us start with thermodynamic aspects: The zeroth order approach treats point defects as structural elements in dilute solution, having a virtual chemical potential of the form

$$\mu_d = \mu_d^o + RT \ln X_d$$

At all practically interesting concentrations one must add a third excess term $RT \ln f_d$; f_d being the activity coefficient. It stems from interactions amongst defects that can be elastic, electrostatic, magnetic or due to a specific chemical bonding. Often the interactions act simultaneously and are interdependent. This immediately explains the difficulty and complexity of a theoretical prediction of defect activity coefficients.

Defect association or the Debye-Hückel-Theory have been applied in order to account for the deviation from the zeroth order approach Fig. 3 gives an example {2}.

The fact that a charged defect is on the average surrounded by a cloud of countercharged defects, results (after calculating the corresponding defect distribution) in an activity coefficient (RT[kcal/mole]):

$$f_d = \exp - \frac{165.2}{RT} \cdot \frac{z_d^2/\varepsilon}{1(\text{Å}) + a(\text{Å})} \qquad (3)$$

Z_d is the effective charge of defects, ε the relative dielectric constant, $1(\text{Å})$ the so called Debye-Hückel length $l_D = \sqrt{D \cdot T}$, $T = \varepsilon/\sigma$ and $a(\text{Å})$ the critical distance between two defects at which a neutral complex is thought to form. In Fig. 3 the experimental vacancy concentrations, taken from a number of authors, are plotted for $Co_{1-\delta}O$ as a function of $p(O_2)$. One can see that the measurements can equally well be interpreted with the help of the aforementioned Debye-Hückel-Theory, or by assuming that doubly ionized vacancies

V''_{Co} exist at low oxygen potentials and low defect concentrations whereas "clusters" of $V''_{Co} + h^o$ exist at high defect concentrations and high oxygen potentials.

Of course, combination of these two first order corrections could be performed, but without a rigorous theoretical justification for the selection of $a(\text{Å})$, and possibly also for the Gibbs energy of association, the interpretation is dubious.

In contrast to CoO, the infinitive dilute solution approach works satisfactorily at the same defect levels in magnetite Fe_3O_4. One notes that the magnetite has a much higher "buffer capacity"

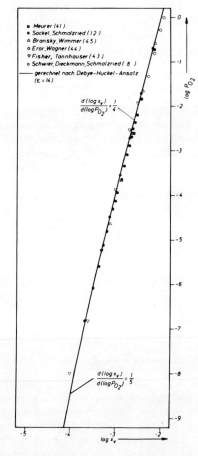

Figure 3: Concentration of cation vacancies as a function of the oxygen component activity for $Co_{1-\delta}O$ at T = 1200°C.

as a consequence of the large number of cations with different valencies, and thus the screening of the effective defect charges is much more efficient. It may safely be said that the detection and the theoretical evaluation of first and second order effects beyond the infinitely dilute solution treatment of point defects is if primary interest in defect chemical thermodynamics nowadays.

To conclude this section, some remarks should be made concerning point defect kinetics. In order to achieve local equilibrium, point defects have to equilibrate. Therefore, knowledge about defect equilibration and relaxation is essential in classifying the solid state reactions. Equilibration can occur in three ways: (1) Reaction between point defects and crystal components taking place at a surface are heterogeneous defect reactions. (2) Reactions between defects and lattice atoms or ions at sites of repeatable growth inside the crystal (at dislocations, small angle grain boundaries etc.) are also heterogeneous defect reactions. (3) Only reactions between point defects themselves are true homogenous reactions. Fig. 4 exemplifies the different situations.

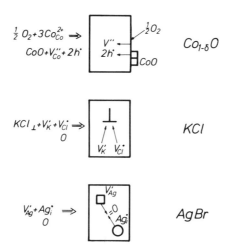

Figure 4:   Three different situations in defect equilibration.

The main problem in connection with equilibrium of point defects is their relaxation time. Clearly, for heterogeneous defect reactions this relaxation time is essentially given by a characteristic length which represents the average distance between sources or sinks at sites of repeatable growth. Since these distances can often deliberately be influenced (for example by sample size or by dislocation density) relaxation times are experimentally accessible and have been determined for a number of heterogenous defect reactions.

The situation is completely different for the case of homogeneous defect reactions and their relaxation times. One may estimate this relaxation time by assuming diffusion control (as for example in calculating the neutralization reaction in aqueous solutions) and then obtains

$$T = \left[8\pi \bar{c}_d \cdot (D_{d_1} + D_{d_2}) \cdot N_o \cdot \frac{E_W}{kT} \cdot a_r \right]^{-1/2} \qquad (4)$$

E_W is the defect interaction energy (in the case of coulombic interaction = $[e_1 \cdot e_2]/[\varepsilon \cdot \varepsilon_o \cdot a_r]$); a_r is the reaction distance (if a<a_r, the defect react to annihilate or to form the complex). Eq. (4) gives for $D_d = 10^{-6} cm^2/sec$, $\bar{c}_d = 10^{-4}$ mole/cm^3 ($\bar{x}_d \simeq 10^{-3}$) and T = 500 K : T $\simeq 10^{-4}$ sec, a time which is quite difficult to determine experimentally in a solid.

Beside the interest in defect relaxation for its own sake, this area of research has quite an influence on all questions related to solid state reactions. This shall be exemplified by a discussion of the interdiffusion of MgO and CoO. Fig. 5 shows the point defect (vacancy) distribution after some time of interdiffusion determined with the assumptions (1) of local equilibrium in the solid solution (Mg, Co)O and (2) if no defect relaxation at all occurs. This means that the initial local defect concentrations are preserved during interdiffusion. In Fig. 6 the corresponding curves for the super-saturated or under-saturated chemical potential of the vacancies which corresponds to the oxygen potential) is plotted. One notes that it differs from the oxygen potential as fixed in the surrounding atmosphere. Clearly, a relaxation process of the defects starts simultaneously with the interdiffusion process; the defect concentration determines the interdiffusion coefficient of the cations. Therefore defect relaxation directly influences the transport coefficients and thus the process kinetics. This example must suffice to illustrate the importance of defect relaxation processes in connection with solid state reactivity.

EXTENDED DEFECTS AND REACTIVITY

Defects with higher than zero dimensionality in crystals are dislocations, planar defects, clusters and inclusions. In a sense they are regions of a different solid phase in the crystal matrix. The "defect phase" is not in equilibrium with the matrix. However, the rate of decay of the "defect phases" is often so small that they may play an important role as parts of the solid during a solid state reaction.

The nonequilibrium character of an extended defect in the crystal is equivalent to the statement that here at least one chemical potential of the crystal components differs from the potential in the

STRUCTURAL DEFECTS AND REACTIVITY OF SOLIDS 245

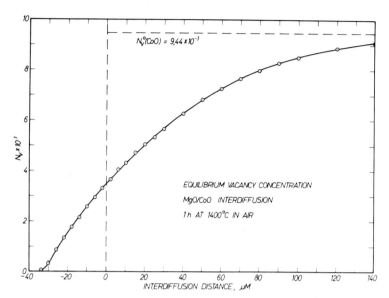

Figure 5: Vacancy concentration (supersaturated) and equilibrium vacancy concentration during MgO-CoO-interdiffusion.

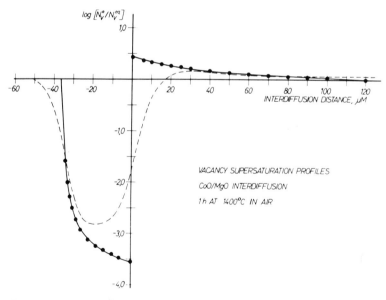

Figure 6: Vacancy supersaturation profiles, no relaxation and partial relaxation assumed.

bulk matrix. Also, the components in general have a different mobility in the defect phase as compared to the matrix. This then gives rise to a great variety of topochemical reactions, which have in part been studied theoretically and experimentally.

The phenomenological problem can be formulated as follows: Given a crystal with a known distribution of fixed extended defects, what is the overall diffusion behaviour of the sample? This problem has been treated in the case of polycrystalline samples, the grain boundaries being the extended defect phase {3}, and recently also in the case of a sample with dislocations {4}. In the latter case one talks about pipe diffusion.

Clearly the solution of the differential equations for transport which can be set up is complicated and depends on the assumptions concerning the geometrical distribution of the defects in the crystal. The theory in the case of a sample with grain boundaries has been worked out and is much in line with the experiment. This is not yet true for a sample containing dislocations with pipe diffusion.

Fig. 7 shows the geometry in the case of rapid diffusion along dislocations. The averaging process brings in unsolved problems, and also the question about the particle mobility as a function of the distance from the dislocation core and the formulation of the proper conditions at the surface of the sample are not yet solved satisfactorily {5}. Work up to now leads to radii of enhanced mobility around a dislocation line which are unrealistically large, if theory and experiment are compared (see Fig. 8). Work is in progress here.

The second important question in this context is as follows: If extended defects are not locally fixed in the lattice as was assumed in the foregoing discussion, can they be responsible for a net transport of matter?

In the case of a conservative movement of dislocations, this movement does not alter the composition of a stoichiometric compound hence matter is not transported if one excludes the drag of defects and foreign particles by way of interaction in the nonstoichiometric crystal {6}.

The situation is of course quite different for extended defects that have been introduced in order to account for the nonstoichiometry of a crystalline compound. Clearly, their movement relative to the lattice frame comprises transport of matter. The question is: do they move in a chemical potential gradient as such, and if so, by what mechanism?

A thought experiment is devised in order to illustrate the consequences:

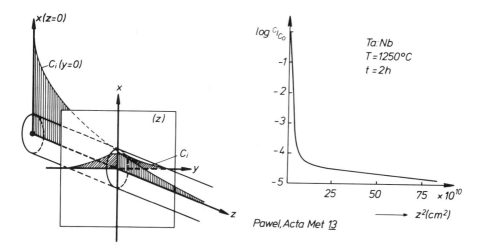

Figure 7:
Pipe diffusion model.

Figure 8:
Concentration profile in the case of simultaneous volume diffusion and diffusion along dislocations.

If a crystalline oxide i.e. $Co_{1-\delta}O$, is brought between different oxygen potentials in a fixed outer frame, as Fig. 9 shows, a macroscopic shift of the oxide relative to the fixed frame is observed. This shift is explained {7} by the drift of cation vacancies in the oxygen potential gradient. The fact that it can quantitatively be calculated from vacancy concentration and vacancy or cation mobilities which are interdependent and have been determined independently, is a strong indication of the dominant role point defects play in solid state reactions. This, of course, is also the basis for Wagner's well known oxidation theory {8} which, without introducing unknown parameters, calculates reaction rates from point defect data. And finally, it is the basis for diffusion controlled heterogenous solid state reactions, as for example spinel formation reactions {9}.

Let us now replace CoO in Fig. 9 by a crystal consisting of one or several phases containing only Wadsley defects to account for the nonstoichiometry. As in the case of CoO, one expects after some time to observe a steady state movement of the crystal surfaces and the phase boundaries in the potential gradient. In a crystal, the nonstoichiometry of which is caused by Wadsley defects only, this implies that extended defects are formed at the surface, that they drift through the crystal in the potential gradient and that they annihilate at the phase boundaries and at the other surface. This then corresponds to a net transport of cations through the lattice towards the side of the higher oxygen potential, since the

Wadsley defects here are supposed to be regions of an increased cation density.

It seems unrealistic or impossible that a Wadsley defect moves in the lattice as a whole in one step. Although mechanistic information is scanty, it is assumed that the lateral movement is a cooperative process. A highly correlated atomic jump sequence would be the consequence, with a correspondingly small mobility. It should also be mentioned, that our thought experiment not only demands the production and annihilation of Wadsley defects at the surfaces, but at the phase boundaries as well, provided they are incoherent phase boundaries.

Formation of a planar defect means nucleation and growth. Nucleation could occur by a collapse of a vacancy disc forming a dislocation loop; growth would be the climbing of that dislocation. Transport of cations could occur either by the lateral movement of planar defects - a sluggish process by nature - or by a correlated cation movement that results in a change of the orientation of the crystal shear plane, also called swinging. This change of orientation seems kinetically preferred as compared to the lateral movement mentioned above. By and large, the present state of knowledge leaves us with the definition of a number of kinetic problems, but the mechanisms concerning reactivity of phases containing planar defects are somewhat speculative at this moment, especially if compared with the well-founded results of the CoO experiment shown in Fig. 9.

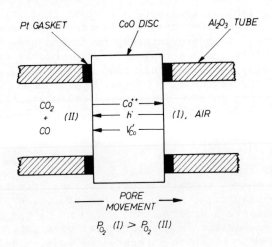

Figure 9: The transport of oxygen through a $Co_{1-\delta}O$ single crystal between two different oxygen potentials {7}.

STRUCTURAL DEFECTS AND REACTIVITY OF SOLIDS 249

This is still more obvious if the question is posed in a different way: what is the acting driving force for the movement of an extended defect, especially of a planar defect, in a chemical potential gradient, since these defects are not in local equilibrium? There is no answer, and therefore a kinetic theory operating with extended defects only, does not exist.

If, however, beside the Wadsley defects which account for the gross nonstoichiometry of the compound in question, there is a small equilibrium concentration of the good old point defect which could be responsible in the well known way for the net transport of matter, then the active role of mobile extended defects is not essential for the reactivity. The extended defects could serve as fast diffusion paths as has been described before. This question is not settled

DEFECTS AND PHASE BOUNDARY CONTROLLED REACTIONS

This final section deals with moving phase boundaries that occur in heterogeneous solid state reactions and in phase transformations of compounds. The gradients of the generalised thermodynamic potentials (T, P, μ_i) make the boundaries move. If one excludes the totally coherent phase boundaries, then these boundaries are structural defects themselves, and the question of their kinetic behaviour constitutes an interesting field of research, related to the question of the mobility of extended defects. It comprises problems such as the atomistic displacement of structural elements in a so-called massive transformation {11}, or, to mention a subject of different quality, the formation and motion of partial dislocation bundles at the phase boundary $Al_2O_3/MgAl_2O_4$ during spinel formation in the reaction $MgO + Al_2O_3 = MgAl_2O_4$, the reactants being single crystals {12}.

Let us start the discussion describing several experiments conducted by O'Briain {13}. The chemical potential drop at the phase boundary Ag/Ag_2S has been determined in the case that silver is crossing the boundary. It was found first of all that the phase boundary reaction is an electrochemical process (equal equivalents of silver ions and electrons cross the boundary independently). Also it was found that the potential drop across the boundary was dependent on the purity and the perfection of the Ag-crystal. Thus there is evidence that structural defects in the matrix of the adjacent phases may decisively influence the kinetics of a phase boundary controlled reaction.

A final point: Often, in phase transformation work, hysteresis is found. Fig. 10 gives an example. It is mainly attributed in the discussion to surface energy and strain, and it is in this context that the concept of the hydrid crystal was introduced {14}. But here is an additional aspect not yet being adequately taken into

account. It is explained in Fig. 11 and includes the action of
structural defects. Any phase diagram of a binary line phase exhibiting a phase transformation is in principle similar to Fig. 11b.
If one induces a phase change by changing the temperature, there
is a multitude of possible reaction paths which depend on the
situation of the Gibbs energy surfaces of the two phases as functions
of T and μ_i (at given P). Since μ_i is a unique function of the point
defect concentration, the individual path depends also on the point
defect relaxation time at and near the phase boundary. This relaxation time can be calculated according to the principles of homogeneous or heterogeneous reaction rate theory, provided the mobility
of the defects and the distribution of the defect sinks and sources
are known {1}. However it is obvious that if point defects, which
are automatically created during the phase transformation of compound crystals, cannot equilibrate instantaneously, hysteresis must
occur. This aspect of hysteresis has not been included in the
discussion. It may be quite important and is amenable to experiment and theory; both are in progress.

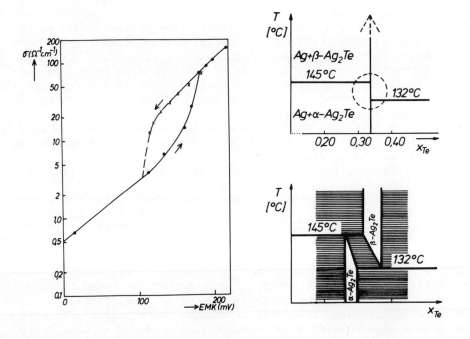

Figure 10:
Hysteresis in phase transformation: Coulometric titration of
Cu_xS in the galvanic cell $Cu/CuBr/Cu_xS/C$ According to Wehefritz {15}

Figure 11:a/b
Section of the phase diagram
Ag-Te.
Enlarged part of Figure 11a as
indicated (Schematic!)

LITERATURE

1. H Schmalzried, Solid State Reactions, Academic Press, Inc., New York 1974
2. R. Dieckmann, Ph.D. Thesis, Techn. University Clausthal, 1975
3. R.T.P. Whipple, Phil. Mag. $\underline{45}$, 225 (1954)
4. J. Mimkes, M. Wuttig, Phys. Rev. $\underline{B2}$, 1619 (1970)
5. P.L. Lin, unpublished, Inst. Theor. Metallurgy, Techn. University Clausthal, 1975
6. A. Gabriel, H. Nägerl, Proceedings, Int. Conference on Defects in Semiconductors, Freiburg 1974
7. G. J. Yurek, H. Schmalzried, Ber. Bunsenges. phys. Chemie $\underline{79}$, 255 (1975)
8. C. Wagner, Z. phys. Chemie $\underline{B21}$, 25 (1933)
9. H. Schmalzried, Progr. Sol. State Chem. (H. Reiss, Editor) $\underline{2}$, p.265, Pergamon Press, New York 1965
10. L.A. Bursill, B.G. Hyde, Progr. Sol. State Chem. (H. Reiss, J.O. McCaldin, Editors) $\underline{7}$, p 177, Pergamon Press, New York 1972
11. T.B. Massalski in Phase Transformations, Amer. Soc. for Metals p. 433, Metals Park, Ohio 1970
12. H. G. Sockel, H. Schmalzried in Mat. Science Res. $\underline{3}$, p 61 (W.W. Kriegel, H. Palmour Editors), Plenum Press, New York 1966
13. C.J. Warde, J. Corrish, C.D. O'Briain, J. electrochem. Soc. $\underline{122}$, 1421 (1975)
14. A.R. Ubbelohde, J. Chim. Phys., p.33 (1966)
15. V. Wehefritz, Ph. D. Thesis, University of Göttingen 1960

STUDIES ON NON-STOICHIOMETRIC STRONTIUM FERRATE, $SrFeO_{3-x}$

Bruce C. Tofield

Materials Physics Division, A.E.R.E., Harwell

Oxfordshire, OX11 0RA, England

PREVIOUS WORK

The oxide perovskites $SrMO_3$, where M is a first-row transition metal, are interesting in that a number of the end members have, unusually, an undistorted simple cubic perovskite structure at room temperature, and also because wide ranges of oxygen non-stoichiometry have been reported for several systems. The most detailed study of the $SrFeO_{3-x}$ system ($0 \leq x \leq 0.5$) reported previously was by MacChesney et al who were the first to prepare the stoichiometric material using high pressures of oxygen, (1). $SrFeO_{3.00}$ was obtained after equilibration at 550°C and 5,000 psi for one week, or at 335°C and 13,000 psi for 16 hours. Samples containing more than 30% Fe^{3+} ($SrFeO_{2.84}$ to $SrFeO_{2.72}$) were reported to have a simple tetragonal distortion of the cubic unit cell, but no superlattice formation was observed. Below $SrFeO_{2.72}$ a two-phase mixture of a perovskite-like phase and a brown-millerite-like phase ($SrFeO_{2.50}$) was found. Gallagher et al assumed from X-ray powder data that $SrFeO_{2.50}$ (and $BaFeO_{2.50}$) were isostructural with $CaFeO_{2.50}$ (space group Pcmn) (2, 3, 4). MacChesney et al also studied the $BaFeO_{3-x}$ system (5), which is complicated by the presence of mixed hexagonal and cubic stacking of octahedra, although, unlike the oxygen vacancy ordering in $SrFeO_{2.75}$, such behaviour is readily revealed by X-ray powder diffraction and has been directly imaged in the electron microscope (6). Several different phases in the $BaFeO_{2.5-3.0}$ region have been observed (7-12) but the only solved structure is the 6H form (cch stacking as in hexagonal barium titanate), determined for $BaFeO_{2.79}$ by powder neutron diffraction (13).

FIGURE 1. Electron diffraction pattern of $SrFeO_{2.78}$ showing the superlattice reflections associated with oxygen vacancy ordering observed in the (100) reciprocal lattice section.

$SrFeO_3$ has an antiferromagnetic ordering transition at 134K, and MacChesney et al noted a steady drop in T_N with increasing non-stoichiometry to 80K for $SrFeO_{2.84}$ (1). Mössbauer spectra of $SrFeO_{3.0}$ showed a single Fe^{4+} line at room temperature and a simple six line spectrum below T_N (2). A spectrum of tetragonal $SrFeO_{2.86}$ showed separate Fe^{3+} and Fe^{4+} lines above and below T_N (2).

No reason for the apparent tetragonal distortion of the primitive unit cell of non-stoichiometric $SrFeO_{3-x}$ was proposed, and certainly no ferroelectric driving force is obvious. Distortions caused primarily by size effects, which might be expected, and which are generally manifested by rotations of octahedra, cause an enlargement of the unit cell, which is almost always observable in powder X-ray diffraction (14). Nevertheless, many similar non-stoichiometric phases have been reported to have primitive cubic or tetragonal cells, for example in the $Ba_{1-y}Sr_yFeO_{3-x}$ system, the $BaFeO_{3-x}-Bi_2O_3$ system and the $SrCoO_{3-x}$ system (15, 16, 17).

PRESENT WORK

We found that we were forced to examine the $SrFeO_{3-x}$ system in more detail because we were unable to reproduce the results of MacChesney et al (1). By reacting $SrFeO_{3-x}$ at 600°C and 15,000 psi, and quenching under pressure, a stoichiometry of only $SrFeO_{2.91}$ was reached. This was cubic as expected from the previous work, however, with a Néel temperature of only 110K, also as expected. However, reaction at >10,000 psi for 7 days

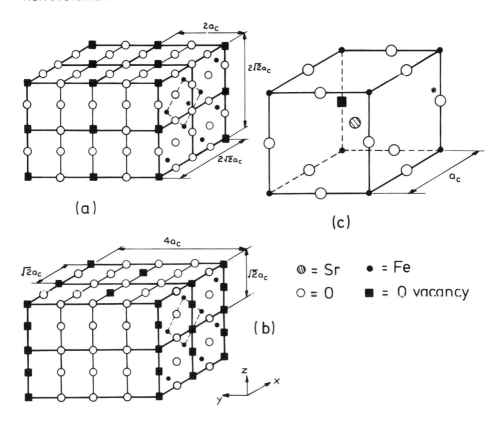

FIGURE 2. (a), a model for the oxygen vacancy ordering in $SrFeO_{2.75}$. The ordering of vacancy strings is related to that in $SrFeO_{2.5}$, (b). The unit cell dimensions are indicated in both cases. The face of a simple perovskite cube is shown by dotted lines in (a) and (b). The cube as found in $SrFeO_{2.75}$ is shown in (c). A particular oxygen atom is denoted by an asterisk to indicate the relative orientations of (a) and (c). The vacancy strings in $SrFeO_{2.50}$ and $SrFeO_{2.75}$ propagate along the (110) direction of the perovskite cube.

at 650°C and 550°C, but without quenching under pressure, gave a higher oxygen content ($SrFeO_{2.93}$) but a tetragonal unit cell. Similar results were obtained at other pressures and the highest stoichiometry reached was $SrFeO_{2.95}$, also tetragonal. All the tetragonal compounds, although non-stoichiometric, had Néel temperatures around 135K, characteristic of stoichiometric $SrFeO_{3.0}$ (1).

On the other hand, a slow cooling of $SrFeO_{2.5}$ in equilibrium

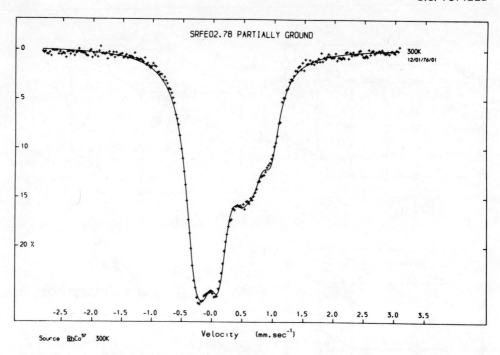

FIGURE 3. The room temperature Mössbauer spectrum of $SrFeO_{2.78}$. The crosses represent the measured spectrum and the solid line is the computed fit assuming two Fe^{3+} and one Fe^{4+} species to be present (Table 1).

TABLE 1. Mössbauer parameters for the iron species determined by fitting to the room temperature spectrum of $SrFeO_{2.78}$.

	$Fe^{3+}(1)$	$Fe^{3+}(2)$	Fe^{4+}
Intensity ratio	1	0.98	2.37
Isomer shift (mm/sec)	0.31	0.48	-0.19
Quadrupole splitting (mm/sec)	1.07	0.20	0.39
Line width (mm/sec)	0.32	0.41	0.41

in oxygen at 1 atmosphere from $1200^{\circ}C$ yielded a cubic product with an oxygen stoichiometry of $SrFeO_{2.78}$. Thus, as well as being unable to prepare stoichiometric $SrFeO_{3.0}$ according to the conditions of MacChesney et al, the apparent symmetry and the magnetic ordering temperature depended on the thermal history as well as the final composition. Such inconsistencies in reactivity indicated possible vacancy ordering, undetected by X-ray powder diffraction and this expectation was strikingly confirmed by electron diffraction. Fig 1 shows the (100) section often observed, with multi-domains causing a doubling of the unit cell along both y and z.

The details of this work have recently been discussed (18). The superlattice appears to be caused by vacancy ordering in a composition, ideally $SrFeO_{2.75}$, which, however, appears cubic in an X-ray powder photograph. The tetragonal phases are in fact two-phase mixtures of this new phase and something close to $SrFeO_{3.0}$; hence the high T_N. The disproportionation temperature, however, is low enough so that a defect cubic phase is easily quenched, and which does have a lower T_N.

At greater than 50% Fe^{3+}, mixtures of $SrFeO_{2.5}$ and $SrFeO_{2.75}$ are formed under normal preparative conditions although electron diffraction indicates occasional intergrowth of the two phases (19). At equilibrium under one atmosphere of oxygen, however, thermogravimetric analysis indicated a single phase region between $SrFeO_{2.5}$ and $SrFeO_{2.75}$ and a first order transition was observed in $SrFeO_{2.50}$ at $847 \pm 2°C$, (18). The structure of $SrFeO_{2.50}$ at room temperature was determined in the course of this work (20), and the body-centred space group Icmm determined in contrast to the primitive space group of $Ca_2Fe_2O_5$ (4).

A tentative model for the vacancy ordering in $SrFeO_{2.75}$ was proposed (Fig 2) which could explain the observed superlattice (18). Although the observation of the new phase rationalises the reactivity anomalies noted above, we gained, however, little definite knowledge about the structure, the iron coordination or the magnetic properties. But knowing that $SrFeO_{2.78 \pm 0.01}$ formed by slow cooling $SrFeO_{2.50}$ in one atmosphere oxygen is apparently almost pure '$SrFeO_{2.75}$' we can now interpret physical measurements such as Mössbauer spectroscopy with more confidence.

More data have recently been collected (21) and Fig 3 shows the spectrum of $SrFeO_{2.78}$ at room temperature. This has been fitted to two quadrupole split Fe^{3+} doublets and a quadrupole split Fe^{4+} doublet (Table 1) and the ratios indicate a composition $SrFeO_{2.77}$, in excellent agreement with that determined by weight changes. By comparison with other systems (eg $CaFeO_{2.5}$ (22)) half the Fe^{3+} would seem to be coordinated by a distorted tetrahedron of oxygens ($\delta=0.31$ mm/sec) and half to be octahedrally coordinated ($\delta=0.48$ mm/sec). The Fe^{4+} site is apparently also distorted. Clearly, therefore, our simple model (Fig 2) is inadequate, although 4- and 6-coordinated Fe^{3+} could arise by a distortion to remove the 5-coordination postulated (18). No compounds with three different types of iron site seem to have been reported in previous Mössbauer investigations of $SrFeO_{3-x}$ or $BaFeO_{3-x}$ (2,3,12). However, the apparent lack at lower temperatures of a phase containing less than 50% Fe^{4+} in $SrFeO_{3-x}$ is in agreement with the conclusions of Jacobson who found that the maximum vacancy concentration in the cubic layers of $BaFeO_{2.79}$ appeared to be $BaFeO_{2.75}$; beyond this, as for $SrFeO_{3-x}$, complete reduction to Fe(III) occurs (13).

The work described indicates that general caution should be observed when interpreting measurements on non-stoichiometric cubic stacked perovskites with apparently primitive cubic or tetragonal unit cells. The many questions still to be answered on the $SrFeO_{3-x}$ system are being pursued using Mössbauer spectroscopy and neutron diffraction.

REFERENCES

1. J. B. MacChesney, R.C. Sherwood and J. F. Potter, J Chem. Phys., 43, 1907 (1965)
2. P. K. Gallagher, J. B. MacChesney and D. N. E. Buchanan, J. Chem. Phys., 41, 2429 (1964)
3. P. K. Gallagher, J. B. MacChesney and D. N. E. Buchanan, J. Chem. Phys., 43, 516 (1965)
4. E. F. Bertaut, P. Blum and A. Sagnières, Acta Cryst 12, 149 (1959)
5. J. B. MacChesney, J. F. Potter, R. C. Sherwood and H. I. Williams, J. Chem. Phys., 43, 3317 (1965)
6. J. M. Hutchison and A. J. Jacobson, to be published
7. S. Mori, J. Amer Ceram Soc. 48, 165 (1965)
8. S. Mori, J. Amer Ceram Soc. 49, 600 (1966)
9. S. Mori, J. Phys. Soc. Japan, 28, 44 (1970)
10. M. Zanne and C. Gleitzer, Bull. Soc. Chim. France, p1567 (1971)
11. E. Lucchini. S. Meriani and D. Minichelli, Acta Cryst. B, 29, 1217 (1973)
12. T. Ichida, J. Solid State Chem, 7, 308 (1973)
13. A. J. Jacobson, Acta Cryst B32, 1087 (1976).
14. J. B. Goodenough and J. M. Longo, in "Landolt-Bornstein, Numerical Data and Functional Data in Science and Technology". New Series (K. H. Hellwege, Ed.), Group III, Vol. 4a, p. 131. Springer-Verlag, New York (1970).
15. M. Zanne and C. Gleitzer, J. Solid State Chem., 6, 163 (1973)
16. M. Zanne, C. Gleitzer and J. Aubry, J. Solid State Chem., 14, 160 (1975)
17. T. Takeda and H. Watanabe, J. Phys. Soc. Japan, 33, 973 (1972)
18. B. C. Tofield, C. Greaves and B. E. F. Fender Mat. Res. Bull, 10, 737 (1975)
19. C. Greaves, D. Phil Thesis, University of Oxford (1973)
20. C. Greaves, A. J. Jacobson, B. C. Tofield and B. E. F. Fender, Acta Cryst., B31, 641 (1975)
21. B. W. Dale and B. C. Tofield, unpublished
22. R. W. Grant, J. Chem. Phys, 51, 1156 (1969)

DISCUSSION

S. MROWEC (Academy of Mining and Metallurgy, Krakow, Poland) Have you made any theoretical calculations on your model of ordering of anion vacancies in strontium ferrate? The free energy of crystals can be dramatically lowered by short and long range ordering. Calculations could help in choosing the most probable configuration of ordered defects

AUTHOR'S REPLY: Such a calculation would be interesting although proper treatment of non-ionic bonding might not be easy. Therefore we have not performed any calculations. By comparison with $SrFeO_{2.5}$, and from the Mössbauer evidence, significant displacements of atoms from regular sites are to be expected and would also have to be determined by the calculation to allow a correct choice between models. Any calculation should probably begin with $SrFeO_{2.5}$ because the atom positions are known and the additional problem of disorder (see Ref. 20) will probably be less severe than in $SrFeO_{2.75}$.

ELECTRON MICROSCOPY OF FERROELECTRIC BISMUTHATE COMPOUNDS

J.S. Anderson, J.L. Hutchison and C.N.R. Rao[*]

Edward Davies Chemical Laboratories, University
College of Wales, Aberystwyth   SY23 1NE   U.K.

[*]Indian Institute of Technology, Kanpur, India

INTRODUCTION

Layered bismuthate compounds of general formula $Bi_2A_{m-2}B_{m-1}O_{3m}$ are ferroelectrics with fairly high Curie temperatures (1,2). They consist of finite perovskite slabs $(A_{n-1}B_nO_{3n-1})^{2-}$ with $\underline{n}$ layers of B-cations (Ti^{4+}, Nb^{5+}, etc.) and $n-1$ layers of A-cations (Bi^{3+} or Ba^{2+}) sandwiched between $\overline{Bi_2O_2}^{2+}$ layers; the square pyramidal BiO_4 groups share edges in such a way that successive perovskite layers are in anti-phase, by a displacement of $\frac{1}{2}[110]$. The Bi atoms in the Bi_2O_2 sheets occupy virtual A-sites of the perovskite lattice, as can be seen in Figure 1; they are displaced from the ideal A-positions by less than 0.05 nm.

Members 1 to 5 of the homologous series $Bi_2O_2(A_{n-1}B_nO_{3n-1})$ were prepared from the constituent oxides (Table 1) and examined in an electron microscope, under optimum conditions for lattice imaging (3).

LATTICE IMAGE CONTRAST

Lattice images were taken with the electron beam parallel to [100] or [010] of the pseudo-tetragonal unit cells. The Bi_2O_2 layers were then projected edge-on and appeared in the images as broad dark lines separated by $\frac{c_0}{2}$. Between these lines fringe patterns corresponding to the positions of the perovskite A-cations were visible. The extent of ordering, domain structures and dislocations could now be studied directly. It was found that all the materials showed perfect ordering of the layers along $\underline{c}$.

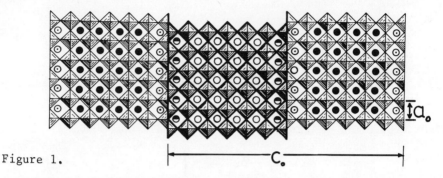

Figure 1.

Schematic projection of $Ba_2Bi_4Ti_5O_{18}$ ($\underline{n}$ = 5) on (010) of pseudo-tetragonal cell. ○, B-cations at y = o; ●, B-cations at y = $\frac{1}{2}$; ◇, A-cations at y = $\frac{1}{2}$; ◆, A-cations at y = o; ⊙, Bi at y = $\frac{1}{2}$ in Bi_2O_2 layers; ◉, Bi at y = o in Bi_2O_2 layers.

DISLOCATIONS

Dislocations were a common feature in these materials and where they were parallel to the electron beam, the image contrast in the core regions could be interpreted directly. Figure 2 shows a dislocation in $Ba_2Bi_4Ti_5O_{18}$ ($\underline{n}$ = 5) which has been analyzed in detail. In the upper part of the micrograph two extra perovskite slabs are present, corresponding to the insertion of a complete unit cell width. These extra layers are accommodated partly by modular side-stepping of the surrounding Bi_2O_2 layers and partly by a widespread compression of the perovskite lattice, over about 30 nm. This compression is about 4% and would involve buckling of the rows of octahedra. Below the dislocation core there is a compensating relaxation of the perovskite lattice, a dilation of about 1% representing stretching of cation - oxygen - cation strings along $\underline{c}$.

Where the Bi_2O_2 layers sidestep progressively, near the core itself, the perovskite slabs become correspondingly narrower and

TABLE 1

n	Compound	Lattice Parameters			Ferroelectric T_c°K
		a	b	c (nm)	
1	Bi_2WO_6	0.5458	0.5438	1.5434	1223
2	$BaBi_2Nb_2O_9$	0.5533	0.5333	2.5550	473
3	$Bi_4Ti_3O_{12}$	0.5448	0.5411	3.283	498
4	$BaBi_4Ti_4O_{15}$	0.5461	0.5461	4.185	668
5	$Ba_2Bi_4Ti_5O_{18}$	0.5514	0.5526	5.0370	590

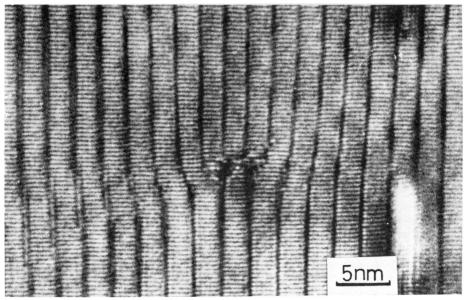

Figure 2.

Lattice Image of $Ba_2Bi_4Ti_5O_{18}$ ($\underline{n}$ = 5) containing a dislocation.

in this way, successive members of the homologous series $\underline{n}$ = 5 → $\underline{n}$ = 1 are generated. These imply measurable composition changes in this region and it is possible that local chemical fluctuations will be involved in the formation of these dislocations.

It has been pointed out by us that these dislocations may be regarded as dislocations of the Bi_2O_2 superstructure imposed on the perovskite sub-lattice and in order to preserve the antiphase relationships across a dislocation core, two extra Bi_2O_2 layers must be incorporated (4). Where only one such layer is present, disorder is also introduced into the perovskite structure.

DIFFUSION PATHWAYS

Because of the large layer spacings, any dislocation movements, or ordering between layers will involve rearrangements of Bi_2O_2 layers over considerable distances. Such rearrangements preserve (or result in) local order and layer periodicity to a high degree and a likely mechanism is here postulated.

It was noted that the Bi atoms occupy virtual A-sites at the edges of the perovskite slabs. In the compounds with n > 2

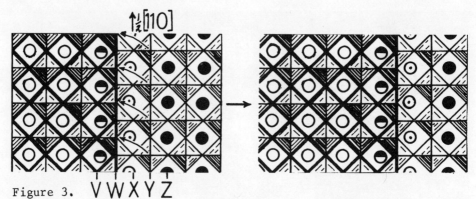

Figure 3.
Proposed atom shifts involved in sideways migration of Bi_2O_2 layers.

they also occupy at least half of the A-sites within the slabs, the remainder being Ba atoms. Two minor displacements of the Bi atoms can now be visualized as in Figure 3:

a) relaxation from a virtual A-site (row V) to the appropriate 'real' A-site at the perovskite slab edge,

b) shift of an available Bi atom from a perovskite A-site to a 'virtual' A-site position, (row Z),

process b) is the reverse of a) but if they occur on opposite sides of a Bi_2O_2 layer, the way is then open for further rearrangement as follows:

c) the Bi atoms and oxygen atoms forming row X are displaced by $\frac{1}{2}$ [110].

d) B-cations (in row Y) migrate to the newly created octahedral positions in row W.

The net effect is to move the Bi_2O_2 layer sideways by one octahedra. It is assumed that these shifts are cooperative as in the mechanism proposed by Galy and Carpy to account for defect diffusion in some closely related structures (5). It was shown on geometrical grounds that B-cation diffusion in perovskites is unfavourable, but this is probably off-set by the simultaneous displacement ((c) above) (6). One immediate implication of this mechanism is that where perovskite A-sites are wholly tenanted by Ba atoms as in $n = 1$ and $n = 2$ compounds, no Bi atoms are available for transfer into Bi_2O_2 configuration, and dislocations of the type described will be chemically pinned in position; long-range diffusion of Bi atoms across the perovskite layers would be required and it is difficult to see how this could take place at ambient temperatures. No information is available on Ba/Bi interdiffusion in such materials.

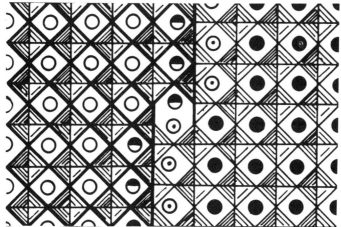

Figure 4.
Creation of a unit step in Bi_2O_2 layer as observed near dislocation cores.

Where Bi_2O_2 layers are free to move, ordering of layers and movement of dislocations, involving relatively simple displacements of atoms, can be visualized. Partial side-stepping of layers, as observed near the dislocation cores, can also be accommodated by the mechanism described. This is shown schematically in Figure 4. The large tunnel created contains two equivalent sets of Bi sites which provide an excellent diffusion path for A-cations, thereby facilitating further dislocation movement.

REFERENCES

1. Schmid, H. in ''Magnetoelectric Interaction Phenomena in Crystals'' eds. A.J. Freeman and H. Schmid, Gordon and Breach, N.Y., (1975).

2. Subbarao, E.C. in ''Solid State Chemistry'' ed. C.N.R. Rao, Marcel Decker, N.Y., (1974).

3. Hutchison, J.L. and Jacobson, A.J. Acta Cryst. B31 1442 (1975).

4. Hutchison, J.L., Anderson, J.S. and Rao, C.N.R. Nature (London) 255 541 (1975).

5. Galy, J. and Carpy, A. Phil. Mag. 29 1207 (1974).

6. Stone, F.S. and Tilley, R.J.D. in ''Reactivity of Solids 1972'', ed. J.S. Anderson, M.W. Roberts and F.S. Stone Chapman and Hall, London (1972).

DISCUSSION

S. MROWEC (Academy of Mining and Metallurgy, Krakow, Poland) Your diffusion path model suggests a highly anisotropic diffusion rate in the bismuthate structure. Did you perform any experiments to examine this?

AUTHORS' REPLY: As far as I know, no diffusion measurements have been made on these compounds, and we do not have the facilities to do the experiments. They would certainly show whether or not the proposed mechanism is valid, and until such supporting evidence is obtained, mechanisms of this type must be purely speculative.

K.J. de VRIES (University of Technology, Twente, Netherlands) You showed that by means of $Bi_2O_2^{2+}$ layers, perovskite materials can be arranged into "plates". Similar observations have been made in the $SrTiO_3.SrO = 1:1$ and $1:2$ systems. In $SrTiO_3$ substituted with $LaO_{1.5}$, SrO can be dissolved in amounts of at least 6 at. % while keeping the material monophasic. Xray diffraction shows that this is accompanied by an increase in unit cell volume (1). It is possible that your suggestion of phase boundary movement could lead to the formation of local structures capable of incorporating the extra SrO? If so, this could also be an annihilation mechanism for specific point defects.
1) J. Bowma, K.J. de Vries, A.J. Burgraaf. Phys.Stat.Sol.(a) **35**, 281, (1976)

AUTHORS' REPLY: To what extent is your material monophasic? X-ray diffraction will give average cell dimensions, but there may be small fluctuations in structure which are not detected. Electron diffraction from very small areas (500 Å diameter) may reveal them; planar boundaries of the type described here, as well as in Galy's paper (these proceedings) would give rise to extra diffraction effects, if present in sufficient numbers. Lattice images may reveal single planar faults. If such boundaries are present in your material, one would expect evidence from electron diffraction and lattice images, if these can be obtained.

SOME OBSERVATIONS ON THE CHEMICAL REACTIVITY OF NICKEL POWDER
MECHANICALLY TREATED BY GRINDING

J. Carrión, J.M. Criado, E.J. Herrera,* and C. Torres

Department of Inorganic Chemistry
and *Department of Materials
E.T.S.I.I., University of Seville, Spain

INTRODUCTION

Lattice imperfections have long been considered to be possible sites of catalytic activity in solids. Considerable research work has been undertaken with the aim of demonstrating such a relationship, but discrepant results have been reported in this field (1 - 4) Nevertheless, in some recent publications (5 - 7) a good correlation has been found between activity of deformed metal sheets and their lattice-defect content, mainly dislocations, as shown by structural investigations. Studies on metal powders are not so numerous. This is mostly due to the difficulty in assessing quantitatively the defect structure of powder catalysts. Also in this sphere, different explanations have been suggested to account for changes in powder activity originated by thermal and mechanical treatment (8 - 9).

In the present work the catalytic activity of a mechanically ground nickel powder for formic acid dehydrogenation has been studied with special reference to the influence of lattice imperfections on activity level. In addition, samples of nickel formates were prepared by passing formic-acid vapour over milled and unmilled powder. The thermal decomposition of these compounds was investigated to determine their possible role as intermediates.

EXPERIMENTAL PROCEDURE

150 g. of a high-purity Ni powder (designated Ni-NG), particle size smaller than 40 microns, was ground for 30 min. in a disk oscillating mill. Samples of milled and unmilled powders were subjected

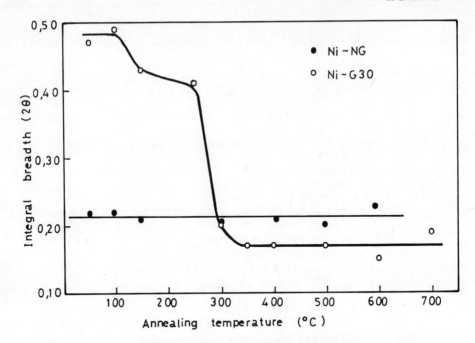

Figure 1. Integral breadth of the (311) X-ray diffraction line versus annealing temperature.

to 1 h. annealing treatments at different temperatures under vacuum protection. All powders were studied, before and after annealing, for their lattice disorder, as indicated by X-ray line broadening. Peak broadening was measured by the integral breadth method, after resolution of the $K\alpha_1$ component by Rachinger's method (10).

TABLE I

Kinetics parameters of catalysts and formates

Sample	α interval	E (Kcal/mol)	A (min^{-1})	A (molec.cm^{-2}seg^{-1})
Ni-NG		8,2		$5,5.10^{22}$
Ni-G30		20,7		$2,4.10^{27}$
NG/HCOO$^-$	0,04-0,90	10,5	$4,9.10^3$	$1,1.10^{24}$
G30/HCOO$^-$	0,1-0,82	30,8	$6,4.10^{22}$	$1,5.10^{33}$

A conventional continuous flow reactor was employed to study the catalytic activity of Ni powders for the dehydrogenation of formic acid in vapour phase, using helium as carrier. The activity was measured in the 300-400°C range; at lower temperatures, blank reaction rate was approximately of the same magnitude as the total one. Decomposition products - CO_2 and H_2 - were analysed by gas chromatography. Working conditions were chosen so that reaction obeyed zero-order kinetics.

Nickel formates were prepared by flowing formic-acid vapour through each type of powder at a temperature of 300°C, without stopping the acid current. A similar method was employed in a previous work[11-12], where the catalytic activity of 3d-metal oxides was studied.

Thermal decomposition was studied by TGA, under nitrogen protection(150 torr). The course of decomposition reaction is not affected by inert-gas partial pressure[13]. Thermal decomposition products are CO_2, H_2 and Ni (12). If the transformation follows the Polanyi-Wigner mechanism[14], the reaction is zero order, just as is the catalytic decomposition. In this case, kinetic analysis of TGA data is based on the equation (15):

$$\ln \alpha - 2 \ln T = \ln \frac{AR}{E\beta} - \frac{E}{RT}, \qquad [1]$$

where α is the decomposed solid fraction, β is the heating rate and the other terms have their usual meaning.

RESULTS AND DISCUSSION

Figure 1 shows the changes in integral breadth followed by the (311) reflection of both powders, after various 1 h. annealing treatments. Peak broadening of the as-received powder is not affected by annealing in the temperature range studied. Line broadening of milled powder, Ni-G30, markedly decreases between 200 and 320°C; at higher temperature it remains stable. B.E.T. specific surface of both types of powders was the same before and after annealing at 400°C(0,40 m^2/g for Ni-NG and 0,20 m^2/g for Ni-G30, respectively). Data in Fig. 1 indicate that though the degree of crystallinity of untreated Ni powder is initially higher than the milled one, annealing-out of crystal defects in the interval in which the catalytic activity was measured makes the previously milled powder a more perfect material than the original one. This fact has been often not taken into account in the literature. On the other hand, the activation energy for formic-acid catalytic dehydrogenation is lower on Ni-NG than on Ni-G30, as contained in table I. The higher activation energy of the ground-and-annealed sample has been interpreted as

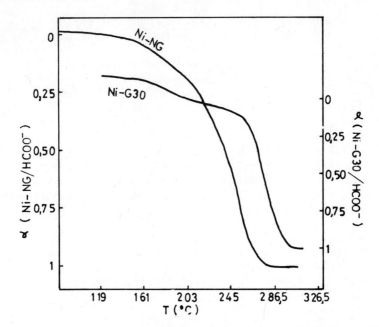

Figure 2.   Thermogravimetric diagram of the decomposition of nickel formates. Atmosphere N_2. Heating rate $12°$/min.

due to its higher degree of crystallinity in the 300-400°C interval, as shown by X-ray studies (Fig. 1).

TGA diagrams for the decomposition of formates obtained on milled and unmilled powders as substrates, are presented in Fig. 2. From this plot were calculated the values of $\ln \alpha - 2 \ln T$, which are, in turn, plotted versus $1/T$ in Fig. 3, according to equation [1]. The good linear correlation found between these parameters is in agreement with the Polanyi-Wigner mechanism for salt decomposition. The activation energies for the decomposition of Ni formates are included in table I. These values of activation energies compared to the ones pertaining to formic-acid dehydrogenation seem to show that the controlling step in the catalytic reaction rate is the decomposition of the intermediate formate. Moreover, these results suggest that when a solid is prepared by a gas-solid reaction, it keeps a memory of the defect structure of the solid matrix.

Concerning the role of lattice defects in catalysis, the previous results point out that many unsuccessful attempts (3-4) to establish a relationship between surface emergent defects in solids and their activity or activation energy for a certain reaction could be ascribed to a lack of structural studies of samples after heating at the same temperature at which the reaction was measured. It may

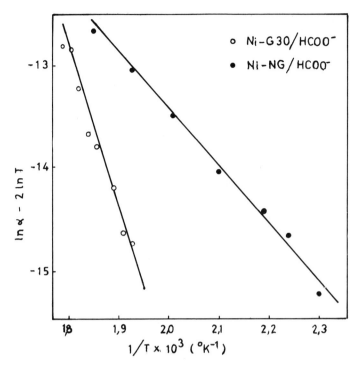

Figure 3. Zero order kinetic analysis of TG decomposition of Ni formates.

well happen, as demonstrated in Fig. 1, that annealing-out of crystal imperfections in the temperature range studied changes the defect solid into a more perfect material than the comparison untreated-specimen.

REFERENCES

(1) J.M. Thomas, Adv. Catal. 19, 293 (1969)
(2) K.B. Keating, A.G.Rozner and J.L.Youngblood,J.Catal.4, 608(1965)
(3) E.M.A. Willhoft, Chem. Commun. 3, 146 (1968).
(4) R. Schmidt, Dissertation, Physikalisch-Chemisches Institut der Universität München (1971).
(5) S. Kishimoto, J.Phys.Chem. 77, 1719 (1973).
(6) J.M.Criado,E.J.Herrera and J.M.Trillo, Proc 5th Int. Congr. on Catal.,North Holland Pub.Co., Amsterdam (1073) p. 541.
(7) J.Carrión,J.M.Criado,F.González,E.J.Herrera and J.M. Trillo, Electrón. Fisc. Apli. 17, 221 (1974).
(8) E.M.Hofer et al., Trans. Faraday Soc.60, 1457 (1964).
(9) R. Schrader et al., Acta Chim.Acad.Sci.Hung.55, 39 (1968).
(10) W.A.Rachinger,J.Sci Instr. 25, 254 (1948).
(11) J.M.Trillo et al., Catal. Rev., 7,51 (1972).

(12) J.M. Criado et al., Rev. Chim. Min., 1041 (1970).
(13) J.M.Criado et al., Thermochim. Acta 12, 337 (1975)
(14) M. Polanyi,E.Wigner,Trans.Farad.Soc.40, 488(1944).
(15) A.W.Coats and J.P. Redfern,Nature 201,68 (1964).

DISCUSSION

A.K. GALWEY (Queens University of Belfast, Northern Ireland, U.K.) Did you measure the surface area of the catalyst following reaction with formic acid? Does the catalytic rate process involving salt intermediate production cause appreciable surface roughening?

Can you give an explanation of the larges value of A in Table 1 and also the compensation effect apparent in this data?

AUTHORS' REPLY: The specific surface of nickel powders was measured before and after annealing at temperatures in which the reaction took place. No change in specific surface was observed. The surface area of the catalyst after the reaction with formic acid was not determined; but no appreciable change in surface-roughening appearance was detected.

Concerning your last question, as is well-known, the pre-exponential term of Arrhenius equation is a complex parameter. The authors have no explanation to justify the highest value of A in Table 1.

M.E. BROWN (Rhodes University, Grahamstown,South Africa) What degree of conversion of nickel powder into formate was achieved?

AUTHORS' REPLY: Formates were prepared as thin layers on nickel-powder substrates. About 2% nickel was transformed into formate.

DEFECT CHEMISTRY OF La-DOPED BARIUM TITANATE

Detlev Hennings

Philips GmbH
Forschungslaboratorium Aachen
Aachen, West Germany

INTRODUCTION

By incorporating small amounts of lanthana in bariumtitanate one obtains in reducing conditions deep black semiconducting perovskite phases, while in oxidizing conditions bright yellow, electrically insulating perovskites are observed. Both can be converted to one another by varying the temperature and the oxygen pressure.

This paper deals with the redox reactions occuring in the perovskite lattice of $(Ba,La)TiO_3$, (BLT), which are governed by equilibria of lattice defects.

The present investigation of defect equilibria in BLT is concentrated on a restricted number of principal lattice defects, the existence of which has been confirmed in other related perovskite materials (1), (2), (3), (4).

In lanthana doped $SrTiO_3$ (1) and $PbTiO_3$ (2) intensity measurements of X-ray diffraction spectra revealed that the La^{3+}-ions enter exclusively the A-sites of the perovskite lattice. In oxydizing atmosphere the lanthanum donors, $La^{\cdot}$, are in these materials compensated by doubly ionized A-site vacancies, V_A''. We can assume that the same holds for BLT.

The yellow oxydized perovskite phase then has the general composition:

$$Ba_{1-1.5x}La_x \square_{0.5x}TiO_3 .$$

The composition of the black reduced phase on the other hand corresponds to the formula
$Ba_{1-y}La_y^{3+}(Ti_{1-y}^{4+}Ti_y^{3+})O_3$, as shown by Eror and Smyth (3). Because of the high electrical conductivity of the reduced material only a rather weak binding between the trivalent titanium and the conduction electrons is assumed so that the Ti^{3+} may be replaced by an electron, e.

In addition, at very low oxygen pressures the formation of single ionized oxygen vacancies $V_O^{\cdot}$ has to be considered (4).

As already shown by the conductivity measurements of Daniels (5) the defect chemistry of BLT is characterized by three ranges which predominate at different oxygen pressures:

(1) $[La_A^{\cdot}] \approx 2[V_A''] \quad ; \quad p_{O_2} \approx 1$ atm

(2) $[La_A^{\cdot}] \approx n \quad ; \quad p_{O_2} < 1$ atm

(3) $[V_O^{\cdot}] \approx n \quad ; \quad p_{O_2} \ll 1$ atm

For each of these lattice defects partial equilibria can be separately determined at certain areas of the oxygen pressure. Thermogravimetric experiments should come to the same result. The above mentioned defects can be combined to a relatively simple equation of neutrality:

$$n + 2[V_A''] = [La_A^{\cdot}] + [V_O^{\cdot}]$$

Other defects as like V_O, $V_O^{\cdot\cdot}$, V_A', V_B'''', etc. are neglected as minorities.

EXPERIMENTS

The experiments were performed with BLT powders, prepared from reagent grade TiO_2, $BaCO_3$ and La_2O_3. The mixed oxides were 15 h calcined at 1150°C and sintered

60h at 1250°C in oxygen. The experiments were carried out with 15 g batches of crushed BLT material in platinum crucibles using a thermobalance. The weight changes were **registered** with an error of $\pm\ 10^{-4}$ g. The gas atmosphere was calibrated by a circulating mixture of CO and CO_2 which was controlled by the emf. of a CaO doped ZrO_2 cell at 650°C. The starting composition of the powders was $Ba_{1-1.5x}La_xTiO_3$. The monophasic composition of the samples was tested by means of X-ray diffraction, light and electron microscopic inspection.

RESULTS

In pure $BaTiO_3$ the weight change measured in dependence on the partial pressure of oxygen directly corresponds to the total number of oxygen vacancies.

$$BaTiO_3 + \delta\ CO \rightleftharpoons BaTiO_{3-\delta} + \delta\ CO_2$$

$$[V_O]_{total} \approx \delta$$

The measured slopes $(\partial \ln [V_O]_{total} / \partial \ln p_{O_2})_T \approx -1/4$ confirm the prevalence of single ionized oxygen vacancies, $V_O^{\cdot}$.

During the reduction of BLT, A-site vacancies are annihilated but nearly no oxygen vacancies are formed. The following reaction is assumed:

$$V_A'' + Ti_B + 3\ O_O \rightleftharpoons 1/2\ O_2 + TiO_2 + 2e$$

The weight loss in moles of oxygen per mol of perovskite corresponds to the concentration of the trivalent titanium or the conduction electrons respectively:

$$\Delta w_O \approx 1/2\ [Ti^{3+}] \approx 1/2\ n$$

Using the measured concentrations of $V_O^{\cdot}$ and n the corresponding mass action constants were calculated for the formation of the single defects. Fig. 1 shows the measured and calculated concentration of conduction electrons in dependence on the oxygen pressure for various La concentrations in $BaTiO_3$ at 1200°C. Fig. 2 shows the concentration of the single defects

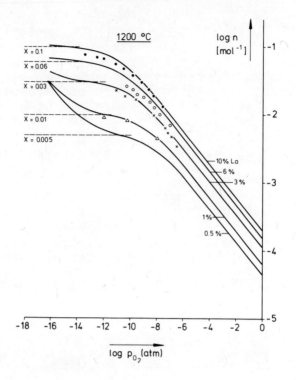

fig. 1: measured and calculated concentration of conduction electrons in BLT at 1200°C

$[V_O^{\cdot\cdot}]$, $[V_A']$, $[La_A^{\cdot}]$ and $[Ti']$ = n calculated in dependence on the oxygen pressure for $BaTiO_3$ containing 1 % La at 1200°C.

The electron concentration calculated on the base of thermogravimetrically determined mass action constants is in good agreement with the concentration directly determined by measurement of the electrical conductivity (5). The results obtained show that the defect chemistry of lanthanum doped $BaTiO_3$ can be sufficiently described over a wide range of oxygen pressures using mass action equilibria of a small number of point defects in the perovskite lattice.

DEFECT CHEMISTRY OF La-DOPED BaTiO$_3$

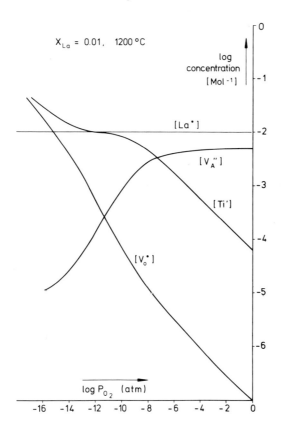

fig. 2: concentration of point defects in BLT containing 1 % La at 1200°C

REFERENCES

(1) T.Y. Tien and F.A. Hummel
    Trans. Brit. Ceram. Soc. <u>66</u>, 233 (1969)
(2) D. Hennings and G. Rosenstein
    Mat. Res. Bull. <u>7</u>, 1505-14 (1972)
(3) N.G. Eror and D.M. Smyth
    The chemistry of Extended Defects in Non-Metallic Solids p. 62-83, ed. Le Roy Eyring, M.O'Keefe, North Holland Publ. Comp. Amsterdam 1970
(4) R.J. Panlener and R.N. Blumenthal
    J. Amer. Ceram. Soc. <u>54</u>, 610 (1971)
(5) J. Daniels - to be published

THE KINETIC PROCESSES OF EQUILIBRIUM RESTORATION IN La-DOPED BaTiO$_3$

R. Wernicke

Philips GmbH
Forschungslaboratorium Aachen
5100 Aachen, West Germany

INTRODUCTION

As shown by the work of Daniels (1) and Hennings (2) the defect-chemistry of La-doped BaTiO$_3$ (BLT) at high temperatures is well understood. On the other hand the explanation of the electrical properties at room temperatures governed by the defect-chemistry is still lacking. The explanation of these properties is of great importance for PTC-resistors and Intergranular-Capacitors. As the diffusion of defects occuring during the cooling cycle after sintering plays an important role (3) investigations about the kinetic processes of equilibrium restoration seem to be useful.

EXPERIMENTAL PROCEDURE

As shown in (4) and (5), dynamic conductivity measurement is a suitable tool for investigating the diffusion of defects. For such measurements the time dependence of the electrical conductivity has to be determined after a sudden change of oxygen partial pressure. From the time constant of equilibrium restoration τ, the diffusion coefficient D can be calculated (6):

$$D = k_1 \cdot \frac{l_D^2}{\tau} \qquad k_1: \text{constant} \qquad [1]$$

The value of l_D is given by the length which the defects have to diffuse to establish equilibrium. For thin plates this diffusion-length l_D is normally determined by the thickness h. With respect to the sample dimensions (15x5x0.5 mm), as a first assumption we set l_D=h. This assumption is only justified if the process is determined by volume diffusion starting from the surface of the sample. Otherwise (e.g. for reaction controlled equilibrium restoration or diffusion of defects along the grain-boundaries) a quadratic dependence of the calculated diffusion coefficient on the sample thickness h should be observed.

RESULTS

Undoped $BaTiO_3$-Ceramic

The results of undoped $BaTiO_3$ samples with different thickness h and different grain-diameters ϕ are given in fig. 1. In the range of oxygen partial pressure (0.1-1.0 atm) at which the experiments have been carried out the electroneutrality is given by $2[\ddot{V}_o] \equiv [A']$ (3), (7). Therefore these diffusion coefficients have to be attributed to the oxygen vacancies $\ddot{V}_o$. The anomalous high values of D found for these defects can be explained qualitatively by the arrangement of oxygen ions in the Perovskite-lattice.

Using eq. [1] and l_D=h we find for all samples - independent of thickness and grain-diameter - the same diffusion coefficients. From these results one can conclude (3)
a) Equilibrium restoration is a diffusion controlled process
b) The diffusion-length l_D determining the time constant of equilibrium restoration is given by the dimensions of the sample. (This holds only for dense materials (3))
c) The equilibrium restoration is governed by the diffusion of oxygen vacancies.

Comparing the diffusion coefficient of oxygen vacancies with tracer-measurements made by Doskocil (8) the experiments show for T ≤ 900°C the same results as found by Paladino (9), (10) for $SrTiO_3$ (see fig. 2).

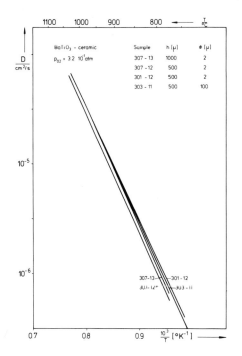
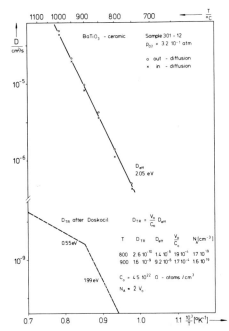

fig.1: Diffusion coefficients of $\ddot{V}_O$ for undoped $BaTiO_3$-ceramic

fig.2: Diffusion coefficients of $\ddot{V}_O$ compared with tracer-measurements (8)

La-doped $BaTiO_3$ Ceramic (BLT)

A quite different behaviour has been found for BLT-ceramic. Using again l_D=h the diffusion coefficients calculated from the time constant of equilibrium restoration are presented in fig. 3a for two samples of different thickness h. This calculation procedure reveales an unexpected change of the diffusion coefficient by the sample thickness h arising from a wrong assumption of the diffusion-length l_D. Due to the formation of Ba-vacancies at the grain-boundaries (as shown below) the diffusion-length in the case of BLT is determined by the grain-diameter ϕ. The correct diffusion coefficients calculated with l_D=ϕ are presented in fig. 3b showing for both samples the same value.

As demonstrated by Daniels (1) the electroneutrality for high oxygen partial pressures is governed by

Ba-vacancies V_{Ba}''

$$[\dot{La}] = 2\,[V_{Ba}''] + n \qquad [2]$$

Therefore the diffusion coefficients in fig. 3b have to be attributed to the Ba-vacancies. During equilibrium restoration these vacancies are formed at the grain-boundaries and diffuse from there into the interior of the grains until the new equilibrium is attained.

The assumption that the diffusion-length and therefore also the time constant of equilibrium restoration is determined by the grain-diameter has been further confirmed by oxidation experiments. These experiments reveal that with fine-grained materials ($\phi \approx 2\mu$) equilibrium can be established at 1100°C within a few hours, while on the other hand with coarse-grained samples ($\phi \approx 100\mu$) no equilibrium could be obtained even after a week. From the measured diffusion coefficients in fig. 3b a time constant of $\tau \approx \phi^2/D =$

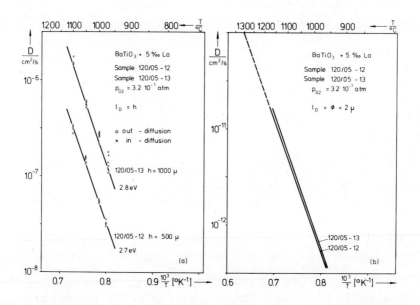

fig. 3: Diffusion coefficient of Ba-vacancies for BLT
a) calculated with $l_D = h$ (wrong values)
b) calculated with $l_D = \phi$ (correct values)

2 hours can be estimated for $\phi = 2\mu$ while for a grain-diameter of $\phi = 100\mu$ the time constant amounts $\tau = 210$ days.

The change of equilibrium in BLT is necessarily involved with the formation of Ba-vacancies. The Ba-ions set free by this process must find new lattice sites. As the diffusion of Ti-ions in the Perovskite-lattice seems to be strongly hampered (11) the reaction of these "free" Ba-ions with second phases (for example $BaTi_2O_5$) is the most likely mechanism of V_{Ba}-formation. As in BLT-ceramic often small amounts of second phases can be observed at the grain-boundaries the formation of V_{Ba} by reactions with these second phases is in good agreement with the results of diffusion experiments. A possible reaction model may be suggested in the following manner:

$$\begin{array}{rcl}
Ba\,(lattice) & \rightleftarrows & \boxtimes_{Ba} + Ba \\
BaTi_2O_5 & + & O + Ba \rightleftarrows 2\,BaTiO_3 \quad [3] \\
O\,(lattice) & \rightleftarrows & O + \boxtimes_O
\end{array}$$

Though the formation of Ba-vacancies is not completely understood up to now it is obvious that the composition and the amount of second phases plays an important role on the properties of these materials. Because of the special formation mechanism the diffusion-length and therefore also the time constant of equilibrium restoration is determined by the grain-diameter.

This influence of microstructure results in severe consequences upon the electrical properties at room temperature. Especially the formation of poorly conducting grain-boundary layers observed in BLT-ceramic can be understood much better in the picture of the kinetic processes (3). These grain-boundary layers playing an important role for PTC-resistors and Intergranular-Capacitors will be formed during the cooling cycle after sintering by diffusion of V_{Ba} from the grain-boundaries into the interior of the grains. With decreasing temperature the diffusion fronts will be frozen in forming a layer in which the lanthanum will be completely compensated by vacancies ($[L\dot{a}] \approx 2[V_{Ba}'']$) while in the interior of the grains only a partial compensation by vacancies takes place ($[L\dot{a}] = 2[V_{Ba}''] + n$). With normal cooling rates ($10°C/min$) a thickness of $0.5 - 1.5\mu$ has been estimated for these layers from the measured diffusion coefficients. It seems likely that these considerations do not only hold for La-dopings but can also be transferred to all other donor-dopes as Sb, Nb, Gd etc.

REFERENCES

1. J. Daniels, K.H. Härdtl, to be published
2. D. Hennings 8th Symp. Reactivity of Solids Gothenburg 1976
3. R. Wernicke Thesis RWTH Aachen (1975) to be published
4. L.C. Walters, R.E. Grace J.Phys. Chem. Solids 28, 245 (1967)
5. K. Kitazawa, R.L. Coble, J. Amer. Ceram. Soc. 57, 250 (1964)
6. W. Jost "Diffusion in Solids, Liquids, Gases" Academic Press Inc. New York 1960
7. S.A. Long, R.N. Blumenthal J.Amer. Ceram. Soc. 54, 515; 54, 577, (1971)
8. J. Doskocil, 4.Konf. Keram. Electron III, 1 (1971)
9. A.E. Paladino, J.Amer Ceram Soc. 48 476 (1965)
10. A.E. Paladino, L.G. Rubin, J.S. Waugh, J. Phys. Chem. Solids 26, 391 (1965)
11. F.S. Stone, R.J.D. Tilley in "Reactivity of Solids" S.262, Chapman and Hall Ltd., 1972

HIGH TEMPERATURE REDOX-REACTIONS OF Co:α-Al_2O_3

A. Schweiger and Hs.H. Günthard

Laboratory for Physical Chemistry, ETH

Universitätstr. 22, CH-8006 Zurich, Switzerland

INTRODUCTION

Cobalt doped sapphire is shown to undergo reversible redox reactions by treatment of the system in hydrogen and oxygen atmospheres above 1100°C. At temperatures above 1550°C irreversible formation of new phases of the spinel type destroy the reversibility of the redox process. The reduction process is found to follow the reaction

$$Co^{+3}:\alpha\text{-}Al_2O_3 + \tfrac{1}{2}H_2(g) \rightarrow (Co^{+2}, H^+):\alpha\text{-}Al_2O_3 , \qquad 1$$

whereas the oxidation may be described by

$$2(Co^{+2}, H^+):\alpha\text{-}Al_2O_3 + \tfrac{1}{2}O_2(g) \rightarrow 2Co^{+2}\alpha\text{-}Al_2O_3 + H_2O(g) \qquad 2$$

RESULTS AND DISCUSSION

The incorporation and extraction of hydrogen according to reactions 1 and 2 may be proved by observation of infrared absorption bands associated with the system $(O\text{-}H\text{-}O)^{-3}$, and by ESR and ENDOR spectra.

Fig. 1 shows polarised infrared absorption spectra of Co doped sapphire after H_2 treatment at 1300°C. O_2 treatment at 1300°C removes the absorption bands in the 3400 - 1900 cm^{-1} range (1). The process may be repeated and reproducibly leads to the appearance and disappearance of the bands. By using deuterium isotope shifts it may be shown directly that they can be assigned to $\nu(OH)$ modes of the fragment OHO^{-3}.

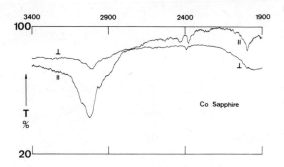

Fig. 1 Infrared spectra of $(Co^{+2}, H^+):\alpha-Al_2O_3$

Until now at least three types of ESR spectra have been reported (2,3). Fig. 2 shows the proton ENDOR spectrum of defect $Co^{+2}:\alpha-Al_2O_3$ first observed by Zverev et al. (2) by means of its ESR spectrum (hereafter called Zb spectrum). This ESR spectrum features axial symmetry and has been associated with Co^{+2} substituted into the interstitial site b (4) with site symmetry C_i (4). There are no features observed so far in the ESR spectrum which would indicate any perturbation of the axial symmetry by the charge compensator, which almost necessarily will destroy local axial symmetry. The ENDOR spectrum clearly demonstrates magnetic hyperfine interaction between the Co^{+2} electron spin and the proton. Under rotation of the crystal around its c axis the ENDOR spectrum exhibits C_3 symmetry and consists of two sets of three spectra which are phase shifted by approximately 12° with respect to each other. It may be interpreted by the defect structure shown in Fig. 3.and should be commented upon as follows:

(i) the protons are substituted in site f between two oxygen ions, one of which belongs to the first coordination octahedron of Co^{+2} (c).
(ii) the two sets of spectra originate from protons located near oxygen ions of the triangles 1,2 and 1',2' respectively which are rotated w.r.t. to each other by 8°.

For all 3 types of ESR spectra of $Co^{+2}:\alpha-Al_2O_3$ it has been possible to show that H^+ may play the role of a charge compensator In the case of the Zc and Zb spectra this has been done by means of ENDOR spectra and in the case of the spectrum reported in (3) by the symmetry of the ESR spectrum. The reduction process with $H_2(g)$ appears to affect all defects at the same time, indicating that it is thermodynamically controlled.

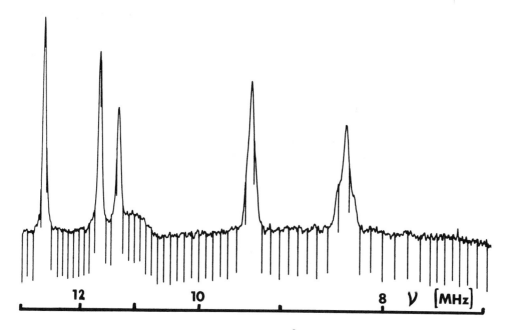

Fig. 2 Proton ENDOR of $(Co^{+2}(b), H_f^+):\alpha-Al_2O_3$

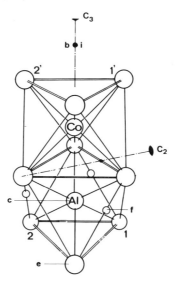

Fig. 3 $\alpha-Al_2O_3$-structure. Positions of the protons are marked by f.

REFERENCES

1. H. Blum, R. Frey, Hs.H. Günthard, Tae-Kyu Ha, Chem. Phys. $\underline{2}$, 262 (1973)
2. G.M. Zverev and A.M. Prokhorov, JETP $\underline{36}$, 647 (1959) Sov. Phys. JETP $\underline{9}$, 451 (1969), Sov.Phys JETP $\underline{12}$, 41 (1961)
3. B. Gächter, H. Blum and Hs.H. Günthard, Chem.Phys $\underline{1}$, 45 (1973)
4. International Tables of X-ray Crystallography Vol $\overline{1}$ (The Kynoch Press, Birmingham 1952), Wykoff Notation

STRUCTURAL EVOLUTION IN THE FeF_3-WO_3 SYSTEM

J.M. Dance, A. Tressaud, J. Portier and
P. Hagenmuller

Laboratoire de Chimie du Solide du C.N.R.S.
Université de Bordeaux I, 351, cours de la
Libération, 33405, Talence, France

INTRODUCTION

The investigation of the oxyfluorides is of great interest since the fluorine for oxygen substitution leads very often to dramatic changes of the physical properties of the resulting materials. This is especially true for those properties characterizing short range order in the structures involved. This type of substitution has been studied recently in various structural types : rutile, perovskite, spinel, garnet, fluorite, etc ... The sharp evolution of physical properties which parallels the increase in the fluorine substitution rate allows a better understanding of the mechanism of the solid state reactions between fluorides and isostructural oxides.

From this point of view the FeF_3-WO_3 system, where both components are derived from the relatively simple ReO_3 structure, is significant: FeF_3 has a weak ferromagnetism below 90°C while WO_3 is diamagnetic and ferroelectric below 223 K.

MATERIALS PREPARATION

Mixtures of variable compositions between FeF_3 and WO_3 were fired, then annealed during 15 h. cycles at 600°C in gold sealed tubes. Before each refiring the mixtures were rehomogenized and analysed.

CRYSTAL CHEMISTRY

The X-ray diffraction investigation of the FeF_3-WO_3 system shows, after a 15 h. firing at 600°C, the existence of a large solid solution domain rich in FeF_3 with the formula $Fe_{1-x}W_xO_{3x}F_{3-3x}$ $(0 \leqslant x \leqslant 0.74)$ (fig.1).

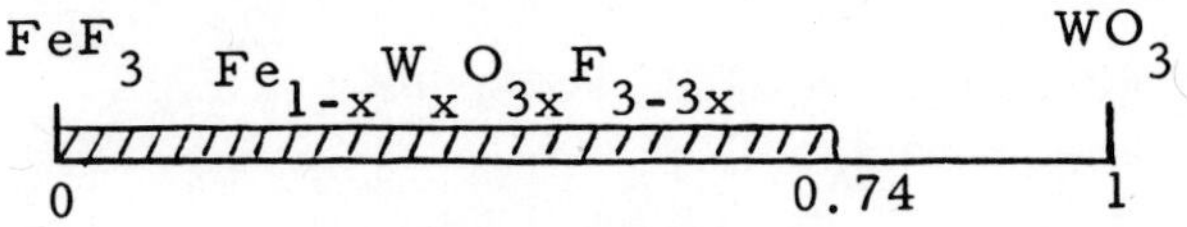

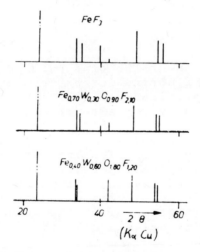

Fig. 1: The FeF_3-WO_3 system at 600°C.

The evolution of the X-ray diffraction patterns with increasing x is shown on Fig. 2. The X-ray lines are sharp and suggest a well crystallized material

Fig. 2: Comparison of X-ray patterns for some compositions of the $Fe_{1-x}W_xO_{3x}F_{3-3x}$ solution.

The indroduction of WO_3 in the rhombohedral structure of FeF_3 leads to a more symmetrical, nearly cubic structure. For $x > 0.74$ the solid solution with the limit composition coexists with WO_3. The variation with x of the parameters of the hexagonal unit cell and of the densities is given on figure 3.

The densities are in good agreement with those calculated for a compensated substitution :

$$Fe^{3+} + 3 F^- \leftrightarrows W^{6+} + 3 O^{2-}$$

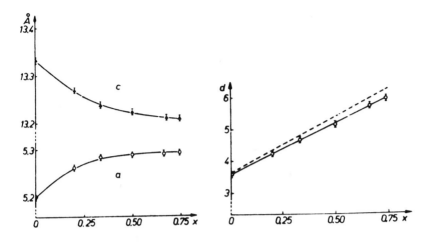

Fig. 3: Variations with x of the hexagonal parameters and (—experimental, --- calculated) densities of the solid solution.

The variation of the parameters of the hexagonal unit cell can be explained by the evolution of the rhombohedral FeF_3 structure towards cubic symmetry of a high temperature FeF_3 variety.

The X-ray diffraction investigation is consistent with the formation of a randomly distributed solid solution, but a physical investigation shows that such a picture is only true after several cycles of firing. After the first firing cycle at 600°C the thermodynamic equilibrium is not reached and the samples contain still homogeneous microdomains of FeF_3 and WO_3, which disappear only after repeated thermal treatments. The X-ray patterns do not change during that thermal procedure.

PHYSICAL STUDIES

The magnetic measurements performed on a series of $Fe_{1-x}W_xO_{3x}F_{3-3x}$ samples heated once for 15 h. at 600°C showed a variation of the Weiss constant θ_p, but no change in the Néel temperature, which remained equal to that of FeF_3 (T_N FeF_3 = 363 K) (table I).

Mössbauer spectra of the same sample showed a central peak corresponding to the resonance of Fe^{3+} ions magnetically diluted, but also a system of six peaks corresponding to the hyperfine structure of the antiferromagnetic FeF_3 (fig. 4a and 4b).

	15 h.	45 h.	
x	T_N (K)	T_N (K)	θ_p (K)
0 (FeF$_3$)	363 ± 1	363 ± 1	−610 ± 20
0,20	363 ± 1	283 ± 5	−440 ± 15
0,33	363 ± 1	215 ± 5	−310 ± 10
0,50	363 ± 1	150 ± 5	−218 ± 10
0,66	363 ± 1	80 ± 5	−150 ± 10
0,74	363 ± 1	62 ± 5	− 98 ± 10

Table I : Néel temperatures and Weiss constants for various values of x after 1 and 3 thermal cycles of 15 h at 600°C.

From these results it is possible to deduce that one firing of 15 h. at 600°C does not allow for the obtention of the completely disordered solid solution, but that microdomains of FeF$_3$ are apparently spread out in an oxyfluorinated matrix where tungsten and iron are statistically distributed. The existence of real FeF$_3$ domains is illustrated by the existence and the position of the corresponding peaks in the Mössbauer spectra and the constant value of the magnetic ordering temperature. These microdomains are too small to be detected by X-ray diffraction.

Although their X-ray diffraction patterns did not change, the magnetic properties and the Mössbauer spectra after three heating cycles show a decrease in the Néel temperature of the solid solution with x. The central Mössbauer peak increases progressively in comparison with the others which disappear completely after the three annealing cycles (fig. 4c).

An equilibrium situation with a random distribution of anions and cations in their sublattices is only obtained after 45 h. The observed Néel temperature corresponds to the value expected for such a solid solution (table I). Local order no longer exists

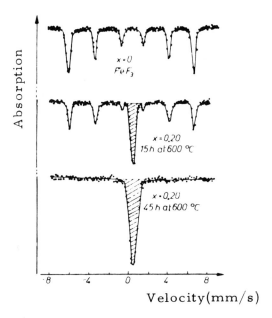

Fig. 4 : Mössbauer spectra at 295 K.

With the help of I.R. spectroscopy, the existence of WO_3 microphases was detected in 15 h heated samples. Figure 5 shows various I.R. spectra obtained after one or three heating cycles, the pure FeF_3 and WO_3 spectra are also given. The latter are characterized by two bands respectively centered at 540 and 800 cm^{-1}.

For the intermediate compositions the spectra show two bands (one for the iron-anion, the other for the tungsten-anion vibrations). For an iron rich composition (x = 0.20) the band intensity relative to tungsten increases with the number of annealing cycles and the band is slightly shifted toward the higher frequencies (900 cm^{-1}) corresponding to an increase in the number of W-F vibrations. The band corresponding to iron decreases and is shifted from 540 to 510 cm^{-1} as a consequence of the fluorine-oxygen substitution. For a tungsten-rich composition (x = 0.66), the results concerning the band intensities are reversed. The tungsten peak centered on 820 cm^{-1} decreases with the number of annealing cycles, whereas the iron peak increases.

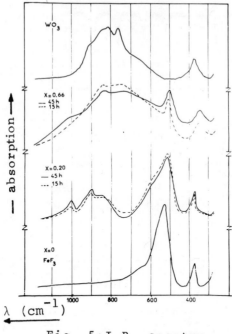

Fig. 5: I.R. spectra

CONCLUSIONS

The relative stability of the microdomains in the investigated solid solution can be explained by the nature of the Fe-F and W-O bonds : the first is essentially ionic, the second covalent. Furthermore, according to Pauling's 3rd rule, the 6+ formally charged tungsten tends to be surrounded by oxygen rather than by fluorine.

It is interesting to compare the length of time needed to reach equilibrium for the $Fe_{1-x}W_xO_{3x}F_{3-3x}$ solid solution, with that to obtain a randomly distributed $Fe_{1-x}Ga_xF_3$ solid solution in the range from FeF_3 to GaF_3 ($0 < x < 1$).

This study seems significant where it concerns certain types of solid state reactions. It illustrates the fact that non-stoichiometry or ion distribution problems cannot be only treated by X-ray diffraction (and even electronic microscopy if the unit cells are too close).

The knowledge of short range ordering phenomena may require an association with appropriate physical properties such as I.R. and Mössbauer spectroscopy.

The authors wish to thank Dr. Ménil for Mössbauer absorption measurements and Dr. Couzy for I.R. spectroscopy measurements.

DISCUSSION

S. MROWEC (Academy of Mining and Metallurgy, Krakow, Poland)
Oxyflouride structures based on oxygen octahedra typical for the WO_3 phase are especially susceptible to shear processes leading to the formation of block structures. Did you observe these blocks or shear planes in your materials?

AUTHORS' REPLY: The structures of oxyfluorinated solid solutions derive from the FeF_3 structure for which no shear planes have ever been observed. X-ray studies did not reveal any block structures, only electron diffraction could answer this point.

DIFFUSION AND REACTIVITY IN PEROVSKITE MATERIALS OF

THE COMPOSITION (Pb,La) (Zr,Ti)O_3

K.J. de Vries, A.H.M. Hieltjes, and A.J. Burggraaf

Twente University of Technology, Dept. of Chemical
Engineering, Laboratory of Inorganic Chemistry and
Materials Science, P.O. Box 217, Enschede, Netherlands

INTRODUCTION

Problems are encountered in the synthesis of homogeneous and dense (lanthana doped) leadzirconate-titanate ceramics. Keizer et al. (1) and Buckner et al. (2) have mentioned inhomogeneities in the distribution of Zr^{4+}- and Ti^{4+}-ions among the available lattice sites. Keizer et al. (1) and Carl et al. (3) have reported the occurrence of a minimum in the ceramic grain size and density after firing at a lanthanum content of 2-4 at%. Haertling (4) and Snow (5) have observed that sinter-reactivity is strongly promoted by adding a surplus of PbO to the reacting powder mixture. The dielectric and ferroelectric properties of these ceramics are dependent on micro structural parameters homogeneity, grain size and density. Hence a better understanding of the reactions that lead to different microstructures is important.

During preparation, homogenizing as well as sintering occur by solid state reactions. In these reactions transport of all ionic species is necessary. Probably the mobility of the highly-charged Zr^{4+}- and Ti^{4+}-ions is smaller than that of the other species and hence is rate determining in the kinetics of the overall transport processes. Therefore the aim of this investigation was to determine the chemical diffusion coefficient $\tilde{D}$ of the Zr- and Ti-ions within the homogeneity range of the perovskite phase as f([La],[PbO]). No single crystals of these solid solutions were available. Hence diffusion couples of ceramic samples were used. The check of their reliability was a part of this investigation. Determination of the predominating diffusion mechanism e.g. solid state-bulk diffusion or liquid film diffusion along grain boundaries was undertaken. If liquid film diffusion were to predominate no strong correlation between diffusion rate and bulk-phase composition can be expected. With a predominating bulk diffusion mechanism it is interesting to know if a minimum in $\tilde{D}$ exists as a function of composition. Defect-structural considerations can be expected to play a role in this case. Hence a description of the phase diagram relevant for the defect chemistry is given.

Lanthana doped lead zirconate-titanate appears as a perovskite phase in the system $PbO-La_2O_3-ZrO_2-TiO_2$. This is drawn schematically in Fig. 1. This phase can be represented by the chemical formula $Pb_{1-\alpha x}La_x(Zr_yTi_{1-y})O_{3+x(1.5-\alpha)}$
Here α, x and y are composition parameters. (αx) denotes the amount of PbO that is removed from lead zirconate-titanate solid solutions on adding x moles of $LaO_{1.5}$
In Fig. 1 the phase boundary on TiO_2 rich side is situated at $\alpha = 1.5$. On the PbO rich side it depends on the value of x,y and temperature. Substitutions of lanthana in lead zirconate-titanate at $\alpha = 1.5$ result in the formation of point defects e.g. lanthanum (III) ions and vacancies at lead ion sites. Compositions with values smaller than 1.5 lie in the homogeneity range of the perovskite phase. They are accompanied by the formation of an additional number of defects.. These are most probably lanthanum (III) ions and vacancies distributed among lead-and titanium ion sites (6).

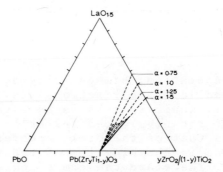

Fig. 1. The homogeneity range of the perovskite phase (hatched area) in the system $PbO-LaO_{1.5}-yZrO_2-(1-y)TiO_2$.

EXPERIMENTAL

Concentration profiles were measured by A.R.L. Electron Microprobe. Diffusion couples were formed by pressing two blocks with different values of y (see Table 1) of dense (> 96%) ceramics against each other, and heating for 6.5 hours at $1240°C$ in a controlled PbO vapour pressure to prevent decomposition of the couple material. The compositions of the couples investigated are given in Table 1. All the couples contained single phase material except the last one. Here excess PbO was added in order that it was present as a second phase. $\tilde{D}$ values were calculated in the Matano-interface using the Boltzmann and Matano method. Diffusion depth is also used as an experimental variable, since this is correlated with the mean value of $\tilde{D}$, and is less sensitive to errors in the determination of the slope of the concentration gradients.

Y	X	α	vs.	Y	X	α
40/60	15	1.5		65/35	15	1.5
40/60	10	1.5		65/35	10	1.5
40/60	3	1.5		65/35	3	1.5
40/60	0	1.5		65/35	0	1.5
40/60	10	1.5		65/35	10	1.5
40/60	10	1.35		65/35	10	1.35
40/60	10	1.2		65/35	10	1.2
40/60	10	0.5		65/35	10	0.5

Table 1. Compositions of the diffusion couples.

RESULTS AND DISCUSSION

In Fig. 2 representative concentration profiles of Zr- and Ti-ions are given as $f([La])$. Concentration gradients are absent in the profiles of Pb- and La- ions. The Matano-interface is found at the original boundary between both parts of the diffusion couples. In Fig. 3 the diffusion depth and the $\tilde{D}$ values calculated at the Matano-interface, are shown as $f([La])$. The error in $\tilde{D}$ is at maximum about 40%. Fig.3 shows that $\tilde{D}$ and the diffusion depth decrease with increasing lanthana content. For diffusion times of 16.5 and 24 hours a quasi plateau in the concentration profiles of Zr- and Ti- at the Matano-interface is developed in lanthana doped samples at a concentration of complete reactions i.e. y = 0.525. The extent of the quasi plateau was increased with diffusion time, An analogous phenomena was described (7) for the diffusion couple system $(La,Sr)TiO_3$.

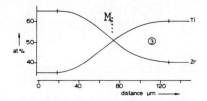

Figure 2. Concentration profiles of Zr- and Ti-ions at 3 at.% La. M indicates the Matano-interface.

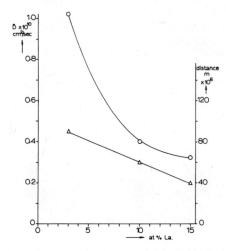

Figure 3. $\tilde{D}(○)$ and diffusion depth (△) as $f([La])$

In Fig. 4 the concentration profiles of Zr- and Ti-ions are shown as obtained in undoped (x=0) diffusion couples. It shows a more extensive diffusion depth compared to doped (x≠0) diffusion couples, with α = 1.5. Moreover two pronounced quasi plateaus are formed, in both profiles. $\tilde{D}$ values calculated in these plateaus are about 100 times larger than those outside. This difference is of the same order as that calculated and measured (8) for diffusion of H^+-ions in TiO_2 where quasi-plateaus occurred too. This model was based on the contribution of local field effects, introduced on the passage of H^+-ions through the TiO_2 bulk phase. The same model may possibly account for quasi plateaus in our doped and undoped systems.

In those cases where the Matano-interface intersects the concentration profiles at the place of a quasi plateau, D contains a relatively large error as a consequence of the small slope of the concentration gradient. Therefore D is also calculated at the value y = 0.50. For this value no quasi plateaus occurred in the concentration profiles of Zr- and Ti-ions of undoped and doped couples heated 6.5 hours at 1240°C. The values of D and the diffusion depth are given in Fig. 5. This Figure shows that no minimum occurs in D of Zr- and Ti-ions as f(La). Hence it is concluded that the minimum in density and grain size (1), (3) is not directly connected with the diffusion rate of Zr- and Ti-ions.

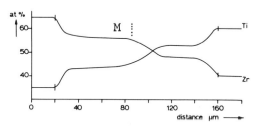

Figure 4. Concentration profiles of Zr- and Ti-ions in undoped (x=0) diffusion couples. The signs indicate the diffusion depth. M the Matano-interface.

Hence it is concluded that the minimum in density and grain size (1), (3) is not directly connected with the diffusion rate of Zr- and Ti-ions.

In monophasic diffusion couples no influence was found of varying PbO content (decreasing α) on $\tilde{D}$ values and diffusion depth. Changes of the defect structure due to the solution of PbO are not significantly reflected in the values of $\tilde{D}$. This may be caused by a coalescence of point defects into layers (9) resulting in a decrease of the point defect concentration in the bulk phase. In diffusion couples with a second phase of PbO a very large diffusion depth was found (>350 μm). Scanning electron microscopy has shown the presence of a second phase layer, identified as PbO, at the grain boundaries. At 1240°C these layers provide a liquid diffusion path of considerable thickness in these couples. From this it is concluded that the diffusion mechanism in the monophasic couples is predominatly solid state-bulk diffusion. This is supported by the dependence on lanthana content as shown in Fig. 3 and 5. These results may also be a proof for the usefulness of ceramic samples in this type of measurements.

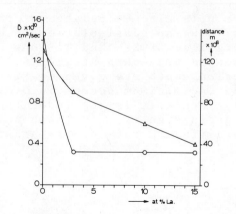

Figure 5. $\tilde{D}(0)$ and diffusion depth ($\triangle$) as $f([La])$ calculated at the concentration y = 0.50.

ACKNOWLEDGEMENT

The authors are indebted to Mr. A.M. Nijssen and Mr. H. Koster for the electron microprobe measurements and to Mr. W.A. van der Heide who contributed to this investigation as a graduate student

REFERENCES

(1) K. Keizer, E.H.J. Janssen, K.J. de Vries and A.J. Burggraaf, Mat. Res. Bull. 8 (1973) 533.
(2) D.A. Buckner and F.D. Wilcox, Ceram. Bull. 51 (1972) 218.
(3) K. Carl and K.H. Härdtl, Ber. Deutsch. Ker. Ges. 47 (1970) 687.
(4) G.H. Haertling and C.E. Land, J. Amer. Ceram. Soc. 54 (1971) 1.
(5) G.S. Snow, J. Amer. Ceram. Soc. 56 (1973) 9.
(6) K.J. de Vries, K. Keizer, J. Bouwma and A.J. Burggraaf, Science of Ceramics 8, Cambridge U.K. (1975).
(7) F.S. Ozdemir and R.J. Schwartz, Ceram. Bull. 51 (1972) 474.
(8) J.W. DeFord and O.W. Johnson, University of Utah, A.D. Report 787667.
(9) J. Galy and A. Carpay, Phil. Mag. 29 (1974) 1207.

DISCUSSION

Hj. MATZKE (Euratom, Karlsruhe, West Germany) You have treated your data in the extreme cases of bulk diffusion and mobility along grain boundaries containing a liquid phase. Did you consider the alternative of transport along normal grain boundaries without a liquid phase?

Also is there a possibility that your results are affected by some reduction of Ti^{4+} to Ti^{3+}, especially while varying the Ti-content?

AUTHORS' REPLY: To determine absolutely the relative importance of grain boundary diffusion, diffusion experiments would have to be performed with extremes of grain size eg $>10\mu m$ and $<0.5\mu m$. Because of the observed influence of dopant La ions on diffusion in ceramic materials with $5\mu m$ grain size, we believe bulk diffusion to predominate.

For your second question, sample colour, Gouy balance and ESR results indicate the absence of reduced ions over the whole concentration range employed.

W.W. BRANDT (University of Wisconsin, Milwaukee, U.S.A.) How do you interpret the plateau's, eg. in Fig. 4? You use the term quasi-plateau and also allude to a trivial reason for their existence.

AUTHORS' REPLY: The quasi-plateau may well be explained by an acceleration mechanism arising from local internal fields. However, as we have not checked whether the unit cell dimension at certain compositions provide obstacles to the diffusion of Ti and/or Bi, a trivial reason may not be completely excluded.

S. MROWEC (Academy of Mining and Metallurgy, Krakow, Poland) Your main conclusion is that diffusion occurs through the bulk rather than along grain boundaries. The dependence of $\tilde{D}$ on La concentration may be insufficient evidence to support this as all your measurements have been made on polycrystalline material whose grain size may vary with dopant concentration.

AUTHORS' REPLY: There are two possible types of grain boundary diffusion, with or without liquid film diffusion. The former is found to be absent in samples whose compositions lie within the homogeniety range. The latter possibility is considered in the reply to Dr Matzke.

FORMATION OF LARGE DEFECTS IN NEUTRON IRRADIATED ALKALI HALIDE CRYSTALS

F.W. Felix and W.E. Montserrat Benavent

Hahn-Meitner-Institut für Kernforschung

D 1 Berlin 39, Glienickerstrasse 100

INTRODUCTION

Several authors have described the formation of large voids in alkali halide crystals, i.e. large enough to be detected by ordinary light microscopy. The earliest experiments by Mollwo and Pohl concerned the formation of large gas bubbles through a chemical reaction with suitable dopants (e.g. by heating KCl crystals doped with CO_3^{--} in Cl_2-atmosphere yielding CO_2 and O_2) (1, 2). The produced gas accumulated in spherical bubbles. The bubbles transformed into parallelepipedic voids, "negative crystals", under prolonged thermal treatment. Most of the gas content of the initial bubbles had then been released into the lattice and had left the crystal by diffusion.

Voids have also been observed in neutron-irradiated alkali halides. Senio has irradiated LiF crystals and found that the nucleogenic gases He and H-3 from the neutron reaction Li-6 (n,α) H-3 precipitated in the form of quadratic voids when heating the crystals (3). Budylin and Kozlov also observed voids in KCl, KBr, KI and the corresponding Na-halides when heating highly irradiated crystals (4).

There is also evidence that smaller voids are formed during electron microscopy of thin crystals (5). The initial process in void formation is believed to be the agglomeration of radiolytically produced halogen interstitials (6).

The results presented here are concerned with void formation induced by reactor irradiation in KI crystals. The interest in these large defects arose during systematic studies of the diffusion of the nucleogenic rare gases in ionic crystals (7). Rare gases have been found to react strongly with lattice defects, as well intrinsic point defects as irradiation-induced defects. The interaction of gases and radiation-induced defects was found even near the melting point. This high stability led to the assumption of extended defects with a need for a direct observation method (8). For this paper this was done by scanning electron microscopy on freshly cleaved $\{100\}$ - planes.

EXPERIMENTAL

The crystal samples were cleaved from commercial material (Dr. K. Korth, Kiel, produced from Merck "suprapur" chemicals) into cubes of usually $3 \times 3 \times 3$ mm^3. Before further treatment the crystals were annealed for 50 h at 550° C in vacuum.

Neutron irradiations were performed in the FRM-reactor Munich to total doses from 1.5×10^{16} to 4.5×10^{18} n cm^{-2} with the crystals sealed in quartz ampoules.

After neutron irradiation, the formation of voids was investigated by scanning electron microscopy (JEOL JSM-35) on freshly cleaved planes coated with gold.

Some preliminary measurements of the formation kinetics of the voids were performed by observing the scattering of light during the heat treatment of the crystals. The influence of water on the defect formation was studied by measuring the decolouring of the crystals when heated in water vapour between 150 and 200°C.

RESULTS

After irradiation the crystals were dark blue in colour. At doses above 5×10^{17} ncm^{-2} as irradiated crystals showed elongated defects with an orientation in $\langle 110 \rangle$-direction, figure 1. These defects have an average length of 3×10^{-4} cm, a volume of order 10^{-12} cm^3 and a concentration of order 10^{9} cm^{-3}. The number of voids was found to increase with irradiation dose.

FORMATION OF LARGE DEFECTS

At about 200°C the crystals became colourless and visually clear. Increasing the temperature above 450°C made a drastic change: the crystals became more and more turbid, finally white and completely opaque. The cleaved planes of such crystals show a non-uniform distribution of larger parallelepipedic voids with $\langle 100 \rangle$ -edges, figure 2. The volume of these voids is of order 10^{-10} cm^3.

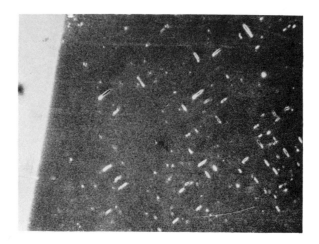

Fig. 1 Cleaved plane of an as-irradiated KI crystal, x 750
Neutron dose 4.5×10^{17} n cm^{-2}

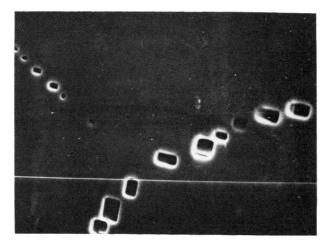

Fig. 2 Irradiated and heat treated KI crystal, x 750
Neutron dose 4.5×10^{17} n cm^{-2}
15 h at 590°C in air

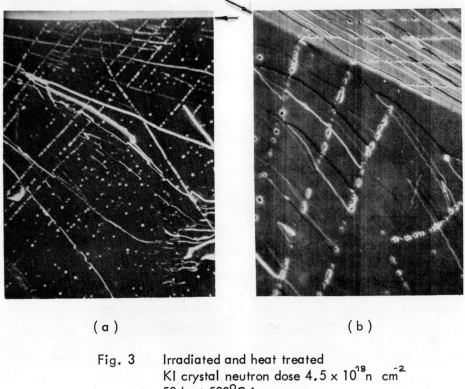

(a) (b)

Fig. 3 Irradiated and heat treated
KI crystal neutron dose 4.5×10^{18} n cm^{-2}
50 h at 500°C in vacuum

(a) x 27 horizontal
(b) x 350, 60° tilted, cleaved edge marked by arrows

It was often found that many voids were arranged on the cleaved plane in parallel rows, e.g. of $\langle 110 \rangle$ -direction as in figure 3a. The average distance between the rows is 3×10^{-3} cm and the length can be several millimeters. The similarity of this linear arrangement with pictures of decorated dislocations is coincidental (9). The voids shown in figure 3a are not only arranged in one dimension, but in $\{110\}$ -planes. This is revealed by cleaving the crystal a second time perpendicularly to the first cleavage. Figure 3b shows that the $\langle 110 \rangle$ -rows continue on the next side as $\langle 100 \rangle$ -rows with apparently the same void density.

From figures 3a and b average values can be deduced for the volume of a single void of 8×10^{-12} cm^3, for the numbers of voids per $\{110\}$ - plane of 1.5×10^6 cm^{-2} and per volume respectively 5×10^8 cm^{-3}.

Preliminary light scattering experiments showed that the voids are formed at 450° C in 5 to 8 minutes. By heating the crystals under water vapour at $\geqslant 10$ mm Hg at 150 to 200° C it was possible to destroy the colour centres quantitatively by diffusing water (10). This treatment was, however, of no influence on the void formation. In unirradiated crystals, voids could not be produced either thermally or by water diffusion.

SUMMARY

The void formation has been investigated in neutron-irradiated KI crystals. A similar behaviour was also found in KBr, RbBr and RbI crystals. The average volume of a single void lies between 10^{-12} and 10^{-10} cm^3 depending on the thermal treatment. It could be established that the void formation is definitely an irradiation induced process.

The observed ordering phenomena have to be studied thoroughly in the future, especially the two-dimensional arrangement of voids in parallel $\{110\}$ -planes. Experiments are in progress which will establish whether this phenomenon depends on preexisting preferential directions in the crystals e.g. the growth axis.

REFERENCES

1) E. Mollwo, R.W. Pohl, Ann. Phys. 39 (1941) 321

2) E. Mollwo, Nachr. Akad. Wiss. Göttingen, Math. Phys. Klasse (1941) 51

3) P. Senio, Science 126 (1957) 208

4) B.V. Budylin, Y.F. Kozlov, Sov. Phys. Solid State 6 (1964) 1237

5) K. Izumi, J. Phys. Soc. Japan 26 (1969) 1451

6) L.W. Hobbs, A.E. Hughes, D. Pooley, Proc. Roy. Soc. A 332 (1973) 167

7) F.W. Felix, J. Physique 34 (1973) C9-149

8) F.W. Felix, M. Müller, Indian J. Appl. Pure Phys., in press

9) S. Amelinckx, Solid State Phys. 8 (1959) 325

10) C. Rühenbeck, Z. Phys. 207 (1967) 446

DISCUSSION

Hj. MATZKE (Euratom, Karlsruhe, West Germany) You saw voids in KI and KBr but not in KCl. It is known that nucleation is aided by very small amounts of rare gases. Although the concentration of gaseous halogens is higher than that of the rare gases we do not know if these are equally effective in the nucleation stage. It is also possible that Ar, formed in KCl, is released whereas Kr and Xe formed in KBr and KI remain and aid nucleation.

AUTHORS' REPLY: The voids are formed within some minutes. After such a short time most of the rare gas atoms, even the most mobile Ar atoms, are still in the crystals. One can therefore assume that the rare gas atoms produced by nuclear reactions do not play a role in the void production, as their concentration is negligible with respect to the concentration of radiolytically produced halogen molecules.

MECHANISM AND CATIONIC REARRANGEMENTS INVOLVED IN THE REDUCTION OF $Co^{2+}[Co^{3+}_{2-x}Cr^{3+}_{x}]O^{2-}_{4}$ SPINELS

P. Bracconi, L. Berthod, L.C. Dufour
Laboratoire de Recherches sur la Réactivité des
Solides, Dept B, Faculté des Sciences Mirande -
21000 - Dijon (France)

INTRODUCTION

This work deals with the Co(II, III)-Cr(III) oxide series (or Co_3O_4-$CoCr_2O_4$ solid solutions) with compositions ranging from Co_3O_4 to $CoCr_2O_4$. These oxides crystallize in the spinel structure and, according to Makkonen, their lattice parameter increases linearly with their Cr content from 8.08 Å for Co_3O_4 to 8.33 Å for $CoCr_2O_4$ (1). They are paramagnetic above $T_c(CoCr_2O_4)$ = 100°K and $T_N(Co_3O_4)$ = 40°K. From their Curie constants we show that the Co^{3+} ions have a zero magnetic moment and, thus, are in a low spin state in octahedral sites (B), the moment of tetrahedral (A) Co^{2+} being 4.6 ∿ 4.7 μ_B. Whence the formula :

$$Co^{2+}[Co^{3+}_{2-x}Cr^{3+}_{x}]O^{2-}_{4}$$

Powders and monocrystals were prepared through various methods and their chemical composition, structure and paramagnetic constants were checked. Their decomposition under oxygen or vacuum or their hydrogen reduction were investigated by T.G.A., X ray diffraction, magnetic analysis, BET-surface area measurement and S.E.M.

The processes involved in such reactions can be divided into 3 groups : a) surface processes leading to oxygen or water molecules formation and evolvement ; b) atomic rearrangements in the solid state ; c) all other processes (e.g. sintering) that can be regarded as subsequent to and independent of the former ones. This paper aim is to find out the nature of the atomic rearrangements involved in the present experimental cases. It gives evidence that they are restricted to cationic rearrangements in the oxygen sublattice of the spinel structure.

Partial reduction of α-Cr_2O_3 in contact with Co metal (above mentioned reactions products below 750°C) and yielding Co-Cr saturated f.c.c. solid solutions is also discussed.

RESULTS

I/ CoO-Co_3O_4 crystallographic orientations in monocrystalline Co_3O_4 decomposition

A small Co_3O_4 monocrystal of octahedral shape, prepared by chemical transport in HCl was decomposed slightly and extremely slowly at 650°C (24 hours). Placed in a quartz tube initially evacuated down to 2×10^{-9} torr, its temperature was raised to 650°C so that the evolved oxygen pressure never exceeded 2×10^{-7} torr. The structural investigation by the rotation and then the Weissenberg methods (rotation around [001], MoK_α and CuK_α radiations) demonstrates that the initial spinel and final f.c.c. CoO structures respectively have all their directions [h00], [0k0] and [00ℓ] parallel (misorientation ± 5°). Particularly the following relationships are verifed :

$$(100)_{CoO} \; // \; (100)_{Co_3O_4} \quad - \quad [001]_{CoO} \; // \; [001]_{Co_3O_4}$$

This, in particular, suggests that the anion sublattices in both structures have been momentarily and locally "end to end", and supports our general conclusion that the structure change occurs via a rearrangement of the Co^{2+} ions in small superficial domains of the solid.

The atomic ordering in the successive planes perpendicular to [100], corresponds to the following sequence :

in spinel ... $A^{2+} \left| B_2^{3+} O_4^{2-} \right| A^{2+} \left| B_2^{3+} O_4^{2-} \right|$ ... B_2O_4 planes 2.02 Å apart

in f.c.c. ... $\left| B^{2+} O^{2-} \right| B^{2+} O^{2-} \right|$ ... planes 2.06 Å apart

The oxygen release and the reduction of Co_B^{3+} according to (in K.V. writing) :

$$O_O^x + 2\, Co_B^x \rightarrow 1/2\, O_2\, (g) + 2\, Co_B' + V_O^{oo}$$

gives the following defect structure :

in spinel ... $A^{2+} \left| B_2^{2+} O_3^{2-} V_o \right| A^{2+} \left| B_2^{2+} O_3^{2-} V_o \right|$ ...

The cation rearrangement involves the two following steps :

a) $Co_A^{2+} \rightarrow Co_B^{2+}$. The resulting defect structure can be regarded as belonging either to the initial spinel structure or to

REDUCTION OF COBALT OXIDE SPINELS

the final f.c.c. structure according to :

in spinel ... $V_A \left| B^{2+}_2 O^{2-}_3 B^{2+}_i V_o \right| V_A \left| B^{2+}_2 O^{2-}_3 B^{2+}_i V_o \right|$ ...

or in f.c.c. ... $\left| B^{2+}_3 V_B O^{2-}_3 V_o \right| B^{2+}_3 V_B O^{2-}_3 V_o \right|$ ...

b) $V_o + V_{Co} \to$ zéro, V_{Co} diffusion step in f.c.c. and vacancies surface annihilation leading to the final structure :

$$\ldots \left| B^{2+}_3 O^{2-}_3 \right| B^{2+}_3 O^{2-}_3 \right| \ldots$$

equivalent to f.c.c. CoO.

II/ Decomposition of powdery Co_3O_4 and solid solutions with $x = 0.005, 0.01, 0.02, 0.05, 0.10 ; 0.25, 0.50, 1.00$

These solutions were prepared from melted nitrate mixtures, evaporated and then calcined, in air, at 850°C for 3 weeks. The total decompositions were carried out at 910°C under 10 torr of oxygen. The magnetic susceptibilities of the products were measured by the Faraday method from 77 K to room temperature in the same apparatus.

For $x = 0, 0.005$ and 0.01, the single final phase formed, which has the CoO f.c.c. structure is purely antiferromagnetic with a Neel temperature decreasing from $T_N = 292.5 \pm 0.5$ K to 291 ± 0.5 K. The Cr^{3+} ions are integrated into CoO structure according to :

$$Co^{2+} \left[Co^{3+}_{2-x} Cr^{3+}_x \right]_B O^{2-}_4 \xrightarrow{O_2(g)} (3 + \tfrac{x}{2}) \left[Co^{2+}_{\frac{6-2x}{6+x}} Cr^{3+}_{\frac{2x}{6+x}} \square_{\frac{x}{6+x}} \right]_B O^{2-}$$

$$\underbrace{\hphantom{Co^{2+} \left[Co^{3+}_{2-x} Cr^{3+}_x \right]_B O^{2-}_4}}_{\text{spinel}} \qquad \underbrace{\hphantom{\left[Co^{2+}_{\frac{6-2x}{6+x}} Cr^{3+}_{\frac{2x}{6+x}} \square_{\frac{x}{6+x}} \right]_B O^{2-}}}_{\text{f.c.c.}}$$

For $x > 0.02$, $CoCr_2O_4$ is detected in the reaction products owing to the sharp sample magnetization change around $T_C = 100°K$. Thus, from an x value comprised between 0.01 and 0.02, Cr-saturated CoO is formed and the excess Cr separates as $CoCr_2O_4$, indicating that the limit solubility of Cr in CoO (= 100 x/3 at %) at 910°C should be comprised between 0.33 and 0.66 at % - compared with 0.5 at % at 1140°C from ref. (2). Thus, here (x > 0.01), the cationic redistribution is inhomogeneous : it requires a segregation of the two kinds of cations but can be described readily by a similar sequence of steps as in § I.

III/ Hydrogen reduction of a solid solution with x = 1.55

Sample prepared at 450°C from a precursor ; grain size : 500 Å ; lattice parameter 8.281 Å.

Below 550°C (P_{H_2} = 40 torr) a partial reduction occurs corresponding to :

- the evolvement of 2-x oxygen g-At (= the number of Co_B^{3+})
- an increase of the spinel matrix lattice parameter up to 8.320 Å and proportional to the number of oxygen g-At evolved.
- the formation of $\frac{2-x}{2}$ Co metal g-At (detected by X ray diffraction and titrated from the sample saturation magnetization).

These three facts together can only be accounted for by assuming that the (2-x) Co_B^{3+} are reduced into Co^{2+} and retained into the spinel structure while (2-x)/2 Co^{2+} are reduced into $Co^°$. This leads, in fact, to an abnormal spinel lattice which can be formally regarded as a solid solution between CoO and $CoCr_2O_4$. This partial reduction is characterized by parabolic type kinetics (activation energy 30 Kcal/M) and no measurable variation of the grain size, indicating that cation diffusion is the rate limiting process.

Above 550°C, by further reduction, a subsequent reordering of the spinel matrix is observed up to the arrangement $Co^{2+}\left[Cr_2^{3+}\right]_B O_4^{2-}$ - as evidenced by a new increase of the lattice parameter from 8.320 to 8.330 Å - and is associated with the reduction of the octahedral Co^{2+} ions. This transformation is immediately followed by and cannot be separated (in dry H_2) from the reduction of $CoCr_2O_4$ into f.c.c. Co and $\alpha-Cr_2O_3$ (3).

IV/ <u>Hydrogen reduction (below 750°C) of $CoCr_2O_4$ into $\alpha-Cr_2O_3$ and f.c.c. Co</u>

Although no direct experimental evidence could be obtained in this case, (by X ray diffractometry), assuming a rearrangement of the non-reduced cations (Cr^{3+}) in the O^{2-} f.c.c. sublattice again here, proves highly helpful in understanding the formation mechanism of the final phase $\alpha-Cr_2O_3$ (corundum structure). The most logical model for this rearrangement should be the $\gamma-Cr_2O_3$ spinel structure (a = 8,36 Å) which would give rise to the equilibrium structure $\alpha-Cr_2O_3$ by associated movements of the oxygen planes and of the Cr^{3+} cations, i.e. by the same shear mechanism that is known to occur in the transformation $\gamma-Fe_2O_3 \rightarrow \alpha-Fe_2O_3$. For this reason and because it requires Cr^{3+} ion in A sites such an arrangement would be highly unstable and could not extend beyond a few atomic planes thick . The discontinuous surface layer of the spinel matrix would thus be transformed "layer after layer" (3).

V/ <u>$\alpha-Cr_2O_3$ partial reduction (above 750°C) in presence of f.c.c. Co</u>

The partial reduction of $\alpha-Cr_2O_3$ observed above 750°C is due to the presence of Co metal but cannot be explained by a catalytic effect of the latter. Indeed, f.c.c. Co-Cr solid solutions are formed and the maximum fraction of Cr_2O_3 reduced at a given temperature is determined by the **solubility limit** of Cr in f.c.c. Co at that temperature : this is shown by the sharp decrease of the latter saturation magnetization **in proportion** to the amount of Cr reduced. This reduction is possible because, contrary to the reduction of pure Cr_2O_3 which would be impossible in the present conditions, it does not require the formation of Cr nuclei (or $Cr-Cr_2O_3$ interface). It stops when the f.c.c. Co lattice, here acting as a host lattice for the reduced Cr atoms, is saturated (refer to Co-Cr equilibrium diagram), i.e., when a new intermetallic phase nucleation would be required so that it might continue (3).

CONCLUSION

The above-experimental results confirm our assumption that cation rearrangements occur as elementary processes in the different reactions investigated.

Concerning Cr-rich solutions, the non-reduced cations rearrange in the spinel structure anion sublattice which is maintained through the successive intermediate reaction steps up to $\alpha-Cr_2O_3$ precipitation. When CoO f.c.c. structure is formed, by decomposition of cobalt rich solutions, the same rearrangements can always take place owing to the fact that the oxygen sublattices in Co_3O_4 and CoO are practically identical.

In every case, mechanical strains apply in these sublattices **as a result of** the different cation orderings in the rearranged zones and in the spinel matrix. Different kinetics are observed in the hydrogen reduction of $Co_{1.45}Cr_{1.55}O_4$ and $CoCr_2O_4$, although the rate-limiting step is always a cation diffusion : this is certainly related to such mechanical parameters. In the case of $CoCr_2O_4$, the mechanical strain relaxation takes place readily owing to $\alpha-Cr_2O_3$ precipitation and separation from the matrix : the **practically linear kinetic law** expresses the transformation "layer after layer" of the initial grains. **Conversely** in the first step of $Co_{1.45}Cr_{1.55}O_4$ reduction, the strains can be withstood and the kinetic law of parabolic type expresses the kinetics of the cationic rearrangement in the whole bulk of the grains.

From this viewpoint, the concept of new solid phase formation in terms of nucleation and growth by diffusion through a two-dimensional interface is advantageously replaced by an analysis in terms of structural rearrangement, mechanical **strains** and precipitation.

1 - R.J. Makkonen, Suomen Kemi B, 35, 230 (1962).
2 - M. Gwishi, D.S. Tannhauser, J. Phys.Chem.Solids, 33, 893 (1972).
3 - P. Bracconi, L.C. Dufour, J. Phys.Chem. 79, 2395, (1975), and
    P. Bracconi, thesis Dijon 1976, C.N.R.S. A.O. 12 128.

DISCUSSION

P.PORTA (University of Rome, Italy) and
F.S. STONE (University of Bath, U.K.) In the oral presentation, details were given of the variation of lattice parameter (a_o) and Curie constant (C) as a function of x in the $Co_{3-x}Cr_xO_4$ spinels.

The following is an extract from the data presented:

x	0	0.006	0.009	0.019	0.200	0.939	1.998
a (Å)	8.086	8.085	8.076	8.070	8.110	8.196	8.330
C	2.70	2.77	2.84	2.90	2.96	4.23	6.39

We wish to point out two interesting effects in this table which are inconsistent with the distribution $Co^{2+}[Co^{3+}_{2-x}Cr^{3+}_x]O^{2-}_4$. These concern the very dilute region ($0 < x < 0.019$).

The end members of the solid solution, Co_3O_4 ($x=0$) and $CoCr_2O_4$ ($x=2.000$), are both normal spinels, so in general one might indeed expect a distribution $Co^{2+}[Co^{3+}_{2-x}Cr^{3+}_x]O^{2-}_4$ for the solid solution, as implied by the authors in the title. Since the ionic radius of Cr^{3+} (0.755Å) is greater than that of Co^{3+} (0.665Å), a_o for the solid solution would be expected to increase linearly with x. In fact a_o decreases with x in the initial region, falling from 8.086Å for Co_3O_4 to 8.070Å for the solution with $x = 0.019$. The results for the Curie constant C also show an apparent anomaly. C increases from 2.70 for Co_3O_4 to 2.90 for the solid solution at $x = 0.019$, which is a much more rapid rate of increase than would be expected for simple replacement of Co^{3+} (d^6, diamagnetic) by Cr^{3+} (d^3, paramagnetic, μ (spin only) = 3.84 μ_B) in octahedral sites.

We believe that these results can be explained by a variation of cation distribution with composition analogous to that recently proposed for $Ni_xMg_{1-x}Al_2O_4$ (1), $Ni_xZn_{1-x}Al_2O_4$, $Cu_xMg_{1-x}Al_2O_4$ and $CoGa_yAl_{2-y}O_4$ (2). We suggest that the initial additions of Cr^{3+} to Co_3O_4 cause some Co^{3+} ions to invert with Co^{2+}, giving a distribution of the form $Co^{2+}_{1-\alpha}Co^{3+}_\alpha[Co^{2+}_\alpha Co^{3+}_{2-x-\alpha}Cr^{3+}_x]O_4$ instead of the simple distribution $Co^{2+}[Co^{3+}_{2-x}Cr^{3+}_{x}]O_4$. Inversion is well-known to decrease a_o (3), and we suggest that the effect is brought about in this case because the octahedral sites occupied by Cr^{3+} are expanded and the adjacent tetrahedral sites are contracted. Electrostatic energy then increases when the smaller Co^{3+} ion replaces the larger Co^{2+} ion in tetrahedral sites: Cr^{3+} is of course 'anchored' in the B sites because of its large crystal field stabilisation. This interpretation also accounts for the unexpectedly large rise in C ince the magnetic moment of Co^{2+} in octahedral sites is ~5.2 μ_B as compared with only ~4.6 μ_B in tetrahedral sites. The inversion process proposed here for $Co_{3-x}Cr_xO_4$ when x is small is simply an electron transfer from Co^{3+} to Co^{2+}. It gives rise to the presence of both Co^{2+} and Co^{3+} on octahedral sites, and thus may lead to a marked increase in electrical conductivity. It would be interesting to examine this.

1. P. Porta, F.S. Stone and R.G. Turner, J.Solid State Chem., 11, 135 (1974).
2. F. Pepe, P. Porta and M.Schiavello, this Symposium p.183)
3. E.J.W. Verwey and E.L. Heilmann, J.Chem.Phys., 15, 174 (1947)

CORRELATIONS BETWEEN THE KINETICS OF DECOMPOSITION AND THE CRYSTAL STRUCTURE FOR SOME INORGANIC AZIDE SINGLE CRYSTALS

H.T. Spath

Institut für Physikalische und Theoretische Chemie

Technische Universität, Graz (Austria)

INTRODUCTION

The kinetics of thermal decomposition of single crystals of a solid must be described in terms of time and space coordinates. An interpretation of the overall fractional decomposition (α) vs. time (t) - functions presupposes information about the geometry of the growing nuclei and the growing interfaces, resp., i.e. their orientation relative to the crystallographic axes. A prominent group of solids where information of this kind is available have been found within the inorganic azides, among others TlN_3, NaN_3, $Pb(N_3)_2$, AgN_3 and $Ba(N_3)_2$.

In TlN_3 (tetragonal) decomposition preferably takes place parallel to (001)-planes, i.e. parallel to the linear N_3^- -ions oriented along <110>, (1). With NaN_3 decomposition takes place parallel to the c-axis of the hexagonal cell, i.e. parallel to the long axis of the N_3^- -ions, (1),(2). The product phase of rather complex composition is separated from the unreacted material by an incoherent interface which is parallel to (hk0)-planes and is moving inwards from the lateral faces of the hexagonal platelets by a diffusion-controlled mechanism (3). In α-$Pb(N_3)_2$ (primitive orthorhombic) there are 4 types of N_3^- -ions and decomposition occurs parallel to the most labile (most asymmetric) N_3^--ions lying along <100>, (1). Decomposition of AgN_3 (bodycentered orthorhombic) exhibits two types of metallic silver nuclei: one type being randomly oriented and found at lattice defects, the other one bearing a direct relation to the lattice structure, (4). In all those examples, with the exception of AgN_3, the geometry of decomposition seems to be strongly related to the N_3^--lattice within the crystal structure.

STRUCTURAL RELATIONSHIPS IN THE DECOMPOSITION OF $Ba(N_3)_2$

The most complete information relating the geometry of decomposition to the lattice structure is available for single crystals of anhydrous $Ba(N_3)_2$. Decomposition starts at the outer (100) and (001) surfaces via randomly distributed discrete nuclei. The growing nuclei of the product phase are separated from the host lattice by planes confining sharp oblique pyramids with diamond shaped bases, (3),(5).

According to (6) the monoclinic unit cell of $Ba(N_3)_2$ (a=9,63 Å, b= 4,41 Å, c= 5,43 Å, ß = 99,56°) contains 2 Ba^{++}-ions in 2 symmetry planes at (x=0,218; y=0,250; z=0,1715) and (x=0,782; y=0,750; z=0,8285). The basal plane projections (001) and (100) for 3 superimposed unit cells are shown in Fig. 1 (the ions are not to scale). Two types of almost linear N_3^--groups (not shown in Fig. 1) lying in the same symmetry planes are oriented perpendicular to <010>. The shortest Ba^{++}-Ba^{++} -distances are enumerated in Fig. 2.

The mechanism of nucleus growth as proposed by Torkar and Spath involves interaction of a Ba-atom with an adjacent Ba^{++}-ion of the lattice to give a more stable complex ($\Box Ba^+ Ba^+\Box$) associated with 2 anion vacancies $\Box$,(3),(5). The species $\Box Ba^+$,placed at some interstitial position at the interface reacquires its atomic properties by electron transfer from N_3^--ions (rate determining step) thus continuing the sequence. In the interaction no considerable displacement of the Ba^{++}ions relative to their lattice positions takes place so that the geometry of a growing nucleus is connected to the Ba^{++}-lattice, whereas there is no direct relation to the N_3^--lattice. The final reaction products (Ba_3N_2/Ba) formed in kinetically irrelevant consecutive steps cause no orienting effect on the rate of growth.

From Figs. 1 and 2 the geometry of the growing nuclei is derived from the Ba^{++}-lattice: Suppose that decomposition at a (100)-plane has reached atom 4 (at y=1/4). Stressing the principle of nearest neighbour interaction, from atom 4 in consecutive steps interactions will occur with ions 3 (1/4), 15(1/4) and 1 (3/4). The next sequence of interactions will involve ions 16 and 8 (both 1/4) and 3 (3/4) etc. Decomposition thus takes place parallel to planes e_1 passing through ions 4,8,12,...(1/4) and 1,5,9,... (3/4) as well as equivalent ions. The total of these planes (analytically written as e_1 : $\pm$ 0,343y + 0,5z = const.), parallel to a, confines an oblique pyramid with a diamond shaped base oriented along the fundamental translations b and c (cf. Fig.1, right hand part). Decomposition proceeds by adding to plane 2 a complete plane 2' of increased area and moving at a constant rate from P_2 along c to P_2' a distance of c= 5,43 Å. At the same time a new plane e_1 starts to grow at P_1'. Any plane between P_1 and P_2 (cf. plane 1 in Fig.1) increases in area by a step of 4,41 Å along b. Fig. 2 shows that in order to proceed from atom 4 (1/4) a distance of 9,63 Å along a to an equivalent atom 8 (1/4) two steps are involved, whereas at the same time decom-

KINETICS OF DECOMPOSITION AND CRYSTAL STRUCTURE

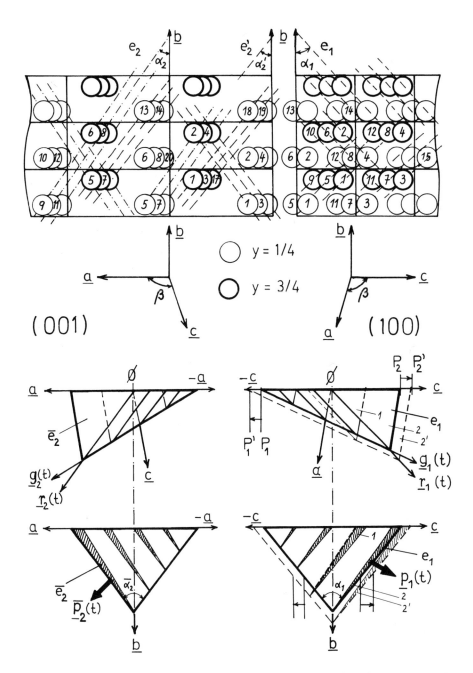

Fig. 1: Geometry of nuclei growing at (001) and (100)-faces as derived from the lattice structure of $Ba(N_3)_2$. Only one half of a nucleus is shown at the lower part of the figure.

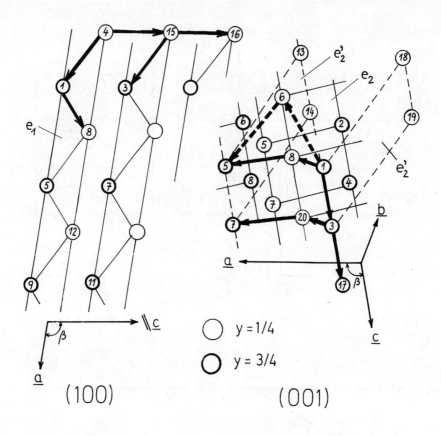

4 (1/4) − 1 (3/4) : 6,41 Å	4 (1/4) − 8 (1/4) : 9,63 Å (=a)
1 (3/4) − 8 (1/4) : 4,83 Å	1 (3/4) − 6 (1/4) : 6,34 Å
4 (1/4) − 15(1/4) : 5,43 Å (= c)	6 (1/4) − 5 (3/4) : 6,37 Å

Fig. 2 : Shortest distances between Ba^{++} − Ba^{++} and interaction sequences in the decomposition of $Ba(N_3)_2$ single crystals. Heavy arrows indicate directions of interaction.

position has reached atom 16 (1/4) along c. The ratio of the rates $k_a : k_c$ is thus expected to be 9,63 : 2×5,43 = 1 : 1,13 as compared to 1 : 1,12 found experimentally, (3),(5). The ratio of the rates $k_b : k_c$ along b and c follows from the intersection of the plane e_1 with the axes b and c, whence $k_c : k_b$ = 1 : 1,19 as compared to 1 : 1,25 , (3),(5). The calculated angle $\alpha_1 = 40,12°$ (Fig.1) is close to $\alpha_1 = 39 \pm 1°$ as found experimentally.

The moving interfaces of a nucleus growing at a (001)-face obey the same principles. However, in this case decomposition takes place parallel to two almost similar planes, e_2 passing through ions 2,4,.....(at y=3/4) and 5, 7 ,...(1/4) and equivalent ions and e_2^T through ions 5,7,...(3/4) and 13, 14,...(1/4) as well as equivalent ions (Figs. 1 and 2). The sequence of interactions, starting from atom 1 (3/4) is seen from Fig. 2. The directions of interaction are the same as those for nuclei at (100)-faces. The heavy broken arrows illustrate a possible alternative involving somewhat larger interaction distances. From the arguments above, the ratio $k_a : k_c$ = 1 : 1,13 whereas the ratio of the rates along $\underline{a}$ and $\underline{c}$ again follows from the intersection-mean of the planes $\bar{e}_2$ and $\bar{e}_2^T$ (i.e. $\bar{e}_2$: 1,5x $\pm$ 0,504y = const.) with the axes $\underline{a}$ and $\underline{b}$, whence $k_a : k_b$ = 1 : 1,38 as compared to 1 : 1,40 found from experiment, (3),(5). The angle $\bar{\alpha}_2$ (Fig.1) is $\bar{\alpha}_2$ = 35,9° (calculated) and $\bar{\alpha}_2$ = 35 $\pm$ 1° (experimental, (5)).

For both types of nuclei the vertex of the pyramide starting from the origin $\emptyset$ moves along the vectors $\underline{r}_1$, $\underline{r}_2$ and $\underline{g}_1$ and $\underline{g}_2$, resp. at a constant rate, whence, with respect to $k_a = 8,9.10^9$. $\exp\{(-26000 \pm 1000)/RT\}$, ref.(5) :

$\underline{r}_1(t) = \underline{r}_2(t) = 1,38.k_a.t.(0,726\underline{a}^o + 0,819\underline{c}^o)$

$\underline{g}_1(t) = 2,31.k_a.t.(0,433\underline{a}^o + 0,976\underline{c}^o)$; origin at $- 1,13.k_a.t.\underline{c}^o$

$\underline{g}_2(t) = 2,13.k_a.t.(0,940\underline{a}^o + 0,530\underline{c}^o)$; origin at $- k_a.t.\underline{a}^o$

($\underline{a}^o,\underline{b}^o,\underline{c}^o$ are unit vectors along $\underline{a}, \underline{b}, \underline{c}$). The interfaces $\bar{e}_1$ and $\bar{e}_2$ may be thought of as equipotential surfaces for the electron transfer from N_3^--ions to $\square Ba^+$ at an energy of 26 kcal/mole. They are moving at constant rates normal to the penetration vectors $\underline{p}_1$ and $\underline{p}_2$, whence (cf. Fig. 1):

$\underline{p}_1(t) = 0,87.k_a.t.(0,129\underline{a}^o \pm 0,640\underline{b}^o + 0,779\underline{c}^o)$

$\underline{p}_2(t) = 0,75.k_a.t.(0,822\underline{a}^o \pm 0,586\underline{b}^o + 0,136\underline{c}^o)$

From this discussion the specific geometry of the growing nuclei and the temperature-independent constancy of the ratios $k_a:k_c:k_b$ = 1 : 1,13 : 1,38 as well as the overall kinetics of decomposition could be definitely related to the Ba^{++}-lattice in $Ba(N_3)_2$.

REFERENCES

(1) P.G.FOX, J.Solid State Chem. **2**, 491 (1970)
(2) R.F.WALKER, N.GANE and F.P.BOWDEN, Proc.Roy.Soc. **A294**, 417 (1966)
(3) K.TORKAR, H.T.SPATH and G.W.HERZOG, in "Reactivity of Solids", ed. by J.W.Mitchell et al., John Wiley (1969), 287
(4) J. SAWKILL, Proc.Roy.Soc. **A229**, 135 (1955)
(5) K.TORKAR und H.T.SPATH, Mh. Chemie **98**, 1712 (1967)
      Mh. Chemie **99**, 118 (1968)
(6) E.M.WALITZI und H.KRISCHNER, Z. Krist. **132**, 19 (1970)

INFLUENCE OF p_{O_2} AND p_{H_2O} IN THE ATMOSPHERE ON THE RATE OF CHANGES IN THE SYSTEM $CaO-SiO_2$

V. Jesenák, Z. Hrabě

Department of Silicates, Slovak Technical University

Bratislava, Checkoslovakia

ABSTRACT

The influence of both the partial pressure of water vapour and the thermodynamic pressure of oxygen in the atmosphere on the rate and mechanism of changes in the system $CaO-SiO_2$ (C-S) was investigated. Pellets pressed from natural α-quartz and calcite in molar ratios C/S = 2/1 and 1/1 were treated in the temperature region 1473-1573 K in different controlled atmospheres, particularly mixtures of air/H_2O and H_2/H_2O. The unreacted portion of both C and S were determined by chemical analysis. Intermediates of the reaction were studied by X-ray phase-analysis and on sandwiches by electron microprobe and optical microscopy. In the studied range of experimental conditions the following intermediates were found : α-cristoballite, wollastonite and dicalcium silicate. No C_3S rancinite, tridymite and glass phase could be detected. The observed values of the conversions of CaO (α_C) and SiO_2 (α_S) as a function of annealing time were computer-fitted to semiempirical kinetic equations. The kinetic constants obtained were correlated with applied p_O and p_{H2O} values and with annealing temperature. It was found that the reactions between CaO and SiO_2 are highly accelerated by both lowering the oxygen pressure and rising the water vapour content in the atmosphere. The former effect dominates at lower, the latter at higher temperatures.

INTRODUCTION

Discussions on thermodynamic equilibrium and on intermediates in the system $CaO-SiO_2$ are unresolved. The equilibrium phases of both the calcium silicates and silica are likely to be influenced by two less investigated factors; trace impurities in the solids and interaction of solids with "inert" components in the atmosphere.

Recent work has reported some observations related to the second problem. Hauffe reported on changes of the electric conductivity of CaO from p- to n- conductivity at $p(O_2) = 1,3$ Pa and 873 K (1). Wagstaff and coworkers observed that the crystallisation rate of fused silica is influenced by the composition of the furnace atmosphere (2). Lee found stable and metastable -OH groups in fused silica when observing the interaction of silica with the atmosphere (3). Burte and Nicholson studying the reaction between CaO and silica monocrystals observed an acceleration of the reaction in the presence of water vapour in the furnace atmosphere (4).

This paper deals mainly with the influence of two "inert" components of the atmosphere i.e. with the influence of oxygen and water vapour on the rate and on the mechanism of the reaction between CaO and silica. The gas components are inert relative to possible chemical changes under the conditions employed and they are interesting from a technological point of view being common components of furnace atmosphere in ceramic technology.

EXPERIMENTAL

<u>Materials preparation</u>. Pure natural α-quartz was ground and refined by HCl extraction. This material contained, according to spectral and microprobe analysis, impurities in concentrations of the order 0,01% or less. The grain size of the material was below 35 μm with 17.5 μm as the predominating fraction.

Calcite - Lachema, CSSR, analytical grade - contained impurities in concentrations 0,005 % or less. The grain size was below 50μm with 19.0 μm as the predominating fraction.

Calcite and silica were mixed, homogenised in molar ratios 2/1 and 1/1, dried and pressed to pellets (d = 10 mm, h = 10 mm) with a pressure of 75 MPa.

<u>Experimental procedure</u>. The samples described above were treated in controlled atmospheres in the temperature range 1473 - 1523 K in a laboratory tube furnace. The studied atmospheres were: air, hydrogen, water vapour and mixtures of these components. (Table 1)

The unreacted portion of both CaO and SiO_2 in the treated samples was determined by chemical analysis: free CaO by the glycolate method and free silica by extracting the bound fraction with concentrated HCl. The composition of the products were controlled by X-ray phase analysis, optical microscopy and microprobe analysis with a JEOL JXA 5 analyser.

Table 1

Investigated experimental conditions

Series	C/S	Temperature (K)	atmosphere composition
1	2:1	1473-1523-1573	air, water vapour, hydrogen
2	1:1	1473-1523-1573	air, water vapour, hydrogen
3	1:1	1573	air, air:$H_2O/9:1, 1:1, H_2O$ $H_2O:H_2/9:1/$, H_2

RESULTS

From the study the following observations were made:
- the reaction rate is in all cases (Table 1) higher in water vapour and in hydrogen than in dry air.
- in the system C/S = 1:1 the reaction is accelerated at 1573 K by hydrogen, at lower temperatures by water vapour,
- in the system C/S = 2:1 the highest reaction rates were observed in hydrogen (series 1.),
- the results of series 3. are given in Fig. 1. The conversion - time dependences could be fitted to semiempirical relations of the type:

$$\alpha_i = \sum_{i=1}^{n} z_i \left[1 - \exp(-k_i t) \right] \quad \{1\}$$

where z_i are weight fractions of the component $\underline{i}$ (C or S) reacting with a rate constant k_i, and t is the reaction time (5).
- from Fig. 1. it follows that a large proportion of both C and S reacts very quickly (in ten minutes conversions >50%) and that the remaining portions react with a very low rate constant,
- correlations of the constants z_i and k_i of the rate equation {1} shows that the effect of H_2 and H_2O is mainly to increase the constant k of the slowly reacting fraction.

DISCUSSION

The accelerating influence of reduced p_{O_2} and of water vapour in the atmosphere on reaction between C and S is very marked. The reaction in the system C/S = 1:1 at 1573 K in hydrogen is accelerated (in comparison to air) by factor of 100 for a conversion α_S= 95% Analysis of the changes of the $CaO.SiO_2$ and $(CaO)_2.SiO_2$ content in the systems studied suggests that the entire reaction proceeds through three partial steps:

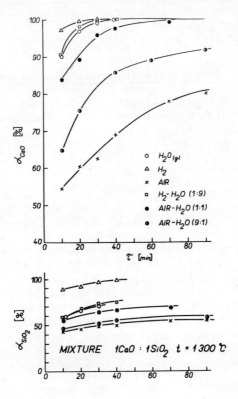

Fig.1. Conversion-time dependence of silica and CaO in different atmospheres at 1573 K

1) C + S = CS
2) 2C + S = C_2S
3) C_2S+ S = 2CS

the last of which seems to be the slowest step. This reaction (3) is realised by diffusion of CaO through the layer of wollastonite and it is principally affected by variations in $p(O_2)$ and $p(H_2O)$. The mechanism of the accelerating effect may be an increase of the diffusion coefficient of CaO through wollastonite either by rising the defectiveness of the oxygen sublattice of wollastonite or by -OH forming in the lattice of wollastonite. The former effect dominates at higher, the latter one at lower temperatures.

REFERENCES

1. Hauffe K. Reaktionen in und an festen Stoffen, Springer Verlag Berlin 1966
2. Wagstaff F.E. Brown S.D. Cutler T.B., Phys.Chem. of Glasses $\underline{5}$, 76 (1964)
3. Lee R.W., Phys.Chem. of Glasses $\underline{5}$, 35 (1964)
4. Buerte A.S. Nicholson P.S., J. Am. Cer. Soc. $\underline{55}$, 469 (1972)
5. Jesenák V., Czechoslovak Conference on Thermal Analysis, Thermanal 73, Oct. 1973, Štrbské pleso, CSSR

AUTOCATALYTIC KINETICS AND MECHANISM OF THE REDUCTION OF CRYSTALLINE COBALT MOLYBDATE IN HYDROGEN

J. Haber, A. Kosłowska, J. Słoczynski

Research Laboratories of Catalysis and Surface Chemistry

Polish Academy of Sciences, 30060 KRAKOW, Poland

ABSTRACT

Reduction of α-CoMoO$_4$ in hydrogen results in the formation of Co$_2$Mo$_3$O$_8$ and Co$_2$MoO$_4$. Its kinetics as measured in a vacuum microbalance and in a X-ray high temperature camera is first order with respect to hydrogen and follows sigmoidal curves. These are well described by the equation of an autocatalytic reaction in a shrinking sphere model $dx/dt = k_1/(1+k_2 x) \cdot p(H_2) \cdot S \cdot (1-x)^{3/2}$ with activation energy 28.2 kcal/mole. The autocatalytic effect is due to traces of metallic cobalt forming in the course of reaction and activating hydrogen, the reduction then proceeding through a spillover effect.

INTRODUCTION

Reduction of simple oxides as examples of reaction types: solid + gas were studied in recent years in a number of papers, whereas only scarce data exists on the reduction of complex oxides such as oxysalts. Many of them, such as molybdates, tungstates, vanadates etc., are known as selective catalysts in the oxidation of hydrocarbons. These reactions proceed via the redox mechanism (1,2), a lattice oxygen ion from the catalyst being inserted into the hydrocarbon molecule, and the catalyst being then reoxidised by oxygen from the gas phase. In the steady state conditions a certain degree of reduction of the catalyst is established, depending on the ratio of the rate of reduction of the solid to the rate of its reoxidation. The steady state degree of reduction may in turn have a pronounced influence on the selectivity of partial oxidation. Information on the kinetics and mechanism of the reduction and reoxidation of such systems is therefore of considerable interest.

EXPERIMENTAL

α-CoMoO$_4$ was obtained by precipitation (3) and annealing at 500° or 600°C. The surface areas determined by the BET method from the adsorption of krypton at liquid nitrogen temperature, were 15.3 m^2/g and 6.7 m^2/g respectively.

Progress of reduction was studied by following the change of mass with a Sartorius vacuum microbalance type 4102, of 1 μg sensitivity. The microbalance was connected to a standard vacuum system (10^{-6} torr) equipped with storage bulbs for gases. The application of two temperature-programmed furnaces enabled the elimination of convection effects, the temperature stability in the range 20-700°C being ±5°C. Measurements were carried out at hydrogen pressures between 4-40 torr. Preliminary experiments have shown that for samples of 0.05-0.2 g the rate of reduction is proportional to the mass of the samples. Thus 0.1 g samples were used in all experiments, the powder of 0.2 mm grain size being spread in the quartz bucket in the forma of a thin layer 1 mm thick.

The experimental procedure consisted in outgassing the sample at 500°C until a constant mass was attained, exposing it to oxygen for 1 hr, cooling to the required experimental temperature, outgassing followed by the introduction of hydrogen at a given pressure. The degree of reduction was calculated on the basis of the stochiometric equation found by Haber and Janas (3,4) and confirmed by Słoczynski (5).

$$4CoMoO_4 + 4H_2 = Co_2MoO_4 + Co_2Mo_3O_8 + 4H_2O \quad \{1\}$$

The second type of measurement of the rate of reduction was carried out in the high temperature X-ray camera. A Rigaku-Denki model D3F diffractometer was used. The finely ground sample was spread in the form of a thin layer of 0.5 mm thick on a 2 cm^2 sample holder made of perforated platinum foil. Temperature was measured by a Pt-PtRh thermocouple placed in the sample. The sample was heated to the temperature of experiment in nitrogen and the rate of reduction was measured in a 1:1 mixture of N_2 and H_2 at a flow rate of 300 ml.min^{-1}. The degree of reduction was calculated from the ratio of intensities of the 3.36 Å line (the strongest line of α-CoMoO$_4$) measured at a given time and in the initial sample. In a special experiment it was checked that the observed rate of reduction was proportional to the surface area of the sample, indicating that there was no limitation by diffusion in the pores.

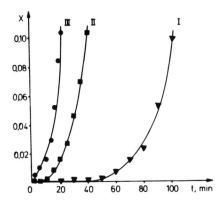

Fig. 1. Reduction of α-CoMoO$_4$ at 480°C in 30 torr of hydrogen, 1 - sample preheated in oxygen, II - sample preheated in vacuum, III - sample of α-CoMoO$_4$ + 5 % CoO.

RESULTS AND DISCUSSION

Fig. 1 shows the kinetic curves of the reduction as measured in the microbalance. Reduction is characterised by the induction time, which depends on the conditions of pretreatment and on hydrogen pressure. It decreases on heating in vacuum and increases after heating in oxygen. All kinetic curves obtained at different temperatures and hydrogen pressures had a sigmoidal shape, showing a strong acceleration after the first period of reaction.

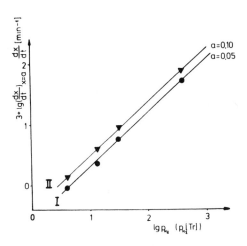

Fig. 2. Rate of reduction of α-CoMoO$_4$ at 500°C as a function of hydrogen pressure at constant degree of reduction x. I-x = 0.05, II-x = 0.1.

In order to determine the reaction order with respect to hydrogen, the rates of the reduction corresponding to the same degree of reduction were plotted as a function of the hydrogen pressure. Fig. 2 shows the data obtained for the degree of reduction x=0.05 and x=0.1 plotted on a double-logarithmic scale. The order of the reaction calculated from the slopes is 0.93, and we may thus conclude that the rate of reduction is proportional to hydrogen pressure.

First order with respect to hydrogen led us to the hypothesis that dissociative adsorption of hydrogen is the limiting step of the reaction. The sigmoidal character of kinetic curves would in such a case by due to the autocatalytic effect of traces of metallic cobalt formed in the course of reduction in the form of highly dispersed clusters. The presence of trace amounts of metallic cobalt in the products of the reduction of α-$CoMoO_4$ was found by one of us in the course of microscopic studies (5). Cobalt is known as a good catalyst for activating hydrogen through its dissociative adsorption. A similar catalytic action of platinum in the reduction of simple oxides has been observed by several authors (6-9).

In order to confirm the autocatalytic model of the reduction of cobalt molybdate, an experiment was carried out in which $CoMoO_4$ was mixed mechanically with 5% or CoO, which is easily reduced to form highly dispersed metallic cobalt. Results of the kinetic measurements carried out in conditions identical to those, in which curves 1 and 2 in Fig. 1 were obtained for pure $CoMoO_4$, are shown as curve 3. In a separate experiment it was shown that under these conditions CoO is completely reduced to metallic cobalt in about 4 min. Results presented in Fig. 1 clearly show that metallic cobalt formed in the first minutes of the experiment accelerates the reduction and completely eliminates the induction period. This indicates that a certain limiting surface concentration of metallic hydrogen activating centres exist, the reduction proceeding through the hydrogen spill-over effect. Such a concentration is easily attained in the circulation reactor, in which dependence of the rate on the mass of the samples is then observed (3), due to poisoning by water.

Microscopic observations of the reduction products indicate that they do not form a compact layer enveloping the $CoMoO_4$ grains, but grow in the form of separate crystallites (5). Thus, there are no diffusional limitations and a shrinking sphere model may be assumed, the decrease of the reaction rate being due to the decreasing surface area of the reduced grains. The rate of reduction may by thus expressed by the equation:

$$dx/dt = k/(a_o + \alpha x) \cdot p(H_2) \cdot S \cdot (1-x)^{3/2} \qquad \{2\}$$

or $\quad Y = \dfrac{1}{(1-x)^{3/2}} \cdot \dfrac{1}{p(H_2) \cdot S} \cdot \dfrac{dx}{dt} = k_1 (1+k_2 x) \qquad \{3\}$

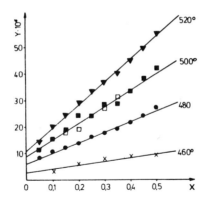

Fig.3. Kinetics of the reduction of α-CoMoO$_4$ in 360 torr of hydrogen at different temperatures, plotted in terms of equation{3}.

where $k_1 = k/a_o$ and $k_2 = \alpha a_o$, k being the rate constant of the reaction, a_o - initial concentration of the catalyst, required to start the reaction, αx - concentration of the catalyst after time t, $p(H_2)$ - partial pressure of hydrogen, S - specific surface area of the solid. Fig. 3 represents the experimental data obtained in flow experiments carried out in the high temperature X-ray camera plotted in the form of Y as a function of x (equation {3}) for different temperatures. A very good fit of experimental points was obtained in a broad range of conversions between 0.05 and 0.5.

In static experiments carried out with the microbalance, deviations from equation {3} were observed at higher degree of conversion, Y assuming higher values. They can be ascribed to heat transfer hindrance, the sample hanging in a quartz bucket of low heat conductivity, isolated by the gas phase. When the reaction attains certain rate, additional acceleration may occur due to rising temperature.

REFERENCES

1. J. Haber, Z.Chem. 13, 241 (1973)
2. J. Haber, B. Grzybowska, J. Catal 28 489 (1973)
3. J. Haber, J. Janas, Proc. Reunion Intern. Cinetique des Reactions dans les Systems Chimiqueś Heterogenes, Dijon 1974
4. J. Haber, J.Less Common Met. 36, 277 (1974)
5. J. Słoczynski, Bull.Acad.Polon.Sci. Ser. sci.chim. (in press)
6. J.E. Benson, H.W. Kohn, M. Boudart J. Catal. 5, 307 (1966)
7. E.J. Nowak, J.Phys. Chem. 73, 3790 (1969)
8. G.E. Batley, A.Ekstrom, D.A. Johnson, J.Catal. 34, 368 (1974)
9. B. Delmon Introduction à la Cinetique Heterogène. ed. TECHNIP, Paris 1969.

DISCUSSION

B. DELMON (Catholic University of Louvain, Louvain-la-Neuve, Belgium)
Two pieces of evidence in this paper suggest an alternative to the
contracting sphere model. As vacuum pretreatment or the addition of
cobalt shortens the induction period and accelerates the first stages
of the reduction, one would expect the nucleation stage to be rate
limiting and hence the number of nucleii to be rather small. The
contracting sphere model would require numerous nuclei to cover
the surface almost completely from the start. Secondly, as curve
III, Fig. 1 shows, metallic cobalt is thermodynamically stable in
your reduction conditions which would suggest that the small metal
clusters that are postulated would grow. In view of these considerations, a nucleation and growth model would be more realistic.

R. HAUL (Technical University of Hannover, West Germany) The autocatalytic process is explained by the intermediate formation of
cobalt clusters which facilitate dissociative adsorption of hydrogen
and subsequent reduction by a spillover mechanism. To what extent
does dissociative chemisorption of hydrogen at the oxide surface
contribute? The mobility of hydrogen on the oxide may however be
rather slow judging from NMR studies on ThO_2 (E. Riensche, Dissert-
tation 1975 Hannover) as compared with the mobility on a metal
surface.

AN IR AND EM STUDY OF THE REACTIVITY OF SOME DIVALENT METAL HYDROXIDES WITH SILICA GEL

T. Baird, A.G. Cairns Smith and D.S. Snell

Department of Chemistry, University of Glasgow

Glasgow G12 8QQ (Scotland)

ABSTRACT

Reactions under reflux conditions between silica gel and solid hydroxides of Ca, Mg, Cd, Cu, Zn and Be were followed progressively by IR and EM techniques and the final products characterised. Identifiable silicates were obtained from the hydroxides of Ca (a tobermorite-like layer silicate C-S-H (I)), Mg (a hectorite-like smectite) and Zn (hemimorphite). Possible reaction mechanisms are discussed.

INTRODUCTION

Granquist and Pollack achieved the synthesis of hectorite by refluxing for some days an aqueous slurry of magnesium hydroxide and silica gel, with a small proportion of lithium fluoride (1). They discussed the mechanism of their reactions, assuming the formation from solution of hectorite "embryos" which crystallised to form solid hectorite. This model, however, did not adequately explain the observed reaction kinetics, and a second, competing reaction was proposed involving the deposition of SiO_2 from solution on to the surface of finely-divided solid $Mg(OH)_2$. A necessary condition of the model was that the competing reactions were independent. Baird, Cairns Smith, and MacKenzie, in a later investigation of this reaction reported the formation of an epitaxially-oriented layer of hectorite-like material on the $Mg(OH)_2$ crystals, thus further demonstrating the existence of the "competing" reaction (2).

The reflux reactions of silica gel with the solid hydroxides of Ca, Mg, Cd, Zn, Cu and Be were studied in this work.

Single polymorphs of these hydroxides can be obtained and unlike the hydroxides of Mn(II), Fe(II), Co(II), and Ni(II) the former hydroxides do not oxidise under reflux conditions. Samples from the reactions were withdrawn at intervals and examined by electron microscopy (EM) and infrared spectrophotometry (IR). The reactions were continued until no further changes were apparent in the spectra, or until an identifiable crystalline silicate was present as the major product. Evidence of the adsorption of silicate ions on to the hydroxide crystals was sought, and an attempt was made to rationalise the observed changes in the early samples from the reactions, in terms of the structures of the hydroxides.

EXPERIMENTAL

The reactions were performed in Pyrex apparatus and were continuously stirred. The source of silica was Mallinkrodt 100 mesh silicic acid. The hydroxides were prepared by precipitation methods. Both ε- and γ-zinc hydroxide were prepared and investigated (3,4). The freeze-dried hydroxides were mixed with the silicic acid in the proportions $M(OH)_2:SiO_2 = 5:6$ (to satisfy the formula of a hypothetical ideal 2:1 trioctahedral layer silicate) and distilled water was then added to form a slurry containing 10% solids. The pH was adjusted to 10 by adding solid LiOH. 50 mg samples were withdrawn at intervals and freeze-dried. The Siemens Elmiskop 1A and Jeol JEM 100C microscopes were used, samples being mounted on carbon film by the droplet technique. IR spectra were recorded on 1 mg samples in 300 mg KBr pressed discs, with the Perkin-Elmer 257, 457, or 221 machines.

The Hydroxide Structures

Ca, Mg, and β-Cd hydroxides share the brucite close-packed layer structure. The cell parameters and the interlayer hydroxyl ion separation vary, but all have (001) cleavage and thus form flat hexagonal crystals. Copper hydroxide is isostructural with lepidocrocite and consists of double layers of distorted octahedra hydrogen-bonded to each other, giving a short interlayer hydroxyl-ion separation, but the crystals still have interlayer (010) cleavage. In γ-zinc hydroxide three ZnO_4 tetrahedra are linked, forming columns along the c-axis. Cleavage between the columns leads to the formation of acicular crystals. By contrast, ε-$Zn(OH)_2$ and β-$Be(OH)_2$ are crosslinked tetrahedral frameworks with no preferred cleavage.

RESULTS

IR spectra of the Mallinkrodt silica gel under various conditions are shown in fig.1a. The band assignments are by Hino and Sato and were based on deuteration studies (5). The only changes in the spectra shown may be ascribed to deprotonation of surface silanol groups of the gel, the Si-O-Si framework remaining essentially undisturbed.

In the early stages of all the reactions, changes were observed in the Si-O-Si stretching region of the spectra. In all cases except $Be(OH)_2$, a new Si-O stretching band was observed between 1000 and 1050 cm^{-1}, and this increased in intensity relative to the original Si-O stretch at 1070 cm^{-1}, as the reaction proceeded. A typical example of this behaviour is shown in fig.1b, the band evolution for the $Mg(OH)_2$ reaction. The absorbance of the new Si-O stretch equals the absorbance of the original after four hours at reflux. This time was different in each reaction, and since it was easily estimated it was taken as a measure of the rate of the early part of the reaction in each case.

The appearance of the new Si-O stretch band was accompanied by a decrease in the particle size of the slurries, a sol showing the Tyndall effect being formed. Electron diffraction studies of the sols were made and all gave fine-grained rings typical of poorly-crystalline phases. In some cases the sol particles could be separately identified in the electron microscope. The sol formed with $Cd(OH)_2$, for example, (fig.2) consists of regular apparently rounded platelets 8 nm across. Only five diffraction rings were observed. The sol from the $Cu(OH)_2$ reaction consisted of tubes about 5 nm in diameter, giving eight diffraction rings. The $Zn(OH)_2$ reactions yielded a sol consisting of irregular particles with a microcrystalline diffraction pattern. The early samples from the $Ca(OH)_2$ and $Mg(OH)_2$ runs differed from the final products only in their lower crystallinity, as far as could be seen in EM, although the IR spectra indicated the existence of a separate phase.

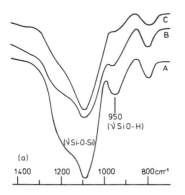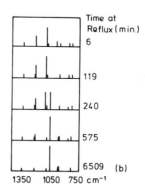

Fig.1a. IR spectra of silica gel; A, air-dry gel; B, 10% slurry at pH 10; C, 10% slurry, pH 10, after 3 days reflux.
1b. IR spectra showing the band evolution for the $Mg(OH)_2$ reaction.

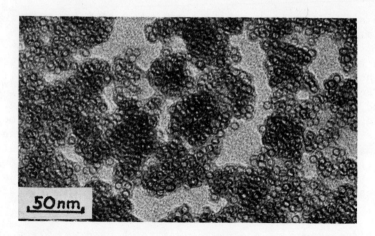

Fig.2. Electron micrograph of the sol particles formed from $Cd(OH)_2$

The Final Products

In the $Ca(OH)_2$ and $Zn(OH)_2$ reactions, the colloidal species recrystallised to form an identifiable silicate. $Ca(OH)_2$ gave the tobermorite-like layer silicate C-S-H(I) (6), $Mg(OH)_2$ gave a hectorite-like smectite, and $Zn(OH)_2$ gave large crystals of hemimorphite. The colloid formed from $Cu(OH)_2$ did not recrystallise. The diffuse electron diffraction pattern is similar to those of natural chrysocollas, and the sol is therefore thought to be a chrysocolla-like copper silicate. The small plates from the $Cd(OH)_2$ reaction also remained unaltered, and the five observed diffraction spacings correspond to none of the known cadmium oxides, hydroxides or silicates. It is thought that this material is a new, poorly crystalline cadmium silicate. At the end of the $Be(OH)_2$ run, the only phases detected were $\beta-Be(OH)_2$ and amorphous silica gel. No reaction between the silica and the beryllium hydroxide was observed.

DISCUSSION

The hydroxides described here form low particle size products on reaction with silica gel. There is some correlation between the times required to form the products, as measured by the IR data described above, and the hydroxyl ion separation distance in the hydroxide crystals (fig.3). This might be interpreted in terms of a mechanism via complete solution of the hydroxide since on the whole, the less soluble hydroxides have shorter OH-OH distances in the crystal. The formation of the low particle size products, however, is particularly slow with the hydroxides without a pre-

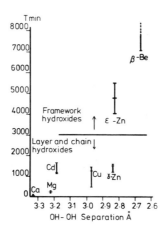

Fig.3. Adsorbate half-times vs. OH-gp separation for $M(OH)_2$

ferred mode of cleavage. In addition, the framework ε-$Zn(OH)_2$ which has a similar solubility to the columnar γ-form exhibits a much longer half-time. We suggest therefore that not only for magnesium hydroxide (2), but generally where there are preferred cleavage modes in a hydroxide, that silica actively tends to break up the crystals: first there is fixation of silicate units on the hydroxide surfaces followed by crystal cleavage, thus imposing fresh surfaces for the process of particle size reduction to continue. Where a crystalline silicate is subsequently formed the sol disappears. Possibly it slowly dissolves to give complex metal silicate ions from which the final product forms through recrystallisation.

REFERENCES

1. W.T. Granquist and S.S. Pollack, Clays Clay Miner., 1959, 8, 150.
2. T. Baird, A.G. Cairns Smith and D.W. MacKenzie, Clay Minerals, 1973, 10, 17.
3. O.K. Srivastava and E.O. Secco, Can.J.Chem., 1967, 45, 579.
4. H.G. Dietrich and J. Johnston, J. Amer.Chem.Soc., 1927, 49, 1419.
5. M. Hino and T. Sato, Bull.Chem.Soc.Japan, 1971, 44, 33.
6. D.S. Snell, J.Amer.Ceram.Soc., 1975, 58, 292.

DISCUSSION

M. FIGLARZ (University of Picardy, Amiens, France) It appears from your paper that it would be difficult to obtain a reaction between Co(II) and Ni(II) hydroxides and silica. These reactions were carried out a long time ago, for example by Cailler or Feitknecht, in pyrex apparatus, without addition of silica, by slight dissolution of the vessel walls. Recently, we have studied the reaction of finely divided Co(II) oxide under reflux conditions when we obtained Co serpentine (J.P. Roussel, Thesis Amiens 1975) The mechanism was considered to be dissolution of the starting materials followed by growth of the disordered silicate phase from the solution to give a turbostratic structure with 001 reflections and hk bands.

AUTHORS' REPLY: We are indeed aware of the work of Caillere, Henin, Feitknecht et al. The divalent hydroxides of Co, Ni, Mn and Fe were avoided, since their structures are unstable under the conditions we used in our present studies. In the absence of oxygen we could, however, predict from Fig. 3 in our paper, "adsorbate half-times" for these hydroxides.

We have also noted dissolution of silica from glass vessels in our studies. We have produced a new magnesium silicate phase by reaction between magnesium hydroxide and saturated potassium carbonate at high pH without prior addition of silica

R.A. KÜHNEL (University of Technology, Delft, Netherlands) What are the swelling characteristics of your synthetic product in comparison with natural smectites and what is the thermal stability of your synthetic product?

AUTHORS' REPLY: Only the hectorite product displays swelling characteristics typical of a swelling smectite. The other products do not appear to be smectites.

All our products lose zeolite water at $120°C$ but no decomposition is observed up to $500°C$.

UNEXPECTED CASES OF REACTIONS BETWEEN SOLID SUBSTANCES AT ROOM TEMPERATURE AND NORMAL PRESSURE. FOUR DIFFERENT EXAMPLES

M.E. Garcia-Clavel, M.I. Tejedor-Tejedor and

A. Martinez-Esparza

Seccion de Termoanalisis y Reactividad de Solidos del

C.S.I.C. Facultad de Quimicas. Ciudad Universitaria,

Madrid-3

INTRODUCTION

We are studying a series of reactions in solid mixtures of oxides with hydrated acids or salts which are prepared in air at room temperature. This paper presents the results of the following equimolecular mixtures $H_2C_2O_4 \cdot 2H_2O/NiO$; $CuSO_4 \cdot 5H_2O/CdO$; $CuSO_4 \cdot 5H_2O/\beta-PbO$ and $CuSO_4 \cdot 5H_2O/\alpha-PbO$. The study has been carried out by X-ray diffraction, IR, TG and DTA techniques.

EXPERIMENTAL

$CuSO_4 \cdot 5H_2O$, $H_2C_2O_4 \cdot 2H_2O$ and NiO, Merck A.R. Crystalline CdO and β-PbO were obtained by calcination of the corresponding carbonates at 300° C/h. heating rate up to 600° and maintaining it at this same temperature for 1 h. These were identified by X-ray. The α-PbO was precipitated from a boiling solution of lead nitrate by means of a caustic soda solution without carbonates. This oxide was indentified by X-ray diffraction

Mechanical equimolecular mixtures of various particle sizes, $H_2C_2O_4 \cdot 2H_2O/NiO < 37\mu$; $CuSO_4 \cdot 5H_2O/CdO < 37\mu$; $CuSO_4 \cdot 5H_2O/\beta-PbO$ and $CuSO_4 \cdot 5H_2O/\alpha-PbO$, were prepared. It was found that no matter what the size of the β-PbO particle, the behaviour of the mixture was indentical. In the case of the α-PbO, the mixture made with particle fractions <60μ showed the same behaviour as the β-PbO mixtures. On the contrary, the mixtures which have been prepared with α-PbO>60μ give place to another reaction.

<u>Apparatus.</u> Chevenard thermobalance; model 93 from Adamel. Photographic register. Heating rate 300°/hr. DTA apparatus constructed in the laboratory. Sintered alumina specimen holder. Chromel/alumel differential thermocouple. Graphic recording Metrohm Labograph E 478. Heating rate 300°/hr. X-ray powder diffraction: Philips P.W. 1010 generator. P.W. 1050 diffractometer. Geiger counter P.W. 1051. Graphic recorder. Cu Kα radiation. Ni filter. Infrared spectroscopy: Unicam spectrophotometer SP1200 Pye from 4000 to 400 cm^{-1} KBr discs.

RESULTS

$H_2C_2O_4 \cdot 2H_2O/NiO$ 1/1 M

The mixture, initially dark grey, becomes humid and turns green very rapidly. The study of the mixture has proved that the reaction is terminated in 48 hrs. Thus, the X-ray diffraction corresponds to the $NiC_2O_4 \cdot 2H_2O$. The TG curve (1-a) corresponds to this, with 3% humidity. Our TG curve does not coincide with the one described by Duval (1). The DTA curve (2-a) confirms the TG results. The reaction produced is the following:

$$H_2C_2O_4 \cdot 2H_2O + NiO \text{ (black)} \longrightarrow NiC_2O_4 \cdot 2H_2O + H_2O$$

This NiO is poorly crystallized, has carbonate and water impurities of 12.8%, which are slowly lost by ignition; it being necessary to arrive at 700° in order to obtain its total loss. The NiO obtained at 700° is green, well crystallized and does not react with oxalic acid at room temperature.

$CuSO_4 \cdot 5H_2O/CdO$ 1/1 M

The mixture shortly after being prepared begins to become humid and several hours later has suffered a complete change in colour, going from reddish brown to blue. The reaction is very rapid, completing the entire process in less than 24 hrs. It is expressed by

$$4CuSO_4 \cdot 5H_2O + 4CdO \longrightarrow Cu_4SO_4(OH)_6 + 3CdSO_4 \cdot 8H_2O + CdO + 9H_2O$$

X-ray clearly indicates the existence of $CdSO_4 \cdot 8/3H_2O$. It has other unidentified lines. The X-ray of the mechanical mixtures $CdSO_4 \cdot 8/3H_2O/CdO$ 1/1, 2/1 and 3/1M, maintained in a water saturated atmosphere, have identified these lines except for the 2θ= 8.00 and 16.03 (fig.3). This fact proves that in the moist mixture, reactions are

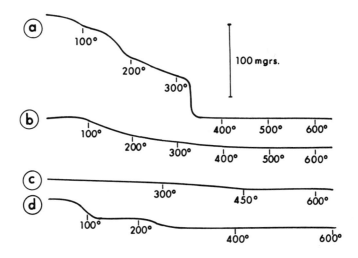

Fig. 1 TG curves of the reacted mixture 1/1 M
a) $H_2C_2O_4 \cdot 2H_2O/NiO$. b) $CuSO_4 \cdot 5H_2O/CdO$. c) $CuSO4.5H/\beta-PbO$
d) $CuSO4.5H/\alpha-PbO$ >60µ.
Weight of samples in mg. a) 322,4. b) 229,3. c) 326,2 d) 342,3.

produced between the hydrated cadmium sulfate, which is the result of the reaction at room temperature, and the excess cadmium oxide. The secondary reactions are confirmed indirectly by the fact that they are not in the $CuSO_4 \cdot 5H_2O/CdO$ 4/3M mixture because of the lack of an excess of CdO. In order to be able to identify the brochantite in the mixture, it was necessary to proceed to the elimination of the $CdSO_4 \cdot 8/3H_2O$, through sucessive washing in cold water. If the elimination of the $CdSO_4 \cdot 8/3H_2O$ is nearly complete (Cu/Cd=60.3(2)), then the diffractogram presents the 8.00 line as very diminished while the 16.03 line has disappeared completely. The diagram of the brochantite appears to be complete. Thus, the $CdSO_4 \cdot 8/3H_2O$ remains trapped in the recently formed lattice of the $Cu_4SO_4(OH)_6$ resulting in the deformation of its crystalline structure.

The results from other techniques confirm the X-ray results. Thus the form of the DTA(2-b) curve becomes similar to that of the poorly crystallized brochantite, given by Pannetier (3). A sample in which the $CdSO_4 \cdot 8/3H_2O$ has been eliminated, has an DTA curve in which the approximation to the well crystallized brochantite curve is very good. Finally the TG curve of the $CuSO_4 \cdot 5H_2O/CdO$ mixture in which we have eliminated the humid water isothermally at 50°, corresponds quantitatively to the stoichiometric formation of brochantite, $CdSO_4 \cdot 8/3H_2O$ and unreacted CdO.

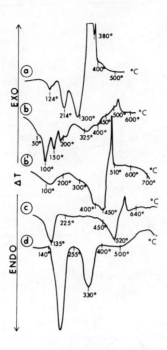

Fig. 2 DTA curves of the reacted mixtures. a) $H_2C_2O_4 \cdot 2H_2O/NiO$ 1/1M b) $CuSO4 \cdot 5H_2O/CdO$ 1/1M b') $CuSO4 \cdot 5H_2O/CdO$ 4/3 washed with water. c) $CuSO_4 \cdot 5H_2O/\beta\text{-PbO}$ 1/1M. d) $CuSO4 \cdot 5H_2O/\alpha\text{-PbO}$ 1/1M >60µ

$CuSO_4 \cdot 5H_2O/\ \beta\text{-PbO}$ 1/1 M

From the moment in which the sulfate and the oxide come into contact, the mixture begins to moisten and to change colour becoming green. This mixture has finished reacting twelve days after its preparation. The X-ray indicates the presence of $Cu_4SO_4(OH)_6$, $PbSO_4$ and β-PbO; the TG gave us the 1-c curve, which is the reproduction of the brochantite TG curve in its zone of dehydration. The 2-c DTA curve fully confirms the TG results. The reaction is:

$$4(CuSO_4 \cdot 5H_2O) + 4\beta\text{-PbO} \longrightarrow Cu_4SO_4(OH)_6 + 3PbSO_4 + \beta\text{-PbO} + 17H_2O$$

The unreacted mol β-PbO, in air, suffers a very slow hydration and carbonation transforming it into $3PbO \cdot H_2O$ and $2PbCO_4 \cdot Pb(OH)_2$.

$CuSO_4 \cdot 5H_2O/\alpha\text{-PbO}$ 1/1 M

In the mixture there is no visible change suggesting that a reaction has taken place. On the contrary, separation of the two phases of the mixture is soon produced, leaving behind the red powder of

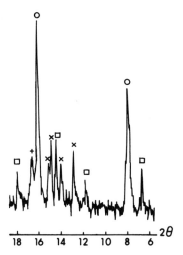

Fig. 3 X-ray diffraction x CdSO4.8/3H$_2$O. ◻ CdSO4.8/3H$_2$O/CdO 3/1 M + Cu 4SO4 (OH)$_6$. o Unidentified lines.

α-PbO in bottom of the container. Nevertheless, the study by X-ray, TG and DTA has revealed that there is a reaction, which is the slow dehydration of the pentahydrate to the trihydrate. The α-PbO did not suffer any transformation. The mixture is maintained in this way for more than 3 months, which the TG and DTA (1-d,2-d) curves prove, these being characteristic of CuSO$_4$.3H$_2$O.

Keeping in mind that CuSO$_4$.3H$_2$O is a very unstable compound, its existence for such a long time in open atmosphere, owed to the presence of α-PbO, must be explained in the following manner: the trihydrate takes water from the atmosphere transforming it into a pentahydrate, but it loses it loses it continuously across the contact zone between the oxide particles and the sulfate particles. Finally the progressive carbonation of the α-PbO (demonstrated by IR) breaks this equilibrium and the trihydrate is transformed back into the pentahydrate.

REFERENCES

1. C. Duval. Inorganic Thermogravimetric Analysis.2d and revis. ed. Elsevier, 362 (1963).
2. G. Schwarzenbach, Las complexonas en al analisis quimico. ed. Atlas, 88 (1959).
3. G. Pannatier et al. Bull.Soc.Chim. 2616 (1963).

DISLOCATIONS AND THEIR ROLE IN THE THERMAL DEHYDRATION OF $Ba(ClO_3)_2 \cdot H_2O$

G.G.T.Guarini and R.Spinicci

Institute of Physical Chemistry, University of

Florence. Via Gino Capponi,9 -50121 FIRENZE- (Italy)

ABSTRACT

The role of dislocations in the initial stages of the dehydration of $Ba(ClO_3)_2 \cdot H_2O$ single crystals is studied. Among the active slip systems, those leading to the formation of probable potential nucleus forming sites are selected.

INTRODUCTION

In a study, by Differential Scanning Calorimetry, of the kinetics of dehydration of $Ba(ClO_3)_2 \cdot H_2O$, some anomalies, mainly concerning the initial portion of the reaction and the behaviour of microcrystals prepared by only slightly differing crystallization procedures, were evidenced. As these could not be explained on the grounds of the thermal method used and of its sensitivity to several parameters, they were assumed to depend on the different content of structural imperfections of the crystals used. (1-4). On the other hand it is well known that the reactivity of solids varies at the point of emergence of line defects on the surface and this enhanced reactivity is often entirely responsible for the early stages of a reaction (5-7). The correspondence between etch-pits and dislocations having been proved beyond any doubt (8,9), we undertook a study, by optical microscopy using the etching technique, of the as-grown and cleavage faces of $Ba(ClO_3)_2 \cdot H_2O$ single crystals with the aim to ascertain the active slip systems and their role in the early stages of the dehydration. The results obtained in this study are reported and discussed in the present communication.

EXPERIMENTAL

Single crystals of $Ba(ClO_3)_2 \cdot H_2O$ were prepared by slow evaporation at room temperature of nearly saturated solutions in bidistilled water. The morphology of the crystals was almost exactly the one indicated by Groth and the interfacial angles, as determined by optical goniometry, compared favourably with both those reported and those computed on the grounds of the structure refinement by Sikka et al. (10,11). Cleavage was performed manually by a razor blade. 96% ethanol was used as etchant, the etching time being 1 to 3 minutes at room temperature. Deformation was performed by compression of opposite faces of the crystal. A Reichert Zetopan microscope equipped with hot stage, interference contrast and interferometer after Nomarski and an Exakta VX 1000 camera were used throughout. (12).

RESULTS AND DISCUSSION

Typical etch patterns for the $\{110\}$ and $\{011\}$ faces are shown in fig. 1,a,b,c,d,e,f; some of the directions of preferential etch pit alignment are indicated. Fig. 1a and 1b show the correspondence of etching on matched (011) cleavage faces, while fig. 1c and 1d show some preferential alignments of etch pits on the (110) cleavage and as grown faces respectively. Some characteristic features of dislocations as revealed by etching are shown in fig. 1e and 1f which refer to the etching of screw dislocations emerging on (011) and to a "pile up", always on (011), respectively. In fig. 2a and 2b alignments of nuclei, on (110) and (011) respectively, are shown, while in fig. 2c and 2d the correspondence of growth nuclei and zones of high etch pit density, on matched (110) cleavage planes, is evidenced.

By the etching technique several preferred etch pit alignments were observed on $\{011\}$ and $\{110\}$ planes. By means of the zone relationships, we determined the crystallographic planes which, upon intersecting the crystal surfaces, gave the directions of the observed alignments. When however some directions of etch pit alignment , say on the (110) face, were justified by a plane which, upon intersecting the (011) face gave a direction for which no alignment was found, the plane was not taken into consideration as a possible slip plane. Among the remaining planes the ones having maximum packing density were chosen according to the rule that glide is easier in high packing density planes and in high packing density directions. A further reduction of the number of possible slip planes was achieved by excluding, on the grounds of a tridimensional model of the structure, the ones whose glide could involve appreciable electrostatic repulsions.

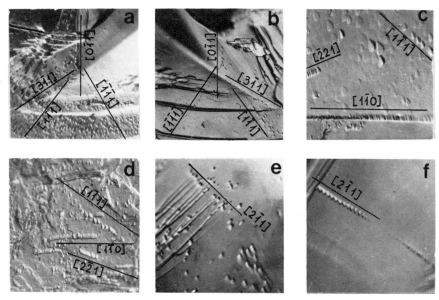

Fig. 1-Etching of $Ba(ClO_3)_2 \cdot H_2O$ crystal surfaces (see text).

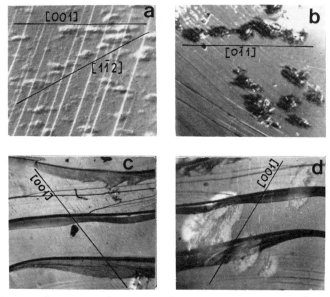

Fig. 2-Dehydration nuclei and their correspondence with high etch pit density zones.

TABLE I
Directions of observed alignements and corresponding slip systems.

Etch pit align. on (011)	(110)	Crystal plane	Slip directions	Nuclei align. on (011)	(110)
[0$\bar{1}$1]	[1$\bar{1}$2]	($\bar{1}$11)	[211]	[0$\bar{1}$1]	[1$\bar{1}$2]
[2$\bar{1}$1]	[1$\bar{1}$0]	(11$\bar{1}$)	[2$\bar{1}$1]	[2$\bar{1}$1]	[1$\bar{1}$0]
[2$\bar{1}$1]	[$\bar{1}$12]	(1$\bar{1}$1)	[21$\bar{1}$]	[2$\bar{1}$1]	[$\bar{1}$12]
[1$\bar{1}$1]	[1$\bar{1}$1]	($\bar{1}$01)	[11$\bar{1}$]	[1$\bar{1}$1]	[1$\bar{1}$1]
[100]	[001]	(010)	[001]		[001]
[$\bar{1}$11]	[$\bar{1}$11]	(101)	[$\bar{1}$11]	[$\bar{1}$11]	[$\bar{1}$11]
[$\bar{3}$11]	[$\bar{3}$31]	(103)	[$\bar{3}$31]		
[$\bar{3}$11]	[$\bar{3}$31]	($\bar{1}$03)	[$\bar{3}$31]		
[$\bar{3}$22]	[$\bar{3}$32]	(203)	[$\bar{3}$32]		
[$\bar{1}$22]	[$\bar{2}$23]	($\bar{2}$12)	[$\bar{1}$22]		
[$\bar{3}$22]	[$\bar{2}$21]	(2$\bar{1}$2)	[$\bar{1}$22]		[$\bar{2}$21]
[$\bar{3}$22]	[$\bar{3}$32]	($\bar{2}$03)	[$\bar{3}$32]		

When large and well defined etch pits were obtained, they were also studied by interferometry and the determination of the planes where the dislocations lie, was performed as suggested by Bengus. (13). The study of dislocation motion caused by heating to a temperature just lower than the decomposition one did not add further information. The results of this work are collected in Table I in which some directions of alignments observed for decomposition nuclei are also reported as well as the probable directions of slip.

The problem of choosing what dislocations were conducive for reaction in our experimental conditions was solved taking into consideration the observed preferential alignments of decomposition nuclei and comparing them with the alignments of etch pits (see Table I and fig. 2a and 2b). From this it appears that only the dislocations $\{111\} \langle \bar{2}11 \rangle$, $\{101\} \langle \bar{1}11 \rangle$ and (010) [001] are effective. The slip on $\{111\}$ planes of $\frac{1}{2}$ of the unit vector along the $\langle \bar{2}11 \rangle$ directions changes, on the (110) crystal plane, the $H_2O...H_2O...H_2O$ sequence along the [001] direction to $H_2O.H_2O...H_2O$. However for the (011) plane, the nucleus forming site might easily be thought the change in the stacking along 001 from $Ba..H_2O..Ba..H_2O$ to $Ba..H_2O..H_2O..Ba$ as brought about by the $\{111\} \langle \bar{2}11 \rangle$ and $\{101\} \langle \bar{1}11 \rangle$ systems.

An interesting feature of the dehydration reaction studies is connected with the shape of growing nuclei. This differs for (011) and (110) surfaces being circular and elliptical, respectively (see fig. 2d for (110) faces); however, a determination of the activation energies for growth in the 117-155°C temperature range shows that while these are almost equal for radial growth on (011) and for transverse and longitudinal growth on (110) (about 10 ± 1 kcal/mole), the longitudinal growth rate is higher. Such a result can only be explained in terms of a higher probability of the growth reaction in the [001] direction on (110) planes as accounted for by the disposition of the water molecules in the same plane.

ACKNOWLEDGEMENTS

The authors are indebted to Prof. E. Ferroni for helpful suggestions and to C.N.R. for financial support.

REFERENCES

1) G.G.T.Guarini and R.Spinicci: to be published.
2) G.G.T.Guarini, R.Spinicci, F.M.Carlini and D.Donati: J.Therm. Anal. $\underline{5}$, 307 (1973).
3) G.G.T.Guarini, R.Spinicci and D.Donati: J.Therm.Anal. $\underline{6}$, 405 (1974).
4) G.G.T.Guarini, R.Spinicci and D.Donati: "Thermal Analysis" Proceedings of the IV I.C.T.A. Vol.I pag. 185, Akademiai Kiado Budapest 1975.
5) J.M.Thomas: Advances in Catalysis: $\underline{19}$, 293 (1969).
6) J.O.Williams, J.M.Thomas, Y.P.Savintsev and V.V.Boldyrev: J.Chem.Soc. A 1971, 1757.
7) V.V.Boldyrev: J.Therm.Anal. $\underline{7}$, 685 (1975); $\underline{8}$, 175 (1975).
8) A.Bergheran and A.Fourdeux: J.Less Common Metals $\underline{28}$, 357 (1972).
9) J.M.Thomas and G.D.Renshaw: J.Chem.Soc. A 1967, 2058.
10) P.Groth: "Chemisches Krystallographie" Vol.II, pag. 114 (Leipzig 1906).
11) S.K.Sikka, S.N.Momin, H.Rajagopal,and R.Chidambaram: J.Chem.Phys. $\underline{48}$, 1883 (1968).
12) G.Nomarski and A.R.Weill: Rev. de Metallurgie $\underline{52}$, 121 (1955).
13) V.Z.Bengus, F.F.Lav'rentev, L.M.Soifer and V.I.Startsev: Soviet Phys. Crystallography $\underline{5}$, 418 (1960).

NEW INSIGHTS ON THE THERMAL DECOMPOSITION OF AMMONIUM PERCHLORATE FROM STUDIES ON VERY LARGE SINGLE CRYSTALS*

P. J. Herley

State University of New York

Stony Brook, N. Y. 11794 (U.S.A.)

P. W. Levy

Brookhaven National Laboratory

Upton, N. Y. 11973 (U.S.A.)

INTRODUCTION

Surprisingly, it appears that past studies on the thermal decomposition of ammonium perchlorate have <u>not</u> included measurements on relatively large single crystals. As described below, studies on crystals larger than 2 or 3 mm provide new insights on the thermal decomposition processes in these and other materials. Also, the information obtained from large crystals obviously applies to microcrystals or powders.

A large fraction of the relevant previous publications describe the thermal decomposition processes in pristine or unirradiated material. A second fraction includes papers on ammonium perchlorate irradiated prior to decomposition. The classical Avrami-Erofeyev kinetics - perhaps with some refinement - appears to provide a relatively good description of the total gas evolution vs heating time curves for unirradiated material (1,2,3,4). Also, the nucleation and growth aspects of this theory appear to correlate well with observations relating reaction sites to etch pits and other dislocation related phenomena (5,6,7). In general, prior irradiation markedly increases the decomposition rate (7). This acceleration has been attributed to a radiation induced increase in the number of decomposition sites while the underlying thermal decomposition reaction appears to be unaffected (7,8).

The relation between decomposition sites and dislocations, and the attribution of the radiation induced acceleration to an increase in decomposition sites, are both strongly supported by the demonstration that dislocations are generated during irradiation (10).

Although an understanding of the ammonium perchlorate decomposition mechanism has improved in recent years, a number of features of this process are not understood. For example, are the mechanisms and kinetics different for the m, c and other faces? Why does the retention, i.e. the material left after the decomposition reaction terminates, vary from one sample to the next?

EXPERIMENTAL

All measurements were made on single crystals cleaved from 20 x 20 x 60 mm crystals grown by slowly cooling well stirred aqueous solutions containing reagent grade material purified by repeated fractional crystallization. The crystals were water clear and free of all visible imperfections. They were thermally decomposed at 227C in the apparatus described previously (8). In every case the decomposition extended beyond the deceleratory period. During irradiation, in water cooled 60-Co sources, the crystal temperature never exceeded 30C. After irradiation and/or decomposition either the entire crystal or cleaved sections were examined with optical or scanning (SCM) and transmission electron microscopes.

RESULTS AND DISCUSSION

A number of conclusions obtained with large, unirradiated decomposed crystals are shown in Fig. 1. Most importantly, all three slabs contain completely unreacted regions. The center portion reacted only within 0.2 - 0.3 mm of the original external faces. Approximately one-half of the original upper and lower m-faces reacted; the remaining parts are free of even partial decomposition. The results for large c-face crystals are illustrated by Fig. 2. Clearly, the entire c-face has been decomposed. This

Fig. 1 Top, middle and bottom portions of a crystal cleaved parallel to the m-face after thermal decomposition.

THERMAL DECOMPOSITION OF AMMONIUM PERCHLORATE

is in accord with a previous report indicating that the c-face reaction proceeds much more rapidly than the m-face reaction (5).

The reacted material adjacent to the external surfaces was studied using SCM techniques. Figure 3 shows an overall and an enlarged area of the surface obtained by cleaving a reacted crystal perpendicular to the m-face. Nearest the original surface is totally reacted material, i.e. residue. The next "layer" contains unreacted material with large multiply connected voids and, with increasing distance from the surface, the voids become smaller, decrease in number, and the connecting channels diminish both in size and number.

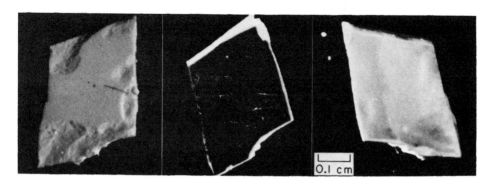

Fig. 2 Top, middle and bottom portions of a crystal cleaved parallel to the c-face after thermal decomposition.

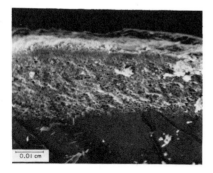

Fig. 3 Scanning electron microscope pictures, at different magnifications, of a section through a thermal decomposition reaction zone on the m-face obtained by cleaving parallel to the c-face.

At the edge of the reacted zone the voids are not connected and have become vanishingly small. Sections through reacted areas on the m and c faces are indistinguishable.

Fig. 4 Top, middle and bottom portions of a crystal cleaved parallel to the m-face after exposure to 10^5R and thermally decomposed.

All measurements described above were repeated with crystals irradiated prior to decomposition. Below 2×10^4R irradiation does not produce observable effects. Between 2×10^4 and 10^6R the crystals are transparent after irradiation but become translucent upon heating, e.g. see Fig. 4. High dose effects were reported previously (11). Namely, after a 10^6R irradiation, the crystals appear milky and are opaque, at 2×10^7R extensive pitting occurs and after 5×10^7R the crystals contain numerous voids.

These results suggest that the application of Avrami-Erofeyev kinetics to both unirradiated and irradiated ammonium perchlorate should be reexamined. Space is available to discuss only a few pertinent points.

First, retentions larger than 70%, and as high as 96% are often observed with large crystals. This large variation in retention is easily explained since the reaction is confined to a layer on the c-face and only part of the m-face. Large retentions occur when part of the crystal does not lie within 0.2 - 0.4 mm of a c-face and/or contains unreacted area on the m-face. Second, the decomposition reaction is clearly surface related, at least on the c-face; more specifically, confined to 0.2 to 0.4 mm from the surface. Thus the assumption that the decomposition proceeds from randomly distributed sites <u>throughout</u> the crystal breaks down for "thick" crystals. It probably applies only to the reacted surface layer permeated by connected voids. Stated alternatively, the usual Avrami-Erofeyev kinetics apparently applies only to crystals with dimensions less than 0.2 or 0.3 mm. Third, the fact that only part of the m-face reacts can be explained in a number of ways. The initial decomposition reaction could be confined to the c-face and the observed m-face reactions represent decomposition commencing on the exposed c-faces of microcracks or cleavage steps. Alternatively, m-face reactions could occur only where dislocation densities exceed a given level; the unreacted part is dislocation free or nearly so. Lastly, consider at least one explanation for the reaction to decrease as it proceeds inward. The unconnected voids at the edge of the reaction zone, in some way, must be influenced by reactions nearer the surface. These voids could be strain related; the reaction occurs at points of high stress. In turn, the stress is produced by trapped gas, differential thermal expansion, etc., associated with the reacting surface. These voids could also be formed by "electronic" or chemical reactions occurring at defects, dislocation pinning points, grain boundaries, etc. The reactions would involve one or more charge and/or energy carriers, such as electrons, holes, protons, excitons, etc., created by the decomposition reaction (12). Such electronic mechanisms explain a number of the radiation produced effects. These mechanisms provide "explanations" for the reaction to diminish as it proceeds. If the strain or the electronic carrier density is too low the reaction is not self-sustaining and diminishes as it proceeds inward.

*Research supported by the Energy Research and Development Adm.

1. P.W.M. Jacobs and H.M. Whitehead, Chem. Rev., 69 551 (1969).
2. D.A. Young, Decomposition of Solids, Pergamon, London, 1966.
3. P.W.M. Jacobs and W.L. Ng, Proc. 7th Int. Symp. Reactivity Solids, Bristol, England, 398 (1972).
4. V.V. Boldyrev, Yu. P. Savintser and T.V. Moolina, Proc. 7th Int. Symp. Reactivity Solids, Bristol, England, 421 (1972).
5. P.J. Herley, P.W.M. Jacobs and P.W. Levy, Proc. Roy. Soc. London, A 318, 197 (1970).
6. P.J. Herley, P.W.M. Jacobs and P.W. Levy, J. Chem. Soc. A, 434 (1971).
7. J.C. Williams, J.M. Thomas, Y.P. Savintser and V.V. Boldyrev, J. Chem. Soc. A, 1757 (1971).
8. P.J. Herley and P.W. Levy, J. Chem. Phys., 49 1493, 1500 (1968).
9. P.W. Levy and P.J. Herley, J. Phys. Chem., 75, 191 (1971).
10. P.J. Herley and P.W. Levy, Proc. 7th Int. Symp. Reactivity Solids, Bristol, England, 387 (1972).
11. P.J. Herley and P.W. Levy, Proc. Conf. Natural and Man-Made Radiation in Space, NASA Tech. Memorandum X-2440, 584 (1972).
12. P.W. Levy and P.J. Herley, Proc. 6th Int. Symp. Reactivity Solids, Schenectady, New York, 75 (1968).

DISCUSSION

A.K. GALWEY (Queens University of Belfast, Northern Ireland, U.K.)
If you remove the outer layer of partially decomposed salt on completion of reaction of a large single crystal and reheat do you get further product formation?

AUTHORS' REPLY: Yes: We decomposed a 5 x 5 x 3 mm single crystal to a $\alpha = 1$ at 227°C in vacuo and then cleaved all faces to expose the pristine unreacted crystal underneath. After decomposition to $\alpha = 1$ this crystal was again cleaved exposing the smaller, unreacted crystal within. Presumably the limitation of unreacted crystal size is a function of the thickness of the reaction product layer, a crystal approximately 0.5 to 0.6 mm in thickness (twice the product layer thickness) is completely reacted throughout.

TRAPPING OF HYDROGEN BY STRUCTURAL DEFECTS IN α-IRON

E.M. Riecke

Max-Planck-Institut für Eisenforschung GmbH

4000 Düsseldorf, Max-Planck-Straße 1, BRD

The anomalous diffusivity of hydrogen in cold-worked iron at low temperatures is due to the attractive interactions between dissolved hydrogen and structural imperfections produced by the deformation. Since hydrogen atoms occupying interstitial sites expand the host lattice, crystal defects such as vacancies, dislocations and interfaces trap hydrogen and affect thus the mobility of hydrogen atoms in the iron lattice. Effusion measurements, therefore, under conditions in which hydrogen diffusion is the rate determining process, yield information concerning the trapping reaction and the binding energy for hydrogen trapped by lattice defects.

The electrochemical permeation technique was used to study this influence of traps on hydrogen diffusion. Pure iron specimens were either cold-rolled (about 60 or 80% reduction in thickness) or recrystallized (24 h at 800°C). Some specimens were cold-rolled and then heat-treated for 24 h at 100°C in order to examine the influence of temperatures up to 100°C on the trap-density. The iron foils were palladium plated on both sides. Hydrogen was introduced into one side of the membrane by means of cathodic polarization at -1 mA/cm^2 in a 0.1 n NaOH solution containing 10^{-3}m As_2O_3, while the other side was held at an anodic potential of 200 mV(E_h). This potential was sufficient to ionize any hydrogen arriving at the surface after passage through the metal. The resulting anodic current is a direct measure of the hydrogen diffusing out.

When steady state permeation was reached, the charging current was shut off and the decay of the permeation current was recorded. During the measurement of the decay transients both sides of the iron membrane were held at 200 mV(E_h). The boundary conditions, pertaining to this method are

$$c_i = c_i^o(d-x)/d \quad \text{for} \quad 0 \leq x \leq d \; ; \; t = 0$$

and $\quad c_i = 0 \quad\quad\quad \text{for} \quad x = 0; \; x = d; \; t > 0$

where c_i is the concentration of dissolved hydrogen on interstitial sites, $c_i^o = c_i(x=0;t=0)$ and d is the thickness of the iron membrane. Hence, the normalized hydrogen flux at the anodic side (x=d) is given by

$$\frac{j}{j_{st}} = \frac{4}{\sqrt{\pi}} \cdot B \cdot \sum_{n=0}^{\infty} \exp\left\{-(2n+1)^2 B^2\right\} \tag{1}$$

where $j_{st} = j(x=d;t=0)$ and $B \equiv d/2\sqrt{Dt}$ if the hydrogen diffusivity D is constant.

Some experimental results are shown in Figs.1 and 2. First, there was no essential change in the run of the decay transients, that means in the density of the effective traps, by heating the cold-rolled specimens up to 100°C, Fig.1. In Fig.2 a series of decay curves is given, measured in the temperature range of 14 to 90°C, using cold-rolled iron foils. The dashed curves represent decay transients calculated according to equation (1) by use of constant values for the diffusivity. Corresponding results are obtained for recrystallized iron. Through parallel displacement of the theoretical curve each point on the experimental curve can be matched with a fictitious diffusion coefficient. The D_s-values so obtained are plotted against the degree of hydrogen effusion in Fig.3. They exhibit in the beginning a certain dependence on hydrogen concentration, approaching then a constant value of D_s^*.

These results can be analysed in terms of a theory developed by A. McNabb and P.K. Foster[1], based on the assumption of local equilibrium between dissolved and trapped hydrogen. Two types of traps are assumed[1,2], one which cannot be removed by annealing and is present in recrystallized iron, the other produced by cold deformation. Let $N_1;N_2$ be the densities of the traps concerned respectively, and $n_1; n_2$ the fractions of traps occupied, then hydrogen effusion can be described by

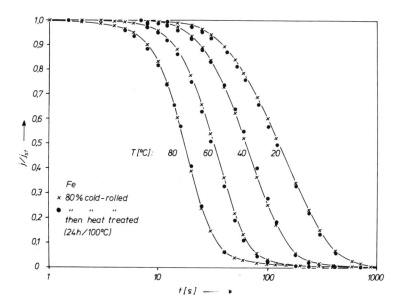

Fig.1 – Decay transients for cold-rolled and for heat-treated pure iron

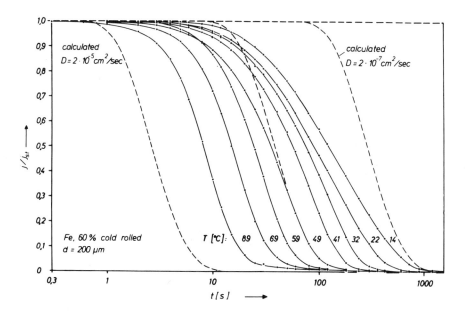

Fig.2 – Decay transients for cold-rolled pure iron in dependence on temperature

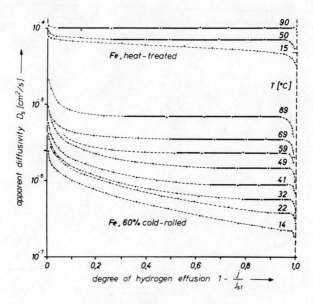

Fig.3 – Apparent diffusivity in dependence on the degree of hydrogen effusion

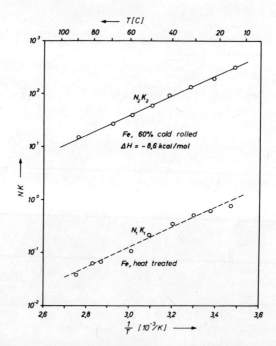

Fig.4 – Product of trap-density and equilibrium constant as a function of reciprocal temperature

$$\frac{\partial c_i}{\partial t} + N_1 \cdot \frac{\partial n_1}{\partial t} + N_2 \cdot \frac{\partial n_2}{\partial t} = D_i \frac{\partial^2 c_i}{\partial x^2} \qquad (2)$$

where D_i is the intrinsic diffusivity of hydrogen in iron. D_i is assumed to be independent on concentration. The rate equations for trapping or hydrogen escape from traps are given by

$$\frac{\partial n_\nu}{\partial t} = k_\nu c_i (1-n_\nu) - p_\nu n_\nu ; \quad \nu = 1, 2 \qquad (3)$$

where k_ν, p_ν are the rate constants and $c_i = n_i N_i$, the concentration of dissolved hydrogen as above, with $n_i \ll 1$. If local equilibrium between trapped and dissolved hydrogen is established at any time during hydrogen effusion the equilibrium constant is

$$K_\nu \equiv \frac{k_\nu}{p_\nu} = \frac{n_\nu}{c_i(1-n_\nu)} ; \quad \nu = 1, 2 \qquad (4).$$

Equations (2) and (4) yield the apparent diffusivity

$$D_s = D_i / \left\{ 1 + \frac{N_1 K_1}{(1+c_i K_1)^2} + \frac{N_2 K_2}{(1+c_i K_2)^2} \right\} \qquad (5).$$

In case of diminishing hydrogen concentration and $c_i K_\nu \ll 1$, the apparent diffusivity becomes constant

$$D_s^* = D_i / (1 + N_1 K_1 + N_2 K_2) .$$

These D_s^*-values can be obtained from the experimental results in Fig.3.

Consider recrystallized specimens, then $N_2 \equiv 0$. Hence, the product of trap-density and equilibrium constant is given by $N_1 K_1 = (D_s^*/D_i) - 1$. D_i-values are taken from literature[3] according to

$$D_i = 4.15 \cdot 10^{-4} \exp(-1.0 \text{ kcal/mol} \cdot RT)$$

The corresponding product $N_2 K_2$ for cold-rolled iron can be calculated in the same way, because $N_1 K_1 \ll N_2 K_2$. In Fig.4 the $N_\nu K_\nu$-values are plotted against temperature. The enthalpy of the trapping reaction is found to be

−8.6 kcal/mol. The similar temperature dependence of $N_\nu K_\nu$ suggests that in both cases one dominant type of trap has controlled the hydrogen effusion process. These traps are assumed to be dislocations and dislocation pile-ups. An estimation yields $N_2 = 8.7 \cdot 10^{19}$ trap-sites/cm^3 or $5 \cdot 10^{11}$ cm/cm^3 for the corresponding dislocation density taking into account 5 trap-sites[4] per plane intersecting the dislocations. Fig. 4 shows that $N_1 = N_2/300$.

Literature

1) A. McNabb; P.K. Foster: Trans. Metallurg. Soc. AIME
        <u>227</u> (1963), S. 618/27

2) A.J. Kumnick; H.H. Johnson: Met. Trans. <u>5</u> (1974),
        S. 1199/1206

3) Th. Heumann; E. Domke: International meeting on hydrogen in metals, Jülich (1972), Vol. 2, S. 492/515

4) R.A. Oriani: Acta Met. <u>18</u> (1970), S. 147/57

CATION DIFFUSION, POINT DEFECTS AND REACTION IN THE CoO-ß-Ga_2O_3 SYSTEM

W. Laqua, B. Küter and B. Reuter

Institut für Anorganische und Analytische Chemie der

Technischen Universität Berlin, Berlin, Germany

INTRODUCTION

The kinetics of a solid state reaction between two oxides AO and B_2O_3 (A, B = cations in the oxidation state +2, +3 resp.), which may form the compound AB_2O_4, can be influenced by the mutual solubility of the reactants as stated by Pelton, Schmalzried and Greskovich (1). Therefore we have to distinguish between so-called reactions of the first and second kind (2); reactions of the first kind take place between pure oxides AO and B_2O_3, while reactions of the second kind occur between the phase AO, presaturated with B_2O_3, and the phase B_2O_3, presaturated with AO.

While the ß-Ga_2O_3 lattice is not able to take up CoO, ß-Ga_2O_3 is readily soluble in the CoO phase (3); therefore in the course of reaction between CoO and ß-Ga_2O_3, there will be a transport of ions not only in the product phase, which is a spinel-type compound of the general formula $Co_{1-y}Ga_{2+2y/3}O_4$ (4), but in the CoO phase itself. In consequence, reaction rate constants of the first kind must be smaller than those of the second kind; the difference is determined by the value of the solubility limit of ß-Ga_2O_3 in the CoO phase, which must be formulated as $Co_{1-z}Ga_{2z/3}O$, and the diffusion flux in this phase, depending on kind and concentration of majority point defects. The situation will be clearer by the schematic illustration given in fig. 1.

The intention of the present paper is to compare experimental determined reaction rate constants with values, calculated on the basis of assumptions made by Pelton, Schmalzried and Greskovich (1) in connection with the more fundamental theory on the formation of ternary ionic crystals developed by Wagner and Schmalzried (5, 6).

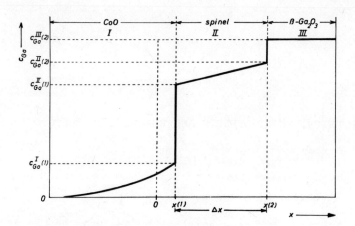

Fig. 1 Gallium concentration profile in a CoO/ß-Ga$_2$O$_3$ sandwich type diffusion couple after a first kind reaction (the initial boundary is located at x = 0)

THEORETICAL ASPECTS

If a solid state reaction proceeds by a parabolic rate law reaction rate constants of the first and second kind $k^{(1)}$ and $k^{(2)}$ resp. are defined by

$$\Delta x^2 = 2 \cdot k^{(1)} \cdot t \quad \text{and} \quad \Delta x^2 = 2 \cdot k^{(2)} \cdot t. \tag{1}$$

In the terminology of Pelton et al. (1) there is:

$$\sqrt{2 \cdot k^{(1)}} = \alpha \quad \text{and} \quad \sqrt{2 \cdot k^{(2)}} = \text{ß}. \tag{2}$$

The following expression is obtained for the ratio R of reaction rate constants of the first and second kind:

$$R = \frac{\alpha}{\text{ß}} = \left[1 + (r_1 f_1)/\alpha\right]^{-1/2}. \tag{3}$$

Here r_1 stands for

$$r_1 = \frac{c_{Ga}^{I}(1)}{c_{Ga}^{II}(1)} \cdot Q_1 \cdot \sqrt{D^{I}(1)} \tag{4}$$

with

$$Q_1 = \frac{1 + 1,5 \left[c_{Ga}^I(1)/c_{Co}^I(1) \right]}{1 - \left[c_{Co}^{II}(1) c_{Ga}^I(1) \right] / \left[c_{Ga}^{II}(1) c_{Co}^I(1) \right]} \qquad (5)$$

The nomenclature is according to fig. 1; as an example, $c_{Ga}^{II}(2)$ means the molar concentration of gallium in phase II (spinel) at the phase boundary 2 (spinel/ß-Ga_2O_3). $D^I(1)$ is the interdiffusion coefficient in the CoO phase, saturated with ß-Ga_2O_3.
The function f_1 is given in a linear approximation by

$$f_1 = 1,1284 + 1,4 \, \eta_1 \qquad (6)$$

for concentration independent diffusion with

$$\eta_1 = \alpha_1/2 \sqrt{D^I(1)} \, . \qquad (7)$$

The constant α_1 describes the rate of phase boundary 1, which obeys a parabolic law (4) $x_1 = -\alpha_1 t^{1/2}$ as does the difference $x_2 - x_1 = \Delta x$ too.

EXPERIMENTAL

Polycrystalline materials were used in all investigations. Because the sintering behaviour of both pure phases and the solid solutions is very good, the samples are of porosities of 5-10 % only with average grain sizes in the range of 300 - 700 µm. Tracer diffusion was measured with the rest activity method by use of $^{60}Co^{2+}$ and $^{67}Ga^{3+}$ radiotracers.

RESULTS

Typical micrographs of reaction couples after first kind and second kind reactions between CoO and ß-Ga_2O_3 and $Co_{1-z}Ga_{2z/3}O$ and ß-Ga_2O_3 resp. are shown in fig. 2. The reaction product formed is the spinel phase $Co_{1-y}Ga_{2+2y/3}O_4$ (4). Platinum markers, originally situated at the interface between the reactants, are embedded in the bulk CoO after first kind reactions, but they are found within the reaction layer after reactions of the second kind.

The parabolic rate law is followed by both reactions of the first and second kind; rate constants are summarized in table 1. The activation energy, estimated from an Arrhenius plot, is in the order of 71 kcal/mol.

A large concentration dependence for gallium and cobalt tracer diffusion rates, which are in the same order of magnitude, has been

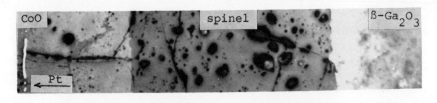

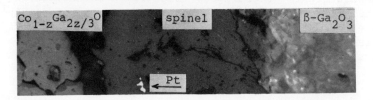

Fig. 2 Transverse section of couples of a.) $CoO/\beta-Ga_2O_3$ and b.) $Co_{1-z}Ga_{2z/3}O/\beta-Ga_2O_3$ after reaction at 1510 °C (200x)

Table 1 Experimental reaction rate constants and calculated values after equation 3 and 7 resp.

T (°C)	$k^{(1)}$ ($cm^2 sec^{-1}$)	$k^{(2)}$ ($cm^2 sec^{-1}$)	$k^{(2)}$(calc.) ($cm^2 sec^{-1}$)	α_1 (calc.) ($cm sec^{-1/2}$)
1260	$1,91 \cdot 10^{-10}$	$4,56 \cdot 10^{-10}$	$4,77 \cdot 10^{-10}$	$-1,61 \cdot 10^{-5}$
1340	$6,15 \cdot 10^{-10}$	$1,50 \cdot 10^{-9}$	$1,57 \cdot 10^{-9}$	$-3,06 \cdot 10^{-5}$
1402	$1,41 \cdot 10^{-9}$	$3,49 \cdot 10^{-9}$	$3,69 \cdot 10^{-9}$	$-4,94 \cdot 10^{-5}$
1510	$5,18 \cdot 10^{-9}$	$1,33 \cdot 10^{-8}$	$1,27 \cdot 10^{-8}$	$-8,46 \cdot 10^{-5}$

found within the spinel phase; an example is given in fig. 3. Within the $Co_{1-z}Ga_{2z/3}O$ phase the concentration dependence of gallium tracer diffusion is negligible, except in the range of very small concentrations (see table 2).

DISCUSSION

The diffusion mechanism within the spinel phase is assumed to be a vacancy mechanism (5). Because the Co^{2+} and Ga^{3+} tracer diffusion coefficients are of the same order of magnitude, spinel must be formed by the Wagner mechanism. The diffusion mechanism within the CoO phase is as follows: complexes are formed in the Co^{2+} sublattice between Ga^{3+} ions substituted for Co^{2+} and single ionized vacancies; because the association degree of these two member complexes is nearly one, the gallium tracer diffusion must be independent of concentration (7).

By setting $D^I(1)$ in equation 7 equal to the Ga^{3+} tracer diffusion coefficient in the high concentration range of the $Co_{1-z}Ga_{2z/3}O$ phase, the calculation of reaction rate constants of the second kind according to equation 3 leads to a very good agreement with experimental results (table 1). Moreover it can be seen,

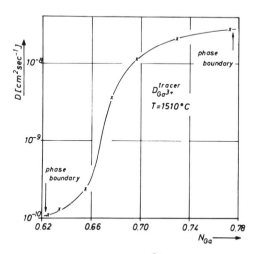

Fig. 3 Concentration dependence of Ga^{3+} tracer diffusion coefficients in the spinel phase $Co_{1-y}Ga_{2+2y/3}O_4$ at 1510 °C (N_{Ga} = cation mole fraction of gallium)

Table 2 Concentration dependence of Ga^{3+} and Co^{2+} tracer diffusion coefficients in the phase $Co_{1-z}Ga_{2z/3}O$ at 1510 °C (N_{Ga} = cation mole fraction of gallium)

N_{Ga}	$D_{Ga}^{tr.}$ ($10^{-7} cm^2 sec^{-1}$)	$D_{Co}^{tr.}$ ($10^{-7} cm^2 sec^{-1}$)
0,0	0,79	1,06
0,0134	1,73	1,67
0,0339	1,71	2,12
0,0690	1,82	2,85
0,105	1,72	2,98
0,143	1,70	3,19

that values calculated for α_1 are negative in sign, which means, that the boundary spinel/CoO moves from the origin (x = 0 in fig. 1) to the right. The cause of this unusual movement is the high diffusion rate of Ga^{3+} within the CoO phase. Thus the position of markers buried within the bulk CoO after first kind reactions can be explained too.

REFERENCES

1. A.D. Pelton et al., Ber. Bunsenges. physik. Chem. $\underline{76}$, 543 (1972)
2. C. Wagner, Acta Met. $\underline{17}$, 99 (1969)
3. W. Laqua et al., J. Solid State Chem., in press
4. W. Laqua et al., Z. Anorg. und Allg. Chem., in press
5. H. Schmalzried et al., Z. Phys. Chem. (Frankfurt/M.) $\underline{31}$, 198 (1962)
6. H. Schmalzried, Z. Phys. Chem. (Frankfurt/M.) $\underline{33}$, 111 (1962)
7. R.A. Perkins et al., Met. Trans. $\underline{4}$, 193 (1973)

THE TEXTURE AND HEAT TREATMENT OF HYDRATED TIN (IV) OXIDES

B. J. Dalgleish, D. Dollimore and D. V. Nowell

The Chemistry Department, Halfield Polytechnic

and The Chemistry Department, University of Salford

ABSTRACT

Various preparations of hydrated tin (IV) oxides are reported. These are classed as α- or β- stannic acids according to their stability in concentrated hydrochloric acid. Precipitation in the cold gives α- stannic acid, whilst preparations at high temperature result in β- stannic acid; the former being of smaller particles than those of the latter.

The surface texture of these preparations have been characterised from their nitrogen adsorption isotherms. They have been subjected to thermogravimetric analysis (TG), differential thermal analysis (DTA) and X-ray diffraction. There is no formation of stoichiometric hydrates in the oxide system but water is adsorbed on the particles which can be identified as small crystallites of cassiterites (SnO_2).

INTRODUCTION

Tin (IV) oxide ignited at high temperatures is rendered insoluble in acids and alkalis in contrast to the sample obtained by low temperature dehydration. This serves to show that many physico-chemical properties of a solid cannot be deduced from its empirical formula alone. Thus different samples of a metallic oxide, each having the same chemical formula, may show large differences in their "activity". The activity reveals itself in such properties as an enhanced rate of reaction with liquid reagents and marked adsorptive properties.

Various methods are available for the preparation of hydrated tin (IV) oxides, the products obtained being termed α- and β-stannic acids. The α- form is prepared by normal hydrolysis reactions in cold aqueous solutions. The β- form is prepared by methods involving relatively high temperatures. The two forms vary in their acid-base stability, the β- form being the more stable.

EXPERIMENTAL

Preparation of Hydrated Tin (IV) Oxides. Except where stated otherwise each product was washed with distilled water 1 day per week for 4 weeks, then filtered off and evacuated in a dessicator. Sample preparation was as follows; sample 1, tin metal (Analar) oxidised with hot nitric acid (8N); 2, as above but using high purity tin foil; 3, precipitation from a stannic chloride solution with ammonia solution in the presence of ammonium chloride (pH varied from 4.2 to 7.1); 4, excess normal sodium hydroxide solution was added to stannic chloride solution and carbon dioxide was bubbled through for 5 hours at room temperature (pH 10.9); 5, homogeneous precipitation using urea, ammonium chloride, hydrochloric acid and stannic chloride solutions and boiling for 24 hours (pH varied from 1.3 to 7.0). Samples 6, 7 and 8 were prepared by acidification of sodium stannate solutions; sample 9, by using a 20% aqueous solution of pyridine to precipitate the hydrated oxide from a weakly acidic solution of stannic and ammonium chlorides at boiling point; sample 10, by hydrolysis of stannic ethoxide using special apparatus (1).

Solubility in Concentrated Hydrochloric Acid. It is not clear from the summarised details but samples 4, 6, 7, 8 and 10 involved precipitation in the cold and produced α- "stannic acid" which dissolved in concentrated hydrochloric acid whilst other samples involved preparations at a high temperature and resulted in β-stannic acids which were insoluble in concentrated hydrochloric acid.

Nitrogen Adsorption Isotherms. Showed that the samples as prepared all had a large surface area (e.g. sample 2, 187 $m^2 g^{-1}$; 5, 96 $m^2 g^{-1}$; 10, 286 $m^2 g^{-1}$). Degassing at room temperature or at 130°C had no effect on the surface area (e.g. sample 6, degassed at room temperature, 173 $m^2 g^{-1}$; degassed at 130°C, 174 $m^2 g^{-1}$). All these samples showed a Type II adsorption isotherm. Heating at 1000°C reduced the surface area to less than 10 $m^2 g^{-1}$ (e.g. sample 8, as prepared 154 $m^2 g^{-1}$; heated at 1000°C, 6 $m^2 g^{-1}$).

X-ray Diffraction Studies. Identified the solid phase of the hydrous oxide as small crystallites of cassiterite SnO_2 in the as-prepared state. In the calcined samples (heated for 2 hours at temperatures up to 1000°C) a progressive increase in crystallite size with increasing temperature illustrates the sintering process. At no stage is there any indication of a stoichiometric hydrate.

Differential Thermal Analysis (DTA) shows a single endotherm indicating a continuous loss of water over the range 140 - 200°C (e.g. peak maxima, sample 4, 169°C; 5, 195°C and 495°C; 6, 138°C; 7, 152°C; 8, 140°C). The anomalous behaviour of sample 5 should be noted.

Thermogravimetric Analysis showed a rapid loss of physically adsorbed water over the temperature range 50° - 200°C, followed by a gradual removal of chemically bound water above ca. 200°C. Derivative thermal analysis (DTG) showed that some samples exhibited a single maxima, e.g. samples 6, 10; others showed a maxima followed by a shoulder, e.g. samples 7, 8; whilst others showed two peaks, e.g. samples 3, 5. The behaviour of samples 3 and 5 is due to impurities. The constitution of the oxide at peak temperature (except for samples 3 and 5) was in the range SnO_2 $0.9H_2O$ to SnO_2 $1.4H_2O$ on the assumption that the weight loss was due to the removal of adsorbed water.

Chemical Analysis indicated that the original as prepared hydrous oxides had formulae in the range SnO_2 $1.6H_2O$ to SnO_2 $2.4H_2O$.

DISCUSSION

The difference between α- stannic acids (precipitated in the cold and the β- stannic acids (precipitated at a high temperature) is that the former are of smaller particle size. This arises from a difference in the state of aggregation of the primary particles. At room temperatures ageing is slow, whilst at elevated temperatures the process is accelerated.

The TG data may show additional peaks due to impurities when plotted as a DTG plot and this can be confirmed by additional lines in the X-ray diffraction studies (e.g. samples 3 and 5). The DTA, TG and X-ray diffraction data show there are no stoichiometric hydrates and in particular that physically adsorbed water is lost between 50° and 200°C. Above 200°C there is a gradual removal of chemically bound water which is located at the surface probably in the form of molecular water held by hydrogen bonds and partly as -OH groups.

The aggregation phenomena leading to particle size differences is only partially reflected by the surface areas determined from nitrogen adsorption data. Stannic oxide gel preparations reported by Goodman and Gregg (2) had a surface area of 172 m² g⁻¹, and a sample studied by Rutledge (3) had a surface area of 173 m² g⁻¹. Sample 10 had an area of 286 m² g⁻¹ which emphasizes the effect on area that experimental variations can cause.

The adsorption isotherms were Type II, the value of the BET C values were over 400. The exceptions were the samples heated at 1000°C. Thus sintered sample 8, had a C value of 2.25 but this really depends upon the range over which the BET analysis was applied. In a method of comparing adsorption isotherms advocated by Gregg the ratio of the amounts adsorbed, are reported at various pressures (4). If the standard adsorption isotherm reported by de Boer et al (5) on alumina is considered as the model for comparison it is seen that the data is reported in terms of t - the statistical thickness of the adsorbed layer. The correlation between t and the amount adsorbed (V_{a_s}) in the reference isotherm at any given relative pressure is:-

$$t = 3.54 \left(\frac{V_{a_s}}{V_{m_s}} \right) \quad \text{or} \quad V_{a_s} = \frac{t V_{m_s}}{3.54}$$

where V_{m_s} is the monolayer capacity of the reference isotherm. If the experimental adsorption isotherm is comparable (V_a indicating the amount adsorbed at the given relative pressure, and V_m the monolayer capacity of the adsorption isotherm under consideration), then:-

$$\left(\frac{V_a}{V_m} \right)_{exp} \bigg/ \frac{t}{3.54} = 1$$

when $\quad \dfrac{3.54 \, V_a}{t} = V_m$

and it is this ratio, $\dfrac{3.54 \, V_a}{t}$ which is used here in place of the direct comparison advocated by Gregg. The value so calculated is nominally the monolayer coverage of the adsorbate. To get all the data on the same graph each of their nominal values is divided by the value $(V_m)_{exp}$ calculated directly from the BET analysis (Figure 1.). This shows identical behaviour for all the isotherms except the material subjected to sintering at 1000°C. The initial comparison indicated that the conventional BET analysis was incorrect (e.g. monolayer coverage, 1.42 cm³ (NTP) g⁻¹, with C = 2.07) and that if an initial small step to the isotherm in the p/p₀ region below 0.05 was considered then the monolayer coverage was smaller (0.324 cm³ (NTP) g⁻¹ and C > 600). The comparisons demonstrate that all the high area samples produced a diminution in the apparent relative value of V_m as the relative pressure increased whilst the sintered sample was exceptional in showing an increase and then a decrease.

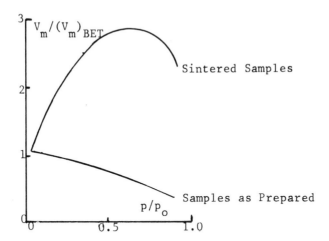

Figure 1. Plot of $V_m/(V_m)_{BET}$ against p/p_o

This would be consistent with the following explanation. First the distinction between the α- and β- stannic acids must be on the basis of the agglomeration of the primary particles. The nitrogen adsorption is taking place, however, on the surface of primary particles which are however, not as small as the BET calculation of the monolayer coverage would indicate. This implies that a surface roughness is present which reduces the effective surface on which later multilayers are adsorbed by a factor of up to 0.5. The sintered sample however, is not agglomerated and its roughness factor is insignificant. Consequently each subsequent layer of adsorbed material is greater than the preceeding layer. This process can continue until at $p/p_o \simeq 0.7$ there is a sufficiently thick adsorbed layer to cause adhesion of particles and a subsequent diminution in the exposed layer upon which further adsorption may occur.

REFERENCES

1. Harris, M. R. and Sing, K. S. W., J.Appl.Chem., 7, 397, (1957).
2. Goodman, J. F. and Gregg, S. J., J.Chem.Soc., 1162, (1960).
3. Rutledge, J. L., Proc.Okla.Acad.Sci., 137, (1965).
4. Gregg, S. J., J.C.S.Chem.Comm., 699, (1975).
5. Lippens, B. C., Linsen, B. G. and de Boer, J. H., J.Catalysis, 3, 32, (1964).

REDOX PROCESSES OF MANGANESE IONS DISPERSED IN OXIDE MATRICES

M. Valigi and D. Cordischi

Centro di Studio (CNR) "Struttura ed Attività
Catalitica di Sistemi di Ossidi" Istituto Chimico,
Università di Roma, Roma, Italy

ABSTRACT

Several oxide matrices (MgO, TiO_2, SnO_2, and Al_2O_3) containing a few per cent of manganese oxide (up to 2 atomic per cent), prepared in oxidizing (air) or reducing (hydrogen) atmosphere, have been investigated by X-ray and ESR techniques.

The results are different for each system. For manganese-doped MgO prepared in hydrogen, the Mn^{2+} in solid solution is oxidised in air to Mn^{4+} starting from 400°C. At the same time Mg_6MnO_8 forms. For specimens prepared in air, the manganese is present predominantly as Mg_6MnO_8. The treatment in hydrogen of these oxidised samples causes the disappearence of this phase and the reduction to Mn^{2+}. In the case of matrices of rutile structure (TiO_2, SnO_2) the manganese ions are incorporated in solid solution as Mn^{4+} when these samples are prepared in air. The dissolved ions are reduced by hydrogen at lower temperature (100-200°C). The redox processes take place in solid solution and are reversible. For the system based on alumina, only a small amount of manganese is detected as Mn^{2+}.

INTRODUCTION

Parameters influencing the valence state of a foreign species dispersed in a host matrix are of great importance in solid state reactions. Indeed, the reactivity of solids may be prevented by low solubility among reactans, which in turn is governed by solid-state chemical processes. In particular, the solubility depends heavily on the charge of the dispersed ions and on the nature of the host matrix. In the present investigation manganese ions, dis persed in several oxide matrices such as MgO, TiO_2, SnO_2 and Al_2O_3 have been examined in order to elucidate the effect of both the host structure and firing atmosphere on the stability of some oxidation states. The oxidation of reduced manganese species and the reduction of oxidised manganese ions are considered.

The results, obtained via X-ray diffraction and ESR techniques, form the main part of this paper. In the final section a comparison between the several matrices is outlined.

EXPERIMENTAL

Manganese-containing MgO samples were prepared by impregnation of magnesium oxide (ex carbonate, 600°C in air) with $Mn(NO_3)_2$ solution. After drying at 110°C, the specimens were mixed, ground, calcined at 600°C in air for 1 hr and ground again. After cooling, each sample was divided into two portions, both were heated at 1000°C the first one in air (5 hr) the second in a stream of hydrogen (10 hr) (purified through a De-Oxo cartridge and liquid nitrogen trap). Samples based on TiO_2 were prepared as described previously (1), as were also the manganese-doped SnO_2 samples. In latter case the specimens were heated at 800°C for 36 hr in air. Alumina-based samples were prepared by the same impregnation procedure starting from γ-Al_2O_3, with a final calcination temperature in air of 1000°C (5 hr). The samples are designated as MM, TM, SM and AM for, respectively MgO, TiO_2, SnO_2, and γ-Al_2O_3 as matrix. The letter a or h after the capital letters specify the firing atmosphere of preparation (a = air, h = hydrogen), whereas the numbers indicate the nominal manganese concentration expressed as manganese ions/100 host cations. For each matrix the manganese concentration in the samples ranged from 0.5 to 2 atomic per cent.

Portions of the samples were submitted to a further heat treatment. Those fired in reducing conditions (hydrogen) were re-heated in air at a set temperature for 5 hr. The specimens prepared in air were fired in a stream of hydrogen for 5 hr.

Lattice-parameter determinations both for MgO (3) and TiO_2 (1) have been described previously. A Debye-Scherrer camera (i.d. 114.6 mm, Cu Kα) was employed to check the crystalline phases present. The ESR procedure has been reported (2).

RESULTS AND DISCUSSION

Manganese Oxide-Magnesium Oxide

<u>Samples prepared in hydrogen (MM-h)</u>. The system MM-h has been studied previously (3). The present results confirm that Mn^{2+} is

substitutionally incorporated in the MgO lattice. Estimation of the total manganese content, from lattice parameter determinations and ESR measurements, indicate that virtually all the manganese ions are in solid solution as Mn^{2+}. The heating in air up to 400°C of these samples produces negligible change in a values and in ESR spectra. By contrast, in the temperature range 400-800°C two processes occur: the progressive oxidation of Mn^{2+} to Mn^{4+} and the segregation of the latter ions in the Mg_6MnO_8 compounds. The first process begins in the low temperature side of this range and is practically complete at 700°C, the second occurs preferentially at higher temperatures. The Mn^{4+} formation in solid solution in MgO is inferred from the marked shrinking of the lattice parameter for specimens heated above 400°C in air and is confirmed by the ESR measurements. The ESR spectra of the specimens treated at 500-700°C are very complex. In addition to the Mn^{2+} sextet, a broad band is present ($\Delta H_{pp} \simeq 40$ G) assigned to associated Mn^{4+} ions, both in Mg_6MnO_8 (detected by X-rays) and in MgO. The Mn^{2+}, in the bulk as isolated ions in cubic symmetry, are very stable to oxidation. After treatment in air at 1000°C almost all the manganese is present as Mg_6MnO_8, only a small amount (0.1 atomic per cent) being incorporated as Mn^{2+} in MgO.

Samples prepared in air (MM-a). In the MM-a samples a small amount of Mn^{2+} is incorporated in the MgO structure (3). The Mn^{2+} ions are easily detected by ESR and from their intensity, a concentration of 0.1 atomic per cent has been derived. Most of the manganese is present as Mn^{4+} in the Mg_6MnO_8 phase. The heating in hydrogen does not affect the samples up to 300°C. At 400°C the Mg_6MnO_8 is present in traces and is no longer observed in the X-ray spectra at higher temperatures. From 400°C an ESR signal consisting of a broad band develops. Superimposed on this band is the typical sextet of Mn^{2+} in MgO. The broad band ($\Delta H_{pp} \simeq 200$ G), attributed to associated Mn^{2+} ions, is the greater part of the signal. At the highest temperature (1000°C) the integrated intensity accounts for the total manganese content. The Mn^{2+} ions formed do not appreciably diffuse in MgO (no variation of the lattice parameter is observed), but are present as a separate phase. Indeed, very weak peaks, partially resolved from those of MgO, are detected in the X-ray diffraction spectra; they are attributed to a concentrated solid solution of Mn^{2+} in MgO, produced from the reduction of Mg_6MnO_8.

Manganese Oxide-Titanium Oxide

Samples prepared in air (TM-a). The incorporation of Mn^{4+} in the rutile structure and its reduction by hydrogen have been studied previously (1,2,4). The results show that in the TM-a samples the manganese is substitutionally incorporated as Mn^{4+} in the TiO_2 structure up to 1.24 manganese atoms/100 Ti atoms. Manganese in excess forms $MnTiO_3$. The heating of the TM-a samples in hydrogen at different temperatures causes the reduction of Mn^{4+} to Mn^{3+} or/and Mn^{2+}. At 300°C the Mn^{3+} is formed preferentially whilst in the temperature range 300-500°C Mn^{2+} is produced. Both the Mn^{3+} and the Mn^{2+} ions remain in solid solution. For temperatures higher than 550°C, Mn^{2+} segregates from TiO_2 and forms $MnTiO_3$. The manganese reduction occurs by hydrogen incorporation in TiO_2 up to 500°C; no loss in weight is observed (4). At higher temperatures water is lost instead.

The additional data given here, concern the behaviour of the reduced samples when re-heated in air. Both the X-ray and ESR measurements show that the reduced manganese is easily oxidised. The samples treated in hydrogen at lower temperatures (up to 500°C)(in which the Mn^{2+} is still in solid solution) are completely re-oxidised at temperatures as low as 100°C. As an example, let us take the samples TM-a 0.5 for which a value of the lattice parameter a = 4.5920 Å has been measured. After reduction at 500°C, $\underline{a}$ is expanded ($\underline{a}$ = 4.5967 Å) due to reduction to Mn^{2+} which is in solid solution. When the reduced sample is again heated in air, already at 100°C (5 hr) the Mn^{2+} is completely transformed to Mn^{4+} and the lattice parameter of the untreated TM-a 0.5 sample is obtained ($\underline{a}$ = = 4,5920 Å). The ESR spectrum of Mn^{4+} in TiO_2 is entirely restored.

Manganese Oxide-Stannic Oxide

<u>Samples prepared in air (SM-a)</u>. A limited solubility of Mn^{4+} is also observed in the system based on SnO_2 (rutile structure). The amount of Mn^{4+} in solid solution is somewhat lower than that of the TM-a system and has been estimated (on the basis of ESR measurements) to be 0.5 atomic per cent. The heating in hydrogen causes the reduction of Mn^{4+} to Mn^{3+} and Mn^{2+}. The formation of Mn^{2+} ions, which give a complex ESR spectrum similar to that described for TM-a samples (4), starts at 200°C a temperature lower than that of the system based on TiO_2. At 300°C the intensity of the ESR signal accounts for the reduction of all the Mn^{4+}. At 400°C SnO_2 is reduced by hydrogen, as shown by a thermogravimetric analysis.

Manganese Oxide-Aluminium Oxide

<u>Samples prepared in air (AM-a)</u>. A complicating feature of this system is the phase trasformation $\gamma \to \delta \to \alpha$-$Al_2O_3$ which takes place, and which is favoured by the presence of manganese. For the more diluted specimens (up to 0.5 atomic per cent) where δ-Al_2O_3 predominates, only a small fraction of manganese is detected by ESR as Mn^{2+} ions. This situation is little different from that found in the manganese supported on η- and γ-Al_2O_3 fired in air (5). After heating the diluted samples in hydrogen a moderate increase of the ESR signal

is observed and, at the highest temperature tested (720°C), a narrowing of the line occurs. This indicates clustering of Mn^{2+}, probably as a surface spinel phase. However, part of the Mn^{2+} formed escapes the ESR observation.

Comparison Between The Different Matrices Containing Manganese

In the MM-h samples the stabilisation of the 2+ oxidation state depends on the dispersion of manganese in the MgO structure. The isolated Mn^{2+} ions, in perfect cubic symmetry in the bulk of the MgO lattice, are not oxidised even on heating in air at 1000°C. By contrasts the Mn^{2+} ions, which are associated, begin to be oxidized at 400°C and become Mn^{4+}. The oxidation takes place with formation of cationic vacancies on the surface and as soon as enough thermal energy is given to the system this unstable situation is removed and Mg_6MnO_8 is formed. The Mn^{2+} oxidation to Mn^{4+} may involve Mn^{3+}. However, for temperatures higher than 600°C the manganese is present chiefly as Mn^{4+}. This result is in agreement with the conclusions based on oxygen and hydrogen absorption measurements (6). For the samples prepared in air (MM-a) most of the manganese is in the +4 oxidation state as Mg_6MnO_8, the remainder being in solid solution in MgO as Mn^{2+}. The hydrogen treatment causes the disappearance of the Mg_6MnO_8 at 500°C. The manganese is reduced to Mn^{2+} but it appears to be still located on the surface. When TM-a samples are considered a different situation is encountered. The manganese is till present in the oxidation states 4+ and/or 2+ but the Mn^{4+} is now in solid solution in TiO_2 and the Mn^{2+} in found as $MnTiO_3$. The Mn^{4+} in solid solution are reduced by H_2. The electronic process takes place at lower temperature with respect to the case of the samples based on the MgO matrix. Mn^{3+} is found in a well-defined range of temperature and predominates at 300°C; at higher temperature Mn^{2+} is formed. Only at temperatures higher than 500°C is the mobility of the species involved high enough to allow water loss and Mn^{2+} separates out as $MnTiO_3$. For TM-a samples the redox process and the formation of $MnTiO_3$ occur in separated ranges of temperature. In the case of SM-a samples, the manganese is incorporated substitutionally in SnO_2 as Mn^{4+} and the action of H_2 is similar; Mn^{2+} is already formed on heating in H_2 at 200°C. For both TM-a and SM-a systems, the reduced manganese ions are oxidised to Mn^{4+} even on heating in air at 100°C. As far as the AM-a system is concerned, the data focus attention on the formation of new phases, which can facilitate the reduction processes.

REFERENCES

(1) M. Valigi and A. Cimino, J. Solid State Chem. 12, 135 (1975)
(2) D. Cordischi, *et al*, J. Solid State Chem. 15, 82 (1975)
(3) A. Cimino, *et al*, Z. phys. Chem. Neue Folge 59, 134 (1968)
(4) M. Valigi, *et al*, J. Inorg. Nucl. Chem. 38, 1249 (1976)
(5) M. Lo Jacono, *et al*, Gazz. Chim. Ital. 105, 1165 (1975)
(6) D. Cordischi, *et al*, Trans. Faraday Soc. 65, 2740 (1969)

STUDIES BY ELECTRON MICROSCOPY OF THE REDUCTION OF MoO_3

G. Liljestrand

Dept. of Inorganic Chemistry, Arrhenius Laboratory

University of Stockholm, S-104 05 Stockholm (Sweden)

INTRODUCTION

During the past years careful examinations of the Mo-O system have shown the existence of many phases in the composition range MoO_3-$MoO_{2.75}$ (1-3). All observed phases in the range $MoO_{2.99}$-$MoO_{2.88}$ are shear phases consisting of largely unchanged slabs of a base structure (MoO_3 or ReO_3) joined along crystallographic shear (CS) planes (fig. 1). The phases Mo_8O_{23} and Mo_9O_{26} which belong to the homologous series Mo_nO_{3n-1} (4) with an ReO_3 type basic structure will not be treated in this article.

Kihlborg found a phase with a composition close to $MoO_{2.89}$ (5). It proved to have a CS structure with $\{35\bar{1}\}$ shear in MoO_3 and to be the member with n = 18 in the homologous series

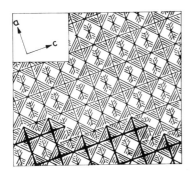 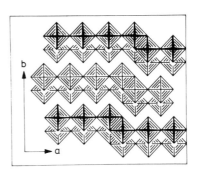

Fig. 1. $\{35\bar{1}\}$ CS in MoO_3, idealized.

Mo_nO_{3n-2}. Additionally a few single crystals were observed with
$\{\bar{2}21\}$ CS (n = 13 in Mo_nO_{3n-1}) and $\{40\bar{1}\}$ CS (n = 26 in Mo_nO_{3n-3}).

Bursill found a phase $MoO_{2.9975}$, a superstructure of MoO_3 with oxygen vacancies disordered due to self diffusion. In this so-called domain structure, $\{120\}$ CS planes began to grow, during exposure in the microscope, parallel to their longitudinal direction (2). Bursill also obtained the members n = 19, 20, 21, and 22 of the $\{35\bar{1}\}$ CS series Mo_nO_{3n-2} (the diffraction spots indicating some disorder) but did not observe any members of the other two series Mo_nO_{3n-1} and Mo_nO_{3n-3}.

The aim of the present study was to clarify the formation conditions determining the CS type formed and to study disorder phenomena in the shear phases. The oxygen pressure during preparation was regulated by a method (previously used by Brusq, Oehlig and Marion (6)), which seemed to offer quite precise control of the formation conditions.

EXPERIMENTAL

In the gas flow apparatus used to prepare the samples, pre-cleaned argon gas, water vapour and hydrogen gas were mixed (7). The quotient p_{H_2O}/p_{H_2} was controlled and kept constant by an oxystat (Hermann-Moritz, Paris), which contained a metal oxide wire as an oxygen probe. The starting material was Matthey Chem. "Spectra pure MoO_3" (Johnson Matthey Chemicals Ltd., London). The relation of the conductivity (σ) of the metal oxide (Nb_2O_5) wire and the oxygen pressure were obtained by calibration with a solid electrolyte cell consisting of ZrO_2 + 15 % CaO kept at a constant temperature (8). The plot of log p_{O_2} = $f(\sigma)$ was linear.

The samples thus prepared were examined in a Siemens Elmiskop 102 equipped with a double tilt-lift goniometer stage. The crystals were crushed in acetone, collected on holey carbon films supported by copper grids and so oriented in the diffraction mode that the superstructure could be observed. The magnifications of

Fig. 2. Lattice image of a Mo_nO_{3n-2} crystal viewed along $\langle\bar{1}12\rangle_{MoO_3}$ ($\langle 100\rangle_{Mo_{18}O_{52}}$).

the microscope were carefully calibrated with the aid of a very well ordered Mo_4O_{11}(o-rh) crystal with accurately known periodicity. The n values in Mo_nO_{3n-2} were determined by photometric measurement of the shear plane spacings on the plate.

RESULTS

Samples, prepared at 614°C and 635°C were mixtures either of MoO_3/"$Mo_{18}O_{52}$" or "$Mo_{18}O_{52}$"/Mo_4O_{11}(o-rh) (analysed by X-ray powder diffraction). Neither Mo_8O_{23} nor Mo_9O_{26} were obtained, in agreement with their reported temperature ranges of formation of 650-780°C and 750-780°C respectively (1). (The formation temperature range of $Mo_{18}O_{52}$ is 600-750°C (1).) The log p_{O_2} values at the sample temperatures were calculated from the pressures measured in the calibration cell and from thermodynamic data reported for the $2H_2(g) + O_2(g) = 2H_2O(g)$ system (9). The equilibrium oxygen pressures of MoO_3/"$Mo_{18}O_{52}$" mixtures at 614°C and 635°C were thus estimated to be $10^{-15.4}$ bar and $10^{-14.7}$ bar respectively. The results are to be compared with $10^{-14.5}$ bar given by Marion et al. as the equilibrium oxygen pressure of MoO_3/$Mo_{18}O_{52}$ at 600°C. In the present study a Nb_2O_5 wire and a H_2O/H_2 buffer were used while Marion et al. utilized a Cr_2O_3 wire and a CO_2/CO buffer.

In the electron microscope the earlier-reported domain structure of very slightly reduced MoO_3 was verified. This phase was sometimes formed by beam heating in the microscope during observation of thin crystals of MoO_3.

{120} CS planes were observed in some crystals, as in Bursill's observations. The interplanar distances corresponded to local compositions in the range $MoO_{2.999}$-$MoO_{2.94}$, with the deficiency in oxygen assumed to be due only to {120} CS.

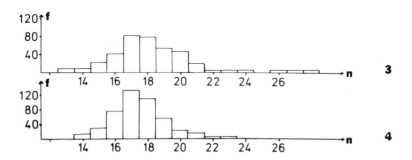

Fig. 3. Distribution of n in Mo_nO_{3n-2} crystals. F is the total frequency. Sample 3 contained MoO_3 + "$Mo_{18}O_{52}$", sample 4 "$Mo_{18}O_{52}$" + Mo_4O_{11}(o-rh).

The diffraction patterns of 47 crystals (very slightly reduced MoO_3 crystals not included) were recorded. Two of these showed $\{1\bar{2}0\}$ CS (see below) and the rest $\{35\bar{1}\}$ CS. The other CS planes reported by Kihlborg, viz. $\{\bar{2}21\}$ and $\{40\bar{1}\}$, were not observed. Values of n of 17, 18 and 19 and in some cases 18 mixed with 19 were found in the patterns of the Mo_nO_{3n-2} ($\{35\bar{1}\}$ CS type).

Most samples were heated approximately 20 hrs. One sample, however, was heated for only ca. 0.5 hr. Of the seven crystals examined in this sample, five crystals showed $\{35\bar{1}\}$ CS, while two others contained $\{1\bar{2}0\}$ CS though their average compositions in both cases were $MoO_{2.84}$ judged from the shear plane spacings. Such highly reduced $\{1\bar{2}0\}$ CS crystals have been found only in this sample. This may indicate that crystals with $\{1\bar{2}0\}$ CS with a composition below $MoO_{2.94}$ (a composition observed both for $\{1\bar{2}0\}$ CS and $\{35\bar{1}\}$ CS, see below) are metastable.

A well-ordered area of a Mo_nO_{3n-2} crystal is shown in fig. 2. The degree of disorder was measured quantitatively in 19 Mo_nO_{3n-2} crystals; the 1772 measured n values (fig. 3) varied between 12 and 35 (values above 27 were very rare, however) corresponding to the local compositions $MoO_{2.83}$ and $MoO_{2.94}$ respectively. Within one sample various crystals could have rather different compositions (these varied from $~Mo_{16}O_{46}$ to $~Mo_{19}O_{55}$) probably due to local variation of the oxygen pressure during crystal growth. The sample heated for only half an hour showed the lowest degree of reduction while the crystals from samples also containing Mo_4O_{11}(o-rh) were most reduced. The uncertainty of the n value measurement was determined by measuring an area in a crystal three times along parallel (but not identical) lines. For a Gaussian distribution of measurement errors, the probability of obtaining the correct n value ± 1 would be 99.8 %.

During the observation in the electron microscope the thinnest crystals of "$Mo_{18}O_{52}$" showed a successive change. The contrast from the shear planes became more and more diffuse while the finer line contrast from the octahedral rows remained. At the same time all reflections except the strongest superstructure spots (those

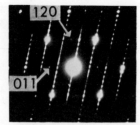 

Fig. 4. Diffraction patterns ($\langle 2\bar{1}1 \rangle_{MoO_3}$ zones) of Mo_nO_{3n-2} crystals. a) After < 2 minutes exposure. b) After approximately 15 minutes exposure to the beam.

nearest each "MoO$_3$ reflection") disappeared from the diffraction pattern, while the positions and the symmetry of the "MoO$_3$ pattern" were invariant (indicating that no tilting of the crystal had occurred) (fig. 4). There seemed to be no change in the shear direction since several zones were observed with the same result. Further studies on this matter are in progress.

In this investigation no members of the proposed homologous MoO$_3$ related series Mo$_n$O$_{3n-1}$ and Mo$_n$O$_{3n-3}$ were observed in spite of several attempts to synthesize them at temperatures near those used by Kihlborg (5). Various oxygen pressures gave mixtures of Mo$_4$O$_{11}$(o-rh)/"Mo$_{18}$O$_{52}$" or of "Mo$_{18}$O$_{52}$"/MoO$_3$. These observations evidently support Kihlborg's suggestion that these alternative kinds of CS in the MoO$_3$ structure appear under metastable conditions. According to the present study "Mo$_{18}$O$_{52}$" tends to be slightly substoichiometric (n$_{mean}$ about 17.5) and so MoO$_3$ should be present if an average n value close to 18 is desired (fig. 3).

In conformity with what is valid for shear in WO$_3$ (10), there seem to be distance-dependent repulsive forces in "Mo$_{18}$O$_{52}$" which permit a considerable disorder among the shear-plane distances but prevent the planes from approaching each other closely.

ACKNOWLEDGEMENTS

I am indebted to Dr. Lars Kihlborg for suggesting this topic and for his continuous interest. I am also grateful to Professor Arne Magnéli for stimulating discussions. Thanks are also due to Tomasz Niklewski for assistance with the gas flow apparatus, Sven Berglund for his oxygen pressure calibration and to Don Koenig for revising the English text. This work has been financially supported by the Swedish Natural Science Research Council.

REFERENCES

1. L. Kihlborg, Arkiv Kemi, 1963, 21, 471.
2. L.A. Bursill, Proc. Roy. Soc. A, 1969, 311, 267.
3. L.A. Bursill, Acta Cryst. A, 1972, 28, 187.
4. A. Magnéli, Acta Cryst., 1953, 6, 495.
5. L. Kihlborg, Arkiv Kemi, 1963, 21, 443.
6. H.L. Brusq, J.J. Oehlig and F. Marion, C. R. Acad. Sc. Paris C, 1969, 268, 1047.
7. T. Niklewski (not published).
8. S. Berglund, Rev. Int. Hautes Tempér. et Réfract., 1971, 8, 111.
9. O. Kubaschewski and E.Ll. Evans, p. 339 in Metallurgical Termochemistry, Thrd Ed., Pergamon, London, 1958.
10. M. Sundberg and R.J.D. Tilley, J. Sol. State Chem., 1974, 11, 150.

INFLUENCE OF CRYSTALLOGRAPHIC, MAGNETIC TRANSFORMATIONS AND STRUCTURAL DEFECTS ON THE REACTIVITY OF SIMPLE AND COMBINED IRON OXIDES

I. Gaballah, C. Gleitzer and J. Aubry

Laboratoire de Chimie du Solide, Université de Nancy I

Case Officielle n° 140, 54037 Nancy Cédex, France

ABSTRACT

Using diverse techniques relatively better experimental conditions for the reduction of simple and combined oxides of iron in the widest possible temperature range, it was possible to show that polymorphic, magnetic transformations and probably point defects and dislocations influence the activation energy (E) and sometimes the rate (V) of the reaction.

INTRODUCTION

In the 1930's, J. Hedvall (1) and H. Forestier (2) found that polymorphic and magnetic transformations increase V and E for solid-gas (3, 4), solid-liquid (3, 5) and solid-solid (1, 2, 6-8) reactions. Table I gives some examples of anomalies in reactivity in the neighbourhood of these changes.

The discussion about the role of dislocations and point defects has been developed in the last three decades (see for example 21-24). But till now there is no definitive answers for questions such as "where does the electron transfer take place ? What is the dynamic behaviour of point defects and dislocations in compounds like oxides as a function of temperature ? Is there a correlation between nucleation sites and points of emergence of dislocations (25) ?..."

On the other hand, it is generally believed that the kinetics of solid-gas reactions may be limited by one or more of the following steps : uniform interfacial reaction, gas diffusion, solid state diffusion, adsorption (26).

Table I

Ref.	Process	Reactant	Product	T°C	Transformations
9,10	Oxidation of Fe	Fe	-	910	α-Fe → γ-Fe
11,12[x]	Reduction of $Fe_{1-x}O$	-	Fe	"	"
13	" Fe_3O_4	-	Fe	"	"
7, 8	Synthesis $CaO+SiO_2$	SiO_2	-	≃900	Q → C
16[x]	Reduction of Fe_2SiO_4	-	SiO_2	870	"
		-	Fe	910	α-Fe → γ-Fe
17[x]	Reduction of $FeAl_2O_4$	-	Fe	"	"
19	Decomposition of NH_4ClO_4	NH_4ClO_4	-	240	Ortho. → Cubic
20	Nitriding of Ca	-	Ca_3N_2	650	Quad. → Cubic
4	Oxidation of Fe_3O_4	Fe_3O_4	-	570	T°c of Fe_3O_4
"	" Fe	Fe	-	770	Tc of Fe
15	" Ni	Ni	-	360	Tc of Ni
14	Reduction of NiO	NiO	-	265	Tn of NiO
3	" "	"	and	"	"
			Ni	360	Tc of Ni
18	Nitriding of alloy (Ni-Co) (12 % Co)	alloy	-	540	Tc of alloy
		Ni	-	360	Tc of Ni

(x) Authors publications ; T°c : Curie point ; Tn = Néel point
Q = Quartz ; C = Cristobalite

The elegant work of J.M. Thomas and al. (27) demonstrated the importance of dislocations in a wide variety of reactions involving solids. Fox (28) in trying to reconcile kinecists' and physicists' view points found that in the decomposition of calcite, four simple slip systems can be activated at different temperatures and so it was possible to relate the decomposition kinetics in different temperature ranges to each type of dislocation system.
X. Duval (29) showed similar results in the oxidation of graphite. So, it seems that the different limiting kinetic steps may be attributed to different kinds of dislocation slip systems.

EXPERIMENTAL

Fe_3O_4, $Fe_{1-x}O$, Fe_2SiO_4 and $FeAl_2O_4$ were prepared from Fe_2O_3,

α-Al$_2$O$_3$ and quartz (Merck), in H$_2$-H$_2$O mixtures (30) at 600, 900, 1000 and 1100°C respectively, and checked by X-ray and measurement of the O/Fe ratios.

The reduction kinetics of these oxides was studied with the following particularities : widest possible range of temperature, relatively better experimental conditions (high flow rate of reducing gas, small quantity of solid (< 100 mg) with specific sur-area (< 1 m^2/g).. etc).

Cahn RG and Setaram MTB 10-8 thermobalances were used for TGA "In situ" X ray diffraction experiments were performed with a hot stage CGR (31). The products are examined by CAMECA SEM and X-ray diffraction.

RESULTS AND DISCUSSION

1. Reduction of Fe$_2$SiO$_4$ with H$_2$: V and E are changed by the transitions quartz-cristobalite and α-γ iron ; they respectively increase and decrease V, and E is successively 55, 63, 66 Kcal/M for T < 870, 870 < T < 910 and T > 910°C (16). The rate is parabolic at T < 870 and linear at T > 870°C (table I).

2. Fe$_{1-x}$O (12) and FeAl$_2$O$_4$ (17) with H$_2$ at 700 < T < 1000°C : V is decreased and increased, respectively, by the α-γ Fe transition.

3. The reduction of Fe$_2$O$_3$ (32) is characterized by change of E at $\simeq$ 425°C (table II). This transition is not influenced by any change of gas composition (10% or 100% H$_2$), gas nature (CO or H$_2$), flow rate (2-4 cm^3/s), nacelle (Cu or Au), mass (10 or 100 mg), experimental techniques (see further). This means that the classical interpretation based on a change of limiting process is doubtful here.

Table II : Reduction of Fe$_2$O$_3$

Gas	cm/sec (x)	Nacelle	Weigh mg	Technique	T range °C	E Kc/M	T range °C	E Kc/M
10%H$_2$+90%N$_2$	$\simeq$3	Cu	10	TGA	340-420	23,5	430-550	9
"	"	"	100	"	"	22,7	"	9,1
100 % H$_2$	$\simeq$4	Au	"	"	220-420	22,5	"	9,8
100 % CO	"	"	"	"	260-420	26,8	430-500	10
100 % H$_2$	$\simeq$2	Fe-Cr-Ni	$\simeq$1000	"in situ" XRD	Detection of wüstite starts at 450°C			

(x) flow rate

Table III : Reduction of Fe_3O_4 (.) with H_2 (+)

T°C (x)	Technique	T range °C	E Kc/M	T range °C	E Kc/M	T range °C	E Kc/M	T range °C	E Kc/M
600	TGA	220-250	50	260-380	17	400-575	11,5	575-620	5
1200	TGA	210-250	17	"	17	400-510	12	510-580	3
600	"In situ" XRD	230-700		Detection of wüstite starts at 390°C					

(.) Nacelle : Au, weight of sample : 100 mg
(+) Flow rate $\simeq$ 4 cm^3/sec for TGA and $\simeq$ 2 cm^3/sec for "in situ" RX
(x) Temperature of preparation

In situ X-ray diffraction proves the presence of quasi stoichiometric wüstite at T > 450°C in the reduction of Fe_2O_3 to Fe via Fe_3O_4. Three possibilities appear :

$$Fe_2O_3 \to Fe_3O_4 \begin{cases} \xrightarrow{T < 450°C} Fe \\ \xrightarrow{T > 450°C} Fe_3O_4 + FeO + Fe \\ \xrightarrow{T > 570°C} Fe_{1-x}O \to Fe \end{cases}$$

V. Romanov (33) found, with Mössbauer spectrometry, stoichiometric wüstite at T $\geq$ 425°C in the reduction of Fe_2O_3 ; the temperature difference probably comes from a better sensitivity of NGR.

4. The reduction of Fe_3O_4 (32) with H_2 is characterized by 3 changes of E at 250, 390 and 570°C (table III). The latter is probably correlated to the Curie point of Fe_3O_4 and not to the stability limit of wüstite (see further). 390°C corresponds to the beginning of wüstite formation, as proved by X rays in situ. 250°C may be attributed, from analogy with metals, to the destruction of defect clusters. If the last hypothesis is true a magnetite prepared at 1200°C will have a different behaviour. Effectively the 250°C transition then disappears (see table III).

Oxidation of Fe_3O_4 (34-36), either powder or pellet, shows exothermic DTA peaks at about 250 and 390°C. The oxidation kinetics present a change of E at $\simeq$ 400°C (37, 38). Compared with reduction these results support the role of intrinsic properties of Fe_3O_4.

The change of E in Fe_3O_4-H_2 or O_2 reactions at 390°C may be explained by the recovery of dislocation or by a modification in the type of involved dislocations. This can occur for Fe_3O_4 at 390°C and for Fe_2O_3 at 425°C, keeping in mind that this difference may be attributed to the fact that Fe_3O_4 generated by Fe_2O_3 reduction is less perfect than when prepared at 600°C with H_2-H_2O.

CONCLUSIONS

Beside polymorphic and magnetic changes which influence the reactivity of solids, it seems that the distribution of point defects and the recovery of dislocations may also influence the reactivity of solids. These changes modify the diffusion paths and the rate of mass and electron transfer by changing the volume of unit cell, type and density of dislocation, point defect concentration and configuration, and spin direction which influence directly the reactivity of solids.

REFERENCES

1. J.A. Hedvall, Z. Anorg. Allgem. Chem.,1924, 67, 135.
   and al.   Z. Electrochem., 1934, 40, 301.
   Reaktionsfähigkeit fester Stoffe, Leipzig, 1938.
2. H. Forestier and R. Lille, Compt. Rend., 1937, 204, 265 & 1254.
   "   "   1939, 208, 891.
3. G. Nury and H. Forestier, 2th ISRS, Gothenburg, 1952, 189.
4. H. Forestier and M. Daire, Akad. Wissenschaften und Literatur, 1966, n° 7, 705.
5. G. Foex and M. Graff, Compt. Rend., 1939, 209, 160.
   G. Foex, Ibid., 1948, 227, 193.
6. H. Forestier, 2nd ISRS, Gothenburg, 1952, 41.
7. J.A. Hedvall and G.Z. Schiller, Z. Anorg. Allgem. Chem., 1934, 97, 221.
8. R. Jagitsch, Z. Physik. Chem. (B), 1937, 36, 399.
9. N.B. Piling and R.E. Bedworth, J. Inst. Met., 1923, 291, 529.
10. K. Fischbeck and al., Z. Elektrochem., 1934, 40, 517.
11. H.K. Kohl and H.J. Engell, Archiv. Eisenhüttenwesen, 1963, 34, 411.
    E. Rieke and al., Ibid., 1967, 38, 249.
12. I. Gaballah and al., Mem. Sci. Rev. Met., 1972, 69, 523.
13. J.M. Quets and al., Trans. AIME, 1960, 218, 545.
14. B. Delmon and A. Roman, J. Chem. Soc., Faraday trans. I, 1973, 69, 941.
15. L. Seigneurin and H. Forestier, Compt. Rend., 1956, 243, 2053.
    H. Uhlig and al., J. Acta. Met., 1959, 7, 111.
16. I. Gaballah and al., Mem. Sci. Rev. Met., 1975, n° 10,
17. I. Gaballah and al., Compt. Rend., 1975, 280, 697.
18. M. Daire and H. Forestier, 4th ISRS, Amsterdam, 1960, 122.
19. L.L. Bircumshaw and B.H. Newman, Proc. Roy. Soc. (London), 1954, A 227, 115 ; 1955, A 227, 228.
    P.W.M. Jacobs and W.L. Ng, 7th ISRS, Bristol, 1972, 398.
20. R. Streiff, Revue de Chimie Minérale, 1967, 4, 707.
21. F.S. Stone, 4th ISRS, Amsterdam, 1960, 7.
    F.C. Tompkins, 5th ISRS, München, 1964, 3.
    P.W.M. Jacobs, 6th ISRS, New York, 1968, 207.
    P.B. Hirsch, 7th ISRS, Bristol, 1972, 362.

22. A.H. Cottrell,"Dislocation and plastic flow in crystals" Clarendon Press, Oxford, 1953.
    W.T. Read, "Dislocation in crystals", Mc Graw Hill Books Cy, New York, 1953.
    J. Friedel,"Les dislocations", Gauthier-Villars, Paris, 1956.
23. W.E. Garner, "Chemistry of the solid state", Butterworths Scientific Publications, London, 1955.
    M.E. Fine, "The Chemical Structure of Solids" Edited by N.B. Hannay, Plenum Press, New York - London, 1974.
    R.W. Whitworth, Advances in Physics, 1975, 24, n° 2, 203.
24. J. Benard,
    "L'oxydation des métaux" Gauthier-Villars, Paris, 1962.
    "Processus de Nucléation dans les réactions des gaz sur les métaux et problèmes connexes" Coll. Internat., CNRS, n° 122, Paris, 1963.
    "Structure et propriétés des surfaces des solides" Coll. Internat., CNRS, n° 187, Paris, 1969.
25. J. Benard, Mem. Soc. Roy. Sci. Liège, 1971, 8, 99.
26. E.T. Turkdogan and J.V. Vinters, Metallurgical transactions, 1971, 2, 3175.
    E. Wicke, Chemie-Ingenieur Technic, 1957, 29, 305.
    L. Bonnetain and G. Hoynant, "Les Carbones", Masson, 1965, tome II, p. 290.
27. J.M. Thomas and al.
    J. Chem. Soc. A, 1967, 2058.
    Proc. Roy. Soc. A, 1968, 306, 53.
    J. Chem. Soc. A, 1969, 2749.
    J. Chem. Soc. A, 1969, 2227.
    J. Chem. Soc. A, 1970, 2938.
    Progress in solid state chemistry, H. Reiss, vol. 6, 1971. Pergamon Press.
28. P.G. Fox, J. Mat. Sci., 1975, 10, 340.
29. P. Magne and X. Duval, Carbon, 1973, 11, 475.
30. F. Marion, Thesis, Nancy, 1956.
    C. Gleitzer, Thesis, Nancy, 1959.
31. N. Gerad, J. Physics, Scient. Instr. E, 1972, 5, 524.
32. I. Gaballah and C. Gleitzer, To be published.
33. V.P. Romanov and al., Izv. Akad. Nauk SSSR, Neorganischeskii Materialy, 1971, 7, n° 3, 450 ;
    Phys. Stat. Sol., 1973, 15, 721.
34. R. Schrader and W. Vogelsberger, Z. Chem., 1969, 9, n° 9, 354.
35. Furuichi Rynsaburo and al., Kogyo Kagaku Zasshii, 1971, 74, 8, 1601.
36. B. Gillot and J. Tyranowicz, Mat. Res. Bull., 1975, 10, 775.
37. D. Papanastassiou and G. Bitsianes, Metal. Transactions, 1973, 4, 487.
38. J.P. Hansen and al., B.F. Coke oven and raw mat. conf., 1960, 185.

NON-QUENCHABILITY OF SOME TRANSITION-METAL CHALCOGENIDES

M. Nakahira and K. Hayashi

Laboratory for Solid State Chemistry,

Okayama College of Science, Okayama, 700, Japan

and

M. Nakano-Onoda and K. Shibata

National Institute for Reseaches in Inorganic

Materials, Sakura-mura, Ibaraki-ken, 330-31, Japan

INTRODUCTION

The establishment of the phase diagrams of transition-metal chalcogenides has so far been based mostly on the quenched specimens. Some of them, however, are not quenchable at all. The present report will briefly summarize the behaviors of those compounds through the comparison between the various measurements of quenched specimens and the in-situ observations at the synthetic temperatures. Attention is mainly directed to the $VS-VS_2$ ($V_{0.50}S$) system.

The structures of the compounds, $VS-VS_2$ system, are based on the fundamental NiAs-type structure. The change in the chemical composition VS through $V_{0.50}$ is accompanied by increasing metal-vacancies within the every second metal-layer of the structure, and eventually $V_{0.50}S$ reaches the structure of the CdI_2-type. The activity of sulphur is equivalent to the partial pressure of sulphur in the atmosphere, so that the chemical composition of solid crystal is determined by the partial pressure of sulphur at the respective temperature. At high temperatures, metal-vacancies will be distributed randomly throughout the crystal, while at low temperatures an ordered arrangement of those vacancies be established. The fact that the quenched specimen, especially in the sulphur-rich region, always exhibits some super-structure indicates the non-quenchability of the high-temperature state. In fact, the equilibrium reaction with the ambient sulphur gas and also the order-disorder transition of the metal-vacancies proceed much faster in the sulphur-rich region. High-temperature X-ray examinations with a single crystal of the V_5S_8 phase sealed in an evacuated quartz capillary exhibit a monoclinic-trigonal(intra-layer disordering)-hexagonal(inter-layer disordering) transition reversibly. In contrast, in the metal-rich region of the present system, the formation reaction and the subsequent ordering of metal-vacancies at lower temperature are very slow, resulting in various degrees of non-quenchability. Lattice image and electron diffraction observations of a single crystal of quenched $V_{0.87}S$ revealed a 4C-3C polytipic transition with no change in the chemical composition through the heating by electron bombardment. In addition, X-ray powder diffractions of the similar specimen in an evacuated quatz capillary at various temperatures revealed a phase separation at above $600^\circ C$. Those observations call for various in-situ measurements at several temperatures to establish the real phase relations.

EXPERIMENTAL

The starting materials were $V_{0.71}S$ and $V_{0.91}S$ powders. Two different approaches were applied to reach the equilibrium between the solid starting material and the ambient sulphur partial pressure in the H_2S-H_2 gas mixture at the respective temperature. In the range VS through V_3S_4 where the reaction rate was low, the specimen was placed at pre-determined atmosphere and temperature to be held for 2 through 24hrs, and quenched (, approach (a)). The composition was determined by oxidizing the sulphide to V_2O_5 at $500^\circ C$. When the same result was obtained from both the $V_{0.71}S$ and $V_{0.91}S$ materials, the equilibrium was regarded to be reached. Preliminary examinations showed that in the range around VS the reaction rate was very slow, requiring about 24hrs of the holding time. The approach (b) was to use a quartz spring balance within the range V_3S_4 through near V_5S_8. The equilibrium was checked by obtaining the same constant weight from both sides. The resulting sulphur partial pressure-composition relations are illustrated in Figs.1(a) and (b).

In the approach (a), the quenched specimen was examined by an X-ray diffractometer with CuK_α radiation. A few remarks should be made here regarding the diffraction patterns of this range. For instance, among those on VS line at $1217^\circ C$, both the hexagonal and monoclinic (similar to V_3S_4) patterns were observed: the former for those with vanadium richer than $V_{0.87}S$ and the latter poorer than that. The characteristic of the latter pattern was in the line-broadening as compared with the pattern of the quenched spec-

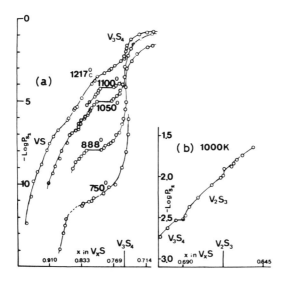

Figs.1(a) and (b). P_{S_2} composition relations.

imen in the purely V_3S_4 region. As shown in Fig.1(a), between 1100°C and 888°C two-phase regions exist. At 1217°C, this region cannot clearly be observed except a small kink at about $V_{0.78}S$. At lower temperarure 750°C, the V_3S_4 phase appears to extend to the vanadium-rich side, but the accuracy of those measurements at low pressure and temperature became very low with the observed points considerably scattered. Small change in the slope of curve could not be detected (refer to the consideration section). In Fig.1(b), a two-phase region at about $V_{0.69}S$ was obtained, indicating the existence of the V_2S_3 phase. The quenched specimens of both sides, however, exhibited very similar X-ray powder patterns.

Under an electron-microscope, the selected area diffractions and the lattice images showed that the quenched specimen of $V_{0.87}S$ was composed of mainly 4C-type hexagonal crystals, but with increasing temperature by electron bombardment this was transformed to the 3C-type. This suggests that, although the major hexagonal framework at high temperature is maintained, the quenched specimen does not necessarily represent the high-temperature state. Moreover, X-ray examination (CuK_α, 500mA) of the powder specimen sealed in an evacuated quartz capillary revealed a phase separation to hexagonal and seemingly monoclinic phases at above 600°C. Upon subsequent cooling, the former seemed to transform to the MnP-type.

One of the authors (M.N.) has already reported the non-quenchability of the V_5S_8 phase, (1). In the present report, therefore, only a figure is reproduced in Fig. 2.

CONSIDERATION

Prior to the discussion of the VS-V_3S_4 region, a brief explanation is given for the V_5S_8 phase (Fig.2). The change in the X-ray intensities for the 555 and 666 reflections (monoclinic cell) with increasing temperature represents the degree of intra- and inter-layer disordering, respectively. At about 800°C, the intra-layer disordering was completed with some short-range order remaining, while the inter-layer one proceeded to about 50% and the subsequent heating up to 1000°C did not cause any further disordering. Rapid quenching to 800°C and room-temperature resulted in the intensities as marked x in the figure, showing the clear reversibility of the transition. This clearly indicates the non-quenchability of the high temperature state in that region.

Regarding the VS-V_3S_4 region, the electron- and X-ray observations of the $V_{0.87}S$ specimen suggest a phase transition and also a certain phase separation at high temperature (probably above 600°C) as described in the experimental section. For $V_{1.00}S$ and $V_{0.95}S$, de Vries and Jellinek have reported a transition of the low-temperature MnP-type structure to the hexagonal NiAs-type one at about 600°C and 350°C, respectively, (2). In the present investigation, a similar transition was observed as mentioned before.

X-ray line-broadening of the seemingly monoclinic patterns of the quenched specimens on the VS line in Fig.1(a) calls for some considerations. As mentioned before, the tail of the V_3S_4 line at 750°C is not reliable at all. It may rather represent various phases which are very much similar to each other in the structure. The difference between the free energies of formation of the closely adjacent phases may be very small, so that the entire sulphur partial pressure-composition curve will be smoothed out. In the process of phase separation, in that case, several mixed-layer-type structures may be formed due to the structural similarity, which cause the characteristic line-broadening of the X-ray powder diffraction lines. Curvatures of the VS lines at 1217°C and especially at 1050°C between $V_{0.91}S$ and $V_{0.82}S$ in Fig.1(a) may suggest an existence of

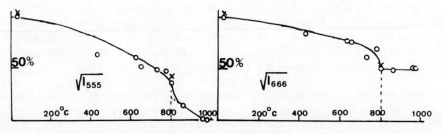

Fig.2. Intra- and inter-layer disordering of V_5S_8 with temperature

some other phases. Further analyses are strongly warranted. Based on the present experiments, the authors are inclined to think that at low-temperature several phases may exist in the range VS through V_3S_4.

ACKNOWLEGEMENT

The electron-microscopic work is of a joint research with Dr. S. Horiuchi at the National Institute for Research in Inorganic Materials, to whom the authors' deep appreciation is shown. Thanks are also due to the Ministry of Education and the Kazuchika-Ookura Memorial Foundation for their Grant-in-Aids which made part of the present work possible.

REFERENCES

(1) H. Nakazawa, M. Saeki and M. Nakahira: J. Less-Common Metals, 40(1975), 57.

(2) A. B. de Vries and F. Jallinek: Revue de Chimie minérale, 11(1974), 624.

ABOUT THE SYMMETRY LOWERING OF SOME APATITES AS EVIDENCED BY THEIR INFRARED SPECTROMETRIC STUDY

J.C. Trombe and G. Montel

Laboratoire de Physico-Chimie des Solides et des Hautes Températures - E.N.S.C.T. 38, rue des 36 Ponts

31078 - Toulouse Cedex - France

INTRODUCTION

The apatite group, with the ideal formula :

$Me_{10}(XO_4)_6Y_2$ with Me = Ca^{++}, Sr^{++} ... ; XO_4 = PO_4^{3-}, AsO_4^{3-} ... ; Y = F^-, Cl^-, OH^- ...

is of considerable importance, because calcium phosphate apatites are found widely as the phosphorite minerals, and are also the main inorganic constituent of calcified tissues.

Apatites generally crystallise in the hexagonal system (spatial group $P6_3/m$) with one unit formula by unit-cell (1,2). Only small departures from the preceding symmetry have been observed, but they are very weak and need precise X ray diffraction studies on single crystals, to be detected (3,4). One interesting feature of the apatite structure consists in the presence of channels running trough the structure, in the c axis direction. This **peculiarity** allows the ions located in channels to be easily substituted, either in aqueous media or by solid state reaction : in the latter case, the substitution of two hydroxyl ions by one bivalent ion ($CO_3^=$, $O^=$, $O_2^=$, $S^=$) associated with a vacancy, has recently been evidenced (5).

The purpose of this paper is firstly to study the effects of this substitution on the structure of the resulting products, by means of X-ray powder diffractometry and infrared spectrometry.

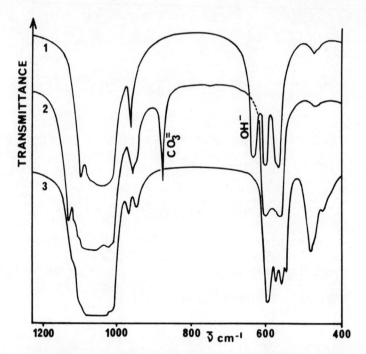

Figure I : Infrared spectra - 1) of hydroxyapatite ; 2) of carbonated apatite ; 3) of oxyapatite

Then the case of apatites containing in their channels two different monovalent ions will be considered.

RESULTS and DISCUSSION

Calcium phosphate apatites **mentioned above,** carbonated apatite, oxyapatite, peroxiapatite and sulfide-apatite, are always slightly defective concerning their respective content of bivalent ion with respect to the ideal formula containing one bivalent ion **per** unit cell. Their X-ray powder diagrams show only an apatitic phase, isomorphous with fluorapatite : all the reflections may be interpreted by considering the spatial group $P6_3/m$.

Differences between these apatites and hydroxyapatite, fluorapatite or chlorapatite are shown by their respective infrared spectrum, and more particularly by the number of bands or shoulders exhibited by these spectra.

The spectrum of hydroxyapatite (Fig. I, spectrum 1) may be interpreted, in terms of site symmetry or factor group analysis, with the spatial group $P6_3/m$. In both cases only nine infrared active bands may be expected for phosphate internal vibrations (Ta-

TABLE I

Fundamental modes	Free PO_4^{3-} Td symmetry	PO_4^{3-} site symmetry Cs	PO_4^{3-} factor group C6h
$\nu 1$	A1 (R)	A'(ir,R)	Elu (ir)
$\nu 2$	E (R)	A'(ir,R)+A"(ir,R)	Elu(ir)+Au(ir)
$\nu 3$	F2 (ir,R)	2A'(ir,R)+A"(ir,R)	2Elu(ir)+Au(ir)
$\nu 4$	F2 (ir,R)	2A'(ir,R)+A"(ir,R)	2Elu(ir)+Au(ir)
ir : infrared active		-	R : Raman active

ble I). This theoretical number is in accordance with the observed number of phosphate bands in hydroxyapatite (Note that some bands are unresolved).

Similar analyses do not fit with the observed experimental spectra of carbonated-apatite (Fig. I, spectrum 2), and oxyapatite (Fig. I, spectrum 3) or of peroxiapatite and sulfide-apatite. Though the spectra of these compounds are not quite similar, they present a common point : the number of bands or shoulders due to phosphate vibrations (all the bands originating from other groups than phosphates are indicated on the spectra) is greater than the theoretical number allowed by an analysis with the spatial group $P6_3/m$. This means that this spatial group, observed by X-ray diffraction, must not represent the true unit-cell symmetry.

In fact, if the substitution, in a unit-cell, of two monovalent ions by a bivalent ion associated with a vacancy is represented (Fig. II), it clearly appears that some symmetry elements (for instance, symmetry plane : m, 6_3 axis) must be suppressed. Thus it is possible to account for the infrared observations which are related to the unit-cell symmetry. However a random arrangement of bivalent ions and vacancies along the channels, allows to maintain these symmetry elements on the scale of many unit-cells ; and then X-ray observations are understandable and do not disagree with infrared ones.

So, in the particular case of apatites containing in their channels bivalent ions associated to vacancies, infrared spectroscopy can provide a better representation of the unit-cell symmetry than X ray diffraction. The perturbations of phosphate vibrations, may be explained as resulting from differences of crystalline field around the six phosphate groups of a unit-cell, according to their respective distance from the bivalent ion and the vacancy. In order to generalise this observation apatites containing simultaneously different monovalent ions have been considered.

The partial hydrolysis of chlorapatite (Fig. III, spectrum 2) results in the broadening of some phosphate vibrations (particularly the band at about 960 cm^{-1}), with respect to that of pure chlorapatite (Fig. III, spectrum 1). Moreover a shoulder at appro-

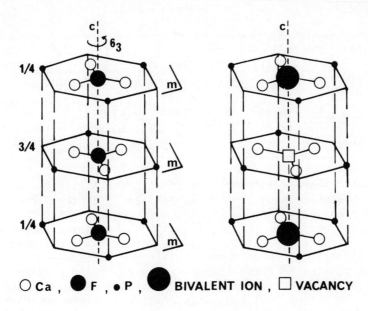

Figure 2 : Atomic arrangement in the fluorapatite lattice or in the lattice of an apatite containing in his channels bivalent ions associated to vacancies (Note that the exact position of bivalent ions along the c axis is not known)

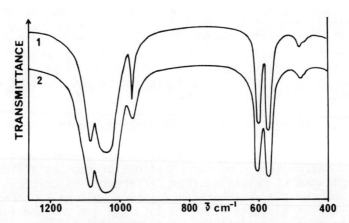

Figure 3 : Infrared spectra - 1) chlorapatite ; 2) chlorapatite submitted to partial hydrolysis

ximatively 1130 cm^{-1} is appearing for chlor-hydroxyapatite while it is absent from the spectrum of the initial chlorapatite. These perturbations, which would mean a lowering of the unit cell symmetry, undiscernible by X-ray diffraction on powders seem, as previously, to result from changes of crystalline field around the phosphate groups of a unit-cell, according to the presence in close proximity of these groups of one or the other monovalent ion.

CONCLUSION

Infrared spectroscopy thus appears as a very **helpful technique**, when studying complex solids like apatites : in some cases, it can provide information at the atomic scale, which are not always evidenced by X-ray diffractometry.

REFERENCES

(1) C.A. Beevers, D.B. Mc Intyre : Miner. Mag., 1946, 27, 254
(2) M.I. Kay, R.A. Young, A.S. Posner : Nature, 1964, 204, 1050
(3) J.S. Prener : J. Electrochem. Soc., 1967, 114, 77
(4) J.C. Elliott, P.E. Mackie : C.R. Coll. Int. Physico-Chimie et Cristallographie des apatites d'intérêt biologique, pp. 69-76
(5) J.C. Trombe : Thèse Toulouse 1972 ; Ann. Chim., 1973, 8, 251 and 1973, 8, 335.

MECHANISM AND KINETICS OF SOLID STATE REACTION IN THE SYSTEM $CuCr_2O_4 - Cu_2Cr_2O_4 - CuO$

J. Haber and H. Piekarska-Sadowska

Research Laboratories of Catalysis and Surface Chemistry

Polish Academy of Sciences, 30060 KRAKOW, Poland

ABSTRACT

The solid state reaction $CuCr_2O_4 + CuO = Cu_2Cr_2O_4 + 1/2\ O_2$ was studied in the temperature range 820-850°C. Electron microprobe analysis showed that reaction consists of decomposition of $CuCr_2O_4$ grains to $Cu_2Cr_2O_4$, transport of chromium in the form of CrO_3 through the gas phase to the surface of CuO grains and growth of $Cu_2Cr_2O_4$ by diffusion across the product layer. In the first stage the rate determining step is removal of chromium oxide from $CuCr_2O_4$ grains, the rate being well described by $kt = 1-(1-\alpha)^{1/3}$, in the second stage the rate becomes diffusion controlled and may be described by the Kröger-Ziegler equation $k \cdot \ln t = \{1-(1-\alpha)^{1/3}\}^2$.

INTRODUCTION

In the previous studies it was shown (1-3) that in the solid state reactions of VIb group metal oxides these oxides always play the role of "attacking" reagent. They are usually rapidly transported either by surface migration or by a transport reaction through the gas phase, covering the surface of grains of the other reagent. The reaction mechanism may be then described in terms of the classical Jander model. When bulk diffusion is more rapid than surface diffusion, the rate constant becomes dependent on the number of intergranular contacts (3). High rates of reaction e.g. with Cr_2O_3 may be explained by the operation of a transport reaction, Cr_2O_3 being rapidly transported through the gas phase in the form of CrO_3 to the surface of the other reactant. It seemed of interest to study whether a similar mechanism operates in case of reactions with more complex compounds of VIb group metals such as oxysalts. The solid state reaction $CuCr_2O_4$ + CuO was chosen because of its importance for the synthesis of selective hydrogenation catalysis.

EXPERIMENTAL

CuO was obtained by precipitation with NaOH from a solution of copper sulphate. $CuCr_2O_4$ was prepared by precipitation with copper nitrate from a solution of ammonium dichromate in the presence of ammonia, decomposition of the precipitate at $400°C$ and dissolution of excess CuO in hot conc. HCl. The preparation was then heated at $650°C$ for 8 hr.

The reagents were pressed into seperate pellets, the pellets were contacted, placed in a furnace at $850°C$ and heated for different periods of time. In one series of experiments the pellets were seperated from each other by a gap 1 mm wide during the heating. After the reaction the pellets were mounted into a resin, cut along their axis and analysed by the electron microprobe. "Camenca MS 46" apparatus was used.

Kinetics of the reaction were followed by measuring the amount of oxygen liberated in the course of the reaction given by the equation :

$$CuCr_2O_4 + CuO = Cu_2Cr_2O_4 + 1/2 \, O_2$$

The reactants in form of powders were mixed together and placed in a quartz reactor connected to a gas burette. The reactor was then inserted into a furnace, previously heated to $850°C$ and the amount of oxygen was determined as a function of time for mixtures of different composition varying from $CuO:CuCr_2O_4$ molar ratio 9:1 to 1:3.

Kinetics of the reactions in the mixtures $CuO + Cr_2O_3$, $MgO + CuCr_2O_4$ and $ZnO + CuCr_2O_4$ were also studied. Products of the reactions were identified by X-ray analysis using a Rigaku-Denki model D3F diffractometer.

RESULTS AND DISCUSSION

As an example Fig. 1 shows the results of electron microprobe analysis along the axis of $CuCr_2O_4$ and CuO pellets, heated for 30 hr at $850°C$. It may be seen that no definite reaction zone can be distinguished, which would be expected in case of diffusion controlled process. Chromium is distributed uniformly along the whole cross-section of the CuO pellet whereas across the whole section of the $CuCr_2O_4$ pellet, grains are visible with increased concentration of copper and decreased concentration of chromium, the compos-

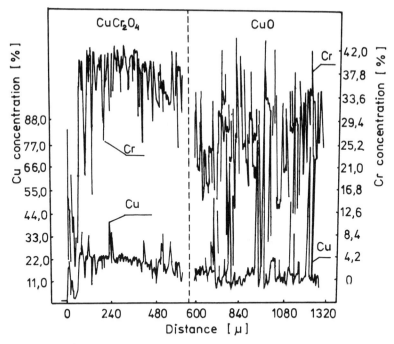

Fig 1. Results of the electron microprobe analysis of $CuCr_2O_4$/CuO pellets. Reaction at 850°C for 30 hr.

ition corresponding roughly to $Cu_2Cr_2O_4$. It may be concluded that the reaction consists in decomposition of $CuCr_2O_4$ to $Cu_2Cr_2O_4$, transport of chromium oxide through the gas phase and its deposition onto the surface of CuO grains, where it reacts further to form $Cu_2Cr_2O_4$. This compound is thus formed in two different parts of the system: in $CuCr_2O_4$ grains by removal of Cr_2O_3 and in CuO grains as a result of the reaction. The transport of chromium oxide through the gas phase was confirmed by the experiment, in which the two reacting pellets have been seperated by a gap of 1 mm. Results of the microprobe analysis were identical to those shown in Fig. 1 except for a smaller concentration of chromium in the CuO pellet.

In the conditions of the experiment the reaction proceeds only when CuO grains are present in the system. Apparently, by binding to Cr_2O_3 they decrease the pressure of chromium oxide in the system below the equilibrium pressure of $CuCr_2O_4$ thus promoting its decomposition. This conclusion is consistent with the observations that in the $2CuO + Cr_2O_3$ mixture the reaction to $CuCr_2O_4$ stops after attaining only about 10% of conversion. No lines of CuO are however visible in the X-ray pattern of the products. Obviously the grains of CuO have been covered with a layer of the product and diffusion across this layer is very slow. When however new portion of CuO is then added the reaction starts again.

A similar mechanism operates on the addition of any oxide to $CuCr_2O_4$ which with Cr_2O_3 forms a compound of dissociation pressure lower than that of $CuCr_2O_4$. This was shown by adding MgO and ZnO which resulted in formation of $MgCr_2O_4$ and $ZnCr_2O_4$, CuO being in this case left after decomposition of $CuCr_2O_4$.

The following mechanism of the reaction of $CuCr_2O_4$ with such oxides as CuO and MgO and ZnO may be thus postulated:

$2CuCr_2O_4 + O_2 = Cu_2Cr_2O_4 + 2CrO_3$ at the surface of $CuCr_2O_4$ grains

$2CrO_3 + 2CuO = Cu_2Cr_2O_4 + 2O_2$ at the surface of CuO grains

further reaction at CuO grains proceeding by diffusion of the depositing chromium oxide across the product layer. It may be thus expected that in the first period the rate of the reaction will be determined by the rate of the removal of chromium oxide from the surface of $CuCr_2O_4$ grains, which may be expressed by the equation:

$$k t = 1 - (1-\alpha)^{1/3} \qquad \{1\}$$

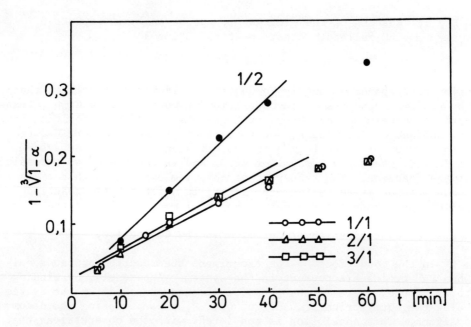

Fig. 2 Conversion as a function of time plotted in terms of equation {1} for $CuO-CuCr_2O_4$ mixtures of different composition, reacted at 850°C.

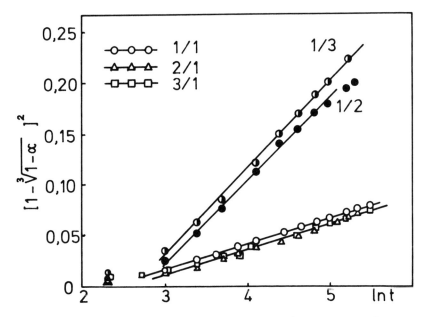

Fig. 3 Conversion as a function of time plotted in terms of equation {2} for CuO-CuCr$_2$O$_4$ mixtures of different composition, reacted at 850°C.

Results summarized in Fig. 2 show that this equation is in fact well obeyed. In the later stages of reaction, when practically all Cu$_2$Cr$_2$O$_4$ originating from decomposition of CuCr$_2$O$_4$ grains had already been formed, further formation of Cu$_2$Cr$_2$O$_4$ at the surface of CuO grains becomes diffusion limited. As indicated by results presented in Fig. 3 kinetics of the reaction may be then expressed by the Kröger-Ziegler equation:

$$k \ln t = \{1 - (1-\alpha)^{1/3}\}^2 \qquad \{2\}$$

REFERENCES

1. J. Deren, J. Haber, Z. anorg.allg.Chem., <u>342</u>, 268 (1966)
2. J. Haber in "Festkörperchemie", Verlag für Grundstoffindustrie, Leipzig 1973, p. 221
2. J. Haber, J. Ziołkowski, Proc. VII Int. Symp. Reactivity of Solids, Bristol 1972, Chapman and Hall, London 1972, p.782.

TOPOTAXY AND SOLID-GAS REACTIONS

J. Guenot, F. Fievet-Vincent and M. Figlarz

Université Paris VII, 2 Place Jussieu
75221 Paris (France)

Université de Picardie, 33, rue St Leu
80039 Amiens (France)

ABSTRACT

A number of solid-gas reactions were carried out on cobalt compounds. Shape, porosity, crystallite size and crystallite misorientation of particles of the reacting solids as well as their changes during the reaction were studied. The results show that topotactic relationships are always present at the beginning of the reaction. As the reaction proceeds either they are maintained to a greater or lesser extent or the particles break up. Influence of the starting solid on growth of the product and occurence of resulting strains followed by their ease of relief seem to be the causes of the observed phenomena.

INTRODUCTION

When a single crystal undergoes a chemical reaction without large change in its external form (pseudomorphous reaction), crystallographic orientation relationships between the starting material and the product are often observed. The occurence of this phenomenon is called topotaxy (1). It is generally explained by a structural similarity and, often, a framework of certain ions is thought to remain unchanged in spite of the reaction. These explanations are certainly true for reactions without real change of structure - like e.g. zeolite dehydration - or with a progressive change - like solid solution development ; the problem remains in question for common solid-gas reactions.

Some time ago, Bernal (2) pointed out two phenomena which might be significant. First, at the very beginning of the reaction, the solid product is hardly able to escape the influence of the starting material and, thereby, must have a growth oriented by the latter. Second, due to the misfit between the two structures, strains arise as the reaction proceeds and the topotactic relationships are not maintained unless the strains are relieved by development of dislocations.

These two factors seemingly have not been sufficiently taken in account in investigations on topotaxy. The aim of the present paper is to show their action by means of examples of solid-gas reactions - which are all carried out from different cobalt compounds - and to draw from them some conclusions on interpreting topotaxy.

RESULTS

Before reporting our results we must emphasize that orientation relationships are never strict because real crystals are never perfect. The model which we use to account for these imperfections is mosaic crystal. A mosaic crystal is formed of parts of perfect crystal - crystallites - more or less misoriented. They are separated by thin zones - boundaries - where dislocations are concentrated.

Our results were all obtained, on solids in powdered form (submicronic small particles of definite shapes), by electron microscopy and diffraction. Crystallite sizes are determined by X-ray powder diffraction. The results are only considered in a qualitative way. More details are to be found in the original papers (3-8). For each solid, the following parameters, as well as their changes owing to the reaction, are studied : shape of the particles ; porosity (shape and size of pores) ; crystallite size ; crystallite misorientation.

The results are summed up in the table.

a) <u>Particle Shapes</u>. The shape of the particles remains unchanged for the greatest number of reactions which are, therefore, pseudomorphous. In two cases - reduction of CoO with small pores to metallic Co and oxidation of $Co(OH)_2$ at room temperature to CoOOH - the particles are broken up.

b) <u>Porosity</u>. $Co(OH)_2$ is made up of non porous particles ; when they undergo the reaction $Co(OH)_2 \rightarrow CoO$, they become porous with tiny pores. For $CoO \rightarrow Co_3O_4$, the porosity remains unchanged.

```
1. CLOSE TOPOTACTIC RELATIONSHIPS

                    Co(OH)₂  ─────────→  CoO      (3)
    hexagonal part. ; non por. ;        pseudomorphs ; small, regular
    large and well oriented cryst.      por. ; small, slighty misorien-
                                        ted cryst.

                    CoOOH    ─────────→  Co₃O₄    (8)
    hexagonal part. ; small, irregu-    pseudomorphs ; small, regular
    lar por. ; medium-sized, fairly     por. ; small, fairly misorien-
    misoriented cryst.                  ted cryst.

                    CoO      ─────────→  Co₃O₄    (5)
    hexagonal part. ; small or large    pseudomorphs ; unchanged por. ;
    regular por. ; small or medium-     unchanged cryst. orientation
    sized cryst. ; slightly misorien-   and size.
    ted cryst.

2. MEDIUM TOPOTACTIC RELATIONSHIPS

        (t > 60°C) Co(OH)₂  ─────────→  CoOOH    (6)
    hexagonal part. ; non por. ; lar-   pseudomorphs ; irregular small
    ge and well oriented cryst.         por. ; medium-sized cryst. ;
                                        fairly misoriented cryst.

                  CoO (160Å) ─────────→  Co      (4,7)
    hexagonal part. ; large regular     pseudomorphs ; unchanged por. ;
    por. ; large cryst. ; well orien-   unchanged cryst. size ; very mi-
    ted cryst.                          soriented cryst.

3. PARTICLE DISRUPTION FOLLOWING TOPOTACTIC RELATIONSHIPS AT THE
   BEGINNING OF THE REACTION

        (room temp.) Co(OH)₂ ─────────→  CoOOH   (6)
    hexagonal part. ; non por. ;        disrupted part. ; non por. ;
    large and well oriented cryst.      large cryst.

                  CoO (85Å)  ─────────→   Co     (4,7)
    hexagonal part. ; small regular     disrupted part. ; non por. ;
    por. ; small slightly misorien-     large cryst.
    ted cryst.

    part. = particles ; por. = pores, porous ; cryst. = crystallites
```

Classification Of Some Topotactic Reactions For Cobalt Compounds

For the reaction $Co(OH)_2 \rightarrow CoOOH$ (t > 60°C), the resulting porosity is irregular with cracks. As for the reaction CoOOH (irregular porosity) $\rightarrow Co_3O_4$, it shows transition from an irregular porosity to a regular one.

c) <u>Crystallite Size</u>. If the crystallite size of the starting solid is large ($Co(OH)_2$ for instance), a much smaller one is obtained for the final product (with a crystallite growth when the temperature rises). If the original crystallite size is small or medium, it changes only a little. For the reactions with particle disruption large crystallites occur.

d) <u>Crystallite Misorientation</u>. A slight increase of crystallite misorientation - accompanying a crystallite size decrease - is observed for the reactions $Co(OH)_2$ (large crystallites) $\rightarrow$ CoO, $CoOOH \rightarrow Co_3O_4$ and CoO (medium-sized crystallites) $\rightarrow Co_3O_4$. The reaction $Co(OH)_2 \rightarrow CoOOH$ (t > 60°C) brings about a rather large misorientation. As for the reaction CoO (large pores) $\rightarrow$ Co, the obtained misorientation is very large, although in no case could it be considered a random orientation. For the reaction $Co(OH)_2 \rightarrow CoOOH$ (at room temperature) which proceeds with particle disruption, tight orientation relationships are very easy to observe at the beginning of the reaction.

DISCUSSION

First of all, it must be emphasized that all the mentioned reactions are topotactic with a greater or lesser crystallite misorientation. For two reactions, the particles split up but only when certain conditions are fulfilled : for $Co(OH)_2 \rightarrow CoOOH$ a low temperature is required and for $CoO \rightarrow Co$, a small porosity of CoO. As far as the first reaction is concerned, orientation relationships are evident at the beginning of the reaction.

A significant influence of close relationships between the solid structures does not appear as limiting crystallite misorientation as it is exemplified by the reactions $CoO \rightarrow Co_3O_4$ and $Co(OH)_2 \rightarrow CoO$ or $CoOOH \rightarrow Co_3O_4$. They all yield a similar misorientation whereas CoO and Co_3O_4 (first reaction) are in close structural relationships.

On the contrary, a more or less easy strain relief is certainly a predominant factor. When it is easy a large texture reordering arises i.e. that small, little misoriented crystallites form. If strain relief is more difficult, the crystallites, although small, are more misoriented. Lastly, large crystallites appear and the particles break up as the reaction proceeds.

The reaction $Co(OH)_2 \rightarrow CoOOH$ ($t > 60°C$) shows that the formation of an irregular porosity is a clue for a difficult texture reordering ; the contrary is true for a regular one.

CONCLUSION

It has been seen that the major factor for the occurence of tight topotactic relationships at the end of a solid-gas reaction is a easy strain relief with formation of a mosaic crystal with small, not very misoriented crystallites. If that strain relief is more difficult the crystallites are more misoriented i.e. the topotactic relationships less close and, finally the reacting crystal may be split up although topotaxy is present at the beginning of the reaction.

Future work remains to be done to explain the occurence or lack of easy strain relief.

REFERENCES

(1) L. S. Dent Glasser, F. P. Glasser and H. W. F. Taylor, Quart. Rev. 16 (1962) 343.
(2) J. D. Bernal, Schweiz. Archiv. Angew. Wiss. Techn. 26 (1960) 69.
(3) M. Figlarz and F. Vincent, Compt. Rend. Acad. Sci. (Paris) 266C (1968) 376 ; F. Vincent, M. Figlarz and J. Amiel in "6th Int. Symp. Reactiv. Solids, Schenectady 1968" (edited by J. W. Mitchell, R. C. Devries, R. W. Roberts and P. Cannon) (Wiley - Interscience, New York 1969) p. 181.
(4) F. Vincent, F. Lecuir and M. Figlarz, Compt. Rend. Acad. Sci. (Paris) 268C (1969) 379.
(5) D. Colaïtis, F. Fievet-Vincent, J. Guenot and M. Figlarz Mat. Res. Bull. 6 (1971) 1211.
(6) M. Figlarz, J. Guenot and J. N. Tournemolle, J. Mater. Sci. 9 (1974) 772.
(7) M. Figlarz, Ann. Chim. 9 (1974) 367.
(8) M. Figlarz, J. Guenot and F. Fievet-Vincent, to be published.

MICROSTRAIN AND THE MECHANISM OF THE REVERSIBLE MONOCLINIC
$\overset{\rightarrow}{\leftarrow}$ TETRAGONAL PHASE TRANSFORMATION IN ZIRCONIA

S. T. Buljan, H. A. McKinstry and V. S. Stubican

Materials Research Laboratory, The Pennsylvania

State University, University Park, PA 16802 U.S.A.

ABSTRACT

The Warren-Averbach method was used to evaluate the microstrain involved in the monoclinic $\overset{\rightarrow}{\leftarrow}$ tetragonal transformation of ZrO_2. As the transformation proceeds microstrain reaches a maximum of 1.9×10^{-3} when 50% of the monoclinic phase has been transformed. Similar behavior was found on cooling. A large hysteresis in microstrain was observed in this transition. The extent of the coherent interface between two phases could explain the behavior of microstrain.

Optical observations have shown that the tetragonal phase is usually twinned on the $(1\bar{1}2)_{bct}$ $(\bar{1}12)_{bct}$ plane and the extent of twinning is influenced by the heating rate. Orientation relations in ZrO_2 transformation are:

$(100)_m \,||\, (110)_{bct}$; $[010]_m \,||\, [001]_{bct}$ and by the virtue of twinning $(100)_m \,||\, (110)_{bct}$; $[001]_m \,||\, [001]_{bct}$.

INTRODUCTION

It is well established that the broadening of x-ray peaks results from the superposition of two effects, one a particle size effect and the other a distortion effect. The distortion effect is dependent on the order of the reflection and arises from small local strains which vary from point to point in the lattice (microstrains). The purpose of this work was to evaluate the microstrain involved in the monoclinic $\leftrightarrow$ tetragonal phase transition of zirconia. Also the twinning and orientation relations in this martensitic type transformation were investigated.

EXPERIMENTAL

In this study, the Stokes - Fourier coefficient method of correcting for instrumental broadening and the Warren-Averbach method of strain evaluation were used to obtain the microstrain and the coherent domain size, (1)(2). The x-ray diffraction studies of microstrain were carried out with a low gradient, spherically wound, Pt high temperature furnace, (3). X-ray data were taken on the Picker (Model 3488) biplane diffractometer using CuKα radiation. Data were taken by step-scanning and a count was taken for the preset length of time at a particular value of 2θ. Changes due to the transformation were observed on a high temperature microscope stage using single crystals of ZrO_2.

To resolve the topotaxial relationship of the monoclinic and tetragonal phases, zirconia crystals were investigated by Laue transmission and Buerger precision methods at elevated temperatures.

RESULTS AND DISCUSSION

Microstrain analysis showed that the zirconia powder contained appreciable amounts of microstrain at room temperature 0.83×10^{-3}. However, the initial strain dropped to a minimum observed value of 0.2×10^{-3} at the temperature just below the transition region. Consequently, for the observation of microstrain in the transformation region, the sample was held at 1105°C for 24 hrs to remove the residual strain before the temperature was raised in the transformation region. Fig. 1 shows microstrain as a function of time for the monoclinic to tetragonal transition. It is evident that the microstrain reaches an equilibrium value after a relatively short time and is almost time independent. Similar results were obtained on cooling for the tetragonal to monoclinic phase transition. The microstrain was measured in two crystallographic directions $\perp(100)$ and $\perp(011)$. Though there is a difference in the magnitude of the microstrain the behavior is similar. Fig. 2 shows that, as the transformation proceeds above 1105°C, the microstrain increases and the root mean square reaches a maximum of $\sim 1.9 \times 10^{-3}$. On cooling again maximum was observed at somewhat lower temperature, but the microstrain does not decrease below the value of 0.8×10^{-3} since two phases coexist to the relatively low temperature of 500°C. These results clearly show the existence of the microstrain hysteresis in the monoclinic $\leftrightarrow$ tetragonal phase transition of zirconia.

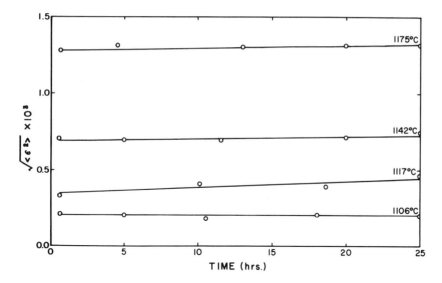

Fig. 1　The root mean square strain perpendicular to the (011) monoclinic plane vs. time of heating in the monoclinic tetragonal transformation range.

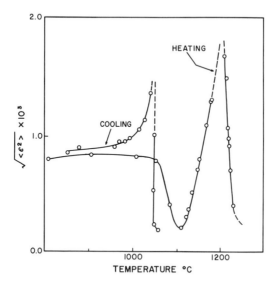

Fig. 2　Microstrain hysteresis exhibited by ZrO_2 monoclinic tetragonal transformation. Microstrain in the direction normal to the (011) monoclinic plane.

Since the maximum in the microstrain occurs at the temperature where 50% of the transformation is completed, it could be interpreted as arising from the maximal coherent interface between two phases. The monoclinic $\rightleftarrows$ tetragonal phase transition in ZrO_2 single crystals was studied at temperature by transmission optical microscopy and x-ray techniques. Optical observations show that on heating, the monoclinic phase transforms to the tetragonal phase by the motion of an interface parallel to the $(001)_m$ plane; in addition, simultaneous twinning occurs behind the advancing interface. The tetragonal phase is usually twinned on the $(1\bar{1}2)_{bct}$ or $(\bar{1}12)_{bct}$ plane, as shown in Fig. 3A and B. The extent of twinning is influenced by the heating rate. Optical and x-ray single crystal work show that orientation relations in the ZrO_2 transformation are:

$(100)_m || (110)_{bct}$; $[010]_m || [001]_{bct}$ and by the virtue of twinning, $(100)_m || (110)_{bct}$; $[001]_m || [001]_{bct}$.

During the cooling the same topotaxial relationships are maintained.

Fig. 4 summarizes the observations of the topotaxial relationship of the monoclinic and texagonal phases and gives the resulting orientations in the monoclinic→tetragonal→monoclinic transformation

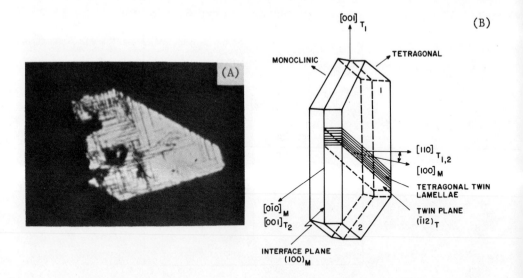

Fig. 3 (A) Twinned crystal of zirconia in tetragonal form at 1200°C. Magnification 90x. (B) Graphical representation of tetragonal twin and interface formation and orientation with respect to crystal morphology.

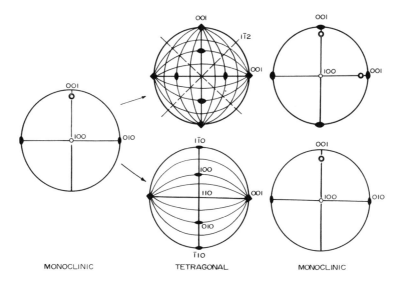

Fig. 4 Resulting crystallographic orientations in the monoclinic→tetragonal→monoclinic transformation of ZrO_2. Upper path results when twinning occurs in tetragonal phase. Representation of twinning in monoclinic phase is omitted for the sake of clarity.

cycle. The upper path shows the case where tetragonal twins are formed and the lower path where there is no twinning in the tetragonal phase, which may possibly be achieved by using a very slow rate of temperature change.

ACKNOWLEDGEMENT

Supported by the National Science Foundation under Grant No. GP-11808.

REFERENCES

1. A. R. Stokes, Proc. Phys. Soc., 61(1948) 382.

2. B. E. Warren and B. L. Averbach, J. Appl. Phys. 21(1950) 595.

3. H. A. McKinstry, J. Appl. Phys. 41(1970) 13 and 5074.

MECHANICAL DAMPING OF IONIC CRYSTALS WITH REGARD TO MgO

J. Kriegesmann, G.H. Frischat, H.W. Hennicke

Lehrstuhl für Glas und Keramik, Technische
Universität Clausthal, Clausthal

auch Sonderforschungsbereich 126 Göttingen - Clausthal

Germany

ABSTRACT

The mechanical damping of doped polycrystalline MgO was measured. A symmetric Debye type peak could only be found in the case of a specimen containing Li^+ with an activation enthalpy of $(2,84 \pm 0,53)$ eV. A Debye type behaviour of damping peaks is excluded when ions with a higher valence state $(+3,+4)$ are introduced.

INTRODUCTION

Substitution of heterovalent doping ions into ionic lattices requires formation of vacancies contrarily charged to the effective charge of doping ions to make the lattice electroneutral. Formation of associates is energetically preferred to homogeneous distribution of vacancies because of coulombic attraction. Associates can be considered as electrical dipoles, if they interact with an applied electric field. These electric dipoles are directional local distortions of electrical homogeneous crystal fields. On the other hand such associates are also characterized by local elastic distortions. Thus an interaction between an applied stress field and associates is equally possible. An electrical field can be explained by a first rank tensor. Therefore this can be considered the main distinction between both types of dipoles.

Interaction between dipoles and applied fields can be demonstrated by measurements of dielectric loss and mechanical damping.

LITERATURE SURVEY

Beckenridge (1) first described electric relaxation in ionic crystals similar to the mechanical relaxation found in metals. But his experiments on electric relaxation of alkali halides could not be reproduced (2). His measurements of mechanical damping (3) were also doubtful because he demonstrated mechanical and electric relaxation on the same specimens. Finally Wachtman (4) proved the postulation that the same defects in ionic compounds caused not only inelastic but also electric relaxation. He used polycrystalline ThO_2 doped with CaO. After Beckenridge's (1)(3) inconsistent conclusions a further attempt was made to find quantitatively evaluated relaxation results for materials with NaCl structure. Dreyfus and Laibowitz (5) first realised mechanical and electrical relaxation of NaCl doped with divalent ions theoretically by supposing a two shell model. Southgate (6) tried to find similar effects with MgO single crystals by doping them with Cr^{3+} and Fe^{3+}. However he could not interpret his results either with a one or a two shell model. Hennicke and Schüssler (7) measured mechanical damping of magnesia bricks. They assumed relaxation of cation pairs at grain boundaries.

Herbst (8) assumed that his relaxation effects found in MgO doped with Ti^{4+}, Fe^{3+}, Al^{3+} and Cr^{3+} were due to jumps near grain boundaries. He referred to the two shell model of Dreyfus and Laibowitz (5).

THEORETICAL BACKGROUND

If additives are able to be dissolved in periclase, simple equations can be formed regarding quadro-, tri-, and monovalent dopants:

$$TiO_2 = Ti_{Mg}^{\bullet\bullet} + 2\ O_O^x + V_{Mg}^{''}$$

$$Al_2O_3 = 2Al_{Mg}^{\bullet} + 3\ O_O^x + V_{Mg}^{''}$$

$$Li_2O = 2Li_{Mg}^{'} + O_O^x + V_O^{\bullet\bullet}$$

In the case of quadrivalent doping ions one vacancy and one foreign ion is needed for formation of a neutral associate while in the case of tri- or monovalent ions one vacancy and two foreign ions are required. However, Wertz and Auzins (9) showed that besides neutral doubly associated vacancies charged single associated vacancies can also exist. If an applied mechanical stress field is acting on dipoles, only singly associated vacancies have the possibility to change their position to energetically more favourable sites.

Doping ions of a single valence state can cause relaxation described by a one shell model. This can happen only if most of the vacancies possess oxygen sites as the nearest neighbours of the doping ions. This is called nn-dipole. It is possible to demonstrate for such a 6 position model (10) that mechanical relaxation behaviour is represented by a single Debye peak. Relaxation becomes more complex when the next nearest oxygen sites (nnn) are also taken into consideration. This two shell model with 14 positions is represented by overlapping of at least two Debye peaks.

The incorporation of tri- or quadrivalent ions into periclase results in a 12 position model for the first shell. Because of this, damping behaviour is characterized by overlapping of two Debye peaks (11). Taking the next nearest neighbour dipoles from the second shell which form a 18 position model together with the first shell, where mechanical relaxation is determined by at least three Debye peaks.

EXPERIMENTAL PROCEDURE

The polycrystalline specimens used for this investigation were prepared in the following manner: MgO powder (purity 99,5 wgt. %) put in Al_2O_3 crucibles was fired at $1480^\circ C$. The resulting bodies were ground in an agate mortar (grains smaller than 63 um). After doping and mixing the substance was pressed hydrostatically at 1800 kp/cm^2 and fired again in an oxidizing medium (air). The specimens were kept at $1500^\circ C$ for 16 hours. Samples of the size $10 \times 15 \times 140$ mm^3 were obtained by cutting and polishing with diamond tools. The dopants used in this paper were ZrO_2, TiO_2, Al_2O_3, Cr_2O_3, Li_2CO_3, and MgF_2 to the amount of 0,3 mol %. Porosity of the samples varied between 10 (doped with Ti) and 22 % (undoped).

Damping was measured by Förster's method (12) using the frequency generator Elastomat 1.024. The samples suspended in a resistance furnace were attached to a piezoelectric vibration system via platinum wires. Resonance frequency of the samples varied from 3000 to 4500 Hz depending on the different samples. Damping could be measured up to $1050°C$.

RESULTS AND DISCUSSION

Small or no peaks were obtained from undoped specimens and specimens doped with ZrO_2 and MgF_2. It could be shown by electrical conductivity measurements (10) that solubility of Zr^{4+} and F^- in periclase is very low. Thus it is not surprising that those specimens show only small effects of damping owing to point defects. All other experiments exhibit special damping peaks. Only in the case of Li^+-doping, however, can a Debye type behaviour be proved. A damping peak of this sample with a maximum at the temperature of $655°C$ and at the resonance frequency of 3406 Hz is presented in Fig. 1, background damping is omitted. The plot of damping times temperature versus reciprocal temperature was approximated by a Debye function according to the equation (10):

$$T (\log.Dekr.) = A \operatorname{sech} (S + P/T)$$

log.Dekr. = logarithmic decrement
A = factor (depending on orientation of crystals)
P = Q/k
Q = activation enthalpy
k = Boltzmann factor

T = temperature
S = $\ln(\omega\tau_o)$
τ_o = preexponential factor of relaxation time
$\omega = 2\pi f$
f = frequency

Tab. 1: Parameters A, P, S, τ_o, τ_{peak}, and Q from Debye peak shown in Fig. 1

A	$P \cdot 10^3$ (K)	S	τ_o (sec)	τ_{peak} (sec)	Q (eV)
13,26	33,12	-35,61	$1,60 \cdot 10^{-20}$	$5,06 \cdot 10^{-5}$	2,84
±0,85	±6,18	±6,64			± 0,53

The activation enthalpy found in this process compared to that of oxygen self-diffusion measured by Oishi and Kingery (13)(2,71 eV) shows a close correlation. Thus we can suppose the 6 position model to be reasonable.
The value of the preexponential factor was found appreciably lower than the normally accepted $10^{-13} - 10^{-17}$ sec for relaxation of point defects. However, it must be considered that the factor S shows a variation of ± 7, that means a limit of variability of three orders of magnitude for the preexponential factor.

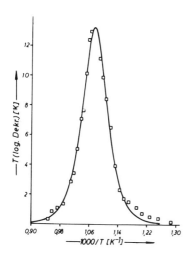

Fig. 1

Damping peak (MgO doped with 0,3 mole % Li_2CO_3)

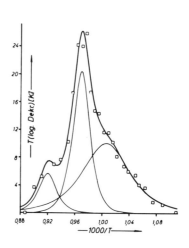

Fig. 2

Damping peak (MgO doped with 0,3 mole % Al_2O_3)

The specimen doped with Al^{3+} gives the most distinct peak. The plot of damping versus reciprocal temperature presented in Fig. 2 indicates that the peak is not symmetric. Generally this means that at least two Debye peaks are overlapping. In this case the experimental results are represented by a summation of three Debye functions. The parameters A, S, P, and Q are listed in Tab. 2.

The values presented in Tab. 2 show that two of the three activation enthalpies are extremely high. Even if we suppose that activation enthalpies must not be exactly comparable with that for diffusion, such high

Tab. 2: Parameters A,S,P, and Q from the Debye peaks shown in Fig. 2

	A	S	$P \cdot 10^3$ (K)	Q (eV)
left curve	5,64	-79,64	86,61	7,46
central curve	20,29	-87,26	90,13	7,77
right curve	10,01	-32,17	32,33	2,77

values cannot be accepted. For this reason we exclude a Debye type behaviour for this peak. A similar result was found when MgO was doped with Cr^{3+}, whereas an even sharper peak was obtained by doping with Ti^{4+}.

The reason for these results, which cannot be compared with those found in literature, is possibly due to the interaction between point defects or with dislocations. The latter can be proved by solid solution hardening in MgO (14). The results found here are not sufficient to satisfactorily explain the unusual effects.

REFERENCES

1 Beckenridge, R.G., J. Chem. Phys. 16 (1948) 959; ibid. 18 (1950) 913
2 Haven, Y., ibid. 21 (1953) 171
3 Beckenridge, R.G., in Imperfections in Nearly Perfect Crystals, John Wiley and Sons
4 Wachtman, J.B., Jr., Diss. University of Maryland 1961; Phys. Rev. 131 (1963) 517
5 Dreyfus, R.W., Laibowitz, R.B., ibid. 135 (1964) A1413
6 Southgate, P.D., J. Phys. Chem. Solids 27 (1966) 1263
7 Hennicke, H.W., Schlüßler, H., Ber. Dtsch. Keram. Ges. 45, (1968) 234
8 Herbst, U., Diss. TU Clausthal 1975
9 Wertz, J.E., Auzins, P.V., J. Phys. Chem. Solids 28 (1967) 1557
10 Kriegesmann, J., Diss. TU Clausthal 1976
11 Haven Y. van Santen, J.H. Suppl. Nuovo cimento 7(1958)605
12 Förster, F., Z. Metallkde. 29 (1937) 109
13 Oishi, Y, Kingery, W.D., J. Chem. Phys. 33 (1960) 905
14 Reppich, B., Mater. Sci. Eng. 22 (1976) 71

FERRITE FORMATION MECHANISM

P.Y. Eveno and M.P. Paulus

Laboratoire d'Etudes et de Synthèse des Microstructures

C.N.R.S. 92190 Meudon-Bellevue France

INTRODUCTION

The most significant information, for the interpretation of the ferrite formation mechanism is provided by methods based on the study of diffusion couples in a semi-infinite medium between two oxides. It is by such experiments that many investigators studied the solid state reaction between Fe_2O_3 and MgO (1,2,3,4,5). However studies of diffusion couples are really convenient only if the examinations lead to quantitative results :

1. To start with, good contact must be achieved between the oxides to avoid any local gas phase transport at the interface. In our diffusion experiments, the specimens were kept in contact through the application of pressure by means of alumina rods. Moreover, marking of the initial interface was achieved by platinum sputtering, forming a fine layer about 500 Å thick. This layer is transformed into globules about 1μm in diameter during heat treatment. This marker technique does not disturb contact and we were able to confirm that it does not interfere with diffusion.

2. Local variations in volume, or material transport, linked to the formation of the intermediate compound, must be brought into evidence at all points of the diffusion couples.

3. Concentration variations throughout the diffusion couple must be known for a correct evaluation of the flux of diffusing elements, and to interpret measurements of spinel quantities formed on both sides of the initial interface.

4. It is necessary to determine the diffusion coefficients of the cations in the same conditions of temperature, partial oxygen pressure and composition as those in which the spinel is formed.

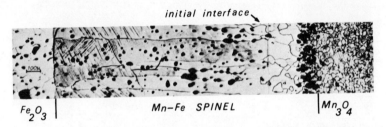

Figure 1 : Micrographic aspect of the diffusion layer in the Fe_2O_3-Mn_3O_4 system (1320° C - 68h - air).

These principles were applied to a study of the mechanisms governing the formation of manganese ferrite and nickel ferrite (6). The essential results about formation mechanisms are given below.

MANGANESE FERRITE FORMATION MECHANISM

Diffusion occurs at 1320°C in air, in Fe_2O_3-Mn_3O_4 couples constituted of sintered oxides. The experimental conditions were selected to take account of the respective ranges of stability of the oxides. Two basic phenomena were revealed. The first was an 11% volume increase in the ex-Fe_2O_3 zone, and the second the formation of high porosity on the ex-Mn_3O_4 side (Figure 1). The volume increase during spinel formation in Fe_2O_3 was detected by the shift in platinum marker planes placed perpendicularly or obliquely to the diffusion axis. This later procedure enables continuous marking of the entire diffusion zone, and we have shown that the volume increase is proportional to the amount of spinel formed on the Fe_2O_3 side. Simultaneously, the porosity rate of

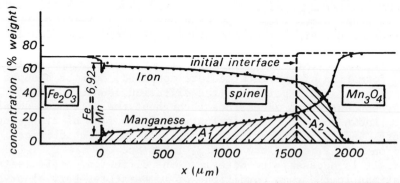

Figure 2 : Cation concentration profiles in the Fe_2O_3-Mn_3O_4 of the figure 1.

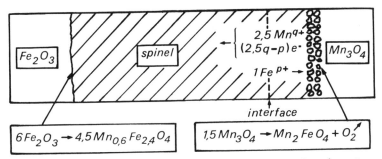

Figure 3 : Schematic mechanism of the manganese ferrite formation between Fe_2O_3 and Mn_3O_4.

this layer remains substantially equal to that of the initial oxide. However, on the ex-Mn_3O_4 side, high porosity is caused by the destruction of part of the manganese oxide. This destruction results from a significant departure of manganese in comparison with the arrival of iron. The analysis of cation diffusion profiles by electron microprobe (Figure 2) shows that the ratio A_1/A_2 of the manganese flux to the iron flux is about 2.5 taking into account the porosity on the ex-Mn_3O_4 side. The electrical neutrality of the flux is respected by diffusion of electrons associated with the manganese. These electrons occur through the liberation of oxygen linked to the destruction of Mn_3O_4. This leads to the formation of pores, which correspond to the annihilation of cation vacancies created by diffusion, with anion vacancies created by the liberation of oxygen.

The diagram in Figure 3 summarizes this mechanism. In order to simplify the presentation, the formulas have been simplified and the cation valencies are expressed by exponents p and q.

This analysis of the results is confirmed by our measurements of self diffusion coefficients. Table 1 summarizes these values. The composition range selected practically corresponds to that of the spinel zone of the Mn_3O_4-Fe_2O_3 couples. The diffusion coefficients of iron and manganese are about 1000 times lower in manganese rich spinel than in iron rich spinel. Hence the flux

spinel composition (Fe/Mn)	5	2	0.5	0.03	0
D^*_{Mn} (cm²/s)	3.3×10^{-8}	7.6×10^{-9}	8.7×10^{-10}	8.4×10^{-10}	
D^*_{Fe} (cm²/s)	4.9×10^{-8}	7.3×10^{-9}	4.3×10^{-10}	3.4×10^{-10}	3×10^{-10}

Table 1 : Selfdiffusion coefficients of Fe55 and Mn54 in iron-manganese spinel at 1320° C in air.

	Nickel ferrite				Wüstite	NiO
Fe/Ni	7	3.5	2.5	2.1	0.25	0
D^*_{Ni} (cm²/s)	1.8×10^{-8}	2.9×10^{-9}	1.9×10^{-10}	5×10^{-11}		4×10^{-10}
D^*_{Fe} (cm²/s)	4×10^{-8}	3.9×10^{-9}	2.9×10^{-10}	3.7×10^{-11}	1.5×10^{-8}	2.5×10^{-9}

Table 2 : Self diffusion coefficients of Fe55 and Ni63 in nickel-iron ferrite (1320° C - air)

ratio is imposed by diffusion coefficients in the manganese rich spinel. The concentration gradients being equal, it may be admitted as a first approximation that the cation flux ratio is equal to the diffusion coefficient ratio. This leads to a ratio of 2.44 between the manganese flux and the iron flux. This value, derived from the diffusion coefficients, is in agreement with that obtained from the concentration profiles. Hence it is the relative value of cation diffusion coefficients, for compositions in which the coefficients are the lowest, which governs the formation of the iron-manganese spinel.

NICKEL FERRITE FORMATION

We have shown (7) that nickel ferrite is only formed in the Fe_2O_3 portion of an Fe_2O_3-NiO couple. The cation fluxes across the initial interface are approximately equal and diffusion occurs between the Fe^{2+} and Ni^{2+} ions. In these conditions, the nickel flux is electrically compensated by the iron flux, and reduction of the iron oxide can only occur through the liberation of oxygen at the Fe_2O_3/spinel interface. The pores are thus formed on the Fe_2O_3 side, in contrast to the Fe_2O_3-Mn_3O_4 system. In order to quantify the suggested mechanisms, we determined the cation diffusion coefficients by means of radiotracers (Table 2). It is at the initial interface, spinel side, that the diffusion coefficients are the lowest and are fairly equal, taking into account experimental errors . This explains why the iron and

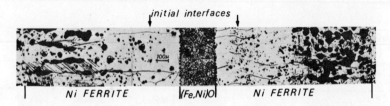

Figure 4 : Micrographic aspect of a Fe_2O_3-NiO-Fe_2O_3 sandwich couple.

nickel fluxes are equivalent. Furthermore, the iron diffusion coefficient in nickelowustite is very high. The electron microprobe analysis shows effectively that iron diffuses very deeply in nickelowüstite. Hence the iron solubility limit cannot be reached and the spinel is not formed on the NiO side. However, in a Fe_2O_3-NiO-Fe_2O_3 sandwich couple such as the diffusion of iron in NiO, occurs in a finite medium. The ferrite grows on both sides of the initial interface (Figure 4). Ferrite formation from the nickelowüstite leads to oxidation of the wüstite with oxygen pick up at the spinel/nickelo-wustite interface and construction of the oxygen lattice. This phenomenon is materialized by a volume increase in the wüstite zone converted into spinel and by pore elimination.

Hence knowledge of the relative values of cation diffusion coefficients permits an interpretation of the phenomena associated with the formation of nickel ferrite.

CONCLUSION

At the conclusion of this comparative analysis, it appears that one of the basic parameters governing the ferrite formation mechanism is the relative value of cation diffusion coefficients since, depending on the case at hand, this leads to oxygen release at the interface with Fe_2O_3 (case of nickel ferrite) or at the interface with the opposing oxide (case of manganese ferrite). Moreover, the more or less significant growth of spinel in the "bivalent" oxide is conditioned by the diffusivity of iron in the combined wüstite.

REFERENCES

(1) R.E. CARTER J. Amer. Cer. Soc., 44, 116 (1961)

(2) P. REIJNEN 5th Intern. Symposium on the reactivity of solids, Munich, G.M. SCHWAB publication, p. 562 (1964)

(3) C. KOOY 5th Interm. Symposium on the reactivity of solids, Munich, G.M. SCHWAB publication, p. 21 (1964)

(4) S.L. BLANK and J.A. PASK J. Amer. Cer. Soc., 52, 669 (1969)

(5) V.R. DERIE and L.A. KUSMAN Ber. Dt. Ker. Ges., 48,119,381(1971)

(6) P. EVENO Thesis, No. 1459 Orsay (1975)

(7) M. PAULUS, and P. EVENO Schenectady. Reactivity of solids Wiley Intercience Publication p585, N.Y. (1969)

PLANAR DEFECTS AND REACTIVITY OF PEROVSKITE-LIKE COMPOUNDS ABO_{3+x}:

THE SERIES $(Na,Ca)_n Nb_n O_{3n+2}$ $(n > 4)$

J. Galy, R. Portier, A. Carpy, and M. Fayard

Laboratoire de Chimie de Coordination du CNRS, BP 4142

31030 Toulouse Cedex (France)

Laboratoire de Metallurgic Structurale des Alliages

Ordonnes, ERA CNRS n° 221, ENSCP 11 rue Pierre et Marie

Curie, 75005 Paris (France)

INTRODUCTION

In the course of studies of the $Ca_2Nb_2O_7$-$NaNbO_3$ system, the existence of a family of phases of composition $A_n B_n O_{3n+2}$ (where A represents Na and/or Ca, $4 \leqslant n \leqslant 6$) was established and investigated by X-ray and electron microscopy methods (1 - 7). In the same area two other families have been studied ; the first one belongs to the same ternary system CaO-Nb_2O_5-Na_2O with the formula : $A_{n+1} B_n O_{3n+1}$ (n = 4 and 5), the structure deriving directly from the K_2NiF_4 type, the other one being $A_2 B_n X_{3n+2}$ (A = Na, B = Nb and/or W, X = O and/or F) (8-9).

RESULTS AND DISCUSSION

Our aim in this paper is to focus the attention on the results obtained in the complex system corresponding to the $(Ca,Na)_n Nb_n O_{3n+2}$ family (fig 1).

These sodium and calcium double niobates are characterized by different stackings of puckered perovskite type sheets of n octahedra in width (n =4, 5 and 6 have been isolated). The cell parameters of the (idealized) structures can be derived from that of

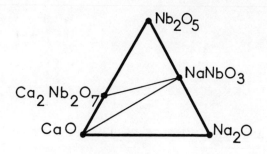

Figure 1

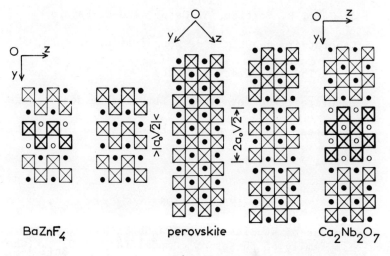

Figure 2

perovskite (a_0) by the relations $a_n, a_n=b_n = n a_0\sqrt{2} + 2K$, $c_n = a_0\sqrt{2}$ where $K = 2.25$ Å is the separation between two neighbouring layers (fig 2) (2).

In fact, $A_n B_n O_{3n+2}$ is a developed ABO_{3+x} formula :

$(Ca,Na)_n Nb_n O_{3n+2}$ $4 \leq n \leq 6$ ⇌ $(Ca,Na)NbO_{3+x}$ $0.5 \geq x \geq 0.333$

the non-stoichiometry being induced by the presence of planar defects (4-5)

Mechanisms have been proposed for the formation and the migration of these planar defects and recently supported by electron microscopy observation (4-10-11).

By means of electron microscopy, a very large number (up to 30) of crystallographically distinct compounds have been studied (4 ⩽ n ⩽4.5. These structures may be described as intergrowths of perovskite type sheets of four or five octahedra in width. Such structures have one-dimensional giant cells, some of these exhibiting a polytypic character (6-7). Figure 3 represents, as an example, the micrograph of the compound $NaCa_8Nb_9O_{31}$ (which corresponds to n = 4.5) or $(Ca,Na)NbO_{3.444}$) and the resulting structure projected on to (100).

The hybrid phases can be synthetized by solid state reaction between the terms n = 4,5, or 6.

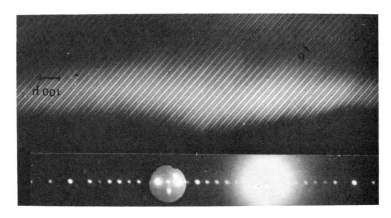

Figure 3

Such studies have been followed simultaneously on samples heated during various periods by comparing X-ray powder patterns and electron microscopy histograms (example in fig. 4 : synthesis of the term n = 4.5 by heating at 1350°C a mixture of $Ca_2Nb_2O_7$ (n = 4) and $NaCa_4Nb_5O_{17}$ (n = 5); (time t = 5mn).

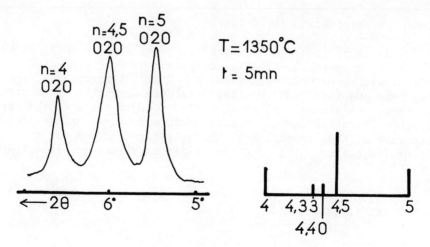

Figure 4

REFERENCES

1. A. Carpy, P. Amestoy and J. Galy, C.R. Acad. Sc. 275C, 883, 1972
2. A. Carpy, P. Amestoy and J. Galy, C.R. Acad. Sc. 277 C, 501, 1973
3. M. Nanot, F. Queyroux, J.G. Gilles, A. Carpy and J. Galy, J. Solid State Chem. 11, 272, 1974
4. J. Galy and A. Carpy, Phil. Mag., 29, 1207, 1974
5. R. Portier, M. Fayard, A. Carpy and J. Galy, Mat. Res. Bull. 9, 371, 1974
6. A. Carpy and R. Portier, C.R. Acad. Sc., 279 C, 691, 1974

ELECTRON DENSITY FLUCTUATIONS AT THE VICINITY OF THE SURFACE

LAYERS

S.Michalak and L. Wojtczak

Institute of Physics, University of Lodz

90-131 Lodz, Poland

ABSTRACT

Electron distribution at surfaces is responsible for the reactivity of solids. Thus, electron density fluctuations can change the chemical properties of a sample at its surface. Fluctuation effects are mainly expected close to phase transition points. Electron density fluctuations lead to dispersion of the dielectric susceptibility. Thus, transition radiation can be applied to the investigation of the reactivity fluctuations at the solid surface.

INTRODUCTION

The solid surface plays an important role in many physical and chemical properties of solids. In particular the electron distribution at the surface, mainly responsible for the reactivity of solids, is determined by the chemical bonding of the system considered in a given reaction. Thus, the description of the electron density is sufficient to know the reactivity of solids.

It should be noted that the electron density can randomly fluctuate and lead therefore to changes in reactivity even at a regular surface. In the case when the lifetime of fluctuations is sufficiently large with respect to the time of chemical reaction on the surface considered the fluctuating behaviour of the local reactivity should be observed. In order to evaluate the conditions under which the reactivity fluctuations play a significant role we consider the dynamic properties as demonstrated by the electron density in the vicinity of the surface layers. In the present note the electron density and its fluctuations are investigated on the basis of the relations between the charge distribution inside a sample and the dielectric properties of the material considered. The electromagnetic radiation produced by charged particles passing through the surface can be applied to the observation giving some information about the electron density fluctuations, and indirectly, about the nature of the reactivity of solids.

MODEL OF FLUCTUATIONS

The idea of fluctuations treated as scattering objects was developed from the critical opalescence theory give by Smoluchowski (1), who assumed that matter at critical points is granular. Investigations of the scattering of neutrons and X-rays by crystals also show that fluctuations play an important role in scattering processes, especially in the critical scattering close to phase transition points (2). Within the theory of fluctuations the inhomogeneous system is divided into homogeneous cells embedded in a thermodynamic reservoir. The interaction between cells corresponding to fluctuations is determined only by the effective potential of the molecular field type.

The model described above can also be applied to the study of the reactivity of solids as referred to the electron density behaviour. In the process of chemical reactions the molecular beam in fact falls on the cell area so that the reactivity of the surface can be different for various cells in which fluctuations of electron density appear. For this reason the reactivity of solids is understood as the average of its local values, related to the electron density fluctuations at the surface. These are expressed in terms of the dielectric susceptibility.

DIELECTRIC SUSCEPTIBILITY

In order to calculate the change of dielectric properties at the surface we consider the correlation functions determined for the polarization vector

$$\bar{P}(\bar{r},t) = \Sigma_\lambda e(\lambda,r)\bar{R}(\bar{r},t)c^+(\lambda,r)c(\lambda,r) \qquad \{1\}$$

given as a local function of its position $R(r,t)$ into a sample with the distinguished surface. $e(\lambda,r)$ denotes the electric charge and $c^+(\lambda,r)$ and $c(\lambda,r)$ are the electron creation and annihilation operators respectively. λ stands for the quantum indices referring to the electron band including the spin orientation. Then, the generalised susceptibility of a given system $\mathcal{H}^{\alpha\beta}(r,r',\omega)$ can be expressed as the Fourier transform of the Green's function for the polarization vector $\bar{P}(\bar{r},t)$ and its temporal derivative $\bar{J}(\bar{r},t)$ interpreted as the shift current conjugated with the polarization field (3), namely

$$\mathcal{H}^{\alpha\beta}(r,r',\omega) = \int_{-\infty}^{+\infty} \langle\langle J^{\alpha}(\bar{r},0) | P^{\beta}(\bar{r},t) \rangle\rangle \exp(-i\omega t) dt \qquad \{2\}$$

α and β are the coordinate components and ω is the frequency of a considered system determined by the Hamiltonian in the form

$$H = \Sigma_{\lambda} \Sigma_{rr'} t_{\lambda}(r,r') c^{+}(\lambda,r) c(\lambda,r) \qquad \{3\}$$

The hopping integrals $t_{\lambda}(r,r')$ are characterised by the geometry of a sample in which the natural boundary conditions at the surface are taken into account (4).

The dielectric susceptibility depends in consequence on the distribution of electrons inside a sample. Its value at the equilibrium state

$$n(\bar{r}) = \Sigma_{\lambda} c^{+}(\lambda,r) c(\lambda,r) \qquad \{4\}$$

influences the frequency dependence of the dielectric constant. Taking into account the fluctuation model in which homogeneity of electron distribution inside a fluctuation is assumed we can write a local density of electrons in a given cell as

$$n(i) = n(1 + \delta(i)) \qquad \{5\}$$

where $\delta(i)$ is the deviation of the electron density in the i-th cell from its equilibrium value. In this case the dielectric susceptibility $\mathcal{H}^{\alpha\beta}(r,r',\omega)$ in the i-th cell is given by $\{2\}$, where the distribution of electrons appearing there is expressed by $\{5\}$.

For the sake of simplicity the considerations of this note are confined to the case of one band model ($\lambda=1$) with a homogeneous density assumed for the susceptibility in the direction perpendicular to the surface ($\alpha=z, \beta=z$). Then the Green's function procedure leads to the result

$$\mathcal{H}^{\alpha\beta}(r,r',\omega,i) = \Sigma_{h} \Sigma_{ss'} A(s,s',h,r,r') (\omega - \omega(s,s',i,h))^{-1} \qquad \{6\}$$

where

$$\hbar A(s,s',h,r,r') = \Gamma(h,r) \Gamma(h,r') \Sigma_{rr'}(h,r) \Gamma(h,r) \langle \tilde{J}^{z}(r,0) \tilde{P}^{z}(r,0) \rangle,$$
$$\tilde{J}^{z}(r,0) = J^{z}(r,0) n^{s}(i), \quad \tilde{P}^{z}(r,0) = P^{z}(r,0) n^{s'}(i), \qquad \{7\}$$

and

$$\omega(s,s',i,h) = t(h) (n^{s}(i) - n^{s'}(i)) \qquad \{8\}$$

The coefficients $\Gamma(h,r)$ and the energy eigenvalues $t(h)$ satisfy a set of homogeneous linear equations

$$\sum_{\bar{r}} t(r,\bar{r})\Gamma(h,\bar{r}) = t(h)\Gamma(h,r) \qquad \{9\}$$

with orthogonality conditions. The summation over s and s' runs over two states $n^+ = n$ and $n^- = 1-n$. In particular when $n=1/2$ the fluctuation effect should appear strongly when

$$\mathcal{H}^{\alpha\beta}(r,r',\omega) = \sum_{\mathcal{M}=0}^{\infty} A(\mathcal{M},r,r',)\delta^{\mathcal{M}}(i)\omega^{-(1+\mathcal{M})} \qquad \{10\}$$

with $A(\mathcal{M},r,r')$ corresponding to terms of a power series expansion of $\{6\}$ with respect to $\delta(i)$.

TRANSITION RADIATION

The transition radiation is produced by charged particles passing through the surface of a medium. Its intensity depends on the difference between the dielectric constants of contacting media. According to the theory (5) the intensity of transition radiation in the unit solid angle Ω per unit energy is given by

$$\frac{d^2 W(i)}{d\Omega d\omega} = \frac{\hbar e^2 \beta^2 \sin^2\theta \cos^2\theta}{\pi^2 c(1-\beta^2\cos^2\theta)^2} \left| \frac{(\varepsilon-1)(1-\beta^2-\beta(\varepsilon-\sin^2\theta)^{1/2})}{\varepsilon\cos\theta + (\varepsilon-\sin^2\theta)^{1/2})(1-\beta\sin^2\theta)^{1/2})} \right|^2$$

$$\varepsilon = \varepsilon(r,r',\omega,\delta(i)) \qquad \{11\}$$

with $\beta = v/c$ where v is the velocity of the incident charged particle beam and c stands for the velocity of light. The particles which are the most frequently used to produce the transition radiation are electrons. The radiation is emitted from the surface under the angle θ determined by the angle between the incident beam and the axis normal to the surface. The dielectric constant is related to the generalised susceptibility as

$$\varepsilon(r,r',\omega,\delta(i)) = 1 - \mathcal{H}(r,r',\omega)/\omega \qquad \{12\}$$

The dielectric constant of the second medium is assumed to be unity. The generalised susceptibility was averaged with respect to its spatial dispersion inside a fluctuation and the dielectric constant of a given fluctuation is a function of the density deviation from its equilibrium value. Let $f(\varepsilon(\omega,\delta(i)),\theta)$ denote the local intensity of the transition radiation given by $\{11\}$ then the total intensity can be expressed as

$$f(\varepsilon,\theta) = (1/N_f)\sum_i f(\varepsilon(\omega,\delta(i)),\theta) \qquad \{13\}$$

where the summation over i runs over N_f fluctuations. Taking into account the power series expansion of $\{13\}$ with respect to $\delta(i)$

we obtain that

$$\left(\frac{d^2W}{d\Omega d\omega}\right)_f = \left(\frac{d^2W}{d\Omega d\omega}\right)_h \frac{1}{2}\left(\frac{\partial^2 f}{\partial \delta^2}\right)_h \langle \delta^2 \rangle \qquad \{14\}$$

where the index f refers to the system with fluctuations and the index h corresponds to quantities calculated in the homogeneous system. The mean square fluctuation $\langle \delta^2 \rangle$ is determined by the probability of fluctuation appearance (2).

CONCLUSIONS

The properties of the transition radiation emitted by a system with electron density fluctuations are different from those for a homogeneous system. According to equation {14}, the study of the fluctuation properties is based on the fact that the plasmon frequency ω_p is shifted due to fluctuations, with the shift proportional to δ^2. Since the transition radiation is generated only in the forming region which extends over few monoatomic layers near the surface (6), observations of the transition radiation characteristics allow us to investigate the properties of fluctuations. They are essentially of a two dimensional character. For this reason the method can be applied to the study of the still unsolved problem connected with the influence of the surface boundary conditions on fluctuation properties, while methods based on the scattering processes lead to the picture of a fluctuation averaged with respect to the whole volume of a sample when the influence of surfaces is negligible.

Investigations of electron density fluctuations are also interesting from the point of view of the surface reactivity. According to the change in electron distribution near the surface its chemical properties are changed in time. In consequence the reaction rate should be of an oscillatory character measureable in the case of an appropriate electrode reaction. Since its rate is rather small the method is more sensitive for electron density changes in the low frequency region. For this reason we suggest in conclusion that considerations of the surface density fluctuations can be confirmed by means of chemical methods.

REFERENCES

1. M. Smoluchowski, Bull. Int. Acad. Sci. Cracovie, A47, 179, 1907; Ann. Phys., 25, 205, 1908.
2. J. Kocinski, L. Wojtczak, B. Mrygon, Acta Phys. Polonica, A43, 425, 1973
3. N.D. Zubarev, Non-equilibrium Statistical Thermodynamics, Moscow, 1971 (in Russian).

4. L. Wojtczak, Acta Phys. Polonica, 36, 107, 1969.
5. F. G. Bass, W.M. Jakovenko, Usp. Fiz. Nauk, 86, 189, 1965
6. S. Mickalak, Post. Fiz., 1, 14, 1968.

HETEROGENEOUS ION-EXCHANGE REACTIONS ON NON-STOICHIOMETRIC TUNGSTEN OXIDE IN AQUEOUS MEDIA

T. Szalay,[*] L. Bartha,[*] T. Nemeth, J. Lengyel

Institute for Physical Chemistry of the University

of Debrecen

[*]Research Institute for Technical Physics of the

Hungarian Academy of Sciences – Budapest

INTRODUCTION

Tungsten oxides particularly those in higher oxidation states can be considered as hydrates in aqueous suspension, and are ion-exchangers in the presence of dissolved metallic ions. The reversible ion-exchange equilibria can be described thermodynamically (1,2) and their characteristic parameters can be determined for systems of different compositions (2,3).

The amphoteric characteristics and behaviour of the solid hydrated oxides depend on the charge density of their surfaces (3) as a function of pH. In the case of dynamic equilibria, such as ion-exchange, the electric charges of the surfaces are not localised. This fact shows itself first in the exchange of multivalent ions.

One of the most important surface characteristics seems to be the isoelectric point (IEP). These parameters are shown for some oxide compounds in Table 1. (4). It can be seen that in aqueous solution oxides of the same composition may adsorb ions with different exchange mechanisms depending on the pH value. Other processes may also take place in the case of non-stoichiometric oxide exchangers in the presence of multivalent ions.

The present work studies ion-exchange of non-stoichiometric oxide compounds demonstrating the phenomena on tungsten oxide of an approximate composition of $WO_{2.85}$, the blue oxide.

<u>Theoretical considerations</u>. An arbitrary tungsten oxide suspended in water can be characterised by a micelle structure as follows:

$$(WO_x)_k \cdot n-y\,(WO_x)_m \cdot H_2O \cdot y(WO_x)_m HO^- \ldots yH^+ \qquad (1)$$

Where $x < 3$, $k > m > n$ and the value y depends on the composition of the solution. Consider a single surface group of the micelle using a symbol ROH to describe it before "dissociation". The possible surface reactions in pure water can be described as follows:

below IEP   above IEP

$$ROH_{2\,(s)}^+ \rightleftarrows ROH_{(s)} \rightleftarrows RO^-_{(s)} + H_2O_{(aq)} \qquad (2a)$$

or

$$R^+_{(s)} + OH^-_{(aq)} \rightleftarrows ROH_{(s)} \rightleftarrows RO^-_{(s)} + H^+_{(aq)} \qquad (2b)$$

The reactions (2a) and (2b) can not be distinguished in practice because the surface charge cannot be distinguished as due to association or dissociation.

In aqueous solution of monovalent cations and anions a regular cation exchange also takes place on the surface of hydrated oxide above its IEP:

$$ROH_{(s)} + Me^+_{(aq)} \rightleftarrows ROMe_{(s)} + H^+_{(aq)} \qquad (3)$$

The reversible ion-exchange equilibrium is reached within a relatively short time and can be described either by an adsorption isotherm of Langmuir-Hückel type (1) or by the mass effect law made valid for heterogeneous systems (3,7).

$$K = \left[\frac{c_{H^+}}{c_{Me^+}}\right]^P \cdot \frac{n_{ROMe}}{n_{ROH}} \qquad (4)$$

In general, the thermodynamic activities of various species should be used in equation (4). However, in the case of dilute solutions of salts containing monovalent ions only, the ratio of activity coefficients ($\gamma_{H^+}/\gamma_{Me^+}$) should be unity and the activities of the solid species can be replaced by their mole-fractions. Applying these assumptions, the value K seems to be the real thermodynamic equilibrium constant and it is useful to determine experimentally the change of standard free energy of the reactions:

$$\Delta G^o = R \cdot 298,16 \cdot \ln K \qquad (5)$$

The value of K can be determined according to equation (4):

$$\lg \frac{n_{ROMe}}{n_{ROH}} = \lg K + p \cdot \lg \frac{c_{Me^+}}{c_{H^+}} \qquad (6)$$

When $c_{Me^+} = c_{H^+}$ at equilibrium then:

$$K = \frac{n_{ROMe}}{n_{ROH}} \qquad (7)$$

Using the thermodynamic activity values of various species, one can make the equation (4) valid to describe the ion-exchange equilibria of multivalent cations, but only in the presence of monovalent anions.

In special cases of multivalent anions an unusual ion-exchange process takes place on hydrated oxides. Besides the regular cation adsorption another reaction takes place with increasing concentration of anions in very dilute solutions. For example, in the presence of SO_4^- ions the reaction (2b) above the IEP becomes reversed ie:

$$ROH_{(s)} + SO_{4(aq)}^{2-} + H^+_{(aq)} \longrightarrow (ROH_2SO_4)^-_{(s)} \qquad (8)$$

This effect is shown in the experimental part by means of the exchange isotherms of multivalent anion-salts.

Another unusual ion-exchange process can be identified on non-stoichiometric oxides, when the solution consists of redox ion-pairs and their ratio Me_{ox}/Me_{red} depends on the pH value. In the reaction of blue tungsten oxide with Fe^{3+}-ions, the ion concentration decreases not only due to the redox process, but also due to ion-exchange process. The former has a heterogeneous and a homogeneous part:

$$x/n\, WO_{(3-n)(s)} + n+1/n\, H_2O_{(aq)} \longrightarrow$$
$$x/n\, H_2WO_{4(s)} + 2x\, H^\bullet_{(aq)} \qquad (9a)$$

$$Fe^{3+}_{(aq)} + H^\bullet_{(aq)} \longrightarrow Fe^{2+}_{(aq)} + H^+_{(aq)} \qquad (9b)$$

On the other hand, the metallic ions can be reduced on the solid surface of the hydrous polytungstate by means of electrochemical reactions (8,9):

$$W_6O_{16}(OH)_{3(s)} \longrightarrow W_6O_{17}(OH)_{2(s)} + H^+_{(aq)} + e^- \qquad (10a)$$

$$e^- + Fe^{3+}_{(s)} \longrightarrow Fe^{2+}_{(s)} \qquad (10b)$$

EXPERIMENTAL

The blue tungsten oxide was prepared by thermal decomposition of solid amonium-paratungstate $(NH_4)_{10}(W_{12}O_{40}(OH)_2) \cdot 4H_2O$ of high purity. The pH values of blue oxide - solution (or water) systems were measured using Ag/AgCl- glass electrodes at room temperature. In some cases radiotracer and pH-metric methods were used in parallel to study the ion-exchange processes in detail. The redox processes were followed by chemical, spectrophotometric and potentiometric methods.

The pH-metric measurements were performed as follows: 2.5 g of blue oxide was suspended in 25 cc of deionised water. The equilibrium was reached within 3 minutes under stirring. After this time 5 cc of solution (concentration of 0.01 - .3 M or M/valency in case of multivalent ions) was added to the suspension. The pH values measured after 3 and 6 minutes were accepted as characteristic of the system in three parallel samples (Fig. 1).

Table 1
The isoelectric points of some oxides

compounds	pH	compounds	pH
WO_3	0.5	TiO_2	6.0
$\sim WO_{2.9}$	1.0	$\gamma-Al_2O_3$	7.6
$\delta-MnO_2$	2.25	ZnO	8.9
SiO_2 (amor.)	3.0	Ag_2O	11.2

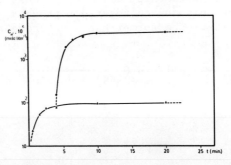

Fig.1. The H^+- ion concentration change in pure water- and solution blue oxide systems.

It should be noted, that the hydrate formation and its dissociation is completed during the first three minutes, while the ion-exchange reaches its equilibrium in the next three minutes. After this time only very slow changes of pH could be measured due to the formation and dissolution of polytungstates.

RESULTS

The regular ion-exchange isotherms of various salts are shown in Fig. 2. Using this data and equation (6), the equilibrium constants (K), the free energy changes (ΔG^o) and the Freundlich empirical constants (p) can be calculated. In the case of multivalent cations, a modified form of equation (6) was used (3,10). The ion-exchange isotherms of salts of multivalent acids (phosphates, sulphates, etc.) show a maximum instead of saturation, Fig. 3.

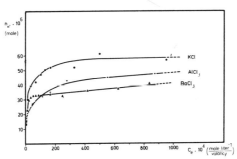

Fig.2. H^+ - Me ion-exchange isotherms of different cations of monovalent anion salts at room temperature.

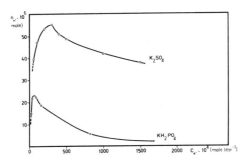

Fig. 3. H^+ - K^+ ion-exchange isotherms in presence of multivalent ions.

This special feature has been investigated using KH_2PO_4 solution with radioactive tracer P(32). From studies of the distribution of P(32), one could conclude that the adsorption of PO_4-ions takes place progressively alongside the cation exchange process. The adsorption of PO_4 ions can be influenced by addition of HCl or KCl to the solution at the same PO_4-ion concentration, Figs 4 and 5.

The effect of these additives however, changed the reaction mechanisms (Fig. 6) This fact is considered as evidence for the anion effect suggested earlier. On increasing the HCl concentration in PO_4-solutions, the adsorption of PO_4-ions will be promoted until the IEP, that is the surface becomes neutral and the PO_4-ions will be present only in H_3PO_4 form. In such a case, the neutral surface will adsorb the maximum number of neutral molecules.

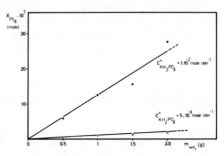

Fig. 4. The adsorption of PO_4-ions measured by radiotracer method in KH_2PO_4 (0.5 HCl M) solutions.

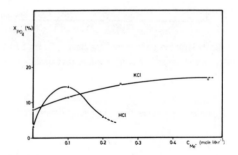

Fig. 5. The adsorbed fraction of PO_4-ions depending on HCl and KCl solute concentration.

In contrast to the pH-effect the addition of KCl to the solution of the same PO_4-ion concentration, H^+-ions will develop continuously as a result of the mechanism of equation (8).

From ion-exchange studies carried out with multivalent redox-cations (Tl^{III}, Fe^{III}, etc.) a new process has been found. The non-stoichiometric hydrated oxide can react with metallic ions of higher oxidation state in a redox process. Fe^{III}- and Fe^{II}-concentrations can be determined in solution by spectrophotometric methods before and after ion-exchange. These data are different in all cases from those determined by radiotracer method (Fe(59)) (Table 11) On the basis of adsorption studies of radioactive ferrous- and ferric-ions it could be concluded that essentially only Fe^{II}- ions are adsorbed by the solid surface (Fig. 7). The Fe^{III}-ions also adsorbed on the surface are reduced to Fe^{II}-ions due to the excess surface electrons. This fact agrees with results on Fe-oxidation states in Fe_xWO_3 compounds (12).

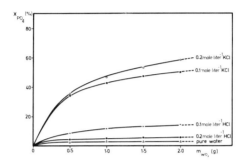

Fig. 6. The adsorbed fraction of PO_4 -ions from solution of different compositions

Table II
Material balance of Fe^{III} and Fe^{II}-ions - blue oxide reactions measured by means of spectrophotometric and radiotracer methods (H^+-concentration 0.5 M in all cases)

quantity of blue oxide (g)	starting concentration $c_o \cdot 10^4$ M			solution concentration at equiln. $c_o \cdot 10^4$ M		
	Fe^{III}	Fe^{II}		Fe^{III}	Fe^{II}	
0.05	6.43	0.19	6.62	3.99	2.52	6.51
0.10	6.43	0.19	6.62	1.68	4.48	6.16
0.20	6.43	0.19	6.62	0.06	6.00	6.06

	decrease of Fe^{III} conc. %	decrease of Fe^{III} conc. due to	
		reduction %	ion-exch.%
0.05	37.95	36.29	1.66
0.10	73.87	66.92	6.95
0.20	99.08	90.62	8.46

The reaction rate of Fe^{III}-ions with non-stoichiometric solids was studied by spectrophotometric methods measuring the specific adsorption of the $Fe(SCN)_3$ complex (Fig. 8). The fast surface processes take place within 10 minutes followed by a slow change in Fe^{III}-ion concentration attributed to the electron exchange among the second, third etc. order surfaces or due to some diffusion mechanism.

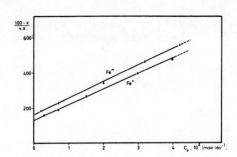 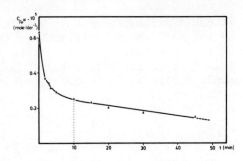

Fig. 7 The ion-exchange isotherms in Fe^{II} solutions according to the Langmiur-Hückel equation
$$\frac{100-x}{v \cdot x} = \frac{1}{z}(C_e - C_o)$$

Fig. 8 The change of Fe^{III}-ion concentration (spectrophotometric data)

CONCLUSIONS

According to the processes described above the following can be concluded:

a. Simple cations are adsorbed by the solid surface as a result of regular ion-exchange processes.

b. Ions of lower oxidation states are adsorbed on the surface in the presence of redox ion-pairs.

c. The adsorption of multivalent anions can significantly change the ion-exchange mechanism.

Thus one can control the simultaneous adsorption of mono- and multivalent cations and anions from solution onto the surface of non-stoichiometric tungsten oxide.

REFERENCES

1. T. Szalay, L. Bartha Z.Phys.Chem 255, 974 (1974)
2. T. Szalay, L. Bartha Z.Phys.Chem 255, 981 (1974)
3. G,E. Boyd, J Schubert, A.W. Adamson J.Am.Chem.Soc. 69 2818 (1947)
4. J.W. Murray J.Coll.Interface Sci. 46, 357, (1974)
5. J. Leugyel unpublished data (1975/76)
6. M.T. Beck Acta Chim. Sci. Hung. 4, 227, (1954)
7. T. Szalay, L. Bartha, T. Németh.Magyar Kemiai Folyoivat 82, 135, (1976)(Hungarian)
8. M.T. Beck private communication
9. M.V. Vojnič, D.B. Sepa, D.S. Ovic Croatica Chemica Acta 44,89, (1922)
10. J. Kielland J.Am.Chem.Soc. 59, 1675,(1937)
11. C.B. Amphlett Inorganic Ionexchangers,Elsevier, Amsterdam (1964)
12. J.P Doumerc, G. Schiffmacher,P. Caro,M. Pouchard,C.R.Acad Sc. Paris 282, 295, (1976)

POLYMERIZATION AND OTHER ORGANIC REACTIONS IN THE CRYSTALLINE STATE

Herbert Morawetz
Department of Chemistry
Polytechnic Institute of New York
Brooklyn, N.Y. 11201

ABSTRACT

Reactions of molecular crystals which do not involve polymerization may exhibit several features reflecting the crystalline order of the reagent, e.g., (a) anisotropy of reactions of the crystals with gases, (b) topotaxy, (c) selection of one of several possible reaction paths, (d) the formation of optically active reaction products from achiral or racemic reagents by the influence of a chiral crystal environment. Solid-state polymerizations are discussed particularly with respect to topotactic effects and the control of the reaction product by the geometric arrangement of monomer molecules in the monomer crystal. The survey deals with (a) the polymerization of trioxane, (b) the polymerization of crystals of vinyl monomers and their clathrate complexes, (c) the "polydimerization" of p-phenylene diacrylic acid derivatives and of distyryl-piperazine and (d) the preparation of large single polymer crystals by the homogeneous solid-state polymerization of diacetylene derivatives.

INTRODUCTION

Although solid-state reactions have been a recognized field of study for the sixty years elapsed since Hedvall's pioneering studies, it is characteristic of a widely shared attitude that a monograph on "Solid State Chemistry" published in 1974 contains no mention of a single reaction in an organic crystal (1).

Scattered information on solid-state organic reactions was first summarized in 1963 and 1965.(2) At that time, the situation was curious in that much more was known about polymerizations than about reactions involving small molecules only. Moreover, in the study of solid-state polymerizations, a great deal of emphasis was placed on kinetic studies. Since polycrystalline samples were generally employed and since the reaction rate may be highly sensitive to crystal imperfections and stresses, the results did not lend themselves to a convincing interpretation. The study of solid-state organic reactions in general, and of polymerizations in particular, is of greatest interest if the results demonstrate topochemical effects or topotaxy and we shall stress these aspects of the field, with special emphasis on recent developments.

Reactions of Small Molecules

A considerable number of solid-state organic reactions which do not involve polymerization have been reported in recent years. Recent reviews of the field have been prepared by Cohen and Green (3), by Paul and Curtin (4) and by Thomas (5), whose extensive treatment is unique in that it discusses the role of structural defects on the reactivity of molecular crystals and deals with organic and inorganic reactions in a unified manner. In addition, Paul and Curtin have reviewed reactions of organic crystals with gases (6) and Cohen (7) has published a summary of photochemical solid-state organic reactions.

In many cases, reactions which can take place both in the liquid and the solid state are much slower in the crystalline phase. Yet, this is not necessarily the case. Pincock et al (8) studied the endo - exo isomerization of compounds I and II and they found that the endo compound reacted with equal velocity in solution and in the crystalline state at temperatures as

much as 40° below its melting point. The high mobility
within the crystals of this substance, which accounts
for its high reactivity, was also reflected in the unusually sharp NMR spectrum and the relatively diffuse
X-ray diffraction. The exo isomer did not exhibit
these features and it was much less reactive in the
solid state. In a related study of the racemization of
III, (9) it was found that the reaction showed no

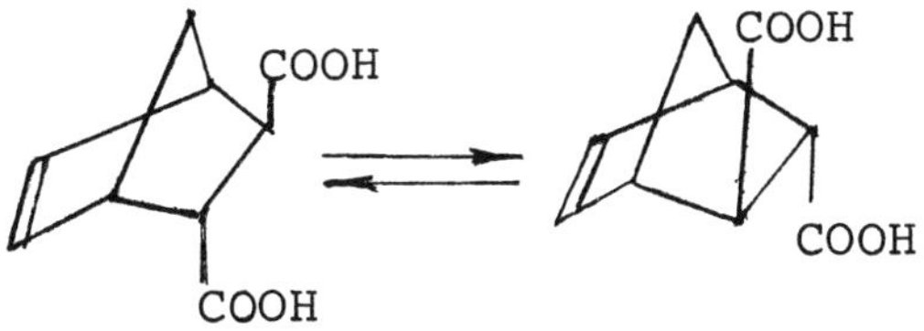

III

evidence of a nucleation step, such as is frequently
characteristic of solid-state processes, but followed
first-order kinetics over its entire course. First-
order kinetics were also observed for the Beckmann-
Chapman rearrangement IV → V. (10)

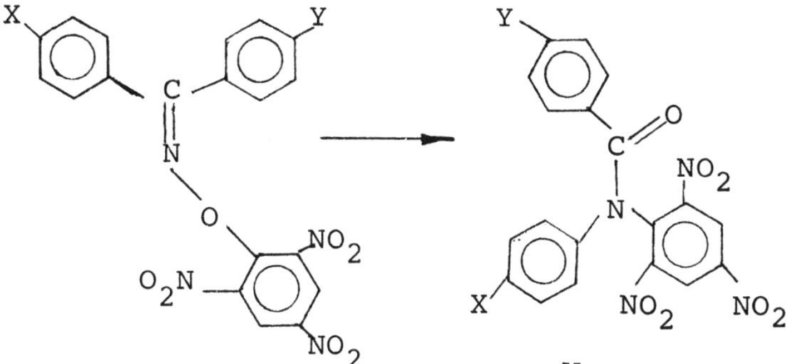

IV   V

The data were interpreted as indicating that the reaction product first formed a solid solution in the reagent crystal; at a conversion of 15-30% it formed
randomly oriented microcrystallites which grew with the
further progress of the reaction.

A remarkable crystallographic study of a chemical
transformation taking place in an organic crystal was
reported by Comer and Trotter. (11) They made the surprising observation that the p-bromophenacyl ester of
hirsutic acid suffers, when exposed to x-rays, an
isomerization in which the product of the reaction remains in solid solution in the parent crystal. The
portion of the molecule involved in the reaction is

represented by VI and VII, with an equilibrium being apparently attained when the two forms are present in similar concentrations. Another pioneering crystallo-

graphic study was carried out on the decomposition of the photo-oxide of anthracene (VIII) to a mixture of anthraquinone (IX) and anthrone (X).(12) This reaction proceeds slowly by a thermal mechanism but is accelerated by exposure to x-rays,

suggesting that the product is preferentially nucleated at crystal defects. A single crystal of the reagent was converted into a mosaic of crystallites of the reaction products whose orientation spanned a range of about 15°.

Topotactic processes might be expected when the reaction involves only small changes in molecular geometry. An extreme case of this kind is the transition from XI to XII which involves merely a change in the intra-

molecular hydrogen bonding.(13) The two isomers are spectroscopically distinct and the conversion of a single crystal of the yellow form XI to a single crystal of the white form XII proceeds in a sharp front along a crystallographic direction. (A similar anisotropy is revealed by a microscopic study of the reaction of single crystals of p-bromobenzoic acid with ammonia (6)). It is much more surprising that the isomerization of XIII to XIV should exhibit topotaxy. In fact, a single crystal of the reagent produces again a single crystal of the reaction product (14,15), although one of the aromatic rings has to flip around, a process which would not be expected in the tightly packed crystal.

XIII   XIV

Some compounds which contain no center of asymmetry form a mixture of levorotatory and dextrorotatory crystals. Thus, the molecules in these crystals are in an asymmetric environment and I suggested some years ago (16) that this could be taken advantage of to produce optically active reaction products from single crystals of non-asymmetric reagents. Such a phenomenon was first demonstrated on the reaction of a single crystal of XV exposed to bromine vapor. As originally reported, addition of bromine across the double bond

XV

yielded a 6% excess of one of the enantiomeric products(17) but it was later found that the optical yield could be increased to 25% by a more careful growing of the chiral crystal.(18) In the photodimerization of various substituted trans-cinnamic acids, studied by Schmidt and his collaborators, the crystal geometry may be utilized to direct the reaction into one of several alternative pathways. Depending on whether the molecules lie parallel or antiparallel to each other, the dimer is β-truxinic acid XVI or α-truxillic acid XVII, respectively.(19) The unsubstituted trans-cinnamic acid may be obtained in crystalline forms with either symmetry and it is then possible to determine the chemical

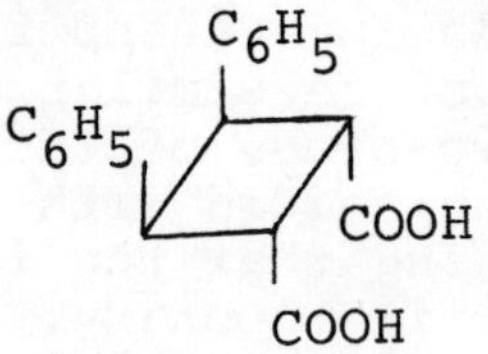

XVI 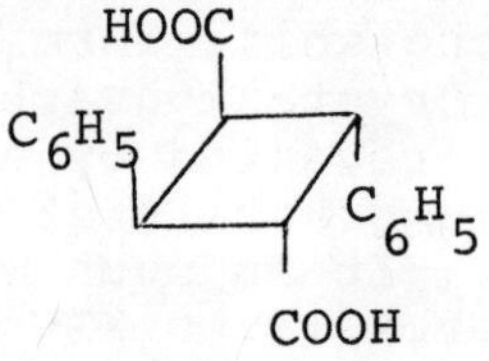   XVII

structure of the reaction product by selecting the appropriate crystalline reagents. In a review of topochemically controlled photodimerizations of crystalline reagents, Schmidt(20) cited a large number of other striking observations illustrating the sensitivity of these processes to changes in crystal structure. For instance, on irradiating compounds of type XVIII dimer XIX is formed if R= -COOH or R= -COOCH$_3$, while R= -COOC$_2$H$_5$ yields a polymer and no reaction results with R= -C$_6$H$_5$. Schmidt formulated some principles on

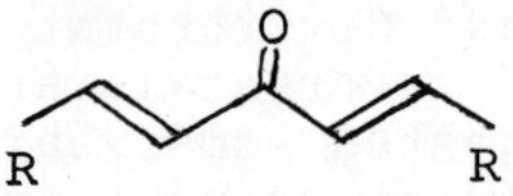

XVIII 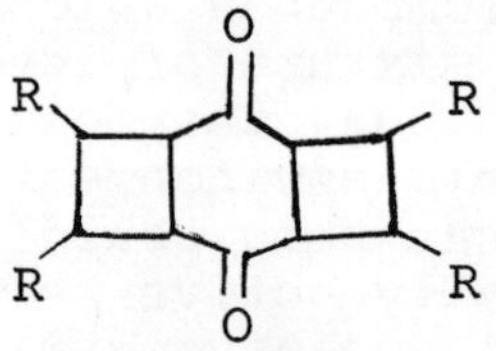   XIX

how to design a compound which will crystallize in a form favorable for a desired solid-state reaction. He called this approach "crystal engineering". For instance, 2,6-dichloro substitution of a phenyl group leads to a characteristic change in molecular packing, which accounts for the fact that XX will photodimerize to XXI, although no dimerization occurs with 1,4-diphenylbutadiene.(21). When a single crystal of XX (which is chiral) containing a small proportion of the analog in which the phenyl is replaced by a thiophene residue was irradiated in the absorption band of the thiophene chromophore, the asymmetric crystal environment led to a mixed dimer which was found to be optically active. (22)   Another interesting example of

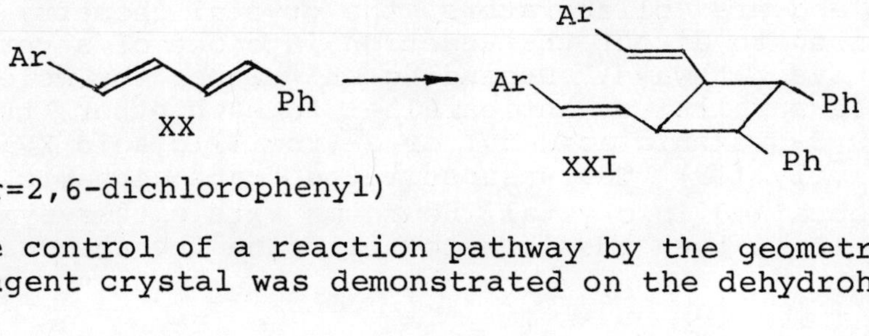

XX ⟶ XXI

(Ar=2,6-dichlorophenyl)

the control of a reaction pathway by the geometry of the reagent crystal was demonstrated on the dehydrohalogenation of crystalline meso β,β' dichloro- and dibromoadipic acid esters by gaseous amines. The only products were esters of the trans, trans-2,4-diene dioic acid, while the same reaction in solution led to a mixture of isomeric products (23).

Recently it was demonstrated that hindered rotation of an atropisomeric compound can take place in the crystalline state and that such a process may be used for the preparation of samples with high optical activities from a starting material containing only a small excess of one of the enantiomers. This result was obtained (24) with 1,1'-binaphthyl, XXII, in which the racemic crystal is thermodynamically unstable with re-

(R)⁻   XXII   (S)⁺

spect to a mixture of the enantiomeric crystals in the temperature interval of 76°C-145°C. If a sample of the racemate is seeded with a small excess of one of the enantiomers, the racemate is gradually converted to the optically active product. A more general method for the isolation of an optically active compound from a mixture containing an excess of one enantiomer was suggested by Lahav et al (25). They prepared 9-anthroyl esters of 1-aryl ethanols and showed that the racemate crystallizes in a form which allows photodimerization while the crystals of the excess enantiomer are photochemically inert and can easily be separated from the photodimer.

SOLID-STATE POLYMERIZATIONS

Ring Opening Polymerization

A number of cyclic monomers, of which trioxane is a typical example, polymerize after exposure of their crystals to ionizing radiation.(26) The reaction occurs, under these conditions, only in the crystalline state at temperatures close to the melting point of the monomer. Moreover, the crystallites of polyoxymethylene formed from a single trioxane crystal have their crystallographic axes highly oriented relative to the axes of the parent monomer crystal.(27) Carazzolo et al(28) showed that the fiber axis of the polymer lies either parallel to the six-fold symmetry axis of the monomer ("Z-crystals") or in three equivalent directions inclined by 76° to this axis("W-crystals"). The theoretical density of the polymer is 1.50 g-cm^{-3} as against 1.40 g-cm^{-3} for trioxane; since the outside dimensions of the crystal do not seem to change during polymerization, voids must form inside the crystal. In the direction of the six-fold axis, the distance between trioxane molecules is 4.175 A° as against a translation of 5.77 A° along the fiber axis of the polymer for three -CH$_2$O- residues, so that growth of a Z-crystal builds up a compressive stress which is relieved by formation of W-crystals.

Although polymerization of pre-irradiated trioxane crystals occurs only in the vicinity of the melting point of the monomer, van der Heijde and Kasteren observed microscopically changes taking place at lower temperatures which influence the later polymerization behavior.(29) They believe that segregation of impurities- particularly traces of water- in cavities formed by radiation damage is a precondition for the spontaneous conversion of trioxane to the polymer. Thorough drying of trioxane invariably led to polymerization without irradiation.(30) NMR spectra of trioxane crystals exhibit a striking narrowing of the absorption peak in the temperature range in which polymerization is observed, indicating a high degree of mobility in the crystalline phase.(31)

Colson and Reneker(32) used α-particle irradiation to initiate trioxane polymerization. The range of these particles is small compared to the thickness of the trioxane crystals and the irradiation dose can be adjusted so that the polymer crystals formed along the track of an α-particle, are well separated from one another. In this way, the development of Z and W crystal morphologies could be followed in detail. The results suggested also that the polymer chains exhibit the folding which is so characteristic of polymer crystallization. However, X-ray diffraction studies(33) indicate that polyoxymethylene obtained by solid-state polymerization of trioxane consists of extended chains, although folded chains result from the solid-state polymerization of tetroxane.

Trithiane behaves in a manner analogous to trioxane, being polymerized, after preirradiation and heating to the neighborhood of its melting point, to polythiomethylene(34). In this case, the monomer is orthorhombic and the polymer chains grow predominantly parallel to the ab diagonals of the lattice of the trithiane crystal. Again, such growth would lead to large compressive stresses and this is presumably the reason for polymer crystal growth in alternative crystallographic directions. (34,35) NMR studies of trithiane crystals do not reveal an increased mobility in the temperature range in which polymerization is favored. However, it appears that highly purified trithiane crystals do not polymerize and that the presence of an impurity is required to initiate the process.(36)

Polymerization of Vinyl Compounds

Although there are many indications that the polymerizability of crystalline vinyl monomers is related to the crystal structure, the geometrical arrangement of the monomer molecules exerts generally little influence on the nature of the reaction. The reaction product is amorphous (except for special cases to be discussed below) and the polymerization proceeds apparently at the interface of a crystalline and an amorphous region.

In spite of this limitation, experiments with single crystals revealed various degrees of topotactic control(37). With barium methacrylate monohydrate, the primary radical formed by the addition of a hydrogen atom to the double bond of the methacrylate ion has an ESR spectrum which changes when a second monomer is added:

$$CH_3 - \overset{COO^-}{\underset{CH_3}{C^{\bullet}}} + CH_2 = \overset{COO^-}{\underset{CH_3}{C}} \longrightarrow CH_3 - \overset{COO^-}{\underset{CH_3}{C}} - CH_2 - \overset{CH_3}{\underset{CH_3}{C^{\bullet}}}$$

and this reaction may therefore be studied kinetically. Moreover, the dependence of the ESR spectrum of the dimer radical on the orientation of the crystal in the magnetic field proves that the dimer radical retains a preferred orientation in the crystal lattice of the monomer.(38,39) By using D_2O in forming the monomer hydrate, it could be established that at least 75% of the hydrogen atoms involved in the formation of the primary radical originate from the water of hydration.(40)

In the case of p-benzamidostyrene, XXIII, the monomer forms a crystal in which the layers with the reactive double bonds lie in planes separated by 28.0A° from each

```
        CH=CH₂
         |
        ⌬
         |
        HN-C=O
             |
            ⌬
```

XXIII

other and the amide groups form a hydrogen-bonded network. This monomer polymerizes spontaneously 80° below its melting point and it can be shown that the infrared dichroism due to the N-H stretching vibration is retained if the polymerization is carried out below the glass transition temperature of the polymer. (41).

The solid-state polymerization of vinyl stearate can be studied by crystallographic methods (42), since the polymer exhibits, in this case, the phenomenon of side-chain crystallization even if the chain backbone is not stereoregular. It was shown that the aliphatic side chains retain their parallel orientation during the polymerization process.

It is unlikely that a high degree of stereoregulation is attainable by the polymerization of crystals of pure vinyl monomers. However, clathrates of vinyl compounds in channel-like cavities of host crystals may provide the geometric constraint for the formation of highly stereoregular polymers in the solid state. Thus, urea adducts of a number of monomers yielded, after radiation-initiated polymerization, products such as trans-1,4-polybutadiene and syndiotactic poly(vinyl chloride)(43). Since the latter polymer is not only stereoregular but is constrained in the clathrate complex to the all-trans conformation, it may be used to aid in the analysis of the infrared spectrum of PVC.(44)

In recent years, a series of interesting investigations were carried out by Farina and his collaborators on the polymerization of clathrate complexes of perhydrotriphenylene.(45-47) Such inclusion leads to polymerization of monomers which would not react, under similar conditions, in the liquid state. It may also lead to the formation of isotactic polymers. The adduct of trans-1,3-pentadiene in optically active perhydrotriphenylene yielded an optically active polymer(45) as a result of the asymmetry of the crystalline environment.

Detailed crystallographic studies have also been reported of the changes accompanying the polymerization of the 2,3-dichlorobutadiene adducts in thiourea(48) and of the butadiene adduct in perhydrotriphenylene. (49)

Polydimerization

When p-phenylene diacrylic acid, XXIV, which contains two functional groups analogous to those in cinnamic acid, is irradiated in the solid state, a dimerization-like reaction takes place at both ends of the molecule, so that a polymer with alternating phenylene and cyclobutane rings (XXV) is formed. A number of phenylene diacrylic acid derivatives behave in the same manner (50) and so does 2,5-distyrylpyrazine, XXVI.(51,52)

The chainlength of the polymer formed in crystals of this type increases with the degree of conversion. Since formation of the cyclobutane rings eliminates conjugation of the double bonds with the aromatic rings, the peaks in the ultraviolet absorption spectrum shift to shorter wavelength. Thus, by irradiating the crystals at a relatively long wavelength, the reaction stops when short oligomers have been formed - these may then be further polymerized by irradiation at a shorter wavelength. However, if the oligomeric product is dissolved and recrystallized, its further polymerization becomes impossible(53); the arrangement of the monomer units characteristic of the monomer crystal becomes apparently unstable during the polymerization, but is essential for the continuation of the process. Also, in contrast with the behavior of trioxane, polydimerizations of distyrylpyrazine (52) and p-phenylene diacrylic acid diethyl ester (54) is favored by a temperature far below the melting point of the monomer.

A crystallographic study of the conversion of distyrylpyrazine to its polymer (55) showed that the monomer and polymer both crystallize in the space group Pbca and the crystallographic density increases by 1% during the reaction. The polymer forms a separate crystalline phase; the polymer crystallites exhibit three-dimensional orientation relative to the parent monomer crystal. If the polymer is dissolved and recrystallized, its diffraction pattern is different from that characterizing the polymer resulting from the solid-state reaction. The space group of the polymer shows also that the substituted cyclobutane rings have the 1,3-<u>trans</u> structure.

When a monomer, analogous to XXIV, forms chiral crystals, polymerization of single crystals may yield optically active polymer though the monomer exhibited no optical activity in solution. A striking demonstration of this effect was recently reported, where chiral single crystals of racemic XXVII were converted to polymers with an appreciable optical activity.(56)

$$\begin{array}{c} CN \\ | \\ C=CH-\bigcirc-CH=CH-COCHC_2H_5 \\ | \quad\quad\quad\quad\quad\quad\quad\quad || \;\; | \\ O=COC_2H_5 \quad\quad\quad\quad\;\; O \;\; CH_3 \end{array}$$
XXVII

Polymerization of Diacetylene Derivatives

A number of diacetylene derivatives have long been

observed to undergo transitions in the solid state to a
polymeric product and the nature of this transition has
been clarified by Wegner and his collaborators.(57-60),
Some of the features of these remarkable reactions are
as follows: (1) Polymerization takes place thermally or
photochemically in the crystalline state, although the
reagents are inert when melted or dissolved. (2) Topotactic control of the reaction is evidenced by the polymerizability of one crystalline form while another crystalline form of the same reagent is inert. Similar conclusions may be reached because of the different behavior
of diacetylene derivatives carrying substituents far from
the reaction center. (3) Single crystals of the monomers
are converted to single crystals of polymers. The polymerization is homogeneous, with the polymer chains distributed, at partial conversion, at random within the
monomer crystal. There is no chain folding.

Crystallographic analysis of the polymer obtained
from 2,4-hexadiyne-1,6-diol-bis(phenylurethane)(59)
showed that the polymerization may be represented by

$$n \ R-C\equiv C-C\equiv C-R \longrightarrow \left[\begin{array}{c} R \\ C-C\equiv C-C \\ R \end{array} \right]_n$$

and this structure of polymers derived from diacetylene
derivatives was later substantiated by Raman spectros-
copy(61). A crystallographic study of the structural
changes accompanying the polymerization process is
generally difficult since the reaction is too fast under
x-ray irradiation. However, with the relatively inert
2,4-hexadiynediol such a study has been shown to be
feasible and it allowed a determination of the changes
in the crystal lattice parameters as a function of the
monomer conversion to polymer(62). The process requires
only minimal motions of the reagent molecules and
Baughman (63) has formulated the geometrical conditions
which have to be satisfied to make possible such conversions of monomers into large, nearly defect-free single
crystals of polymers.

Poly-p-xylylene

When [2,2']paracyclophane is pyrolyzed, the diradical
p-xylylene is formed which polymerizes spontaneously when
condensed on a cold surface. Wunderlich and his
collaborators have subjected this process to detailed
calorimetric(64) and crystallographic(65) study. Since

$$\underset{\text{CH}_2-\text{CH}_2}{\overset{\text{CH}_2-\text{CH}_2}{\bigcirc\bigcirc}} \xrightarrow{600°C} \dot{C}H_2-\bigcirc-\dot{C}H_2$$

$$\xrightarrow{t<80°C} (-CH_2-\bigcirc-CH_2-)_n$$

the polymer is highly crystalline and polymer crystallization is generally arrested below the glass transition temperature (which lies, in this case, at 50-80°C) it was concluded that crystallization must accompany the polymerization process. The calorimetric data suggest that at -196°C the vapor condenses to the crystalline monomer which polymerizes at somewhat higher temperatures. So far, no single crystals of the monomer have been obtained, so that we do not know whether the direction of chain growth is controlled by the crystal structure of the monomer.

Polycondensation

In a polycondensation reaction a small molecule is eliminated during the conversion of a bifunctional monomer to a high polymer. Yet, even such processes have been reported to take place in the crystalline state. Even more surprisingly, a single crystal of ε-aminocaproic acid (ACA) heated 30° below its melting point was found to be converted to crystallites of polycapramide (Nylon-6) whose crystallographic axes were highly oriented with respect to the parent monomer

$$H_2N-(CH_2)_5COOH \longrightarrow [-NH-(CH_2)_5-CO-]_n + H_2O$$

$$\text{ACA} \hspace{4cm} \text{Nylon-6}$$

crystal, although the polycondensation involves 17% contraction in the fiber direction (66,67). The polymer chains grow in two alternative directions as required by the space group of the monomer. The polymerization involves an interaction of the monomer crystal with its vapor; in the presence of an inert gas which would compete with the monomer gas adsorption, the process was strongly retarded and no polymer is formed if the ACA vapor is continuously removed under high vacuum. Three phases of the process could be distinguished, i.e., (a) an induction period, (b) a period during which monomer was converted at a constant rate (determined probably by its vapor pressure) to polymer (c) a continuing condensation of the polymer chains with each other after exhaustion of the monomer. Epitactic polymer growth was observed on exposure of (100) faces of KCl to ACA vapor (67-68). Surprisingly, the polymer formed simultaneously in two different crystalline modifications, as spherulites of the α-form in which neighboring chains are antiparallel and as single crystals of the γ-form, in which the chains are parallel to each other.

CONCLUDING REMARKS

Studies of reactions in molecular crystals have led to the realization that molecular mobility is in some cases surprisingly large. The most important conclusions from work in this field may be summarized as follows:

(1) The crystal structure may impose restrictions directing a chemical reaction into one of several alternative pathways.

(2) The asymmetric environment of a crystal may be utilized for the synthesis of chiral products from achiral or racemic reagents.

(3) Polymerization in the crystalline state may be topotactic, allowing the preparation of oriented polymer from crystalline monomers.

(5) In favorable cases, topotactic polymerization proceeds in a single phase and single crystals of a monomer can then be converted into macroscopic single crystals of a polymer.

Acknowledgement

I am indebted to the National Science Foundation for financial assistance for the preparation of this paper by their Grant DMR 75-05234. I should like to thank Professor M.D. Cohen of the Weizmann Institute of Science for making available to me manuscripts of papers from his laboratory prior to publication.

REFERENCES

1. C.N.R. Rao, "Solid State Chemistry", Marcel Dekker, New York (1974).
2. H. Morawetz, in Physics and Chemistry of the Organic Solid State, D.Fox, M.M. Labes, A.Weissberger,

eds., Interscience, New York, vol. 1(1963) pp. 287-328; vol.2(1965) pp. 853-872.
3. M.D. Cohen and B.S. Green, Chem.Brit., **9**,490(1973).
4. I.C. Paul and D.Y.Curtin, Acc.Chem.Res.,**6**, 217 (1973).
5. J.M. Thomas,Phil.Trans., A**277**,251 (1974).
6. I.C. Paul and D.Y. Curtin, Science, **187**, 19 (1975).
7. M.D. Cohen,Angew.Chem.,Internat.Ed., **14**,386 (1975).
8. R.E. Pincock, K.R. Wilson and T.E. Kiovsky, J.Am. Chem. Soc., **89**, 6890 (1967).
9. R.E. Pincock, M.M. Tong and K.R. Wilson, J. Am. Chem. Soc., **93**, 1669 (1971).
10. J.D. McCullough, D.Y. Curtin and I.C. Paul, J. Am. Chem.Soc., **94**. 874 (1972).
11. F.W. Comer and J. Trotter, J. Chem.Soc., B,**11** (1966).
12. K. Lonsdale, E. Nave and J.F. Stephens, Phil. Trans., **261**, 1 (1966).
13. S.R. Byrn, D.Y. Curtin and I.C. Paul, J.Am.Chem. Soc., **94**, 890 (1972).
14. J.Z. Gougoutas and J.C. Clardy, Acta Crystal., **B**, **26**, 1999 (1970).
15. J.Z. Gougoutas, Pure Appl.Chem., **27**,305 (1971).
16. H. Morawetz, Science, **152**, 705 (1966).
17. K. Penzien and G.M.J. Schmidt, Angew. Chem., **81**, 628 (1969).
18. M.D. Cohen, in"Topochemistry and G.M.J. Schmidt", Verlag Chemie,(1976).
19. G.M.J. Schmidt, J. Chem. Soc., 2014 (1964).
20. G.M.J. Schmidt, Pure Appl. Chem., **27**,647 (1971).
21. M.D. Cohen, E. Elgavi, B.S. Green, Z. Ludmer and G.M.J. Schmidt, J.Am.Chem.Soc., **94**,6776 (1972).
22. A. Elgavi, B.S. Green and G.M.J. Schmidt, J.Am. Chem. Soc., **95**, 2058 (1973).
23. G. Friedman, M. Lahav and G.M.J. Schmidt, J. Chem. Soc., Perkin Trans. II, 428 (1974).
24. K.R. Wilson and R.E. Pincock, J.Am.Chem.Soc., **97**, 1474 (1975).
25. M. Lahav, F. Laub,E. Gati, L. Leiserowitz and Z. Ludmer, J. Am. Chem. Soc., in press.
26. S. Okamura and K. Hayashi, Makromol. Chem., **47**, 230, 237 (1961).
27. (a) J. Lando, N. Morosoff, H. Morawetz and B. Post, J.Polymer Sci., **60** S24 (1962); (b) S. Okamura, K. Hayashi and M. Nishii, J.Polymer Sci., **60** S26 (1962).
28. G. Carrazzolo, S. Leghissa and M.Mammi, Makromol. Chem., **60**, 771 (1963).

29. H.B. van der Heijde and P.H.G. Kasteren, Phil.Mag., 13, 1039 (1966).
30. H.B. van der Heijde, Phil.Mag., 13, 1055 (1966).
31. A. Komaki and T. Matsumoto, Polymer Lett., 1, 671 (1963).
32. J.P. Colson and D.H. Reneker, J.Appl.Phys., 41, 4296 (1970).
33. G. Wegner, A. Munoz-Escalona and E. W. Fischer, Ber.Bunsenges, 74, 909 (1970).
34. J.B. Lando and V. Stannett, J. Polymer Sci., Pt. A, 3, 2369 (1965).
35. G. Carazzolo, M. Mammi and G. Valle, Makromol.Chem., 100, 295 (1967).
36. J.E. Herz and V. Stannett, Polymer Lett., 4, 995 (1966).
37. H. Morawetz, Pure Appl.Chem., 12, 201 (1966).
38. J. H. O'Donnell, B. McGarvey and H. Morawetz, J.Am. Chem. Soc., 86, 2322 (1964).
39. M. J. Bowden and J.H. O'Donnell, Macromolecules, 1, 499 (1968).
40. M.J. Bowden, J.H.O'Donnell and R.D. Sothman, Macromolecules, 5, 269 (1972).
41. S.Z. Jakabhazy, H. Morawetz and N. Morosoff, J. Polymer Sci., Pt. C, 4, 805 (1964).
42. N. Morosoff, H. Morawetz and B. Post, J. Am. Chem. Soc., 87, 3035 (1965).
43. D.M. White, J.Am.Chem.Soc., 82, 5678 (1960).
44. S. Krimm, J.J. Shipman, V.L. Folt and A.R. Berens, Polymer Lett., 3, 275 (1965).
45. M. Farina, G. Audisio and G. Natta, J. Am. Chem. Soc., 89, 5071 (1967).
46. M. Farina, U. Pedretti, M.T. Gramegna and G. Audisio, Macromolecules, 3, 481 (1970).
47. M. Farina, G. Audisio and M.T. Gramegna, Macromolecules, 4, 265 (1971); 5, 617 (1972).
48. Y. Chatani, S. Nakatani and H. Tadokoro, Macromolecules, 3, 481 (1970).
49. A. Colombo and G. Allegra, Macromolecules, 4, 579 (1971).
50. F. Suzuki, Y. Suzuki, H. Nakanishi and M. Hasegawa, J. Polymer Sci., 7, 2319 (1969).
51. M. Hasegawa, Y. Suzuki, F. Suzuki and H. Nakanishi, J. Polymer Sci., 7, 743 (1969).
52. H. Nakanishi, Y. Suzuki, F. Suzuki and M. Hasegawa, J. Polymer Sci., 7, 753 (1969).
53. T. Tamaki, Y. Suzuki and M. Hasegawa, Bull.Chem. Soc. Japan, 45, 1988 (1972).
54. H. Nakanishi, F. Nakanishi, Y. Suzuki and H. Hasegawa, J.Polym.Sci., Polym.Chem.Ed., 11, 2501 (1973).

55. H. Nakanishi, M. Hasegawa and Y. Sasada, J.Polymer Sci., 10, 1537 (1972).
56. L. Addadi, M.D. Cohen and M. Lahav, J.Chem.Soc., Chem. Commun., 471 (1975).
57. G. Wegner, Makromol. Chem., 154, 35 (1971).
58. G. Wegner, Polymer Lett., 9, 133 (1971).
59. E. Hädicke, E.C. Mez, C. H. Krauch, G. Wegner and J. Kaiser, Angew.Chem., 83, 253 (1971).
60. J. Kaiser, G. Wegner and E.W. Fischer, Kolloid.Z.- Z.Polym., 250, 1158 (1972).
61. A. J. Melveger and R. H. Baughman, J. Polymer Sci., Polymer Phys. Ed., 11, 603 (1973).
62. R. H. Baughman, J. Appl. Phys., 43, 4363 (1972).
63. R. H. Baughman, J. Polymer Sci., Pt. A-2, 12, 1511 (1974).
64. G. Treiber, K. Boehlke, A. Weitz and B. Wunderlich, J. Polymer Sci., Polym. Phys., 11, 1111 (1973).
65. R. Iwamoto, R.C. Bopp and B. Wunderlich, J. Polym. Sci., Polym. Phys., 13, 1925 (1975).
66. E.M. Macchi, N.Morosoff and H. Morawetz, J. Polym. Sci., Pt. A-1, 6, 2033 (1968).
67. E. Macchi, J. Polym. Sci., Pt. A-1, 10, 45 (1972).
68. E. Macchi, Macromol. Chem., 165, 313 (1973).

LASER INDUCED REACTIONS IN DOPED POLYMETHYLMETHACRYLATE AND THEIR

CORRELATION WITH RESULTS OBTAINED BY X-RAY AND UV IRRADIATION.

D.J. Morantz and C.S. Bilen

London College of Printing, London, S.E.1. UK

ABSTRACT

Emission spectra and kinetics of induced phosphorescence in doped PMMA are reported for laser irradiation; compared with previous studies, using X-ray and UV irradiation, of induced phosphorescence and thermoluminescence. Possible mechanisms are proposed.

INTRODUCTION

Thermoluminescence and room temperature induced phosphorescence are observable for doped polymethylmethacrylate (PMMA), pre-irradiated for 20 minutes with X-ray or UV light, or heated for 10 hours at 473 K, in the absence of air, (1,2). However, no induced phosphorescence resulted on exposure to a CW Argon Ion laser beam. The absence of effect may be due to the inability of low energy visible light photons (below 4.5 e.v.) of the laser to produce the species responsible for induced phosphorescence or to the absence of significant absorption of blue or green light by the samples; the latter would also diminish the possibility of a non-coincident two-photon process. The use of a secondary blue/green light absorbing dye was tried and, as expected, laser induced phosphorescence was obtained in the dyed, doped PMMA samples; the kinetics were, however, found to be different.

EXPERIMENTAL

PMMA (B.D.H. Chemicals) was of low and high molecular weight of average particle size of 170 and 270 μ respectively. Triphenylene (purity 99.9%) was used as the primary dopant and eosin (laboratory reagent grade) as a secondary light absorber. In some qualitative experiments, dibenzothiophene as primary dopant and rhodamine or fluorescein as secondary dopants gave positive results. The samples were as described elsewhere (1,2) with the addition of secondary dopants of concentration 10^{-3} Molar in the sample mixture and the use of laser light instead of UV and X-ray sources. The full beam of an Argon Ion laser, with output powers up to several watts, was used without a filter. Individual exposures ranged from 1 to 8 minute periods and the integrated exposure totalled up to 120 minutes.

RESULTS AND DISCUSSION

The emission spectra of triphenylene doped PMMA, containing eosin as the secondary dopant, are shown in Figure 1 before and after various laser irradiations. The emission spectra before irradiation show only fluorescence peaks ($S_1 - S_0$) whereas, after laser irradiation, afterglow peaks corresponding to $T_1 - S_0$ transitions are also present. The laser induced phosphorescence is in the whole thickness of the sample, as in the case of X-ray irradiation, and the induced phosphorescence lasts for 3 - 4 days. This contrasts with UV irradiation where the effect is near the surface lasting for 7 - 10 hours. Typical curves for induced phosphorescence intensity versus irradiation time are shown in Figure 2, for laser, X-ray and UV irradiation. The X-ray irradiation curve has an induction period and all three differ at longer exposures, with saturation for X-rays and decline for intense UV.

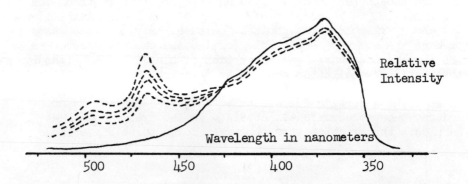

Figure 1. Emission spectra of PMMA doped with both triphenylene and eosin: before irradiation (solid line); and after laser irradiation of 11,22,33,44 minute periods (showing increasing intensities).

The curves for two different laser powers for the induced phosphorescence effect are shown in Figure 3, and the high power curve approaches saturation after a rapid linear increase. The temperature dependence of this effect was shown in experiments where the same total laser irradiation period (32 minutes) was achieved by individual exposures of 2, 4 or 8 minutes. The eight minute exposure series, at a higher average temperature, resulted in lower phosphorescence intensities, compared with the four minute series, and the two minute series was the most intense. We have also found that, using a range of laser powers, the nett induced phosphorescence intensity, after two minutes irradiation, increases monotonically with laser power up to about 600 mW (T_{max} = 328 K) where it reaches a plateau; after which there is an indication of some decrease. Under such conditions, the centre of the irradiated spot gave less bright induced phosphorescence than the outer ring (where the light intensity of the incident laser beam is relatively low). These effects were also observed for spots with more than optimum UV irradiation.

The existence of species, similar to those obtained for X-ray irradiation, derived from PMMA, in the laser irradiated doped samples which also contained dyes is suggested by the presence of induced phosphorescence of the dopant, and its subsequent decay

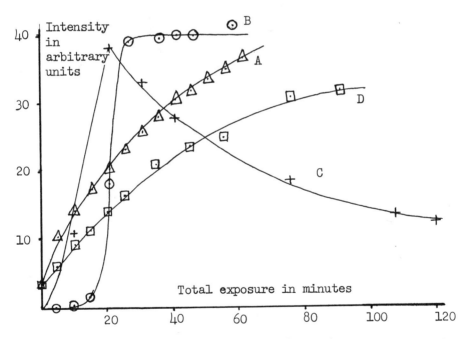

Figure 2. Induced phosphorescence intensity v. exposure for triphenylene doped PMMA for: (A) laser (sample also includes eosin); (B) X-ray: (C) intense UV; and (D) low intensity UV irradiation.

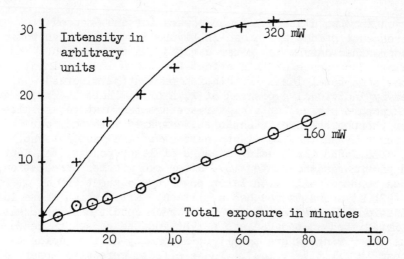

Figure 3. Induced phosphorescence build up of triphenylene doped PMMA containing eosin compared at two laser powers.

kinetics. In the case of X-ray and UV light (below 370 nm) the photon energy, primarily absorbed by the matrix itself, is sufficient to cause chain scission, producing radicals. For laser irradiation, the primary energy input is due to absorption by the dye molecules, thus requiring an energy transfer process to pass the energy to the matrix. With intense UV irradiation, total absorption of the light beam by the PMMA occurs near the surface and the concentration of radical species produced is sufficient for secondary reactions to cause a decrease of induced phosphorescence after the rapid initial rise. Laser irradiation induced phosphorescence, resulting from dye absorption, appears throughout the depth of the sample for the low dye concentrations we used. At higher laser beam intensity, or for samples having higher secondary dopant (dye) concentration, we may expect curves similar to those for higher UV intensity.

The behaviour of radicals, produced by gamma irradiation of PMMA with temperatures ranging from 110 K to 320 K, was demonstrated by esr, (3). UV produced PMMA radicals were also identified by esr, (4). We found that, with UV and X-ray irradiated samples, there was good correlation in the kinetics of thermo-luminescence and of induced phosphorescence properties, (2). Radicals, involved in energy transfer, were indicated by an anomalous reversible decrease and increase of induced phosphorescence in doped PMMA observed during liquid N_2 cooling and rewarming to ambient, (5). Thus, the presence of PMMA radicals in laser irradiated samples is also a possibility.

An intriguing question concerns the photo-chemical action by photons of low energy, i.e. 2.5 ev, which is below the 4.5 ev requirement for radical production in PMMA. Two-photon coincident absorption, yielding 5 ev, is ruled out as the energy density is insufficient for a non-linear absorption process. Multiphoton, non-coincident, absorption is a possibility and this was reported for the laser photolysis of a pyrene solution, where energy states more than twice the ingoing photon energy were obtained, (6). However, the Q-switched laser used was of the order of a megawatt during a short pulse of the order of nano-seconds, whereas the laser we used is not comparable having an intensity of 6 orders of magnitude less.

Alternatively, the necessary energy (4.5 ev) states may be populated by absorption from higher vibrational levels of dye or of matrix molecules. Such transitions have been studied using two laser beams, (7). One laser, in the infra-red, caused molecules to go into higher vibrational levels where the photons from a second laser beam in the UV region led to dissociation or ionisation. Thus the molecules normally transparent to high energy photons from UV laser beams were rendered absorbing in the presence of infrared photons. In our case only one, high intensity CW visible, laser beam is used and we believe that, similarly, higher vibrational levels result from nearly adiabatic absorption by the dye molecules. An energetic level sufficient for reaction (4.5 ev) may then be reached by further photon absorption in the dye, followed by energy transfer to the doped matrix. Alternatively, through energy transfer processes from the vibrationally excited dye, the matrix may become capable of absorbing the visible light, to which it is normally transparent, and react.

We acknowledge an ILEA Research Fellowship to one of us (C.S.B.) and assistance by Mr. A.T. O'Hare.

REFERENCES

1. D.J. Morantz and C.S. Bilen, 7th Int'l. Symp., "Reactivity of Solids", Bristol, 525, 1972
2. D.J. Morantz and C.S. Bilen, Polymer, 16, 745, 1975
3. G. Geuskens and C. David, Makromol. Chem., 165, 273, 1973
4. P.F. Wong, Polymer 15, 60, 1974
5. C.S. Bilen and D.J. Morantz, Nature, 258, 66, 1975
6. J.T. Richard, G. West and J.K. Thomas, J. Phys.Chem. 74, 4137, 1974
7. R.V. Ambartzumian and V.S. Letokhov, Appl. Opt. 11, 354, 1972

THERMAL ISOMERISATION OF CISAZOBENZENE IN THE SOLID STATE

H.K. Cammenga, E. Behrens and E. Wolf

Institut für Physikalische Chemie, Techn.
Universität Braunschweig, Braunschweig, GFR

INTRODUCTION

Reversible photoisomerisation in molecular crystals may eventually become a new means of information storage. A certain loss of information can occur by thermal re-isomerisation. Furthermore thermal isomerisation is a reaction of type A → B other than a phase-transformation, which type has not often been studied so far. For these reasons we investigated the thermal cis → trans isomerisation of azobenzene.

Azobenzene (M = 182.23 g/mol) can exist as two geometric isomers (fig. 1), first isolated in pure form by HARTLEY[1]. The cis-isomer, m.p. 71.6 °C, is thermodynamically unstable and, depending on conditions, more or less slowly rearranges to the stable trans-isomer, m.p. 68.3 °C. From crystallographic X-ray analysis it is known that in the solid the trans-molecules exhibit a nearly planar conformation [2, 3], whereas in the cis-molecules the planes of the two phenyl rings show an angular deflection of 53° to each other due to steric hindrance of the H-atoms in the 2,2'-position [4, 5]. The nonplanarity results in an appreciably lower resonance energy for cis-azobenzene. The dipole-moment of cis- is 3.0 D compared to 0 D for the trans-form [6].

SUBSTANCES AND ANALYSIS

Trans-azobenzene (trans-a) was crystallised from pure methanol, dried, zone-melted, sublimed in HV (to remove traces of hydrazobenzene) and again zone-melted under argon. Benzo(c)cinnoline was prepared by photoisomerisation of azobenzene in conc. H_2SO_4, and purified as trans-a; m.p. 157.3 °C.

A new method for the photochemical synthesis of larger amounts of cis-a was developed. A solution of pure trans-a in dry pure hexane was circulated at -10 °C through a filter, a radiation cuvette protected from IR radiation of the high pressure Hg lamp by a water shield and a large column of Al_2O_3 (Act. II, neutral). Cis-a formed is adsorbed until nearly all trans-a in solution has been converted. The rest is eluted with hexane and thereafter by elution with hexane/benzene cis-a is recovered and crystallised. Crystals are obtained at -20 °C from hexane.

It has been found that 0.01 per cent of trans- in cis-a and vice versa may be detected by thin layer chromatography, a method applied to test the purity of every sample before use. To follow the kinetics of the isomerisation reaction spectrophotometric analysis of a solution of the sample or a fraction thereof in hexane was employed. The 315 nm absorption band was used, which corresponds to a $\pi-\pi^*$ transition, intense in trans- and very weak in cis-a. In the concentration region used the LAMBERT-BEER-law was followed both for the pure components and the mixture. The mole fraction of trans-isomer in a sample could be calculated from

$$x_t = \frac{\varepsilon - \varepsilon_{co}}{\varepsilon_{to} - \varepsilon_{co}}, \qquad (1)$$

where ε measured molar extinction coefficient of sample and ε_{co}, ε_{to} molar extinction coefficients of pure cis- and trans-a.

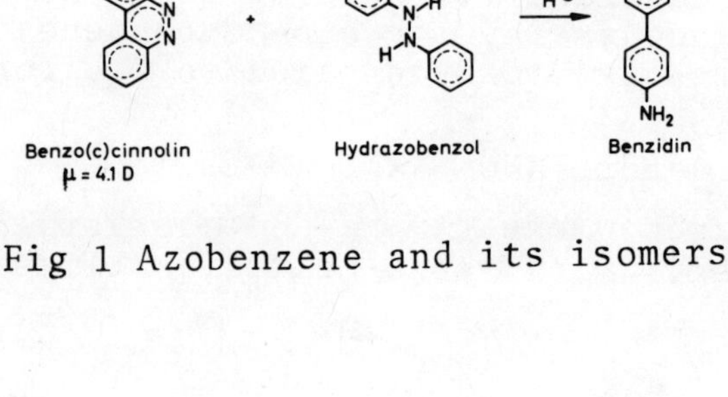

Fig 1 Azobenzene and its isomers

KINETICS OF ISOMERISATION

To follow the kinetics of the cis →trans isomerisation several crystals of the pure cis-isomer were placed in small, clean glass ampoules and these were sealed off while the crystals were kept cool. At t = 0 8...10 of these ampoules were placed in a constant temperature bath, equipped with a precision thermometer, and held at constant temperature in the dark for extended periods. A dummy probe beneath the other probes could be observed with a long focus microscope at intervals under faint illumination. At intervals samples were taken from the temperature bath and rapidly cooled in ice. The ampoule was broken, the content weighed on a microbalance and transferred into a volumetric flask. The sample was quickly dissolved in hexane (Spectranal, Merck), analyzed immediately and the mole fraction x_t or reaction extent α calculated. Measurements were made in the range 54.2 to 65.3 °C. The α, t curves are shown in fig. 2 up to α = 0.5. At higher temperatures and $\alpha > 0.5$ fractional melting may be observed and these results have not been taken into consideration.

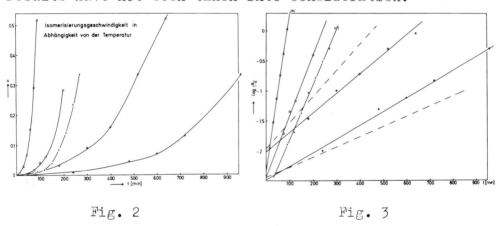

Fig. 2   Fig. 3

Two regions may be seen on fig. 2. During the induction period nuclei of trans-a are formed, preferentially at defects at the crystal surface as observed microscopically. The induction period is followed by an acceleration region of growth of the new phase. As seen from fig. 3, the kinetics of the reaction follows rather closely the law deduced by PROUT and TOMPKINS [7]

$$\ln \frac{\alpha}{1-\alpha} = k_3 \cdot t + C. \qquad (2)$$

From the temperature dependence of k_3 (fig. 4) an apparent activation energy of 215 kJ/mol is deduced, more than twice the mean value for unimolecular isomerisation in solution (98.4 kJ/mol) found for various solvents [8, 9] . Previous measurements of the solid state isomerisation rate of cis-a by TSUDA and KURATANI [10] lead to higher rates and an activation energy of only 131 kJ/mol, but these measurements were made on finely ground powder of cis-a, mixed with KBr and pressed into disks for IR-spectrophotometric analysis of the reaction rate. The authors state that their rates were dependent on the extent of grinding. Furthermore the application of pressures of 10 tons/cm^2 may raise the temperature within the disks by several ten degrees and together with the decreasing molar volume (cis: 150.2, trans: 148.0 cm^3/mol) favour enhanced nucleation.

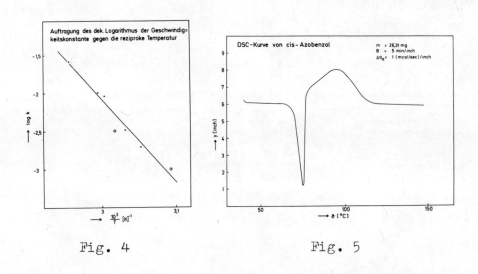

Fig. 4   Fig. 5

THERMODYNAMICS OF THE REACTION

Appreciable difference exists between values of the enthalpy of isomerisation as given by several authors. HARTLEY using a crude calorimetric method found ΔH_{iso} = -50 kJ/mol at 21 °C [1]. CORRUCINI and GILBERT deduced ΔH_{iso} = - 41 kJ/mol from the difference of the standard enthalpies of formation determined by bomb calorimetry [11]. Using the HMO method ADRIAN calculated -64.9 whereas LJUNGGREN and WETTERMARK obtained only -10.9 kJ/mol by a CNDO/2 calculation [13].

We used differential scanning calorimetry (DSC) to

measure ΔH_{iso}. As the reaction rate below the eutectic point, 41°C, is too small to allow direct determination of ΔH_{iso} totally in the solid state, ΔH_{iso} and the enthalpy of fusion ΔH_f are determined in the same experiment, cf. fig. 5. First ΔH_f for the trans-isomer and for benzo(c)cinnoline were measured and found to be 22.7 and 20.9 kJ/mol respectively. The molar heat capacities of cis- and trans-a were measured, and it was calculated that ΔC_p = 10.3 J/K mol. On assuming that the enthalpy of melting of cis-a is equal to the mean of the enthalpies of trans-a and benzo(c)cinnoline, which seems to be reasonable in view of the very similar molecular structure, space group and lattice energy (see below) and using the measured ΔC_p, the enthalpy of isomerisation at STP is obtained as $\Delta H_{iso,st}$ = $-$ (48.0 $\pm$ 1.5) kJ/mol, which is close to the approximate value of HARTLEY.

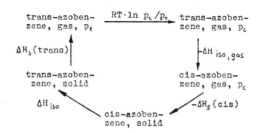

Fig. 6 The cycle at T = 298.15 K

The theoretically calculated isomerisation energy as obtained by ADRIAN and by LJUNGGREN and WETTERMARK from the difference of the computed resonance energies should in fact apply to the free molecules of the isomers, i.e. within the gaseous phase. We have calculated this value by the isothermal cyclic process at 298.15 K shown in fig. 6. The vapour pressures and the enthalpies of sublimation of the isomers are obtained from ^{14}C tracer effusion measurements: ΔH_s (trans) = (93.84 $\pm$ 0.12) kJ/mol, p_t = 8.01·10^{-4} Torr, ΔH_s (cis) = (92.92 $\pm$ 0.12) kJ/mol, p_c = 6.11·10^{-4} Torr. With these data the enthalpy of isomerisation in the gas phase at 298.15 K

$$\Delta H_{iso}(\text{gas}, 298.15) = \Delta H_{iso,st} + \Delta H_s (\text{trans}) + RT \cdot \ln p_c/p_t - \Delta H_s (\text{cis}) \quad (3)$$
$$= -47.8 \text{ kJ/mol}.$$

and is found to be not far from the value in the solid state, in contrast to the arguments of LJUNGGREN and WETTERMARK, raised by them to support their low calculated value. Following the energetics of similar solid state reactions may yield other accurate data, with which results of theoretical calculations may be compared. As the accuracy of such data obtained directly is much higher than that from bomb calorimetry (small difference of large numbers) these methods of calculation may be improved to give better accordance with experiment.

REFERENCES

1. G.S. HARTLEY, J.C.S.(1938)633; 2. J.J. De LANGE, J.M. ROBERTSON and I. WOODWARD, Proc.Roy.Soc. A171(1939) 398; J.C. BROWN, Acta Cryst. 21(1966)146; 4. G.C. HAMPSON and J.M. ROBERTSON, J.C.S.(1941)409; 5. A. MOSTAD and C. RØMMING, Act.Chem.Scand.25(1971)3561; 6. G.S. HARTLEY and R.W. Le FEVRE, J.C.S.(1938)867; 7. E.G. PROUT and F.C. TOMPKINS, Trans. Faraday Soc.40(1944)488; 8. R.W.J. Le FEVRE and J. NORTHCOTT, J.C.S.(1953)867; 9. J.HALPERN, G.W. BRADY and C.A. WINKLER, Can.J.Res.28 (1950)140; 10. M.TSUDA andK.KURATANI, Bull.C.S.J.37(1964) 1284;11. R.J.Corrucini and E.C.Gilbert, JACS61(1939)2925 12. F.J. ADRIAN, J.Chem.Phys.28(1958)608;13. S.LJUNGGREN and G. WETTERMARK, Acta Chem.Scan.25(1971)1599.

ENERGY TRANSFER IN THE SOLID STATE PHOTOPOLYMERIZATION

OF DIACETYLENES

G. Wegner, G. Arndt, H.-J. Graf and M. Steinbach

Institut für Makromolekulare Chemie der
 Universität Freiburg
7800 Freiburg im Breisgau, West Germany

INTRODUCTION

The photoreactivity of solid crystalline compounds containing two or more conjugated triple bonds per molecule was observed as early as 1882 by A. V. Bayer who synthesized N,N'-diacetyl-O,O' diaminodiphenyldiacetylene as an intermediate in one of his famous attempts to synthesize indigo (1). Several years later, Straus (2) gave the first systematic report on photoreactive diacetylenes but he was unable to elucidate the structure of the polymer. Among others, Seher (3) and Bohlmann (4) in Germany and Jones, Armitage and Whiting in England (5) came across this peculiar photoreactivity which is shown by the monomer crystals only. In solution or melt, diacetylenes are quite stable toward uv- or high energy radiation. In the solid state, however, the same compounds are extremely light-sensitive and their colour deepens upon irradiation or annealing below their melting point from bright red for diacetylenes, via pink for triacetylenes, blue for tetraacetylenes, green for penta-acetylenes and to black for hexaacetylenes (4).

REACTION MECHANISM

The first attempt to explain this photoreactivity of crystalline conjugated polyynes was made by Bohlmann (6), who discovered that the photoreactivity is closely related to the packing distance between the adjacent polyyne "rods" and thus depends to a large extent on the particular chemical constitution of these compounds and on their packing properties.

If a bulky group such as t-butyl-group is linked to the polyynes no reactivity is observed, while polyynes with substituents such as n-paraffin and substituted benzenes exhibit photoreactivity. As far as di-t-butylpolyynes are concerned Bohlmann suggested that the polyyne rods are too far apart from each other because of the bulky endgroups and therefore a reaction between neighbouring polyynes cannot occur . Although Bohlmann recognized the basic relationship between the solid state reactivity and packing properties he didn't study the polymer products and thus this reaction remained unexplored until 1968.

Based on x-ray evidence and general knowledge of packing properties of organic molecules G. Wegner (7) recognized in 1968 that the photoreaction of crystalline diacetylenes obeys the rules of topochemical reactions which were established by G. M. J. Schmidt (8). After a basic investigation on a broad variety of diacetylene derivatives Wegner proposed that diacetylenes polymerize according to equation [1] in order to form a polymer I with completely conjugated backbone (7). As was shown by subsequent exact x-ray structure analyses (9,10) Wegner's proposal had to be completed by a second mesomeric structure II of the polymer backbone which turned out to be more stable and hence more common. The polymer backbone is therefore better described as a sequence of triple single and double bonds with a trans-arrangement of the substituents with regard to the double bonds.

$$R-C\equiv C-C\equiv C-R \qquad \begin{array}{c} R\diagdown C=C=C=C\diagup \cdots \\ \cdots \diagup \qquad\qquad \diagdown R \end{array} \quad I \qquad [1]$$

$$\begin{array}{c} R\diagdown C-C\equiv C-C\diagup \cdots \\ \cdots \diagdown \qquad\qquad \diagdown R \end{array} \quad II$$

Besides x-ray structure analysis Raman spectroscopy of the solid polymers proved to be a valuable tool in structure determination (11). The Raman spectra are clearly explained in terms of structure II. Although the structure of the polymer has been established without any doubt and although it has been demonstrated by Wegner and his coworkers (12,13) as well as by others (14,15) that polymerization of diacetylenes is an extremely versatile method to produce perfect macroscopic single crystals of polymers with conjugated backbone and consequently interesting electrical and optical properties (16) the mechanism of the polymerization is not yet clear. In order to clarify the mechanism of this polymerization consideration should be focused on the change of the geometrical and electronical structure of the diacetylene group in the monomer respectively polymer matrix during polymerization. In the following we discuss a model of the reaction mechanism and report some spectroscopic experiments which have been performed in this direction.

ACTION SPECTRA

A series of experiments was performed to study the polymerization mechanism (17,18,19). Especially the dependence of the rate of polymerization within single crystals as a function of the wave lengths of the incident monochromatic radiation and as a function of the particular chemical constitution of the endgroups R was measured and will be discussed in this paper. A selection of some photoresponse spectra, which contain the main characteristic results is given with the corresponding absorption of the same compounds in solution in figures 1, 2, 3 and 4.

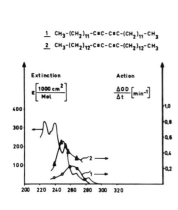

Fig. 1 Spectra of compounds 1,2

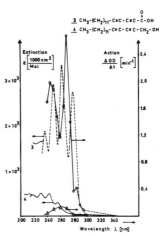

Fig. 2 Spectra of compounds 3,4

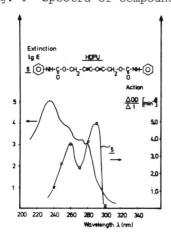

Fig. 3 Spectrum of compound 5

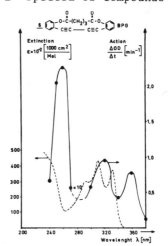

Fig. 4 Spectrum of compound 6

Photoaction of diacetylenes in the solid-state and uv-absorption of the same compounds in solution

As a most surprising result, the maximum of photoreactivity is in many cases not found in the absorption range of the triple bond system but rather in the absorption maximum due to the endgroups. Generally speaking, if a reactive modification of a given compound is found at all, the action spectrum resembles approximately a distorted form of the absorption spectrum. This can be unumbigiously seen if one compares the photoresponse spectra with the absorption spectra of the different diacetylenes in the figures 1, 2, 3 and 4. Although the spectra are not yet understood in all details it is obvious that in the case of compounds 1 and 2 with paraffin endgroups a maximum of photoreactivity can only be observed at about 250 nm since the aliphatic residues do not absorb at higher wave lengths. A similar result is also observed with compound 3. On the other hand the cyclic compound 6 as well as compounds 4 and 5 exhibit maxima of photoreactivity at wave lengths up to 360 nm, which clearly correspond to their absorption spectrum. One of the maxima of photoreactivity is generally found between 280 and 290 nm, that is in the long wave length tail of the absorption maximum of the conjugated triple bonds which is found around 250 nm in monomer solutions. Even if one assumes a bathochromic shift by going from the dissolved to the solid state, it seems possible that photopolymerization occurs also upon irradiation outside the absorption band of the isolated triple bond system. Therefore, the question arises whether photoreactivity is due to photochemical excitation of the isolated triple bonds at all or rather due to excitation of photoexcitable groups and radiationless transfer of excitation energy to the reaction center. To answer this question in a hypothetical manner attention should be given to the previously proposed mechanism (19) as shown in figure 5.

The only way to form a polymer with the indicated backbone structure is a reaction between C_1 and C'_4 of adjacent monomers which leads to a linear unsaturated polymer chain in the monomer matrix. Since in the polymer the translation period between successive substituents (R and R' or R' and R'' is 4.9 Å, polymerization is not possible if the space needed by the substituents is larger than 4.9 Å in chain direction. This polymerization mechanism thus clearly indicates the required packing properties of the monomers in order to form a polymer and correlates reactivity with monomer packing in a simple manner.

Following the proposed mechanism one must consider a small rotation of the diacetylene group around its center of gravity because the distance of the Van der Waals bonded monomers between C^1 and C^4 is shortened from around 3 Å to 1.36 Å to form a double bond. Although this movement has not yet been experimentally proved it is indicated by the small shortening of the lattice dimensions perpendicular to the polymer chain direction.

Since the diacetylene-group is linear and rodlike it must be bent in order to form bondangles of 120° which is required by the double bond in the polymer backbone.

Thus a coupling of electronic excitation with thermal vibrations of the crystal seems to be of basic importance and interpretation in terms of energy transfer in the crystalline materials is supposed to clarify the polymerization mechanism. As far as experimental evidence is concerned it is unambigious that energy transfer must take place in the case when the polymerization is initiated by photoexcited sidegroups, since the reaction takes place at the diacetylene groups as described above. Further on the observed absorption peaks at 600 nm in the very beginning and at 540 nm in the early stages of the reaction and the rather broad absorption peak of the final polymer centered around 490 nm or 560 nm depending on the substituents indicate that several stable intermediates appear during photopolymerization and disappear to give the final polymer at the end of the reaction.

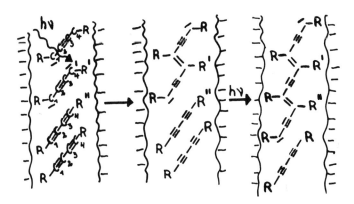

Fig. 5 Molecular mechanism of the solid-state polymerization of diacetylenes "carbene polymerization"

REFERENCES

1) A. v. Baeyer and L. Landsberg, Ber. dtsch. chem. Ges. 15,61 (1882)
2) F. Straus et al., Ber. dtsch. chem. Ges. 63, 1868,1886 (1930)
3) A. Seher, Liebigs Ann. Chem., 589, 222 (1954)
4) F. Bohlmann, Angew. Chem. 69, 82 (1957)
5) J. B. Armitage, E.R.H. Jones, M. C. Whiting et al. J. chem. Soc. (London) 1952, 1998
6) F. Bohlmann, E. Inhoffen, Ber. 89, 1276 (1956)
7) G. Wegner, 2. Naturforschg. 24b, 824 (1969)

8) G. M. J. Schmidt, Photochemistry of the Solid-State in: Reactivity of the Photoexcited Organic Molecule, Wiley New York 1967
9) E. Hädiche et al., Angew. Chem. Intern. Ed. 10, 266 (1971)
10) D. Kobelt, H. Paulus, Acta cryst. B30, 232 (1974)
11) A. J. Melveger, R. H. Baughman, J. Polymer Sci. A-2, 11 603 (1973)
12) G. Wegner, Makromol. Chem. 154, 35 (1972)
13) J. Kiji, J. Kaiser, R. C. Schulz, G. Wegner, Polymer (London) 14, 433 (1973)
14) R. H. Baughman, E. Turi, J. Polymer Sci., A-2, 11, 2453 (1973)
15) D. Bloor, F. H. Preston, D. J. Ando and D. N. Batchelder, in: Structural Studies of Macromolecules by Spectroscopic Methods, Ed. K. J. Ivin, J. Wiley & Sons (1975)
16) W. Schermann, G. Wegner, Makromol. Chem. 175, 667 (1974)
17) D. Bloor, D. J. Ando, F. H. Preston and G. C. Stevens, Chem. Phys. Letters, 24, 407 (1974)
18) D. Bloor, L. Koski, G. C. Stevens, F. H. Preston, D. J. Ando, J. Material Sci. 10, 1678 (1975)
19) G. Wegner, Paper presented at the 169th ACS-meeting, Philadelphia 1974

SOLID-GAS REACTIONS. PART V. BROMINATION OF ORGANIC SOLIDS

E. Hadjoudis

Department of Chemistry

Nuclear Research Center "Demokritos"

Athens, Greece

INTRODUCTION

The subject of organic solid-state reactions has been reviewed and Schmidt[1] stressed the high degree of topochemical control in photochemical processes the products of which are determined to an important extent by the structure of the crystal. We now explore the range and limitations of the topochemical concept in these two-phase, two-component reactions.

Schmitt[2] showed that solid cinnamic acid reacts with bromine vapour to give cinnamic acid dibromide (reaction 1).

$$\text{PhCH=CHCOOH} + \text{Br}_2 \longrightarrow \text{PhCHBr·CHBr·COOH} \quad (1)$$
$$\text{(solid)} \quad \text{(vapour)} \quad \text{(solid)}$$

Many examples of this type of reaction have since appeared[3-6], but the generality of such reactions was not established. In view of the experimental simplicity of the reactions and the possibility to obtain stereoselective products, it was of interest to establish its general applicability; we also hoped to find examples suitable for more detailed analysis of the various processes involved in reaction (1). Our previous results combined with the present data call for a mechanism which is presented here.

EXPERIMENTAL

The quantitative experiments in which rates of bromine uptake were followed as a function of time were carried out on samples (average weight 50mg) sieved to constant mesh size (350 mesh) and placed on a microscope coverslip. These samples were introduced into a modified wash bottle containing a standard amount of liquid bromine, and immersed in a thermostat at $25\pm0.2°$. The samples were removed after measured intervals and introduced into a Cahn R-G Electrobalance and weighed. The balance was attached to a water pump and the loss of weight during evacuation was displayed on a recorder. In a number of cases the rate of absorption of bromine was measured by means of a quartz spring balance. In these cases the reaction tube and the bromine container were immersed in a thermostat at the same temperature, keeping thus the vapour pressure of bromine saturated at the reaction temperature. The change in weight was always expressed as the molar ratio of organic solid:bromine.

Powder photographs were taken on a Guinier camera with Cu-Kα radiation.

RESULTS AND DISCUSSION

In our previous[5] work on the reaction between bromine vapour and solid olefins it was shown that the reaction is stereoselective. Thus the majority of compounds studied, cinnamic acids, stilbenes, dibenzoylethylenes and arylidene acetophenones (chalcones), yield products (dibromides) which are trans-adducts. We have found so far, as the only exceptions to the above rule the solid cis-stilbene and the solid cis-1,2-di-p-methyl-benzoylethylene yielding the same dibromide as the trans-isomers (meso-dibromides). This behaviour was attributed to a cis-trans isomerisation prior to or during bromine addition. The possibility of solid-state cis-trans isomerisation was demonstrated by treatment of olefins with iodine vapour, which was found to depend on the density of solid material[7,8].

Further, we made the hypothesis which viewed the reaction taking place through the formation of a "complex", as an intermediate step. This suggestion is supported by the appearance of colours during the bromination process and by the fact that light slows down the rate of reaction of gaseous bromine with solid cinnamic acids by destroying the intermediate "complex"[9]. The possibility of a weak complex between gaseous bromine and crystalline organic compounds is evidenced by the reaction between para-iodo-benzoic acid and gaseous bromine. Benzoic acids were chosen as the organic counterparts mainly for two reasons: (i) they are structurally close to cinnamic acids which had been studied before[5,9], but lacking the >C=C< group where bromine adds leading to stable dibromides and (ii) structural information and behaviour with ammonia gas were available[10]. The studied benzoic acids were the ortho, meta, and para-derivatives of the Cl, Br and I substituents. The interaction of these solids with gaseous bromine was negative except in the case of para-iodo-benzoic acid, which takes up bromine (1:0.5 M ratio) and it is shown in Fig. 1. The bromine uptake is completed at about 8 minutes and it is followed with a quartz spring balance. The colour o the solid acid from white turns to red-orange. The brominated compound can be degassed completely (turning again to white) either by suction with a water pump or by leaving the sample in the open air. The behaviour was the same when a single crystal of the acid was used instead of polycrystalline material.

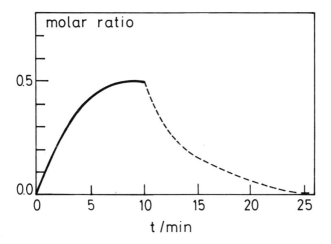

Fig. 1. Weight uptake of para-iodo-benzoic acid at RT. Broken curve corresponds to the degassing process.

It was desirable to clarify if the brominated red-orange product was a new phase and if the degassed white material was still para-iodo-benzoic acid or had been subjected to some change. In order to answer these two questions we took X-ray Guinnier photographs of the pure acid, the brominated and the degassed acid. An examination of the d-spacing of these photographs shows that in the brominated sample (1:0.5) new lines appear, suggesting the appearance of a new phase. The new lines disappear on degassing and the remaining lines have the same pattern with that of para-iodo-benzoic acid, showing thus a reversible process.

In our previous experiments we also had noticed that, while the stereo-structure of the product was independent of crystal structure of the organic compound, significant differences in rates of bromine uptake were shown by the crystal modification of dimorphic compounds and by chemically related compounds crystallising in different packing arrangements[5]. This is supported by new findings. Thus the two forms of para-chloro-chalcone (form-I, m.p.=114-115°, form-II, m.p.=111-112°) give the same dibromide (m.p.=180-183°) but with different rates of bromine uptake. The chemically related compounds pyridyl-2-, pyridyl-3-, and pyridyl-4-acrylic acid also show different rates of bromine uptake. The time needed for these solids to complete the reaction with gaseous bromine (1:1) is 3, 3½ and 6½ hours respectively (experiments similar to those described before[5]). The differences between rates of addition to the double bond in polymorphic forms or in closely related compounds are unlikely to be due to different rates of reaction of the molecular species, which therefore are attributed to differences in rates of bromine diffusion into the crystal lattice.

Other workers[3] have concluded that the reaction with bromine evidently takes place in an absorbed phase on the surface of the crystals, in a film of solution formed by the aromatic compound dissolved in liquid bromine, or a combination of both. Our experiments tend to indicate that these alternatives indeed exist though not necessarily in combination; in other words, a compound undergoes a true gas-solid reaction while another tends to form a liquid film of bromine in which an homogeneous reaction may possible take place. The true nature of the first step of the reaction can only be appraised by working with small specimens and short exposure time to bromine.

In conclusion we would summarize that our results point towards a mechanism in which gaseous bromine: (a) penetrate the solid, (b) forms a "complex", (c) adds stereoselectively by trans-addition. An abnormal direction of addition to some cis-compounds is due to cis-trans isomerisation prior to or during addition to erythro- or meso-dibromides.

ACKNOWLEDGMENT

The author would like to thank Mrss. C. Mallerou and S. Filippakopoulou for valuable assistance.

REFERENCES

(1) G.M.J. Schmidt, "Photochemistry of the Solid State" in "Reactivity of the Photoexcited Organic Molecule", Wiley, N.Y., 1967, p. 227.

(2) A. Schmitt, Annalen, 127, 319 (1863).

(3) R.E. Buckles, E.A. Hausman, and N.G. Wheeler, J.Am.Chem.Soc. 72, 2494 (1950).

(4) M.M. Labes, H.W. Blakesbee, and J.E. Bloor, J. Amer. Chem. Soc. 87 4251 (1965).

(5) E. Hadjoudis, E. Kariv, and G.M.J. Schmidt, J. Chem. Soc., Perkin II, 1056 (1972).
(6) R.S. Miller, D.Y. Curtin, and I.C. Paul, J. Am. Chem. Soc., 93, 2784 (1971).
(7) E. Hadjoudis and G.M.J. Schmidt, J. Chem. Soc., Perkin II, 1060 (1972).
(8) E. Hadjoudis, Isr. J. Chem., 11, 63 (1973).
(9) E. Hadjoudis, Isr. J. Chem., 12, 981 (1974).
(10) R.S. Miller, D.Y. Curtin, and I.C. Paul, J. Am. Chem. Soc., 93, 6340 (1974).

MECHANISMS OF SOME ROOM TEMPERATURE PHOSPHORESCENCE PHENOMENA IN DOPED SOLID POLYMER MATRICES

D.J. Morantz, C.S. Bilen and R.C. Thompson

London College of Printing, London, S.E.1. UK

ABSTRACT

Room temperature phosphorescence of certain doped crystalline, thermoplastic and thermosetting organic systems are compared. Results are given for a novel doped particulate amino-resin system. It is suggested that these phosphors depend on enhanced energy pathways to the triplet.

INTRODUCTION

Work on organic phosphor systems has emphasised photo-inert low temperature glasses. The theory is well established, accounting for the behaviour of emission parameters including intensity, lifetimes and energy transfer efficiencies, (1,2). Examples of room temperature phosphorescence are few and their explanation incomplete, (3,4). One of these, based on amino-phosphors, was developed for postal applications, (5). We have found a variety of room temperature phosphor systems including: thermoplastic resins with varying amorphous/crystalline ratios; doped single crystals of cyclododecane; and induced phosphors, prepared by pre-irradiation with UV or X-rays, of doped polymethylmethacrylate (PMMA), (6). We have investigated new thermosetting, particulate, materials based on doped amino-resins, obtaining intense room temperature phosphorescence. The induced phosphorescence systems appear to depend on transient radical species which enable energy absorbed by the matrix to be transferred, efficiently, to the dopant triplet. Such UV generated radicals in PMMA were identified using esr, (7) and the induced phosphorescence was correlated with thermoluminescence, (8). Our emission and absorption spectroscopy data for particulate doped amino-resin phosphors also suggest the presence of species (not radicals), other than dopants, involved in similar energy transfer.

EXPERIMENTAL

The particulate room temperature phosphor comprised a melamine-formaldehyde type resin, with carbazole sulphonic acid (4%) as primary dopant and para-aminobenzonaphthalene (1%) as secondary dopant. Other secondary dopants were also effective. Phosphorescence emission spectra of these systems were recorded using an emission spectrophotometer, reported earlier, (8), but with a rotating chopper incorporated to distinguish phosphorescence from total emission. Degradation of phosphorescence emission was studied by exposing samples to an unfiltered UV source (HBO 200 mercury lamp, AC operated). Cumulative changes in emission spectra were recorded over a range of time intervals until the phosphorescence had dropped to less than 10% of the initial peak intensity. Degradation of phosphorescence, for samples held at elevated temperatures (483 K for periods up to 115 hours) was also investigated.

RESULTS AND DISCUSSION

We recall that induced phosphorescence of PMMA was found to deteriorate on exposure to oxygen, prolonged UV or heat, consistent with the transient nature of the species responsible for room temperature phosphorescence, (8). Similarly the amino-resin particulate phosphor, exposed to sunlight or heat, showed deterioration but was not significantly affected by oxygen. UV exposure reduced the emission intensity in the phosphorescence region, 500 nm, whereas the fluorescence band, peaking at 390 nm, was relatively unaffected. Curves A and B of Figure 1 show total luminescence (fluorescence and phosphorescence) compared with the phosphorescence emission of the same samples; Curves C to H show the effects of

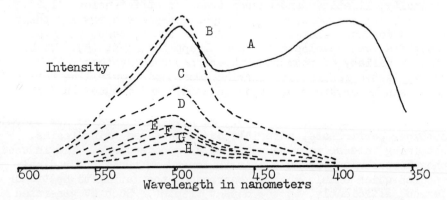

Figure 1. Emission spectra of doped particulate resin samples at various stages of UV irradiation. Curve A: total emission, Curve A: phosphorescence only. Curves C to H: increasing UV exposure of 1,3,5,7,10 and 20 minutes.

increasing UV exposure on the phosphorescence emission. Figure 2, curve A, shows the decrease of the normalised phosphorescence intensity with increasing UV irradiation time for a sample exposed at 160 mm from the source. Graphs of reciprocal phosphorescence intensity against exposure period result in linear plots suggesting the involvement of bimolecular processes. In contrast with thermal degradation, the UV degradation is found to be a surface rather than a bulk effect. Figure 3 shows the effect of storing particulate melamine-formaldehyde phosphor samples at 483 K. The fall in room temperature phosphorescence intensity with time suggests a rapid initial destruction of the species responsible for the energy transfer to triplet state levels. A closer examination of Figure 3, however, shows that prolonged heating after 20 hours resulted in a small restoration. A similar initial decay was also seen in earlier (unpublished) work for X-ray induced phosphorescence in doped PMMA. Samples stored at various temperatures (below the glass transition) showed that phosphorescence decreased more rapidly with increasing temperature.

The particulate aminoresin phosphor, at room temperature, gives a strong green afterglow, peaking at 500 nm and characteristic of the secondary dopant. In addition, a weaker phosphorescence shoulder is also apparent at 425 nm, corresponding to the triplet of the primary dopant. At 77 K only a low intensity blue afterglow (425 nm) remained. An anomalous decrease of phosphorescence with reduction of temperature was also observed for induced, room temperature phosphorescence in PMMA, (9), contrary to the _increase_ expected for the well known low temperature phosphor systems, (1). The existence of strong room temperature

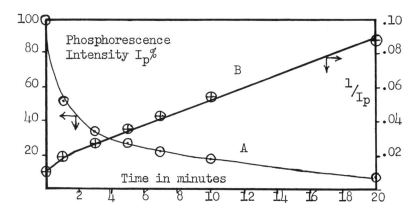

Figure 2. Curve A: Normalised phosphorescence intensity versus UV exposure. Curve B: Reciprocal of phosphorescence intensity versus UV irradiation time.

phosphorescence in the particulate phosphor may be accounted for, by analogy with the PMMA phosphor, as due to energy transferring species in the matrix. Moreover, this explanation is consistent with the gradual reduction of the green phosphorescence as the temperature is reduced, where there is a substantial energy gap between the 500 nm emission and the 366 nm excitation band, and the energy transfer enhancement of the triplet is quenched as the temperature is reduced. Transfer, over the smaller energy gap, from 366nm to the blue triplet at 425 nm is less affected, as we would expect in line with our explanation. These findings not only support the role of the energy transferring species, for populating the emitting triplet levels, but also underline the importance of thermal activation in populating the vibrational and rotational states of such species. Reduction of the temperature, thus, decreases the overlap of cascade levels and there is a consequent reduction in energy transfer efficiency. Thus, "enhanced" phosphorescence is diminished at lower temperatures.

The dopants used in this work, when dissolved in other, rigid, polymer matrices, do not phosphoresce at room temperature, due to non-radiative deactivation of the triplet. In the light of present results and earlier studies in these laboratories, (6,8), we propose that room temperature organic phosphorescence may take place when there are suitable energy transferring species, within the matrix system. These must provide energy to the triplet state of the dopant at a rate which is faster than that of natural non-radiative deactivation. Such "energy-transferring" species should have energy levels extend-

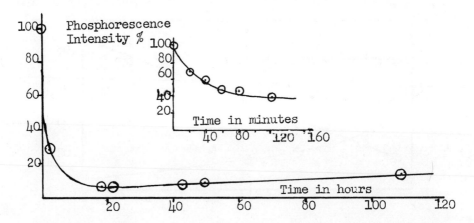

Figure 3. Effect of prolonged storage at 493 K on normalised phosphorescence intensity (main curve); small insert shows expanded time scale up to 120 minutes.

ing between the emitting triplet levels and the absorption levels of the matrix and/or the primary dopant. Furthermore, such species should not be significantly radiative, otherwise energy will be emitted, and not passed on to the dopant triplet state. Thus in the presence of such species, during UV excitation, sufficient numbers of triplet levels may be populated for room temperature phosphorescence to prevail, despite competing non-radiative deactivations. Species, suitable for transfer of energy to triplet levels may include trapped or transient radicals associated with certain matrices (e.g., PMMA); these are produced by irradiation, prolonged heating, or (under certain conditions) laser irradiation. In the case of the particulate phosphor, the species may arise in preparation, during catalysed polymerisation and cross-linking of doped intermediate resins. The decrease in room temperature phosphorescence, through intense UV irradiation or by heating of the particulate phosphor system, appears to be due to destruction of the trapped species. These species are not apparently radicals, and may be free chain ends with or without ionic character. The room temperature phosphorescence for doped crystalline and semi-crystalline hosts cannot be readily accounted for through the intervention of a matrix species. We observed a positive monotonic correlation between crystallinity and room temperature phosphor intensity for doped thermoplastic alcohol/acid systems and also freshly doped sorbitol. An explanation, which we believe will also require additional energy pathways, may also involve consideration of some characteristic properties of crystals.

Thus, for the systems we have considered, additional energy pathways to the triplet (and external to the singlet-triplet manifold), are required for strong room temperature phosphorescence. These may be provided by various energy transferring species, depending on the nature of the phosphor substrate.

We acknowledge an ILEA Research Fellowship to one of us (C.S.B.)

REFERENCES

1. G.N. Lewis and M. Kasha, J. Amer. Chem.Soc., $\underline{66}$, 2100, 1944
2. A.B. Zalhan, Ed. The Triplet State, Camb.Univ.Press.,Lond., 1967
3. C.F. Forster and E.F. Rickard, Nature, $\underline{197}$, 1199, 1963
4. G. Oster, N. Geacintov and A.U. Khan, Nature, $\underline{196}$, 1089, 1962
5. J.C. Harrison, et al, Proceeding of the Institute of Mechanical Engrs., $\underline{184}$ Part 11, 52, 1969/70.
6. D.J. Morantz and C.S. Bilen, 7th Int'l.Symp., "Reactivity of Solids", Bristol, 525, 1972.
7. P.K. Wong, Polymer, $\underline{15}$, 60, 1974
8. D.J. Morantz and C.S. Bilen, Polymer, $\underline{16}$, 745, 1975
9. C.S. Bilen and D.J. Morantz, Nature, $\underline{258}$, 66, 1975

REACTIONS IN VITREOUS SOLIDS

W. Vogel

Otto-Schott-Institute der Sektion Chemie

Friedrich-Schiller-Universität Jena (DDR)

INTRODUCTION

Increased knowledge of the activity and reactivity of vitreous solids has opened completely new fields of application over the past 20 years. As examples, mention may be made of the use of glass as an optical medium in high efficiency optics, as laser material, as a photosensitive or photochromic medium, as a light or wave conductor (e.g. in fibre optics), as an electric insulator or semiconductor and finally as glass ceramic with minimal thermal expansion coefficient having a high resistance or machinability analagous to that of metals.

All these developments have only become possible after better knowledge of the correlations between structure and properties of glasses by means of new and modern methods of investigation. A great number of authors have contributed to this, e.g. Bray with studies on nuclear resonance, (1,2), Maurer with studies on light scattering, (3), Porai-Koshits with studies on small angle X-ray scattering, (4,5), Kurkjian with Mössbauer-spectroscopic investigations (6), Douglas, Paul and Bates with spectroscopic investigations in the range of visible wavelengths (7,8), and Stevels, Konijnendijk, and Neuroth, by IR- and Raman spectroscopic investigations (9,10). A special part in the elucidation of microstructure formation processes and the resulting fine structure of glasses was played by electron microscopy (12).

RESULTS OF ELECTRON MICROSCOPIC INVESTIGATIONS INTO MICROSTRUCTURE FORMATION PROCESSES IN GLASSES

Electron microscopic studies of glasses of different composition have shown that, according to the network or crystallite theory, glass structures represent border-line cases. During the solidification phase to a vitreous solid microphase-separation processes take place in most glass melts. As a consequence most glasses are of micro-heterogeneous structure with a droplet-like glass phase being embedded in a glass matrix. In clear glasses the dimension of the droplet regions is 20 - 150 Å, in opal glasses it rises by a factor of 10 - 100.

The following results were obtained from glasses with relatively large microglass phases. Therefore details of the microstructure formation processes became more readily visible. Until now it has been generally accepted that glasses with microglass phase separation derive from melting systems with subliquidus-regions-of-segregation.

As the subliquidus-regions-of-segregation in a melting system are always to be found between well-defined chemical compounds, the microglass phase developed in an unmixing process show compositions which are more or less the same as those of well defined chemical compounds. This process is decisively influenced by the cooling speed of the melt.

Thermodynamic interpretation of unmixing phenomena in glasses

The theoretical treatment of unmixing processes in glasses as expressed by Cahn and Charles (13,14) as well as by Haller (15,16) was a considerable contribution to the understanding of these processes on the basis of thermodynamic considerations. It has to be pointed out, however, that due to experimentally demonstrable segregation structures the correlations seem to be much more complicated than can be expressed in a thermodynamic interpretation of an ideal case. A positive free mixing enthalpy ΔG_M is a thermodynamic prerequisite for segregation.

Gibbs' necessary stability criteria, were applied to vitreous systems for the first time by Cahn and Charles. For a system with a miscibility gap they are:

Coexistence boundary between
stability and metastability:
$$\left(\frac{d^2 \Delta G_M}{dx^2}\right)_{p,T} > 0$$

region of instability:
$$\left(\frac{d^2 \Delta G_M}{dx^2}\right)_{p,T} < 0$$

spinodal (boundary between
metastable and instable regions):
$$\left(\frac{d^2 \Delta G_M}{dx^2}\right)_{p,T} = 0$$

Figure 1 shows the free mixing enthalpy ΔG_M and the resulting miscibility gap of a binary system. Because of the above mentioned stability criteria, mixtures within the coexistence curve and the spinodal are metastable, and below the spinodal are unstable with respect to thermodynamic fluctuations.

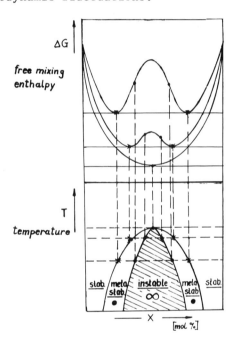

Figure 1. Free mixing enthalpy ΔG_M as a function of glass composition and the resulting miscibility gap in a binary system.

Phase separation in the sense of a nucleation and growth mechanism takes place in the metastable region of the miscibility gap. Fluctuations raise the free enthalpy and do not lead to a phase separation. In order to get from the metastable to the stable state an energy barrier in form of the nucleation energy for the formation of a phase boundary surface has to be overcome. After reaching a critical size the nuclei grow under a decrease of the free enthalpy. In this way, a droplet phase develops in the matrix phase in glasses. A typical example is shown in Figure 2.

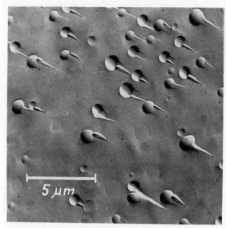

Figure 2. Electron micrograph of a barium borosilicate glass after vacuum fracture and replica preparation. After a simple unmixing process a droplet phase developed in a matrix phase

In the unstable or spinodal range fluctuations lead to a decrease of the free enthalpy and thus the thermodynamic conditions for spontaneous phase separation automatically arise.

By solving the diffusion equation for the spinodal range Cahn and Charles found that the growth of this fluctuation leads to the formation of a composite structure. According to Haller and coworkers (15,16) as well as according to our own research, a composite structure also occurs due to the growth process of the droplet if the volumes change correspondingly. Figure 3 shows a typical composite structure. Two continuous phases can clearly be seen.

Figure 4 shows three glasses, whose compositions lie to the left, exactly at the maximum, and to the right hand side of the unmixing maximum respectively. The change of the functions of the microphases is clearly to be seen

Kinetic interpretation of unmixing phenomena in glasses

Up until now only the formation of two coexistent phases has been discussed, representing the result of a simple primary unmixing process. Not until 1958 was an unmixing process in glasses discussed as taking place in two steps (17). This was proved experimentally

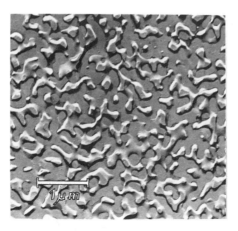

Figure 3. Electron micrograph of a sodium borosilicate glass made visible by a single step replication method. The micrograph shows a typical composite or penetration structure. In order to make the phase boundaries more distinct the glass was treated with water for a short time.

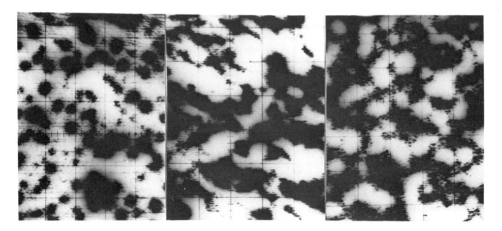

Figure 4. X-ray BaLα scanning micrograph of a barium borosilicate glass whose composition lies (a) to the left, (b) at the maximum, (c) to the right of the unmixing maximum. A change of the functions of the microglass phases can be seen, ie the droplet phase takes over the matrix function depending on the composition of the glass. The maximum of unmixing shows the typical penetration structure.

in 1958/59 (17,18,19,20). According to this model glasses consist of three microglass phases, two droplet phases of very different dimensions embedded in a matrix phase. Subsequently other examples of secondary unmixing effects and of three phase glasses could be demonstrated experimentally. Already these examples showed that the structure formation processes taking place even when the simplest liquid glass melts are cooled, are much more complicated than expected, because kinetic processes are of decisive importance in them.

Thus after stepwise unmixing, up to 8 different microglass phases could be demonstrated by us for the first time in the $BaO-B_2O_3-SiO_2$-system by electron optics (21,22,23). Figure 5 shows 6 different microglass phases.

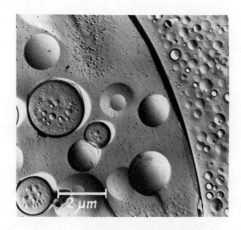

Figure 5. Electron micrograph of a barium borosilicate glass made visible by the replication method. After stepwise unmixing, 6 microglass phases of different compositions were formed.

Figure 6 shows the diagram of a stepwise unmixing process in glasses which leads to the formation of more than two microglass phases. Due to a primary unmixing process two phases of different viscosity develop. During the cooling process these behave as if they were separate systems with different unmixing processes. An equalisation of concentration in the system as a whole in accordance with the co-nodes becomes very complicated and is possibly only in the separated primary phases. This leads to suppression nonequilibria or to local equilibria phases.

The diffusion processes during equalisation of concentration can lead to shell formation around the microglass phases. This can

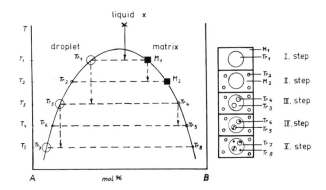

Figure 6. Diagram of a stepwise unmixing process in glasses which leads to the formation of more than two microglass phases. On the right hand side: Diagram of the resulting microstructures which can be clearly demonstrated by electron optics.

be demonstrated experimentally. Thus Figures 7 and 8 clearly show the diffusion courts around the big droplets. As the temperature of the glass melt was falling a limited ion diffusion was still possible in the matrix phase. The necessary equalisation of concentration according to the co-nodes of an unmixing dome was possible only within the courts. The components diffusing in the direction of the big droplets were deposited as a shell on the droplets, as these were already solidified. Thus an SiO_2-shell developed. Further away from the big droplets the SiO_2-part has undergone secondary separation from the matrix in the form of little droplets.

It is also possible for the diffusion zone to be within the droplet. This is the case when the melting composition is on the other side of the unmixing dome and when the phase in which diffusion processes readily occur has a droplet rather than matrix. Figure 9 demonstrates such a process. Due to the diffusion from the big droplet in the direction of the matrix, a diffusion court has developed within the big droplet.

At the same time another diffusion court can be seen around the secondarily formed droplet phase C in the big droplet. This can be readily understood because the matrix phase A and the secondarily formed droplet phase C (according to the scheme in Figure 4) are similar but not identical.

Multiple unmixing phenomena can also appear in glasses which are at the maximum of the unmixing dome and which show primarily penetrating structures. This can be seen in Figure 10. Both penetrating phases A and B are secondarily unmixed.

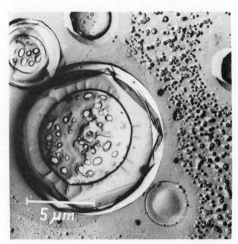

Figure 7. Electron micrograph of a barium borosilicate glass made visible by the replication method, following HNO_3-etching. In the matrix phase diffusion courts are clearly visible around the droplet phase. The big droplet phase is surrounded by a SiO_2-shell.

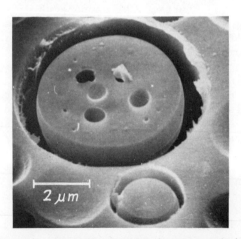

Figure 8. Electron micrograph of a barium borosilicate glass made visible by the replication method, following HF-etching. The SiO_2-shell is dissolved and the droplet phase nearly isolated.

REACTIONS IN VITREOUS SOLIDS 513

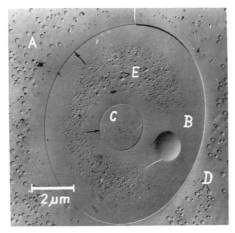

Figure 9. Electron micrograph of a barium borosilicate glass with multiple separation. The droplet phase is the phase where place exchanges are more readily achieved. This is shown by the stepwise separation of the big droplet into four different phases and by the diffusion courts in the big droplet.

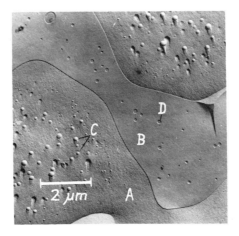

Figure 10. Electron micrograph of a barium borosilicate glass made visible by the replication method. The glass composition is at the unmixing maximum and shows a penetration structure. In both the primary microglass phases secondary unmixing processes has occured.

Such microstructure as for instance in opal glasses have become completely intelligible only after the process as a whole has been elucidated to a high degree.

Distribution and incorporation of heavy-metal ions into unmixed base glasses (24,25,26,27)

According to present understanding of unmixing phenomena in glasses derived both from experimental work and theoretical considerations another question must be answered. It is the question concerning the distribution of small additions of 3d-elements (e.g. Co, Ni, Cu and other colour indicators) on the microglass phases of a base glass. This is relevant to such practical topics as the development of colour glasses, laser glasses, photosensitive or photochromic glasses, of glass ceramics etc. In all cases a slight addition of certain ions to the base glass produces an enormous effect, and leads to the development of a new property of the glass.

Concerning the distribution of e.g. Co-, Ni- or Cu- ions in the microglass phases of a base glass, certain distibution equilibria could have been expected. Experimental findings showed a completely different result. All glasses studied showed almost 100% enrichment of colour indicators in the network modifier rich microglass phase. Figures 11 and 12 may serve as examples. They show barium borosilicate glasses.

Figure 11 shows an electron back-scattering micrograph of a barium borosilicate glass doped with NiO (1.89 Mol%) consisting of 4 microglass phases. A barium borate glass phase ($\simeq$ BaO $4B_2O_3$) is embedded in the form of large droplets into a B_2O_3/SiO_2-matrix phase. In the droplet phase smaller B_2O_3/SiO_2-droplets separated immediately due to secondary unmixing. Their composition is not completely identical with the matrix phase. The seperation of a second very small droplet phase also took place in the matrix phase after a secondary unmixing process, which is neglected here because a micro probe examination cannot be carried out due to the very small size of the droplets.

Figure 11 also shows the $Ba_{L\alpha}$ X-ray intensity curve. The electron ray was led along the white horizontal line across the sample. The intensity curve verifies the above mentioned data concerning the Ba-distribution in the micro phases. While the Ba-intensity of the matrix is very low, it rises sharply in the large droplet phase and decreases in the same way in the small droplet phase. The intensity value, however, is a little higher than in the matrix according to the stepped unmixing mechanism as shown in Figure 6.

Figure 12 shows the same electron back-scattering micrograph with the superimposed $Ni_{K\alpha}$ X-ray intensity curve. The course of

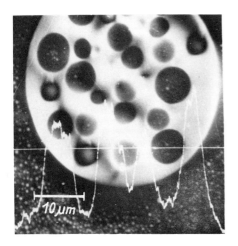

Figure 11. Electron back-scattering micrograph of a barium borosilicate glass doped with NiO, with superimposed $Ba_{L\alpha}$ X-ray intensity curve. The electron ray was led along the white horizontal line over the sample. The intensity curve shows the enrichment of barium in the large droplet phase.

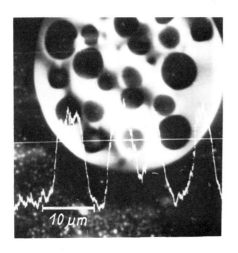

Figure 12. Electron back scattering micrograph as in Figure 11 with superimposed $Ni_{K\alpha}$ X-ray intensity profile. The X-ray intensity curve clearly shows the complete enrichment of nickel ions in the BaO-rich droplets.

the curve corresponds to a high degree to that of the Ba-intensity profile. Ni was found neither in the matrix nor in the small droplet phase.

In further experiments two barium borosilicate glasses, in which the barium glass phase has a matrix or a droplet function, were doped with CoO. In both cases cobalt enrichment took place in the basic glass phase. In one case, colourless droplets could be seen in a deep blue matrix glass phase. In the other case deep blue droplets were embedded in a colourless matrix glass phase, as could be demonstrated by light microscopy.

ON FURTHER INFORMATION OF MICROSTRUCTURE FORMATION PROCESSES IN SOLIDS

In spite of complete transparency, in most cases optical or technical glasses have a microheterogeneous structure. The dimension of the microglass phase however is small and therefore it is often difficult to explain the formation processes. For the electron-optic studies described, those glass systems were selected which have a strong tendency towards unmixing and whose structure formation processes can easily be studied. This applies to certain model glasses. According to Goldschmidt, beryllium fluoride are doubly weakened models for silicate glasses. In these the chemical binding of the glass building units is reduced to 1/4 of that in the original glasses, the silicate glasses. Such glasses provided a lot of information which could not be gained from silicate glasses. The results allowed conclusions on structure formation processes, which had obviously already started in the melt. Therefore it is absolutely necessary to find possibilities and methods which allow the study of formation processes and reactions of the glass components already in the liquid melt. This applies to study of the early stages of the unmixing processes as well as of nucleation and crystallisation processes.

Another question remains to be answered in glass structure research: Until now the phenomenon of microglass phase separation in glasses with existence of subliquidus-unmixing domes in the respective thermodynamic equilibrium has been discussed. Numerous examples demonstrate however, that under rapid cooling even melts of purely eutectic systems solidify as microheteregeneous glasses i.e. as glasses consisting of two glassy microglass phases

Numerous results of phase separation in glasses gained over the last 20 years show that phase separation is already possible when, in a solidifying glass melt, two stable structural elements with different volumes are formed. Quite a few phenomena can be explained in such a way

FINAL REMARKS

Recent results in glass structure research have lead to a new structural concept of glasses. New knowledge gained from a broad experimental basis has considerably improved the understanding of sometimes barely understandable variations in the properties of glasses during their production. They are a valuable basis for the manufacture of glasses with completely new properties. Measures for the control of phase separation in glasses in the sense of promotion or suppression are prerequisites for improving the chemical stability of chemical glass or for the production of Vycor glasses. A decrease of phase separation is a prerequisite for the increase of light transmittance of optical glasses, because the losses due to scattered light are decreased. An enrichment of certain additional components in microglass phase lowers the degree of work necessary for nucleation and crystallisation. Thus controlled crystallisation, i.e. the development of new glass ceramic materials becomes possible.

In spite of the great achievements of the last 25 years in glass research it is necessary to state that we are still at the beginning compared with the results of other fields of research in solids.

REFERENCES

1. P.J. Bray, J.G. O'Keefe.Phys. Chem. Glasses 4, 37, 1963
2. P.J. Bray, Wiss. Zeitschrift der FSU, Jena 23, 267, 1974
3. R.D. Maurer J. Chem. Phys. 25, 1206, 1956
4. E.A. Porai-Koshits, N.S. Andreev.Nature 182, 335, 1958
5. E.A. Porai-Koshits.Phys. Chem. Glasses 16, 385, 1975
6. C.R. Kurkjian J. Non-Cryst. Sol. 3, 157, 1970
7. A. Paul, R.W. Douglas Phys.Chem. Glasses 10, 133&138, 1969; 6, 207, 1965
8. T. Bates.Modern Aspects of the Vitreous State. Ed. J.D. Mackenzie. Vol.2, p195. Butterworth London 1962
9. W.L. Konijnendijk, J.M. Stevels.J. Non-Cryst. Sol. 20, 193,1976
10. W.L. Konijnendijk.Philips Res. Rep. Suppl. No 1. 1, 1975
11. N. Neuroth.Glastechn Ber. 41, 243, 1968
12. W. Vogel.Struktur und Kritallisation der Gläser. VEB Deutscher Verlag für Grundstoffindustrie. Leipzig. 1st. Auflage 1965 2nd Auflage 1971
13. J.W. Cahn, R.J. Charles,Phys. Chem. Glasses 6, 181, 1965
14. J.W. Cahn. Trans. Inst. Min. Met. 242, 166, 1968
15. W. Haller. Chem. Phys. 42, 686, 1965
16. G.R. Srinvasan, I.Tweer, P.B. Macedo, A. Sarkar, W. Haller.J. Non-Cryst. Sol. 6, 221, 1971
17. W. Skatulla, W. Vogel, H. Wessel. Silikattechnik 9, 51, 1958
18. W. Vogel Silikattechnik 9, 323, 1958
19. W. Vogel Z. Chem. 3, 271, 1963
20. W. Vogel Silikattechnik 15, 282, 1964
21. W. Vogel, W. Schmidt, L. Horn Z. Chem. 9, 276, 1969
22. W. Vogel, W. Schmidt, L. Horn Z. Chem. 9, 401, 1969
23. K. Gerth, A. Rehfeld Silikattechnik 20, 227, 1969
24. W. Vogel, A. Rehfeld Silicates Industriels 5, 1, 1967
25. W. Vogel, A. Rehfeld , H. Ritschel Symp. on Coloured Glasses ICG. Prague 1967. Comptes Rendues 114
26. W. Vogel Lecture. Annual Meeting of the Belgian Chemical Society Brussels 1973
27. W. Vogel Wiss. Zeitschrift der FSU, Jena 23, 241, 1974

THERMAL DECOMPOSITION OF RHODIUM OXIDE GELS

E.Morán(*), M.A.Alario-Franco(**), J.Soria(***) and
M. Gayoso(*)

(*) Universidad Autónoma de Madrid, Departamento de
Química Inorgánica. Cantoblanco. Madrid. Spain

(**) Instituto de Química Inorgánica "Elhuyar"
C.S.I.C. Serrano 115. Madrid. Spain

(***) Departamento de Catálisis del C.S.I.C.
Serrano 115. Madrid. Spain

INTRODUCTION

The system $Rh-O_2-H_2O$ is a very interesting one showing no less than six phases, five of these are crystalline and consequently, well characterized: RhO_2 (1,2), rhombohedral and orthorhombic Rh_2O_3 (3), high pressure Rh_2O_3 II (4) and RhOOH (5).

Rhodium oxyde pentahydrate is an "amorphous" substance of a fairly constant composition quoted as $Rh_2O_3 \cdot 5H_2O$. As with most other amorphous gels our knowledge of this one is quite imperfect and the aim of this work was to try to characterize this material as well as to study its thermal decomposition under different conditions. Although Rh(III) is always diamagnetic (6), EPR spectroscopy was used in order to find out whether rhodium in oxidation states different from that one was present, by analogy with other gel systems (7)

EXPERIMENTAL

Materials

Rhodium oxide hydrate gel was prepared at room temperature by mixing, under controlled conditions, a Rh(III) solution with another of potassium hydroxide. The gel obtained was washed until no potassium could be detected (3). It was then filtered to a yellow gel which dried to a temperature of 100°C in air became a brown mass of composition $Rh_2O_3 \cdot 5H_2O$.

Two different solutions of Rh(III) were used and both were obtained by disolving rhodium metal in melted potassium bisulphate and dissolving the obtained mass on hot water. The metal used was, in one case, 99,9% rhodium metal from RIC/ROC, in the other it was from an unknown source and certainly of lower purity.

Methods

The thermal decomposition experiments were performed on a Mettler Thermoanalyzer instrument in still air, vacuum and dynamic atmospheres of oxygen, nitrogen and water vapour. The heating rate was commonly 10°C/min and calcined alumina was used as reference material.

X-ray diffraction was effectuated on a Debye-Scherrer camera, with molybdenum $K\alpha$ radiation filtered with zirconium on a Phillips PW1051 instrument.

Electron microscopy and diffraction were performed on a Siemens Elmiskop 102 instrument with the corresponding samples deposited on carbon coated grids.

EPR experiments were made on a Jeol "JES-PE" spectrometer with a modulation of 100 KHz.

RESULTS

Although the gel was "amorphous" to the usual X-ray diffraction with copper radiation, molybdenum radiation did show, after some 10 hours exposure, the presence of three very weak lines corresponding to spacings of 3,34; 2,21 and 1,92 Å. By means of electron microscopy it was seen that the gel was, in fact, a two-phase mixture in which the, by far, more abundant component was an amorphous solid made up by clusters of very small particles together with some very small single crystals which, so far, we have not been able to indentify.

The BET surface area of the pentahydrate sample was 130 m^2/g and the presence of hysteresis in the isotherm suggested the existence of porosity.

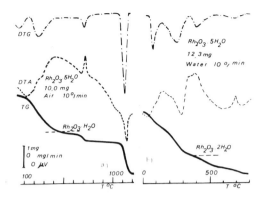

FIG. 1: Thermograms of $Rh_2O_3 \cdot 5H_2O$, a) in air, b) in water vapour

Thermal decomposition results

We have plotted in Fig. 1 a) and b) the TG, DTG and DTA plots of $Rh_2O_3 \cdot 5H_2O$ in air and water, respectively. It can be seen that there are several steps each one characterized by a change in the slope of the weight loss plot and a thermal effect.

It can also be appreciated from the weight loss data that a monohydrate and a dihydrate seem to be stable products. Both the dihydrate and the monohydrate were black, and morphologically and crystallographically similar to the brown initial pentahydrate, that is, they were essentially amorphous with a small amount of crystalline component.

EPR Results

Spectra were obtained on the three different hydrates as well as in anhydrous Rh_2O_3.

$Rh_2O_3 \cdot 5H_2O$: At room temperature, the spectrum seems to be formed by two superposed signals: signal I, a rather symmetric line with $\Delta H \simeq 800$ gauss and $g \simeq 2,07$, and signal II an asymmetric line (Fig. 2 a). At liquid nitrogen temperature only signal II was observed. This being a line typical of a paramagnetic ion on a field of axial symmetry, in this case $g_\perp = 2,17$ and $g_{||} \simeq 2,05$ (Fig. 2 b).

Also $Rh_2O_3 \cdot 5H_2O$ did show a similar spectra at room temperature, that is a line with essentially the same characteristics of signal I above: however at liquid nitrogen temperature, this signal dissapeared and no other could be observed.

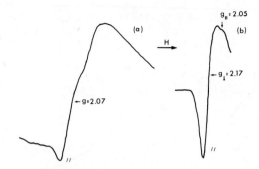

FIG. 2: EPR spectra of $Rh_2O_3 \cdot 5H_2O$. a) room temperature b) liquid N_2 temperature.

On the other hand no signal could be detected for the crystalline anhydrous Rh_2O_3.

DISCUSSION

The TG results indicate that, in air as well as in vacuum, the decomposition happens in three steps:

$$Rh_2O_3 \cdot 5H_2O \rightarrow Rh_2O_3 \cdot H_2O + 4H_2O \qquad (I)$$

$$Rh_2O_3 \cdot H_2O \rightarrow Rh_2O_3 + H_2O \qquad (II)$$

$$Rh_2O_3 \rightarrow 2Rh + 3/2\, O_2 \qquad (III)$$

However, in the presence of flowing water vapour it seems to happen according to:

$$Rh_2O_3 \cdot 5H_2O \rightarrow Rh_2O_3 \cdot 2H_2O + 3H_2O \qquad (IV)$$

$$Rh_2O_3 \cdot 2H_2O \rightarrow Rh_2O_3 + 2H_2O \qquad (V)$$

The only effect of a flowing oxygen atmosphere being a slight delay of process (III), while in a nitrogen atmosphere the thermograms were practically identical with those obtained in air.

Both the DTG and DTA results show that reaction (I) above does happen in, at least, two steps which are both characterized by endothermic effects. However, reaction (II) is a clearly exothermic process. All this seems to be an indication than the water is bonded in three different ways. On the other hand, the exothermicity of process (II) is influenced by both heating rate and sample history; it is nevertheless always exothermic and commonly very much so; it is, in this respect, analogous to those proceses known for other gels and which are denominated as the "glow phenomenon".

Work in other gel systems, e.g. Cr_2O_3 gels (7), has shown that chromium(IV) in the form of CrO_2 is formed in the calcination of the hydrated chromic gels (8), so it was thought that in view of the analogies existing between Cr(III) and Rh(III) (8) a similar situation could operate on the rhodium gels. In this respect, Rh(IV) and Rh(II) were the most likely candidates. Signal II characteristic of the pentahydrate is similar to those found for Rh(II) in an AgBr matrix (9) and also in $ZnWO_4$ (10). On the other hand, the only published data of Rh(IV) that we are aware of (11), place its signal at $g \simeq 1,7$. Consequently we do attribute signal II to Rh(II) located in an axial field with the unpaired electron delocalized from the $4d_{z^2}$ level of the rhodium atom. On the other hand, the fact that this signal does not change with decreasing the temperature is indicative of a stable environment. The presence of Rh(II) on the samples can be attributed to incomplete oxidation of the metal by the potassium bisulphate in the solubilization process.

However, as we have seen above, Fig. 2, signal II dissapears on heating the sample in air at 250°C (conditions under which one can obtain the mono- or dihydrate depending on the atmosphere surrounding the sample). This can be attributed to the oxidation to the more stable Rh(III).

Signal I, on its hand, observed in all the samples except the crystalline anhydrous Rh_2O_3, value of $g \simeq 2,07$ is close to $g_{iso} = 1/3 \ (g_{||} + 2g_{\perp})$ of Rh(II). The lineshape does not change with heating but the signal dissapears at low temperatures. This behaviour can be attributed to a change in the surrounding field at liquid nitrogen temperature, this producing a broadening of the line that makes it undetectable. It appears then that Rh(II) is present in both a well ordered and a disordered environments which can perhaps be identified with the crystalline and amorphous materials seen on the electron microscopic observations.

If this reasoning is valid it is interesting to remark that the Rh(II) present in the well crystallized material which produces signal II, seems to be easier to oxidize than the Rh(II) present in the amorphous component responsible of signal I, which does not dissapears until after the glow. This is a somewhat surprising result in view of the common connection between the amorphous character of some materials and their reactivity (12).

In the same context it is interesting to point out that the monohydrate, being amorphous, is still stable up to $\sim 700°C$ in air. Crystalline RhOOH, with the same stoichiometric composition, does decompose at a clearly lower temperature and in a very different way:

$$RhOOH \xrightarrow{\sim 480°C} RhO_2 \xrightarrow{\sim 900°C} Rh_2O_3$$

which is reminiscent of the behaviour of orthorhombic CrOOH (13).

The fact that the sample is always crystalline when anhydrous, that is, after the glow, suggest that this is due to a short range-order to a long-range order tranformation in which an important amount of the lattice energy of rhombohedral Rh_2O_3 is released.

We have not seen any trace of the rhombohedral to orthorhombic Rh_2O_3 phase transformation which has been reported as sluggish (2,3).

The rhombohedral material always decomposed endothermally to the metal plus oxygen before the transformation at a constant heating rate of 10° to 25°/min. Nevertheless, we did obtain the orthorhombic sesquioxide by oxidation of rhodium metal during 72 hours at 1000°C; however this product was still contaminated with up to 10% of unreacted rhodium.

REFERENCES

(1) R.D. SHANNON, Solid State Comm. 6, 139-143 (1968).
(2) O. MULLER and R. ROY: J. Less-Common Metals: 16 129-146(1968).
(3) A. WOLD, J. ARNOTT and W. CROFT: Inorg. Chem. 5 972 (1963).
(4) R.D. SHANNON and C.T. PREWITT: J. Solid State Chem. 2 134-136 (1970).
(5) J. CHENAVAS, These, Grenoble (1973). J. Solid State Chem. 6 1-15 (1973).
(6) S.E. Livingstone: "The platinum metals" in Comprehensive Inorganic Chemistry, J.C. BAILAR et al (Editors)Pergamon, Oxford 1973.
(7) M.A. ALARIO-FRANCO, J. FENERTY and K.S.W. SING: Reactivity of Solids, 7 327 (1972), J.S. ANDERSON et Al (Editors) Chapman and Hall, London (1973).
(8) F.A. COTTON and G. WILKINSON, Advanced Inorganic Chemistry, 3rd edition. John Wiley, New York (1973).
(9) R.S. EACHUZS and R.E. GRAUBS: J. Chem. Phys. 59(4) 2160 (1973).
(10) M.G. TOWSEND: J. Chem. Phys. 41 3149 (1964).
(11) I. FELDMAN, R.S. NYHOLM and WATTON: J. Chem. Soc. 4724 (1965).
   A.D. WESTLAND: Phys. Letters 32A 30 (1970).
(12) H. REMY: Treatise on Inorganic Chemistry. Elsevier, London (1967).
13) M.A. ALARIO-FRANCO and K.S.W. SING, J. Thermal Analysis 4, 47 (1972).

POLYMERIZATION EFFECTS DURING THE CRYSTALLIZATION OF SILICATE GLASSES

J. Götz, D. Hoebbel,* and W. Wieker*

Joint Laboratory for Silicate Research of the

Czechoslovak Academy of Sciences and the

Chemical University Prague, Czechoslovakia

*Central Institute for Inorganic Chemistry of the

Academy of Sciences of the German Democratic

Republic, Berlin, GDR

During recent years, much indirect evidence has been gathered indicating that many silicate glasses may contain polymeric groupings consisting of silicon and oxygen atoms. To study the configuration of these silicate groupings, three chemical methods have been developed, which enable the direct determination of silicate units in crystalline and glassy silicates.

The molybdate method measures the speed of formation of β-molybdatosilicic acid; this speed characterizes the type of silicate anion. A special procedure for paper chromatography of silicate anions minimizes the possibility of polycondensation or hydrolysis. (1,2). Finally, the recently described method of direct trimethylsilylation converts silicate anions into stable TMS-derivatives, which may be separated by GLC chromatography (3,4,5).

All three methods have been tested on crystalline silicates with known constitutions. This has helped to improve experimental proceedure and to learn more about the special features of the individual techniques. It was found that the molybdate method alone is not suited for systems containing more than two different silicate groupings. Paper chromatography does not separate the linear anions $[Si_2O_7]^{6-}$, $[Si_3O_{10}]^{8-}$, $[Si_4O_{13}]^{10-}$ etc. The TMS method fails to detect higher polymerized silicate groupings e.g. polysilicates or phyllosilicates. However, by simultaneous application of all three methods it is possible to eliminate many inherent shortcomings of the individual methods and to obtain new information on the configuration of silicate groupings in such crystalline and glassy materials which are soluble in 0,1 N NCl.

According to new X-ray identifications (6), the lead silicate $2PbO.SiO_2$ crystallises in four polymorphs; rapid quenching of the corresponding melt easily produces the glassy $2PbO.SiO_2$. Heat treatment of this glass develops the low temperature modification $T-Pb_2SiO_4$, which converts later to the first medium temperature modification $M_1-Pb_2SiO_4$. Higher temperature turns $M_1-Pb_2SiO_4$ into the second medium temperature phase $M_2-Pb_2SiO_4$ which at the highest temperatures yields the stable $H-Pb_2SiO_4$. The mutual relationship between these polymorphs is expressed by the following scheme:

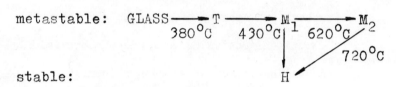

Silicate anion analysis demonstrates that this conversion of the glass via several metastable crystalline modifications into the stable high temperature phase H is connected with polymerisation of silicate anions (7). The glassy $2PbO.SiO_2$ is characterised by a whole array of very different silicate units, in which no single type prevails. Thermal vibrations induced by heat treatment of the glass cause fragmentation of the largest silicate groupings and thus promote the formation of dimeric units $[Si_2O_7]^{6-}$. These polymerise into tetrameric cycles $[Si_4O_{12}]^{8-}$ as the T phase turns into the M_1 phase. The second medium temperature modification M_2 is an intermediate between M_1 and the stable H phase, which contains predominately high molecular polysilicate chains $[SiO_3]_n^{2-}$ (fig 1).

So far, no information on the details of polymerisation is available. It is probably complicated and may include fragmentation and various polymerisation steps. Nevertheless the general tendency outlined above, towards the formation of higher polymerised silicate groupings has been confirmed by our recent results on the crystallisation of glassy $3PbO.SiO_2$ (8). As for the structure of the glasses studied, our results support a concept suggested by Hägg (9), according to which glassiness is connected with the existence of very many structural groupings of different size and shape. In such a system the formation of a three-dimensional crystalline lattice is hindered and consequently glassiness promoted.

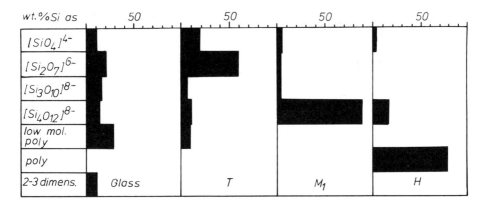

Fig. 1 Distribution of silicate anions in $2PbO \cdot SiO_2$.

REFERENCES

1. E. Thilo, W. Wicker and H. Stade, Z.anorg.allg.Chem. 340, 261, 1965
2. W. Wieker and D. Hoebbel, Z.anorg.allg.Chem. 366, 139, 1969
3. J. Götz and C.R. Masson, J.Chem.Soc. A, 2683, 1970
4. J. Götz and C.R. Masson, J.Chem.Soc. A, 686, 1971
5. F.F.H. Wu, J. Götz, W.D. Jamieson and C.R. Masson, J.Chromat. 48, 515, 1970
6. R.M. Smart and F.P. Glasser, J.Am.Ceram.Soc. 57, 378, 1974
7. J. Götz, D. Hoebbel and W. Wieker, J.Non-Cryst.Solids 1976 in press
8. J. Götz, D. Hoebbel and W. Wieker- to be published
9. G. Hägg, J.Chem.Phys 3, 363 1935

THE USE OF LASER RAMAN SPECTROSCOPY IN THE STUDY OF THE FORMATION OF OXIDE GLASSES

H. Verweij and H. van den Boom

Philips Research Laboratories

Eindhoven, The Netherlands

SUMMARY

Laser Raman Spectroscopy (LRS), which has been found to be a powerful tool in the study of the structure of glasses (1, 2, 3), has been used to investigate the process of the formation of glassy material from batch components. The model glass system which is considered in this paper is a $30K_2O.70SiO_2$ (molar composition) glass, formed from a batch containing 30 mole percent K_2CO_3 and 70 mole percent SiO_2 (quartz). Raman spectra were recorded of samples taken from the incompletely reacted batches. Glassy and crystalline intermediate products were identified by the well-known "finger print" method.

INTRODUCTION

The investigations presented in this paper are part of a research program in which reactions during glass melting are studied. The melting process can be roughly divided into three parts: Reaction of the batch components, fining (making the glass bubble-free), homogenisation. In order to achieve refinement of the glass (which is a very important part of the process) a "fining agent" is usually added to the batch. Frequently used refining agents are: Na_2SO_4, sometimes in combination with carbon, and As_2O_3 and Sb_2O_3, sometimes in combination with nitrates.

Few direct techniques are available to follow the glass-forming process. Of the direct techniques other than LRS, only

X-ray diffraction has been extensively used. This technique has the disadvantage that only crystalline materials can be identified, whereas in the glass-forming process both crystalline and glassy intermediate products are formed. Furthermore it is impossible, using X-ray diffraction, to follow the reaction products of small amounts of As_2O_3, Na_2SO_4 etc. added to the batch (concentrations of about 1%).
DTG and DTA measurements have been used extensively. The conclusions made by different authors are not consistent. Most studies were done in the system $Na_2CO_3-CaCO_3-SiO_2$.
Kröger et al. studied the equilibria, occurring in the systems alkali oxide - CO_2 - SiO_2. They used their results to explain the melting reactions in an indirect way (4). A survey of the literature of the glass making process is given in ref. 5.

The three mentioned processes: melting, refining and homogenisation, have important interactions with each other, so that it is desirable to have one single measuring technique with which it is possible to study the whole process including the development of gases from the melt and the homogenisation. LRS appears to be a useful technique in studying the glass formation from the batch components.

- LR spectra of the crystalline and glassy components are in most cases very well resolved; the peak positions and relative intensities, together with the line widths of the peaks can be used for identification of the intermediate reaction products.
- The reaction path of small amounts of Na_2SO_4 and As_2O_3 can be followed easily. It should be noted, however, that other fining agents such as Sb_2O_3 and NaCl give mainly reaction-products which have a very weak Raman-effect.
- The development of CO_2 gas can be followed by measuring the Raman-peaks of the CO_3^{2-} still present in the batch. It was possible to detect H_2O and H_2 in vitreous silica (7) and there are indications that N_2 and O_2 can be measured too.
- It is possible to use LRS to give quantitative information about the reaction products.

LRS has been used as a finger print technique. By analogy with IR spectroscopy LRS can give information about the molecular structure of glasses (1,2,3). LRS spectra of glasses are much better resolved and more characteristic than IR spectra.
Knowledge of the structure of the glasses, is important for an understanding of the glass-melting process, as will become clear below.
We started our study of the glass forming process with the composition $30K_2CO_3.70SiO_2$. Konijnendijk and Buster investigated LRS well-refined glasses of the same composition with various amounts of fining agents added and melted in various atmospheres (6).

LRS STUDIES OF STRUCTURES OCCURRING IN THE GLASSY SYSTEM $K_2O-SiO_2-CO_2$

There is a strong resemblance between the LR-spectra of glassy and crystalline compounds in the case of silicates and borates (1,2,3). Crystalline $K_2O.2SiO_2$ (potassium disilicate) shows LR-peaks which are also found in "$K_2O.2SiO_2$" glass at the same positions (1101, 529 and 486 cm^{-1} ref. 1,3).
Glasses prepared from $xK_2CO_3.(1-x)SiO_2$ (with x = 0,30-0,50) show peaks of crystalline $K_2O.SiO_2$ (960 and 590 cm^{-1}) of crystalline $K_2O.2SiO_2$ (1101, 529 and 486 cm^{-1}); and of crystalline K_2CO_3 (1068, 1410 and 690 cm^{-1}). Some of the spectra are given in fig. 1.
The peaks (at energies > 200 cm^{-1}) in the LR-spectrum of $K_2O.2SiO_2$ are tentatively ascribed to vibrations of a SiO_4-tetrahedron forming part of a two-dimensional network (disilicate plane). The peaks in the LR-spectrum (at energies > 200 cm^{-1}) of K_2O-SiO_2 are tentatively ascribed to vibrations of a SiO_4-tetrahedron, forming part of a one-dimensional network (metasilicate chain). The peaks (at energies > 200 cm^{-1}) in the spectrum of K_2CO_3 are tentatively ascribed to vibrations of the CO_3^{2-} triangle.

The spectra given in fig. 1 indicate that large quantities of CO_2 remain dissolved in the glasses of low SiO_2 content in the form of "glassy carbonate". A more complete description and interpretation of the studies in the system $K_2O-SiO_2-CO_2$ will be published in the near future.

INVESTIGATIONS INTO THE FORMATION OF $30K_2O.70SiO_2$ GLASS FROM A $30K_2CO_3.70SiO_2$ BATCH

LR-spectra were taken of incompletely reacted mixtures of $30K_2CO_3$ and $70SiO_2$. The mixing conditions of the batch and the amount of batch material were kept constant. The batches were heated at different temperatures during varying periods of time, after which they were quenched in quartz-vessels which could be evacuated subsequently. The samples were ballmilled and the LR-spectra of the powders were taken.
The spectra given in figure 2 show how the reactions in the batch proceed under isothermal conditions. The reaction has been followed at temperatures from 800°C to 950°C. After firing for 1 hr at 800°C some crystalline $K_2O.2SiO_2$(d) has been formed. Furthermore a glass is observed, which also occurs in the system $K_2O-SiO_2-CO_2$ (kcs). This glass resembles one of the glasses that can be melted from $50K_2CO_3.50SiO_2$ at low temperatures. It is concluded therefore that part of the crystalline K_2CO_3 (mpt.

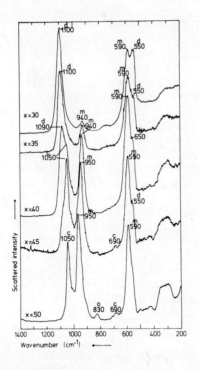

Figure 1.
Raman spectra of glasses molten from $xK_2CO_3 \cdot (1-x)SiO_2$. The glasses with $x = .30$ and $.35$ are heated during 1 hr at $1200°C$; the glasses with $x = .40, .45$ and $.50$ are heated during 1 hr at $1100°C$. The peak positions in the spectra are given in cm^{-1}. The meaning of the symbols is: $c = CO_3^{2-}$ ion; d = disilicate plane; m = metasilicate chain; o = orthosilicate. The polarization direction was perpendicular (z) to the optical plane formed by the direction (x) of excitation and the direction (y) of collection of scattered light in brief $x(zz+zx)y$, ref.[8]).

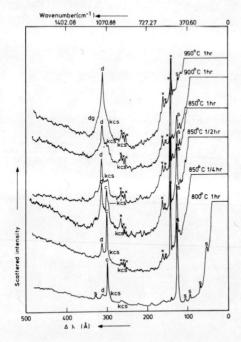

Figure 2.
Raman spectra of some incompletely reacted $30K_2CO_3 \cdot 70SiO_2$ batches. A more sophisticated apparatus was necessary due to the high scattering level of the powders. The scale is linear in wavelength difference. The meaning of the symbols is: x = laser plasma line dg = disilicate glassy, d = disilicate crystalline; kcs = $K_2O-CO_2-SiO_2$ glass; s = quartz; c = carbonate (crystalline). Optical measuring geometry: $x(zz=zx)y$.

891°C) is converted into a "glassy" state at 800°C.
Experiments at lower temperatures, not given here, indicate that crystalline $K_2O.2SiO_2$ is the first reaction product which is formed. At 850°C the crystalline K_2CO_3 (c) has completely disappeared after 1 hr heating, whereas SiO_2 (s) is still present. The spectra also show a strong formation of cryst. $K_2O.2SiO_2$ (d) and $K_2CO_3-SiO_2$ glass (kcs).
At 900°C the crystalline K_2CO_3 (g) has disappeared after ¼ hr heating and a large amount of CO_3^{2-} is present in a glassy form. When the reaction at 850°C proceeds, the SiO_2 has reacted almost completely after 1 hr heating and the proportion of crystalline $K_2O.2SiO_2$ (d) has grown. The quantity of $K_2CO_3-SiO_2$ glass (kcs) reaches a maximum during the isothermal treatment. Additional experiments at 950°C indicate that at this temperature a mixture of crystalline $K_2O.2SiO_2$ (d) and glassy $K_2O.2SiO_2$ (dg) is present and perhaps some $K_2CO_3-SiO_2$ glass (kcs) after one hour heating.

DISCUSSION OF THE RESULTS

From the experimental results, given in the preceding section, we can give a qualitative description of the glass-melting process of a $30K_2CO_3-70SiO_2$ batch in the temperature region from 800°C to the melting point of cryst. $K_2O.2SiO_2$ (1015°C).
When the reactions take place, the SiO_2 grains are attacked by K_2CO_3 and different phases are formed in layers around the SiO_2 grains (fig. 3). The potassium concentration decreases from the $K_2CO_3-SiO_2$ glass layer towards the center of the SiO_2 grain. During the melting process the $K_2O.2SiO_2$ layer reacts with the SiO_2 grain. The potassium and oxygen ions needed for this process are supplemented by the $K_2CO_3-SiO_2$ glass layer; K_2CO_3 transports K^+ ions and CO_3^{2-} ions to the glass layer.
When all the crystalline K_2CO_3 has reacted the remaining CO_3^{2-} disappears from the glass since the CO_3^{2-} solubility is decreasing due to the out-diffusion of the K^+ ions. At 950°C (below the mpt. of $K_2O.2SiO_2$) the first formation of $K_2O.2SiO_2$ glass is indicated. This glass is probably formed by the out-diffusion of K^+ and O^{2-} ions from the alkali-rich $K_2CO_3-SiO_2$ glass.

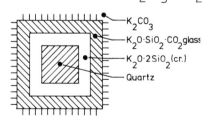

Figure 3. Schematic drawing of the reaction products formed around a SiO_2 grain during the reaction.

In the present paper only a qualitative description of the reactions is given. Experiments are underway to obtain more quantitative data by using the peak intensities in the LR-spectra.

CONCLUSIONS

The results obtained are not compatible with the general view (ref. 5). Generally it is thought that $K_2O.SiO_2$ is the first phase which is formed after which $K_2O.2SiO_2$ is formed. At elevated temperatures an eutectic melting should take place.
In our experiments it has been shown that in the temperature region of $800°C - 950°C$ in a $30K_2CO_3.70SiO_2$ batch under isothermal conditions the concentrations of K_2CO_3 and SiO_2 decrease, that of $K_2CO_3-SiO_2$ glass reaches a maximum and that of $K_2O.2SiO_2$ (glass or cryst.) increases. It is found that $K_2O.2SiO_2$ is the first phase which is formed.
It is concluded, that the development of CO_2 gas at temperatures above $800°C$ is causes by the out-diffusion of CO_3^{2-} from the $K_2CO_3-SiO_2$ glass and possibly by the reaction of K_2CO_3 with the glass layer.

LITERATURE

1. Konijnendijk, W.L.; "The structure of borosilicate glasses". Philips Res. Repts Suppl, 1975, no. 1 (thesis).
2. Bril, T.W.; "Raman spectroscopy on crystalline and vitreous borates", Philips Res. Repts Suppl 1976 (in press, available April 1976).
3. Brawer, S.A., White, W.B.; J. Chem. Phys., 63 (1975), 2421-2432.
4. Kröger, C.; Glas techn. Ber., 15 (1937), 371-379.
5. Glass-making (melting and fining). Bibliographic review Ed. Union Scientifique Continentale du Verre, Charleroi, Belgium (1973).
6. Konijnendijk, W.L., Buster, J.H.J.M.; to be published.
7. v.d. Steen, G.H.A.M., v.d. Boom, H.; to be published.
8. Damen, T.C., Porto, S.P.S. and Tell, B.; Phys. Rev. 142, 570 (1966).

RELATIONS BETWEEN PROPERTIES AND STRUCTURAL EVOLUTION OF SOME

Si, Ge, Sn TERNARY CHALCOGENIDES

E. Philippot, M. Ribes and M. Maurin

Laboratoire de Chimie Minerale C, Universite des

Sciences et Techniques du Languedoc, Place E.

Bataillon 34060 Montpellier - Cedex (France)

INTRODUCTION

The "oxo" compounds of Si, Ge and Sn with alkaline and alkaline-earth metals have been studied, both for structural analysis as well as physical properties. On the other hand the homologous sulphide compounds are not yet fully understood.

SYNTHESIS AND REACTIVITY

In our department we have developed the study of the $MS-XS_2$ and M'_2S-XS_2 systems (with M = Ba, Sr, Ca, Pb; M' = Na, K and X = Si, Ge, Sn). During the course of the systematic syntheses, as a function of temperature, for different MS or M'_2S/XS_2, we have isolated a number of ternary chalcogenides. All of the compounds were synthetized under vacuum, in sealed silica tubes by reaction in the solid state. Each compound has been characterized by X-ray powder diffraction, infra-red, Raman and , for tin compounds, Mössbauer spectra.

STRUCTURAL ANALYSIS

For each new stoichiometry, a detailed structural analysis by crystal X-ray diffraction allowed us to determine on the one hand the coordination of Si, Ge and Sn atoms by the sulfur atoms and on the other hand the geometry of the various anionic groups.

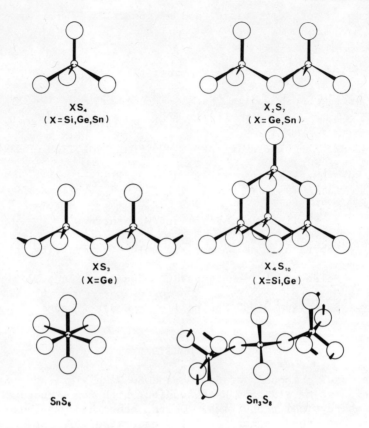

Fig. 1. Different types of coordination in ternary chalcogenides

This study shows a great variety of stereochemical environments of group IV b elements summarized in figure 1.

Atom in tetrahedral co-ordination :

XS_4^{4-} — discrete anions as in Ba_2GeS_4, Sr_2GeS_4, Ca_2GeS_4, Ba_2SiS_4 and Na_4SnS_4 compounds

$X_2S_7^{6-}$ — two XS_4^{4-} tetrahedra sharing a corner as in $Na_6Sn_2S_7$, $Na_6Ge_2S_7$ and $Ba_3Sn_2S_7$ compounds

$X_4S_{10}^{4-}$ — four XS_4^{4-} tetrahedra, each of them sharing three corners with three other different tetrahedra as in $Na_4Ge_4S_{10}$, $Na_4Si_4S_{10}$ and $Ba_2Ge_4S_{10}$ compounds

$(XS_3)_n^{2n-}$ — infinite chain of XS_4^{4-} tetrahedra, each tetrahedron sharing two corners with two other different tetrahedra as in Na_2GeS_3 and $PbGeS_3$.

Atom in both tetrahedral and trigonal bipyramidal coordination:

- as in $Na_4Sn_3S_8$

Atom in octahedral coordination:

- as in $PbSnS_3$ and in the MDO structure of Na_2SnS_3

PHYSICAL PROPERTIES

Within the framework of our work on structure and physical property relationships, some alkaline-earth silicates doped by europium were found to be good phosphors. Thus we studied the emission characteristics of phosphors based on alkaline and alkaline-earth chalcongenides as host lattices for Eu^{2+} and Ce^{3+}. The table 1 summarizes the cathodoluminescence results for $x = 0,007$.

TABLE 1

System	Colour
$BaS - SnS_2$	
$\quad Ba_{2-x}Eu_xSnS_4$	zero
$M^{II}S - GeS_2$ (M^{II} = Ba, Sr)	
$\quad Ba_{2-x}Eu_xGeS_4$	zero
$\quad Sr_{2-x}Eu_xGeS_4$	zero
$M^{II}S - SiS_2$ (M^{II} = Ba, Sr)	
$\quad Ba_{2-x}Eu_xSiS_4$	turquoise blue (strong)
$\quad Ba_2 SiS_4 : x\ Ce$	blue
$\quad Sr_{2-x}Eu_xSiS_4$	yellow (strong)
$\quad Ba_{1-x}Eu_xSi_2S_5$	green (very strong)
$\quad BaSi_2S_5 : x\ Ce$	blue
$\quad Sr_{1-x}Eu_xSi_2S_5$	turquoise blue
$M^I_2S - SiS_2$ (M^I = Na, Li)	
$\quad Na_{4-0,06} Eu_{0,03} SiS_4$	blue
$\quad Na_{2-2x}Eu_xSi_2S_5$	green (weak)
$\quad Li_{2-2x}Eu_xSi_2S_5$	red (very weak)

For those doped compounds where the cathodoluminescence results were positive, we carried out a quantitative luminescence study, Table 2.

TABLE 2

	$X^{II}_{1-x} Eu_x Si_2S_5$, $X^{I}_{2-2x} Eu_x Si_2S_5$ (A)								
X →	Ba			Sr			Na		
	λ	I	Δλ	λ	I	Δλ	λ	I	Δλ
	5075	3,74	650	4900	3,85	625	4875	0,010	625
	$X^{II}_{2-2x} Eu_x SiS_4$, $X^{I}_{4-2x} Eu_x SiS_4$ (B)								
	4925	1,63	563	5450	4,09	625	4600	0,013	575
T ↓	Ba_2SiS_4 : x Ce					$BaSi_2S_5$: x Ce			
	λ_1	λ_2	I_1	I_2	Δλ	λ_1	λ_2	I_1	I_2	Δλ
300°K	4350	4650	29,05	25,6	825	4600	5050	14,1	16,5	1125
78°K	4300	4710	20,4	17,9	724	4550	5000	sh.	16,5	1000

PM = 400 V, λ = wavelength in Å, x = 0,007, I = intensity in mV, Δλ = half-width in Å.

This study of luminescence curves (figure 2) shows the relations between stoichiometry, structure, doping nature and colour.

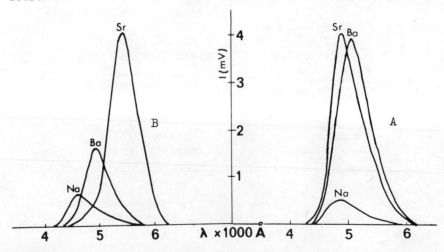

Fig. 2 Luminescence curves

The germanium and silicon compounds can be obtained very easily as glasses. A preliminary study of the ionic conductivity of these compounds, indicates that the properties of Na_2SiS_3 glass are comparable with the homologous Na_2SiO_3 compounds.

ACKNOWLEDGEMENT

The authors wish to thank Dr J. Olivier-Fourcade and Dr J.C. Jumas who participated actively in this research.

HIGH AND LOW TEMPERATURE FORMS OF LEUCITE

L. Hermansson and R. Carlsson

Swedish Institute for Silicate Research

Gothenburg, Sweden

ABSTRACT

Phase transitions of leucite and solid solutions of leucite, obtained by crystallization of glasses in the K_2O-CaO-Al_2O_3-SiO_2 system, have been investigated by means of high temperature X-ray diffraction and DTA.

The unit cell dimensions of leucite (a=13.08 Å, c=13.75 Å) change continuously with temperature up to the transition between the tetragonal and the cubic form at approximately 625°C, where a discontinuous change is observed. Solid solutions of the low temperature form of leucite (a=13.12-13.15 Å, c=13.69-13.74 Å) showed the same transition behaviour. The discontinuous change, however, occurred at a lower temperature, in one case as low as 300°C. For two samples a temperature interval of approximately 100°C occurred within which low and high leucite co-existed.

Two samples yielded the cubic form of leucite at room temperature (a=13.43 Å). The linear thermal expansion coefficient of high leucite was determined to be $(11.7-12.8) \cdot 10^{-6}$ (C^{-1}) (25-600°C).

INTRODUCTION

Some investigations of crystal structure and phase transitions of leucite are described in the literature (1,2,3). According to Wyart (1) the tetragonal low temperature form of leucite changes continuously into a cubic high temperature form. The transition is finished at 625°C. During the present investigation on controlled crystallization in the K_2O-CaO-Al_2O_3-SiO_2 system, solid solutions of leucite existing in both high and low temperature forms at room temperature have been obtained. In connection with determinations of the thermal expansion of these solid solutions of leucite, the phase transitions of leucite in this system have been studied.

EXPERIMENTAL

Only a few investigations of crystallization in the SiO_2-Al_2O_3-K_2O system have been described (4,5). Glasses of this system are very difficult to crystallize (6). In order to obtain significant crystallization, addition of CaO is necessary as well as a nucleating agent. In some samples ZnO has been added. The main phase is leucite but low percentages of sphene also exist as well as a glassy phase. Small changes of the composition cause a drastic change in the crystallization. A paper on crystallization in the SiO_2-Al_2O_3-K_2O system is in preparation (7).

The investigations of phase transitions of leucite have been performed for five glass compositions crystallized for 24 h at 900-950°C, samples 2-6 in Table I. Synthetic leucite, sample no 1, was prepared from fine grained SiO_2 (quartz), Al_2O_3, and K_2CO_3. The firing conditions were 1300°C and 16 days.

The phase transitions, unit cell dimensions, and coefficients of thermal expansion have been determined by means of high temperature X-ray diffraction. Corrections for the movement of the sample holder have been made by measuring the diffractions for Pt(1,1,1) and by utilizing Campbell's expansion tables (8). Differential Thermal Analyses have also been made.

Table I: Chemical composition of samples 1-6

Sample No	Composition (wt.-%)					
	SiO_2	Al_2O_3	K_2O	CaO	TiO_2	ZnO
1	55.06	23.36	21.58	-	-	-
2	62.75	12.35	17.00	1.42	2.04	4.44
3	52.00	24.00	13.00	3.00	8.00	-
4	59.32	12.10	15.00	2.85	4.08	6.65
5	63.07	13.56	12.60	2.85	4.08	4.44
6	60.00	16.00	13.00	3.00	8.00	-

RESULTS

The unit cell dimensions of stoichiometric leucite (a=13.08 Å, c=13.75 Å) change continuously with temperature up to the transition between the low and high temperature forms at approximately 625°C, where a discontinuity is observed ($a_{625°C}$=13.51 Å). The phase transition is shown in detail in figure 1a.

Three of the crystallized samples, nos 2-4, yielded solid solutions of low leucite (a=13.12-13.16 Å, c=13.69-13.74 Å) and two samples, nos 5-6, yielded high leucite (ss) at room temperature. Solid solutions of low leucite change into high leucite as the temperature is raised. The transition is similar to that of stoichiometric leucite. The transition, however, is completed in the temperature range 300-530°C, depending on the different compositions of the original glasses. The figures 1b-d show the transition in detail. For samples 3 and 4 a temperature interval of approximately 100°C was obtained, during which high and low leucite co-existed (figure 1c and 1d).

Samples 5 and 6 yielded high leucite (ss). The linear thermal expansion coefficient of leucite for these samples was determined to be 11.7 10^{-6} and 12.8 10^{-6} °C^{-1}, respectively (25-600°C). The thermal expansion is small in the temperature range 600-1000°C (figure 2). The unit cell dimension of high leucite (ss) is a=13.43 Å (25°C) for both samples.

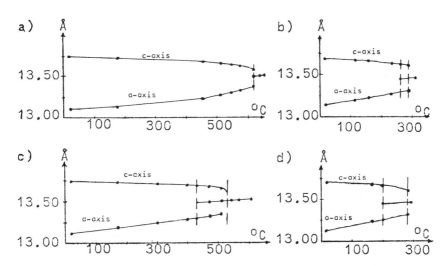

Figure 1: Phase transition behaviour of leucite, a) stoichiometric leucite, b) sample no 2, c) sample no 3, d) sample no 4.

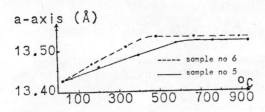
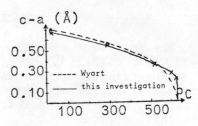

Figure 2: Unit cell dimensions of high leucite at different temperatures.

Figure 3: Differences in unit cell dimensions at different temperatures.

DISCUSSION

Our findings concerning phase transition of stoichiometric leucite are not in complete accordance with those of Wyart (1) who assumed a continuous change without a terminating discontinuous change at 625°C. The discrepancy, however, may be attributed to an arbitrary interpretation of measured data. Wyart has only reported a few values of unit cell dimensions over the final temperature range. In figure 3 the differences between unit cell dimensions (c-a) at different temperatures are shown according to Wyart's investigation and the present one.

As is apparent from the nature of the crystallization samples 2-6, it is possible not only to lower the phase transition temperature of leucite but also to eliminate the formation of low leucite. These modified transitions are attributed to formation of defective crystal structures. Some foreign atoms can be included in the structure during crystallization and slightly disordered structures develop. If high leucite (ss) is exposed to heat treatment above the formation temperature of high leucite, a structure will develop that at lower temperatures will change into a low temperature form of leucite. In the temperature range up to the formation temperature, however, high leucite (ss) seems to be stable. Stabilization or meta-stabilization of high temperature modifications has previously been reported in connection with crystallization of glasses (9-12) Greig (13) also reported high cristobalite crystals embedded in a glass matrix.

Table II: Peak-broadening of leucite (ss)

Sample No	Half-height width (2θ)	
	004 reflexion	400 reflexion
1	0.20	0.26
2	0.25	0.27
3	0.29	0.32
4	0.29	0.31

The co-existence of high and low temperature forms of leucite in samples 3 and 4 is attributed to the presence of a whole series of solid solutions of leucite in each sample. Diffractograms of these samples show peak-broadening (Table II). Samples 1 and 2 show a considerably more limited temperature range of co-existence. This co-existence is attributed to temperature gradients within the samples.

Measurements have been performed by means of two different types of high temperature X-ray diffractometer, one using powder and the other a suspension of the sample in alcohol. Consistent results were obtained between both diffractometers.

REFERENCES

1. Wyart,M.J., Bull. soc. Min. 61,228(1938).
2. Wyart,M.J., Bull. soc. Min. 63,5(1940).
3. Faust, Schweiz min. petr. Mitt. 43,165(1963).
4. Carlström,E. and Hermansson,L., Silikatrapport 75-3(1975).
5. Akagi et.al., Pat. 7408,921; CA 81:81713.
6. McMillan,P.W., 'Glass-ceramics' 79(1964).
7. Hermansson,L., in preparation.
8. Campbell,W.J., U.S. Bur. of Mines. Inf. Civ. 8107(1962).
9. Eppler,R.A., J. Am. Ceram. Soc. 46,97(1963).
10. Beall,G.H. et.al., J. Am. Ceram. Soc. 50,181(1967).
11. Petzoldt,J. Glastechn. Ber. 40,385(1967).
12. Skinner,B.J. and Evans,H.T., Am. J. Sci 258A,312(1960).
13. Greig,J.W., J. Am. Ceram. Soc. 36,389(1953).

DESORPTION OF Kr (OR Kr^{85}) AND SF_6 FROM VITREOUS AND CRYSTALLINE SiO_2, AND OF Kr FROM B_2O_3 AND GeO_2 GLASSES

W. W. Brandt and H. W. Ko

Department of Chemistry and Lab. for Surface Studies, U. of Wis.-Milwaukee, Wisconsin, USA

INTRODUCTION

The diffusion and solubility coefficients of He in various inorganic glasses are well known, while only few data for Ne and Ar are available (1-6). Kr and SF_6, a fairly inert and nearly spherical molecule of large size, have been used very little in studies of this sort. It is not clear to what extent the larger molecules can reach subsurface layers or interior regions of typical glasses.

Very often, the Arrhenius equation for the diffusion coefficient, $D = D_0 \exp(-E/RT)$, is used to evaluate typical sorption, desorption, or permeation rate data (all symbols having their customary meaning) and one finds a linear correlation to exist between most Log D_0 and E pairs of values, reported in the literature (7). In spite of this apparent simplicity, there are various reasons for suspecting that E and Log D_0 are subject to wide distributions, in a given glass-inert gas system, since there may well be very different and competing diffusion paths available in random structures which can lead to perfect, limited, or inhibited percolation of the bulk, depending on the "fit" of the various diffusion paths present and the diffusing species (8).

EXPERIMENTAL

Table (1) shows the average grain radii and specific surface areas of the samples used in this work.

Most samples were degassed and then exposed to Kr or to SF_6 at elevated temperatures and for periods of time listed in Table (2). They were then quenched and transferred to a clean vycor vessel which was connected to a Hitachi-Perkin Elmer Model RMU-6E mass spectrometer. After evacuation, the temperature of the desorption cell was increased slowly and at a constant rate. The resulting desorption transients were evaluated as described in more detail earlier (9); using the transient peak positions, T_{max}, and half-widths, ΔT, as primary data; these also are listed in Table (2).

One sample, (E), was exposed to Kr^{85} at 1021°K for 19.5 hrs, quenched and then chemically etched in 1 $\underline{M}$ NaOH at 40°C. After suitable etching time intervals this sample was treated for 20 min in concentrated HCl, was filtered, dried, weighed, and its γ-activity measured under reproducible conditions.

RESULTS AND DISCUSSION

The half widths, ΔT, and temperatures of maximum desorption rate, T_{max}, measured by mass spectrometry

Sample	Type	Average Grain Radius r (cm)	Specific Surface Area (cm^2/g)
A	Fused SiO_2	3.8×10^{-4}	3.0×10^3
B	"	5.0×10^{-4}	2.3×10^3
C	"	7.2×10^{-4}	1.6×10^3
D	"	0.61	1.5
E	"	3.6×10^{-3}	3.2×10^2
F	"	0.45	2.5
G	"	0.12	9.6
H	"	4.1×10^{-3}	2.8×10^2
I	Natural Quartz	4.9×10^{-2}	36
J	"	3.6×10^{-3}	3.2×10^2
K	"	~1.0	~1.6
L	Vitreous B_2O_3	$~2.7 \times 10^{-3}$	7.5×10^2
M	Vitreous GeO_2	4.9×10^{-3}	1.5×10^2

Table 1. Samples used in this work. The grain radii of the powder samples were obtained microscopically, as described previously (1). The specific surface areas of these samples were obtained from the average r values, assuming spherical shapes. Samples (A) through (H) are identical in composition, Samples (D) and (F) being identical with (D) and (F) of Ref (1). Samples (I), (J), and (K) originated from a single piece of natural quartz from Rock Springs, Arkansas. The Sample (L) is similar to (G) of Ref (4), while (M) resembles (F) and (G) of Ref (3).

DESORPTION OF Kr and SF$_6$ FROM GLASSES

	Sample	No. of Experiments	Time (hr)	Temp (°K)	Pressure (atm)	T_{max} (°K)	ΔT (°K)	E (kcal/mole)	Log D_0 (a)	Log D_0/r^2 (b)	(Q/pw)10⁷ (c)	(Q/pA)10¹⁰ (d)
Kr:	A	11	20	1165	0.624	1114	166	53	-1.6	5.3	13	5
	B	3	27	949	0.432	1182	182	54	-1.6	5.0	16	7
	C	3	18	869	0.457	1051	(256)	--	--	--	~47	~30
	E	2	19	1056	0.553	1113	213	46	-0.4	4.5	6	18
	G	2	16	1004	0.774	1109	203	43	2.5	4.3	2	200
	I	2	16	999	0.612	1059	267	29	-2.8	1.8	.2	8
SF$_6$:	C	1	15	469	0.987	451	229	6.3	-7.6	-1.3	.6	.4
	H	2	~24	484	0.993	495	194	8.5	-6.0	-1.3	1.2	4
	F	2	26	484	1.00	434	205	6.6	-2.7	-2.0	.7	280
	D	2	26	483	1.00	434	(96)	--	--	--	~.1	~60
	J	2	18	669	0.967	633	260	12	-5.5	-0.6	0.8	3
	K	2	~11	456	0.961	570	306	7.3	-2.5	-2.5	0.02	13
	L	2	9	508	~1.0	399	115	10	-3.4	-1.8	--	--
	M	2	26	727	~1.0	632	286	10	-5.5	-0.9	--	--

Table 2. Desorption of Kr and SF$_6$ from fused and natural quartz and from vitreous B$_2$O$_3$ and GeO$_2$. T_{max} and ΔT are, respectively, the temperature of maximum desorption and the half-width of the desorption transients. E and Log D_0, the parameters of equation (1), obtained from ΔT and T_{max} (9).
Samples A-H--Fused Quartz; I-K--Natural Quartz; L--Vitreous B$_2$O$_3$; M-- Vitreous GeO$_2$.
a_{D_0} in (cm^2 sec^{-1}); b_{D_0}/r^2 in (sec^{-1}), r is the average grain radius. c_Q the total amount of gas desorbed (in cc (STP)), w the sample weight in (g). ^dA is the surface area of the sample in cm^2.

are listed in Table (2) as well as the Arrhenius parameters E and D_0, obtained as described earlier (9). Also, the very approximate gas uptakes per unit sorption pressure and sample weight, Q/pw as well as the uptakes per unit sorption pressure and surface area, Q/pA, are listed. (Reporting E and D_0 is convenient, but does not imply that one is dealing largely or entirely with bulk diffusion processes.)

As a first important result one notes that Log D_0 increases markedly with grain size, in all cases where this comparison is possible, while the uptake per unit sorption pressure and weight, Q/pw, tends to decrease. These trends are of course expected if the desorption involves only, or largely, the surface-near regions of the samples.

At this point, one might be concerned that incomplete saturation of the glass samples with Kr or SF_6 could lead to these observed trends. To eliminate this possibility, the sorption times and temperatures were varied in the individual experiments, and (Q/p) was found to be independent of sorption time or temperature, within the ranges used.

The E values obtained for SF_6 appear to be very low for bulk diffusion controlled processes; SF_6 molecules are larger than Kr and one would expect $E(SF_6) > E(Kr)$ if only bulk diffusion were important. Low apparent E's have been reported for various other surface desorption processes, using the present data treatment methods (10).

The relative desorption rate coefficients D_0/r^2 (in sec^{-1}) listed in Table (2) as well as (Q/pA) should be independent of the grain size if the desorption rate processes were entirely surface related which is not the case. It appears that surface and bulk regions are involved in all these processes, to varying degrees.

The relative γ-radioactivities of partly etched samples are shown in Figure (1), as a function of the (decreasing) average grain radius. It turns out that approximately one half of the Kr^{85} is removed after about 2 μm of the external surface layer is dissolved. A uniform distribution of Kr^{85} throughout the grains would correspond to 50% being stored in a layer of 2.9 μm.

Ion implanted Kr^{85} lodges even closer to the surface, depending on the ion energies. For example, for 50 keV ions the "50% layer" is approximately 300 Å thick, while 10 keV ions lead to a layer of 50 or 100 Å (11).

A relatively simple model of the glass network structure can be described which is consistent with the slow transition from bulk to surface processes as one proceeds from He to Kr and then to the much larger SF_6, and from very fine grained samples to large pieces: It appears that a certain small portion of the interstices or pores used by the migrating gas atoms are interconnected, throughout the solid, while many others lead into, but do not traverse, the remaining regions of the random network structure. Obviously, the average internal dimensions of the preferred diffusion paths are

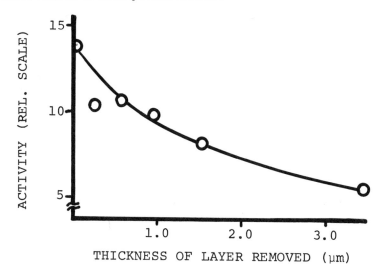

Figure 1. γ-Ray count of vitreous SiO_2 (Sample (E)) after etching. The count is corrected for the self-absorption. The layer thicknesses are calculated from the weight losses after successive etchings.

very different for the above gases, and the total number of paths open to Kr atoms is expected to be much larger than that available to SF_6 (see Sample (C) Table (2).

In terms of percolation theory, some regions of the (nearly) random networks can be assumed to have a finite percolation probability, with respect to the transport of Kr and SF_6, while other regions, constituting a disperse phase in a continuous matrix, are more compact and constitute barriers to percolation (8).

The model described here corresponds to very wide distributions of E and Log D_0 parameters, which are now characteristic of short segments of the available diffusion paths through the bulk or near the surface; this is undoubtedly due to a wide distribution of the interstices which are utilized by the migrating atoms.

REFERENCES

1. M. Abe, B. Rauch, W. W. Brandt, Z. Naturf., $\underline{26a}$, 997 (1971).
2. W. G. Perkins, D. R. Begeal, J. Chem. Phys., $\underline{54}$, 1683 (1971).
3. W. W. Brandt, B. Rauch, J. J. Wagner, Z. Naturf., $\underline{27a}$, 617 (1972).
4. W. W. Brandt, T. Ikeda, Z. A. Schelly, Phys. Chem. Glasses, $\underline{12}$, 139 (1971).
5. (a) J. E. Shelby, S. C. Keeton, Report SLL-73-5269, Sandia Laboratories, Livermore, Cal. (Sept. 1973); (b) J. E. Shelby, J. Appl. Phys., 44, 3880 (1973), and references cited therein.
6. W. W. Brandt, M. Abe, Reactivity of Solids, Chapman and Hall, Ltd., 1972, p. 210.
7. W. W. Brandt, W. Rudloff, Z. Phys. Chemie (Frankfurt), $\underline{42}$, 201 (1964), Figure (2).
8. W. W. Brandt, J. Chem. Phys., $\underline{63}$, 5162 (1975).
9. W. W. Brandt, Int. J. Heat Mass Transfer, $\underline{13}$, 1559 (1970).
10. Ref (8), Table (2), and references cited therein.
11. P. Lukač, Radiochim. Radioanal. Letters, $\underline{18}$ (#4), 203 (1974).

SURFACE CRYSTALLIZATION IN THE SiO_2-Al_2O_3-ZnO SYSTEM BY DTA

Z. Strnad and J. Šesták
Research and Development Institute for Technical Glass, 11000 Prague, Czechoslovakia, and Institute of Solid-State Physics of the Czechoslovak Academy of Sciences, 162 53 Prague, Czechoslovakia

INTRODUCTION

Easy attainable DTA measurements provide information about the conversion of glass into glass-ceramics.

However, the direct relation between the DTA curve and the absolute values of nucleation and crystallization rates obtained from optical measurements has not been established as yet. There were a few attempts to use the shift of the top of DTA peak for the estimation of the effect of the nucleating agent or preliminary heat treatment (1,2). A simplified mathematical evaluation of DTA peak was also used for the determination of the activation energy (3,5) but it was noticed that these data are not necessararily related to those established from the rate of radial growth of individual crystals (3). If glasses have a different disposition for bulk and surface crystallization it is possible to estimate their mutual relation from the shift of the top of DTA peak of coarse and fine particles assuming that the shift represents the change in the activation energy (6). However, these activation energies are complex because the surface and the volume crystallization often occur simultaneously. It would be of interest to study the volume and surface crystallization seperately, particularly to examine the effect of different grain size of particles on the surface crystallization which is the object of our contribution.

SURFACE CRYSTALLIZATION

Using the 70 SiO_2-10 Al_2O_3-20 ZnO glass which exhibits no volume crystallization we made a series of DTA runs for powdered samples (weighing about 0.3g) prepared by crushing and separating particles by sieve analysis, see FIG.1. This glass characteristically crystallizes from the surface at relatively low temperatures below $1000°C$. X-ray powder diffraction of the crystalline surface showed that the crystal phase was β-quartz solid solution only. Owing to the large surface area of the powdered sample the number of surface nuclei can be very large and thus the beginning of the exotherm on DTA curves can occur at a lower crystallization rate as the grain size of the granulated glass decreases. For a simple characterisation of individual peaks, the value of the Steepness of the advancing part of DTA exotherm suits best. Using the DTA equation (7) and a modified simple method for the estimation of the activation energy (4), see Fig.2, it was found that the value of E changes with the size of particles, (Fig. 3). A similar result was obtained for DTA study of the crystallization of pure diopsite glass which also exhibits a distinct surface crystallization (8)

DISCUSSION

From FIG.3 two parts of E can be distinguished, the first which can be found by extrapolation to the flat surface and falls to about 50 kcal/mol. It can be assumed as independent of curvature and related merely to the crystal growth. The second part is a function of grain size and can be explained on the basis of nucleus formation on a curved surface. If the contact angles stay constant between the phases α-β and γ-β (Fig. 3.), it is necessary

SURFACE CRYSTALLIZATION BY DTA

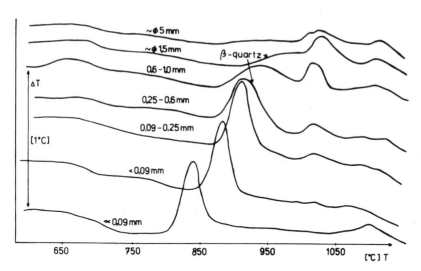

FIG.1. DTA curves of the surface crystallization of powdered samples of the glass 70 SiO_2-10 Al_2O_3-20 ZnO (recording sensitivity of 1K/150mm for T and 1K/15mm for ΔT by the apparatus NETZSCH, 5°K/min)

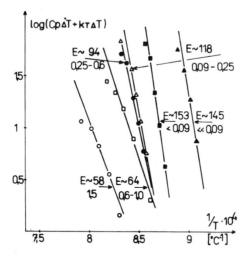

FIG.2. Evaluation of apparent activation energies from the advancing parts of DTA peaks in Fig. 3. (C_p = 36 cal K^{-1} mol^{-1} s^{-1}, K_T = 5 K^{-1} mol^{-1} s^{-1}, T scanned from the electronic derivative).

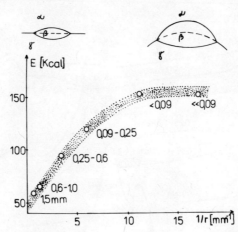

FIG.3. The plot of apparent activation energy vs. grain size of the sample (compare Figure 3 and 4)

to assume the formation of nucleii of a different size as a function of curvature of the surface γ-α. It means that equally stable nucleii in the form of a spherical cap are formed by changing their lenticularity as graphically for a planar and curved boundary in Fig 3. It results in the change of E which also includes nuclei formation. It is in accordance with the knowledge that the concave and convex surface decreases and increases the work of nuclei formation, respectively, when the γ-phase is a solid one which is not readily deformable (9).

The validity of our conclusion is certainly limited by the extent to which we can fulfil the requirements necessary for a simple evaluation of DTA curves. The increasing amount of the generated heat yields relatively a higher temperature gradient and heat exchange which may contribute to the DTA peak formation. This can be neglected because for the different heating rates the value of E varies only within $\pm$ 15%. The different size of the DTA peak has also no effect in the logarithmic representation in FIG.2 if the peaks are geometrically similar. A small deviation of the sample temperature from programmed linear increase (=0.5°K) is also negligible with regard to the temperature interval of the peak (=20K).

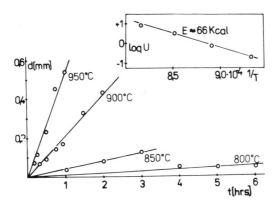

FIG. 4, Linear growth (d) as a function of time (t) and corresponding Arrhenius plot for the surface crystallization of β-quartz solid solution obtained by optical observation.

The extrapolated value of E can be examined in terms of crystal growth. For comparison we can use optical measurements (10) of the motion of the reaction interface on the planar surface and the associated E obtained through an ordinary Arrhenius plot, E = 66 Kcal/mol, (Fig. 4.). We can see that E_{DTA} = 50 Kcal/mol is lower than E calculated from direct observation. This is in agreement with the formalism of the nucleation dependant three dimensional growth (1.U^3) where the value of E is composed of E_1 and 3 E_U. If to a first aproximation we neglect E_1, then for a derivativ evaluation according to (4) 3/4 E_U = E_{DTA} and hence E_U = 66 Kcal/mol.

LITERATURE

(1) Thakur R.L.,Takizawa K.,Skaino T.and Moriya T., Centr. Glass and Cer. Res. Inst. Bull. 11 (1964).
(2) Sack W.and Sheidler H., Glastech.Ber. August 1970,p.322
(3) Clinton D.,Mercer A.and Miller R.P., J.Mat.Sci. 5/1970/ 171
(4) Šesták J.,Phys.Chem.Glasses 15 /1974/ 137
(5) Buri A.,Marcota A.and Orsini P.,Proc.9th Inter.Congr. Glass, Volume 1,Section A 1.3,p.343

(6) Thakur R.L. in Advances in Nucleation and Crystallization in Glasses by Amer.Cer.Soc.,Ohio 1971,p.166
(7) Nevřiva M,Holba P.and Šesták J.,Silikáty 20 /1976/ in press
(8) Voldán J. in Bulletin State Res.Inst.Hradec Králové 1972, No 4,p.40
(9) Fine M.E., Introduction to Phase Transformations in Condensed Systems, MacMillan, New York 1965, p.33
(10) Strnad Z.,Špirková B. and Dusil J., Silikáty 20 /1976/ in press

SOME RECENT DEVELOPMENTS IN THE INVESTIGATION OF SOLID-STATE AND SURFACE REACTIONS

Eric G. DEROUANE

Facultés Universitaires de Namur, Laboratoire de Catalyse

61, rue de Bruxelles, B-5000-Namur. Belgium

ABSTRACT

The last few years have witnessed an uncommon growth in the design of new techniques and the improvement and application of more conventional methods for the investigation of solid-state, thin films, and surface reactions.

Following a brief discussion of the modern instrumentation which is offered to experimentalists, for which several critical reviews have been proposed, the attention is focussed on the application of the rather conventional but often underestimated magnetic techniques as used in dynamic, i.e., resonance, conditions (E.P.R, N.M.R, F.M.R.)

Specific applications are discussed in relation with the mobility of labile oxygen in transition metal oxides, the dissolution and adsorption of hydrogen by metals (Pd,Pt,Cu,W,V), the binding of olefins to oxidic surfaces, and the application of ferro- and ferri- magnetic resonance to the investigation of solid-state processes. All of these are presented as examples showing the potentiality of these methods, reflecting the momentary interest (and knowledge) of the author.

INTRODUCTION

It is rather a cliché to state that the development of reliable ultrahigh vacuum techniques has prompted scientists to invent a bewildering array of electron, ion, atom, and X-ray spectroscopic techniques for the characterization of surfaces and thin films and for the investigation of their dynamic behavior, i.e., their reconstruction mechanisms and their reactivity towards outer phases.

However, apart from few fundamental research laboratories, these "surface" methods have only been slowly appreciated probably because they were dealing almost exclusively with so-called clean or ideal surfaces of rather limited interest to the "practical" scientist. The developing awareness that these techniques can provide composition profiles in the submicron scale and the pioneering work of several physicists and chemists [1] led to their extensive use for the study of solid-phase reactions in thin films[2] and surface reactions on metal [3] and oxide [4] surfaces.

Many reviews in this area have been proposed by various authors with different viewpoints [5-13]. Hence, we will not deal further with these methods, but for quoting some specific results when necessary and briefly discussing some of their main features.

Figure 1 compares the main spectroscopic techniques which are available today for surface and thin film investigations with respect to their optimum detection sensitivity (in ppm. atomic) and their probing depth (in layers, one layer being estimated at 0.25 nm). It is immediatly apparent that, in the monolayer coverage range, a variety of techniques are of potential use, the capabilities and limitations of which have been discussed in detail by Honig [5]. A great number of major considerations determine the choice of one of these for the study of any specific problem. If however one is mostly concerned by surface reactivity, and if we define "surface" as the outermost layer of atoms bounding the solid, only two techniques appear extremely promising. These are Secondary Ion Mass Spectromety (SIMS) [14] and Ion Scattering Spectrometry (ISS) [6]. In view of the very difficult quantitative analysis and of the complex interaction between the incident beam and the surface atom(s) that occur in SIMS, there is a fair possibility that ISS will become one of the most quantitative surface spectroscopies because of the inherent simplicity of the scattering process.

Diffraction techniques will, similarly, not be considered in this lecture. Recent reviews have been proposed[15] and most of the scientists are aware of their great capabilities.

The more conventional techniques which have developed more slowly along the past years, are proving now to provide extremely interesting information on surfaces and their reactivity.

Such is Infrared Spectroscopy (IR). The advent of differential cells [16], very simple in their principle, or that of wavelength modulated reflection-adsorption IR [17,18], rather sophisticated and similar in principle to the modulation methods used in AES, enable one to get information on the very first adsorbed layer exclusively. Ellipsometry has proved to be very successful in the study of chemisorption by semiconductors and sometimes by metals[19], and the same can be said about Inelastic Electron Tunneling Spectroscopy (IETS). It allows detection of vibrational energy changes in organic molecules adsorbed as dopants on the surface of thin films of metal oxides [20]. Polarized infrared internal reflection spectroscopy (IRS) showed existence of various CO species at the surface of Pd films[21]. In addition to the well-known linear

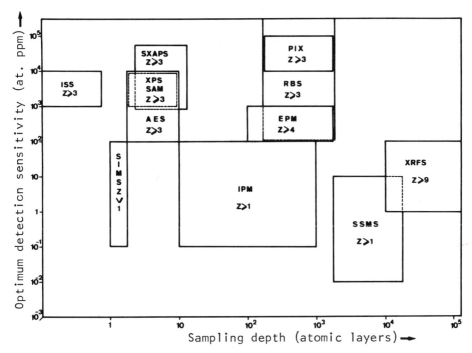

Fig. 1 : Comparison of spectroscopic methods for surface and thin film analysis

```
XRFS : X-ray fluorescence spectroscopy
EPM  : Electron-probe microanalysis
SSMS : Spark-source mass spectroscopy
ISS  : Ion scattering spectrometry
SIMS : Secondary ion mass spectrometry
IPM  : Ion-probe microanalysis
AES  : Auger electron spectrometry
SAM  : Scanning Auger microanalysis
XPS (ESCA) : X-ray Photoelectron spectroscopy
SXAPS: Soft X-ray appearance potential spectroscopy
PIX  : Proton induced X-ray analysis
RBS  : Rutherford Back-scattering
Z refers to atomic number of elements that can be
  detected in principle.
```

(ν > 2000 cm^{-1}) and bridged (ν < 2000 cm^{-1}) structures, a particular form of CO with high adsorption energy (> 40 kcal.mol^{-1}) and low vibration frequency (ν < 1800 cm^{-1}) was found to involve bonding of both carbon and oxygen to Pd surface atoms on a low index face of Pd crystallites. Such "flat" CO species have also been reported recently [22] in XPS studies of the low coverage adsorption of CO on stepped noble metal surfaces. Further investigation of those is of primary interest as they could eventually act as precursors in the dissociation of CO which results in the coking of the metallic surface under particular conditions.

Following this brief general survey, our attention will now be focussed on the too often under-estimated "magnetic" techniques applied in their "dynamic" resonance aspect (Electron paramagnetic Resonance, EPR; Nuclear Magnetic Resonance, NMR; Ferro- or Ferrimagnetic Resonance, FMR). We will restrict our discussion to few typical examples for which we think some progress has recently been made or which should merit closer attention in the very near future, such as :

a. the mobility and reactivity of labile oxygen in transition metal oxides;
b. the use of Nuclear Magnetic Resonance (NMR) for studying the adsorption and dissolution of hydrogen by metals and investigating the interaction of olefins with oxidic surfaces;
c. the use of Ferro- and Ferri-magnetic Resonance (FMR) in the investigation of solid-state reactions.

THE REACTIVITY OF OXYGEN AND OXIDES

Based on adsorption, kinetic and physical measurements, various oxygen species have been postulated (and sometimes observed) as intermediates in different oxidation reactions, among which several show paramagnetism (O^-, O_2^-, O_2^+, O_3^-, etc...).

Electron Paramagnetic Resonance (EPR) has proved to be a helpful technique in catalysis [34] and for the study of the formation, mobility, and reactivity of surface species [35]. An exhaustive review of the EPR of adsorbed oxygen species confirmed this about three years ago [23]. In the latter, most of the attention was focussed on the properties and characterization of the superoxide (O_2^-) and ozonide (O_3^-) ions.

In a recent paper [28], relying on XPS chemical shift analysis for various metal oxides, it was pointed out that, in all cases where comparison could be made with lattice oxygen, the binding energy, E_B, of the O(1s) lattice oxygen level was lower than the corresponding value for chemisorbed oxygen. This was interpreted as some indication for chemisorbed (atomic ?) oxygen species closer to O^- than O^{2-}.

Although the "atomic" character of these species was not further evidenced, O^- radicals deserve nevertheless more attention.

Indeed, O⁻ species have been identified by EPR on MgO[24] and found to react easily with simple hydrocarbons such as ethylene [25] and alcohols [24]. O⁻ radicals were also detected using V_2O_5 supported on silica gel [36], by reacting N_2O with Mo^{5+} on silica gel [37], or by UV-irradiating a ZnO surface covered by N_2O [38]. O⁻ ions produced on supported MoO_3 and V_2O_5 were found to be relatively stable in the 100-300°C temperature range [37].
It is clearly seen from Table 1, which reports the EPR parameters of the various O⁻ radical species, that high energy separation of the $2p_z$-($2p_y$,$2p_x$) levels (ΔE) seems to indicate good oxidation catalysts. Whether this observation is fortuitous or not should be discussed in view of more extensive results on a comprehensive series of oxides such as the transition metal oxides of the fourth period.

Table 1 : EPR parameters for O⁻ radical on various oxides

Oxide	Species	$g_\parallel$ (a)	$g_\perp$ (b)	ΔE (c) (eV)	Ref.
MgO	$(O^-)_s$	2.0016	2.041	0.85	24
MgO	V^-	2.0032	2.0385	0.92	24
V_2O_5/SiO_2	$(O^-)_s$	2.003	2.023	1.62	36
MoO_3/SiO_2	$(O^-)_s$	2.002	2.019	2.00	37
ZnO	$(O^-)_s$	2.0023	2.021	1.79	38

(a) $g_\parallel \simeq g_e$
(b) $g_\perp \simeq g_e [1 + \lambda/\Delta E]$
(c) E = separation between upper $2p_z$ orbital and lower $2p_x, 2p_y$ doublet for axial symmetry.
λ for O⁻ was taken equal to 135 cm⁻¹ [39].

Indeed, it is well known for the latter and many authors have tried it with more or less success in the past [32,40], that correlations seems to hold between the metal oxide bond energies and their catalytic activity. By screening the recent literature, one finds a variety of parameters which can be used as estimates of the M-O bond strength. Some of these are compared in Table 2 and Fig. 2 for the fourth period transition metal oxides. One is the enthalpy per oxygen equivalent of the formation of oxide, ΔHe [33]. It is also

Table 2 : Metal-oxygen interaction data for first-series transition metal oxides

Oxide	ΔH_{ads}(a)	E_d(b)	$E_B(O_{1s})$(c)	ΔE_M(d)	$-\Delta H_e$(f)	E_A(e)
TiO_2	104	76	531.5	4.9	56.4	–
V_2O_5	60	57,5	529.6	4.2	37.1	–
Cr_2O_3	36	–	529.3	3.1	–	21.9
MnO_2	23	30	529.6(g)	–	31.0	28.3
Fe_2O_3	35	39	530.5(h)	2.0(j)	32.8	35.9
Co_3O_4	15	37	529.75(i)	2.0(j)	26.6	24.5
NiO	21	44	529.6(j)	1.5(j)	28.6	26.2
CuO	23	48	529.6	1.3(j)	18.8	28.7
ZnO	65	83	530.6	–	41.6	–

(a) Heat of adsorption from isochore method; in Kcal.mol^{-1}, from ref. 26.
(b) Desorption activation energy (state 1) from flash-desorption; in Kcal.mol^{-1}; from ref.27.
(c) in eV from Fermi level; O_{1s} binding energy unless otherwise indicated from ref. 28
(d) Chemical shift for $2p_{3/2}$ metal level in oxide, in eV from Fermi level; from ref. 28 unless otherwise indicated
(e) Activation energies for propane oxidation, in Kcal.mol^{-1} from ref. 32
(f) Enthalpy per oxygen equivalent of formation of the oxide; in Kcal.mol^{-1}, from ref. 33
(g) for β-MnO_2, see ref. 29
(h) for Fe_3O_4, see ref. 30
(i) CoO = 529.6; Co_2O_3 = 529.9; see ref. 31   (j) see ref. 31

possible to estimate the binding energy between adsorbed oxygen and the surface of the catalytic oxide either by the isochore method[26] (ΔH_{ads}) or by flash-desorption [27] (E_d). The latter often show distinct states of labile oxygen. State (o) is weakly bound oxygen, state (1) and state (2) are relatively strongly bound and state (3) corresponds to the beginning of dissociation. Among those we will only consider state (1) which corresponds to the oxide standard state.

One can also try to estimate the oxygen species binding characteristics from XPS data. Indeed as covalency effects in ionic oxides are usually understood to mean some sharing of valence electrons instead of complete transfer, they can account for the gradation of binding energy values, E_B, as measured by XPS, lower values corresponding to more covalent oxides [28]. O(1s) binding energies and $2p_{3/2}$ level chemical shifts, ΔE_M, for metals in various metal oxides have been considered in Table 2 and Fig. 2.

In order to characterize the surface reactivity, activation energies for the oxidation of propane, E_A, have been used [32]. It is known indeed that, for this case, the reaction kinetics on various oxides are virtually first order in hydrocarbon and zeroth order in oxygen. Hence, the weakness of hydrocarbon adsorption seems to favor an oxygen - covered surface and E_A data are probably reasonably adequate in this case to reflect the surface reactivity of the oxides.

Fig. 2 shows a plot of the metal-oxygen bond parameters for various oxides. It is clearly seen that heats of adsorption [26] and desorption activation energies [27] vary in a parallel way. The same can be said of ΔE_M and $-\Delta H_e$ and this is not surprising as both parameters measure directly M-O bond strength. However, one sees immediatly that major discrepancies will be observed when correlating these parameters with the propane oxidation reaction activation energies, and this for $Cr_2O_3(d^3)$, $Fe_2O_3(d^5)$ and $Co_3O_4(d^6-d^7)$. Better agreement, however, is observed with the O(1s) level binding energy, strongly suggesting that this parameter could be used successfully for such correlations.

Our suggestion is, therefore, that more attention should be paid in XPS to the accurate determination of binding energies, free of charging shifts and/or non-stoichiometry effects. Such "absolute" data could well take over the presently used "thermodynamic" values which sometimes are shown to be inadequate correlating parameters.

Similarly, we would urge for more extensive and comprehensive EPR studies of O^- species in various (transition) metal oxides in order to gather systematic data on their behavior and reactivity. This would complement the scarce data that we have summarized in table 1 and eventually lead to most interesting conclusions regarding the possibility of O^- type ions being intermediates in oxidation reactions on catalytic oxides.

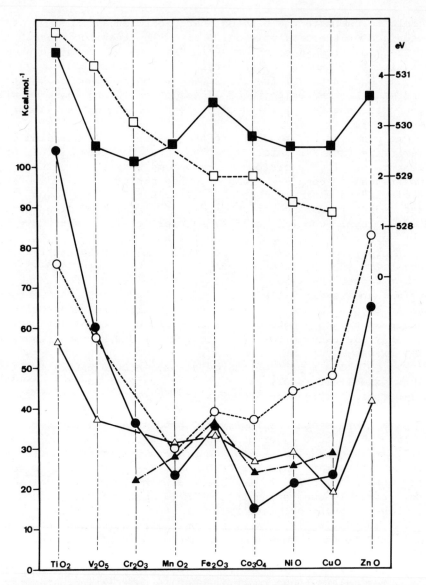

Fig 2 : Metal-oxygen interaction data for the first series transition metal oxides

- ● ΔH_{ads} : Heat of adsorption [26] (Kcal.mol^{-1})
- ○ E_d : Desorption activation energy [27] (Kcal.mol^{-1})
- ■ E_B : O_{1s} binding energy [28-31] (eV)
- □ ΔE_M : Chemical shift for $2p_{3/2}$ metal level in oxide [28-31] (eV)
- ▲ E_A : Activation energy for propane oxidation [32] (Kcal.mol^{-1})
- △ $-\Delta H_e$: Enthalpy per O equivalent of oxide formation [23] (Kcal.mol^{-1})

SOLID-STATE AND SURFACE REACTIONS

HYDROGEN ADSORPTION AND DISSOLUTION BY METALS

Investigations of hydrogen in metals have received renewed interest in the recent years in view of possible industrial applications such as selective gas filtering, hydrogen storage as hydrides and development of new superconducting materials. The metal surface-adsorbed hydrogen interaction as well as the interference between surface (sorbed hydrogen) and bulk (dissolved hydrogen) states are also of primary importance for the understanding of catalytic phenomena.

In view of its importance as a hydrogenation catalyst, the adsorption and absorption of hydrogen by palladium have been extensively studied. It has been reported recently [41] that chemisorbed hydrogen was affecting the "solubilization" of hydrogen in Pd in a way similar to that of Ag in Pd-Ag alloys. This was explained in the rigid band approximation by supposing that each solute atom was contributing its "s" electron to the unfilled "d" band of the metal. De Haas-van Alphen effect measurements [57] show that Pd has 0.36 holes atom^{-1} in the 4d-band. The decrease of susceptibility caused by alloyed silver or dissolved hydrogen [58] seems to indicate that about 0.55 electrons are used to fill up the 4d-band of the Pd alloys [42]. Moreover, two Pd-H phases α and β coexist at room temperature. The hydrogen concentration in the α-phase is low (0 < x < 0.03) while it is high in the β-phase. Hence, as they have different lattice parameters, strains will appear in mixtures of both phases as shown namely by NMR[45] from measurements of concentration dependent activation energies for proton diffusion.

Pt, W and Cu also deserve some interest. Pt and W are known to be good adsorbent for hydrogen and different adsorption states have been detected on these metals while Cu is generally considered to be an intermediate adsorbent for H_2.

Table 3 summarizes some data as they can be obtained from NMR studies of metals, hydrogen on metals, and metal hydrides. The mobility of adsorbed hydrogen species is readily evaluated by measuring activation energies for hydrogen diffusion, E_A, as derived from the proton linewidth dependence on temperature.
A quantitative indication on the binding mode of hydrogen to the metal surface is obtained by studying the shift of the resonance line with respect to a diamagnetic compound containing the same element. This is the so-called Knight-shift (K) which for transition metal(s) (alloys) is expressed as follows[59], n_e being the electron concentration and T the temperature :

$$K(n_e,T) = K_s(n_e) + K_d(n_e,T) + K_{VV}(n_e) + K_{dia}$$

K_s is the classical contribution due to the Fermi contact interaction of the nucleus with the spins of s conduction electrons and it is then positive. K_d arises from the spin polarization of the inner s-levels because of the spin paramagnetism of the d-conduction elec-

Table 3 : NMR data for metals, hydrogen on metals, and metal hydrides

Metal (M)	x (for MH_x)	E_A (a) (kcal.mol^{-1})	K (b)	Ref.
Pd	0	–	^{105}Pd: – 1.8 ± 0.014	46
	0.025	–	^{105}Pd: – 1.49 ± 0.015	46
	0.014	–	^{1}H : –0.0093 ± 0.0008	46
	0.035	–	^{1}H : –0.0050 ± 0.0005	46
	0.62	–	^{1}H : –0.00135 ± 0.00005	46
	0.75	–	^{1}H : –0.0002 ± 0.00005	46
	> 0.7	≃ 3.0	–	45
	< 0.7	f(x) – <3.0	–	45
Pt	0	–	Pt : –3.53 ± 0.01	49
(b-state)	adsorbed	0.5 ± 0.2 (L.T.)	^{1}H : –0.004 ± 0.001	47
(a-state)	adsorbed	4.3 ± 1.0 (H.T.)	^{1}H : ≃ 0	47
Cu	0 or ads.	–	^{63}Cu: 0.214 ± 0.002	50
	adsorbed	2.0 ± 0.3	^{1}H : 0.0094 ± 0.0005	50
W (a-state)	adsorbed	–	^{1}H : 0.0008 ± 0.0002	54
(b-state)	adsorbed	–	^{1}H : –0.005 ± 0.002	54
V	0	–	V : –0.58 ± 0.001	49
	0.2	≃ 4.0	–	55
	0.3	≃ 2.5	–	52

(a) E_A : activation energy for hydrogen diffusion
   L.T. low temperature
   H.T. high temperature

(b) Knight shift in % ($\frac{\Delta H}{H}$) for nucleus indicated

trons and contributes negatively to K. K_{VV} is caused by the Van-Vleck orbital paramagnetism of the d-conduction electrons; it is positive or negative and can be omitted for protons. K_{dia} is always negative as it results from diamagnetic screening, being small it can be neglected as compared to the other contributions.
Hence, dominant contributions in K are K_s and K_d. These being of opposed signs compensate each other and the value and sign of K will then depend closely on the relative amount of s and d conduction electrons.

For the sake of discussion, we have summarized in Table 4 some qualitative XPS data on Pd, Pt, Cu and Ag[56]. The negative shifts observed for pure Pd and Pt and the positive shift for Cu are readily explained by the Fermi level positions in these three metals with respect to the high density of states d-band and the low density of states sp-band. In addition, one can also note that K absolute values decrease with increasing width of the Fermi level edge and that particle size effects are eventually observed (Cu,ref.51).

Table 4 : XPS data for various metals[56]

Metal	Position of Fermi-level	(a) Δ (eV)	Fermi-level "width"[b] (eV)
Pd	4d-band	–	≃ 0.76
Pt	5d-band	–	≃ 0.60
Cu	4s-band	≃ 1.94	≃ 0.96
Ag	5s-band	≃ 3.96	≃ 0.48

(a) Distance from top of d-band to Fermi-level

When hydrogen is sorbed (or solved for Pd) by these metals, proton Knight shifts are observed which are of the same sign as those observed for the pure metal (with one exception for a particular state of H_2 in the case of W [54]) their magnitude being in some cases functions of hydrogen content (Pd[46]).

The following quantitative conclusions can be drawn. For the palladium-hydrogen system, the ^1H and ^{105}Pd shifts depend linearly on hydrogen content in α - PdH_n. Hence, the main contribution to K is due in this case to the high d-electron density of states which progressively decreases with filling of the d-band by the hydrogen electrons. For pure β-PdH_n, it is assumed according to the rigid band model that the 4d band is completely filled, that the 4d susceptibility is almost zero, and that the Fermi level is positioned in the 5s band. This increases the 5s electron density and possibly, to a small extent, the 5s electron density of states. Hence, this accounts for the K increase for β-$PdH_{x>0.7}$. This difference in the behavior of d density of states for the α- and β-phases is also reflected in the experimental g-factors (measured by EPR) for the α- and β-phase resonances of Mn-doped Pd [43,44]. Remembering that the g-shifts are dominated by d-electron contributions, one understands the progressive decrease of the α-phase g-factor with increasing H/Pd ratio (g_α (Pd) = 2.134; decreasing rate = 0.168 per one unit H/Pd), while the β-phase g-factor stays constant (g_β = 1.984).

Another interesting case is that of hydrogen adsorbed on Pt [47]. Two different adsorption states are detected in agreement with previous data [60]. Below 290°K, a first state (b) is detected with a negative Knight shift and a relatively small activation energy for the migration of H atoms. Above 290°K, another state (a) is shown to be present with nearly zero Knight shift and E_A of about 4 kcal. mol^{-1}. Both states correspond to atomic hydrogen. The difference of shifts and activation energies suggest differences in Pt-H adsorption bonds. Considering that K results from the contributions of K_s (>0) and Kd(<0), we propose that the 6s electron contribution is higher for the (a) state, resulting in a stronger Pt-H bond than for the (b) state (note that the same explanation could hold to explain the two H adatom states on W [54]).
It might be useful to note, in addition, that further evidence for the low-activation energy b-state has recently been gained from Neutron Inelastic Scattering Spectroscopy [48], a peak at 20 meV indicating the presence of a potential barrier of about 0.5 kcal. mol^{-1}.

Completely different is the adsorption of hydrogen by Cu as the Knight shift is now positive because of the dominant role played by the 4s electrons, the 3d levels lying now at about 2 eV below the Fermi level.

It is our opinion that these results demonstrate most convincingly the potential use of NMR as a means for studying chemisorption by metals.
Our recommendations would be to extend these investigations to other metals (and alloys) and other adsorbates (mostly ^{13}C labeled molecules such as ^{13}CO, $^{13}C_2H_2$, etc.). Quantitative interpretation of Knight shifts, and their correlation with available XPS data, would give a complementary view of the system as NMR provides simultaneously information about the bonding and the motion state of the adsorbate.

SOLID-STATE AND SURFACE REACTIONS

THE INTERACTION OF OLEFINS WITH OXIDIC SURFACES

Until quite recently, very little significant contribution had been made by high-resolution NMR spectroscopy to the understanding of surface and catalytic processes [73]. The recent advent of Fourier Transform ^{13}C-NMR Spectroscopy stimulated a great number of studies dealing with adsorption of hydrocarbons, mostly olefins, on various oxidic surfaces [61-70].

Our intention is not to propose here a full review of those [70] including our own results, rather to point out the evidence for the identification of some possible intermediates and/or binding modes to the surface when considering the adsorption of olefins. 1-butene and 2-butene (cis- and trans-) will be used as examples and a possibility of intermediates classification and nomenclature will be proposed and discussed.

Table 5 summarizes some of the available experimental data. Figs. 3 and 4 are stick-diagrams showing the relative changes in chemical shift values for 1-butene and the two 2-butene isomers respectively. Numbering of the C-atoms is as follows :

1-butene $C_{(1)}H_2 = C_{(2)}H - C_{(3)}H_2 - C_{(4)}H_3$

(cis-trans) 2-butene $C_{(1)}H_3 - C_{(2)}H = C_{(2)}H - C_{(1)}H_3$

Fig. 5 shows the proposed structures for a possible rationalization of the various intermediates into four classes A, B, C, and D, in view of the results quoted in Table 5.

Type A intermediates are physisorbed or weakly chemisorbed species with almost no electron transfer to or from the surface. Δ, the difference between the sum of all C shifts per adsorbed molecule and the corresponding sum for the free substrate is very small. Although rather non-specific adsorption is more or less achieved at the level of the double bond. Such species are observed on silica and on Tl-exchanged X-zeolite ; in the latter case, the high ionic radius of Tl^+ (1.44 Å) probably prevents the occurence of a strong surface-substrate interaction.

Type B intermediates are also characterized by low values of Δ and weak electron transfer occuring upon adsorption. It is seen, however, that adsorption is now highly specific. A small positive charge is localized on $C_{(2)}$ while the other C-atoms share the corresponding negative charge. Such a structure is in the line of some recent theoretical predictions [72] although the latter have been made for extreme configurations. Type B intermediates can reasonably be attributed to a vynilic type of adsorption.

Type C intermediates are observed when adsorption is made on an Ag-exchanged X or Y zeolite. Δ values are clearly negative. This

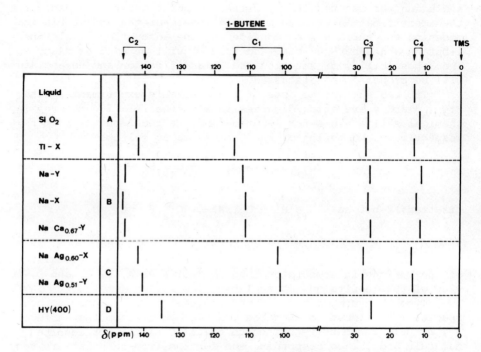

Fig. 3 : Stick-diagram showing chemical shifts of 1-butene species adsorbed on various supports. X and Y stand for zeolite type. Values in parentheses indicate either the exchange level or the evacuation temperature.

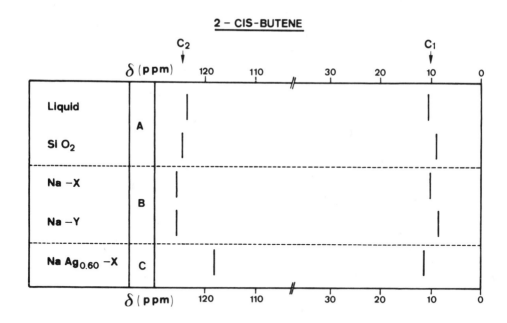

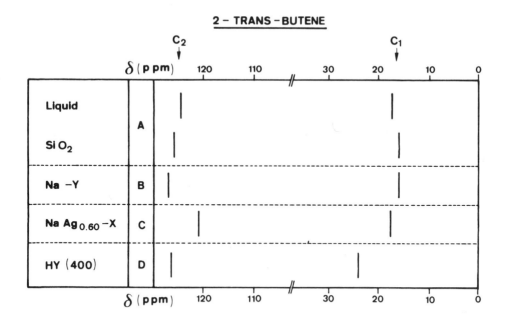

Fig. 4 : Stick-diagram showing chemical shifts for cis- and trans- 2-butene species adsorbed on various supports.

a. 1-BUTENE

A. Physisorption or weak chemisorption

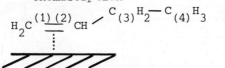

B. Vinylic-type adsorption

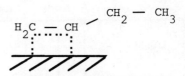

C. Coordinative adsorption

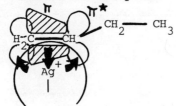

D. Cyclic adsorption

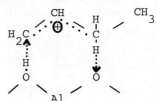

b. 2-BUTENE (cis and trans)

A. Physisorption or weak chemisorption

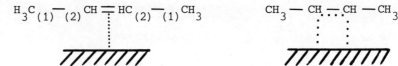

B. Vinylic-type adsorption

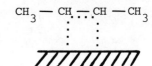

C. Coordinative adsorption

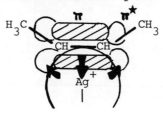

D. Cyclic adsorption

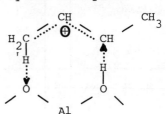

Fig. 5 : Proposed structures and adsorption types for olefin-intermediates on various supports.
  a. 1-Butene
  b. 2-Butene (cis- and trans-)
  (A,B,C,and D are the typical structures referred to in Table 5 and in the text; similar structures are probably valid for other olefins).

Table 5 : Chemical shifts of adsorbed butene species (a)

Compound	Support (c)	θ	T(°C)	Type	C_1	C_2	C_3	C_4	Δ (b)	Ref.
1-butene	–	–	–	A	113.3	140.1	27.1	13.3	–	61,70
	SiO_2	0.99	33	A	112.7	142.4	26.5	11.7	- 0.5	61
	Tl-X	0.70	60	A	114.4	141.7	27.1	13.3	2.7	66
	Na-Y	0.90	60	B	111.9	145.5	26.0	11.2	0.8	64,70
	Na-X	0.75	57	B	111.1	146.2	26.5	12.6	2.6	69
	NaCa(0.67)Y	0.82	60	B	111.1	145.6	25.7	11.5	0.1	65
	NaAg(0.60)X	0.85	60	C	101.8	141.6	27.7	13.9	- 8.8	66
	NaAg(0.51)Y	0.90	60	C	101.1	140.3	27.8	13.5	-11.1	66
	HY(400)	≃ 0.7	-70	D	25.3	134.9	–	–	+26.6	70
2-cis butene	–	–	–	A	10.6	123.3	–	–	–	62,63
	SiO_2	0.80	33	A	9.1	124.6	–	–	- 0.4	61
	Na-X	0.80	40	B	10.2	125.5	–	–	3.6	69
	Na-Y	0.60	60	B	8.4	125.5	–	–	0	64,70
	NaAg(0.60)X	0.55	54	C	11.3	118.1	–	–	- 9.0	66
2-trans butene	–	–	–	A	17.3	124.5	–	–	–	62,63
	SiO_2	0.80	33	A	15.9	125.9	–	–	0	61
	Na-Y	0.76	60	B	16.0	126.9	–	–	2.2	64
	NaAg(0.60)X	0.75	80	C	17.6	121.0	–	–	- 6.4	66
	HY(400)	≃ 0.7	-70	D	24.0	126.4	–	–	+17.2	70

(a) All values in ppm from TMS (see quoted references for detail). (b) difference between the sum of all individual C-shifts per adsorbed molecule and corresponding sum for free molecule. (c) X and Y refer to zeolite type. Numbers in parentheses correspond to exchange level or activation temperature.

type of adsorption is essentially characterized by a small, but noticeable, transfer of electrons from the substrate onto the surface which then becomes slightly negative. The adsorption center is essentially $C_{(1)}$ for 1-butene and both $C_{(2)}$'S for 2-butene. We ascribe the formation of this type of intermediate to coordinative adsorption as shown in Fig.5. It corresponds to the formation of Ag^+ olefin complexes of approximately the same type as those which have been described previously and, at length, in the literature [71]. Bonding occurs through σ and π bonds. The π back-bonding is less important than the σ-bonding and the adsorbate will become slightly positive. In the particular case of 1-butene, which can be considered as ethyl-ethylene, the reaction center $C_{(1)}$ changes from sp^2 to sp^3 hybridization and a high field shift is observed. One particular difficulty however remains in interpreting C-13 shifts in a straightforward manner. While ^1H shifts are mainly influenced by the diamagnetic term (depending on charge) the C-13 shifts include the paramagnetic term as the principal contribution. It has been shown [71] in the case of olefin-Ag^+ complexes, that the paramagnetic term dominates the shift.

The last type of intermediate that we will refer to, type D, will be formed by a concerted "cyclic" type adsorption. Here, Δ values are found to be positive and relatively high. Considering the facts that these intermediates occur on supports which are known to be strongly acidic and that only two different C atoms are observed, one has to propose cyclic structures such as those shown in Fig. 5. Both 1-butene and 2-trans-butene lead to the same type of species (which is rather close to 2-trans-butene) but for slight differences in charging, which results from the different attacking site of the surface proton. Hence, the adsorption mechanism must involve two sites acting in a concerted manner, a surface proton (from an acid site) is partially added to the more "negative" pole of the butene molecule while a hydrogen from the adsorbed butene interacts with a basic surface oxygen. Such cyclic intermediates could well explain the rather easy isomerization of 1-butene to 2-butene as observed on some acidic catalysts [74,75].

Further work is in progress [70] in order to ascertain the validity and generality of this classification, to investigate the possibility of progressive interchange between these intermediates, and to uncover their role in catalytic exchange and isomerization reactions of olefins.

FERROMAGNETIC AND FERRIMAGNETIC RESONANCE STUDIES OF SOLID STATE REACTIONS

Although Mössbauer spectroscopy has been applied to study the structure of Fe and Sn compounds for many years, it is only recently that it has been used for the study of solid state, surface, and/or catalytic reactions oxides [76-79], metals [80,81] and alloys [82,83].

As recently shown, ferro- and ferri-magnetic resonances (FMR) are able to provide complementary information on bulk and surface structure changes if ferro- or ferri-magnetism is present. This is of particular interest for iron containing catalysts. Metallic iron is ferromagnetic as well as magnetite Fe_3O_4 while a large variety of materials, some of the iron (III) oxides and the so-called ferrites, also show ferrimagnetism.

Our attention has been focussed recently [89] on the investigation of the production and stabilization of small iron particles on magnesium oxide as obtained by decomposition of a Fe(III) containing basic magnesium carbonate [79]. As evidenced by Mössbauer spectroscopy, fully reduced samples were found to consist of MgO (as support) and two iron-containing dispersed phases as Fe metal and FeO-MgO [79]. The stable metallic iron particle size was in the range 1.5-30 nm, increasing with higher Fe-content. The same precursors were used in our study (8 and 16 % Fe). They were either calcined in air at 800 K or progressively decomposed in vacuo or reduced, the latter according to the procedure described in ref.79. Thermoferromagnetic curves (as represented in Figs. 6 to 9) have been obtained for the different samples after various pretreatments by plotting the reduced intensity of the FMR signal, $\frac{I}{I_o}$ (where I_o is the intensity at the lowest recording temperature) as a function of temperature. Such thermoferromagnetic curves can give us information on the nature of the ferro- or ferri-magnetic phases which are present (from the value of T_c) and on their state of dispersion (from the observation of superparamagnetism in small particles or bulk-type ferro- or ferri-magnetism). Table 6 shows the various iron species which are ferro- or ferri-magnetic as well as their Curie temperature, T_c, and eventual decomposition products.

It is readily seen, from Fig.6, that very different situations occur depending on the thermal treatment.

Calcination in air leads to the formation of $MgFe_2O_4$ exclusively. This phase must be in a very high dispersion state as it shows a strong superparamagnetic behavior.

When the precursor is progressively evacuated (same thermal treatment as for the reduction, but in vacuo) [79] the final product consists of two iron containing phases which are respectively Fe_3O_4 and $MgFe_2O_4$. After reduction (max.temp.700 K) and evacuation at 670 K, two different iron phases are present. One is metallic iron, the other being $MgFe_2O_4$.

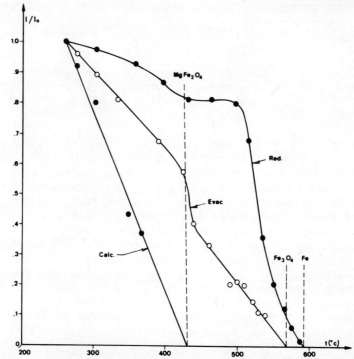

Fig. 6 : Thermoferromagnetic curves for 16 % Fe/MgO samples as obtained by calcination, thermal decomposition in vacuo and reduction

Table 6 : Ferromagnetic and Ferrimagnetic Iron compounds

Compound	T_C (°C)	Decomposition product	Ref.
Fe	770	-	84
Fe_3O_4	585	γ-Fe_2O_3 at 493C for particles < 300 nm	84,88
$MgFe_2O_4$	320-440	-	84,85
α-Fe_2O_3	685	-	86
γ-Fe_2O_3	675	α-Fe_2O_3 at 483 C	87

It is of major importance to note, at this point, that the presence of FeO cannot be ruled out on the basis of the FMR results, as it is not ferro(i)- magnetic.

But for the absolute signal intensity, calcination in air of the 8% Fe precursor leads to the same observation. Hence, the iron phase which is always produced by calcination is $MgFe_2O_4$ and not $\alpha-Fe_2O_3$ or $\gamma-Fe_2O_3$.

When the precursor is treated in vacuo, however, differences appear between the 8 % Fe-and 16 % Fe-samples as shown in fig. 7, where contributions of Fe_3O_4 and $MgFe_2O_4$ are seen. Lower iron content leads to the formation of more Fe_3O_4 with respect to $MgFe_2O_4$. In addition, the Fe_3O_4 phase shows higher superparamagnetic behavior than the $MgFe_2O_4$ phase. These effects can be explained by taking into account the slightly reducing character of thermal treatments in vacuo and considering that its effect will be more pronounced for the smaller particles and lower Fe contents. Hence, Fe_3O_4 domains, which contain iron reduced as Fe^{2+}, must obviously be the smallest, superparamagnetic, and of increasing relative importance with decreasing iron loadings.

Figure 8 compares the thermoferromagnetic curves of the fully reduced 8 % and 16 % Fe/MgO samples. Besides metallic iron, of which the Curie temperature is strongly lowered as a result of an important interaction with the support, one still sees the presence of unreduced iron as $MgFe_2O_4$, the contribution of the latter being more important as the iron concentration increases. From this, one would be tempted to conclude that the fraction of iron in the metallic state decreases with higher Fe content in strict contradiction with previously reported data [79]. Therefore, we have to consider the presence of FeO, which is not seen by FMR, as suggested by Mössbauer and X-ray diffraction data.

The question, however, arises whether FeO is formed by H_2 reduction of Fe^{III} containing compounds or by reaction between Fe and magnesium ferrite which is present. Although H_2 reduction is certainly occuring, the latter process can be evidenced by comparing the thermoferromagnetic curves of samples progressively re-evacuated at high temperature. From fig. 9, one sees that $MgFe_2O_4$ disappears with further evacuation treatments at 870K. This could be attributed to the formation of Fe_3O_4 by :

$$4MgFe_2O_4 + Fe \rightarrow 4MgO + 3Fe_3O_4$$

or MgO-FeO, according to :

$$MgFe_2O_4 + Fe \rightarrow MgO + 3FeO$$

As no Fe_3O_4 is identified by FMR, the second possibility must be preferred.

Our results are then found to be in very good agreement with those obtained previously; the strong iron-support interaction is

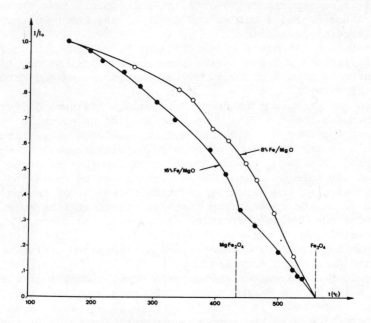

Fig. 7 : Thermoferromagnetic curves for 8 % and 16 % Fe/MgO samples as obtained by thermal decomposition in vacuo

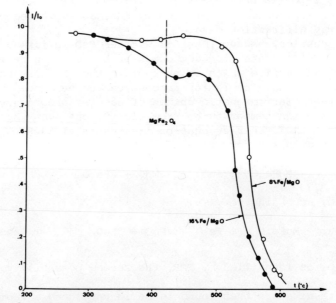

Fig. 8 : Thermoferromagnetic curves for 8 % and 16 % Fe/MgO samples obtained after reduction.

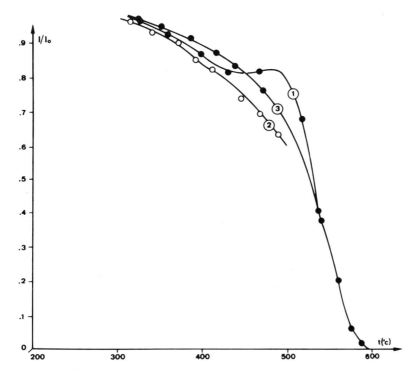

Fig. 9 : Thermoferromagnetic curves for 16 % Fe/MgO samples
1. after reduction at 700 K and evacuation at 670 K
2. after reduction at 700 K, evacuation at 670 K and reevacuation at 870 K for 90 min.
3. after reduction at 700 K, evacuation at 670 K and reevacuation at 870 K for 3 hrs.

confirmed and additional evidence is brought for the formation of FeO.

These observations are summarized in Fig. 10. They are complementary to those reported from Mössbauer, X-ray diffraction, or chemisorption studies and they put forward the major role played by the $MgFe_2O_4$ phase which is observed after all three pretreatments.

These results show clearly the interest of applying FMR for the investigation of surface and bulk processes. The former example was dealing with oxidic phases, but one should emphasize that FMR is equally good for studies on metals and alloys whenever those show ferromagnetism.

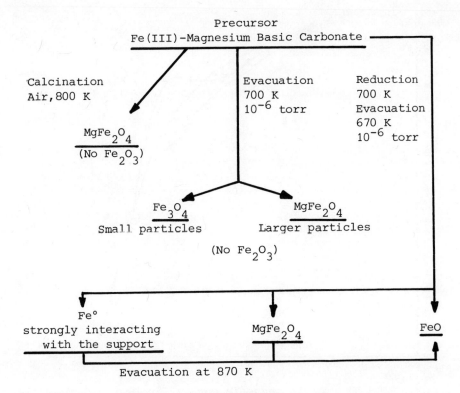

Fig. 10 : Schematic decomposition of the precursor Fe(III)-magnesium basic carbonate for various pretreatments (air, vacuum, hydrogen) (89).

CONCLUSIONS

Writing a comprehensive review on the subject of the present lecture would lead to writing a monograph in addition to those already existing. Restricting our presentation to our own results would have been equally inadequate, by limiting the list of concerned people to those directly interested in our work.

Hence, we preferred to cover some sort of a neither too narrow nor too wide area more or less centered on magnetic resonance techniques and uniquely defined by our present interest (and knowledge). This sure implies a sometimes arbitrary (and eclectic) selection of examples and we apologize to those whose work has not been referred to in this lecture... as well as to those who know of better examples.

In view of the recent technical progress and the growing interest of an increasing number of laboratories, it is our opinion that magnetic resonance techniques, of which some applications have been discussed in this paper, will prove in the near feature to be of increasing interest for the study of solid state and surface reactions, mostly when used in combination with other spectroscopic or non-spectroscopic techniques.

ACKNOWLEDGEMENTS

We would like to thank Dr. J. B.Nagy for his cooperation in collecting and discussing the ^{13}C-NMR data mentionned in this paper. We are also gratefull to the "Laboratoire de Spectroscopie Electronique" of the F.U.N.D.P. for permission to use their X.P.S. data.

BIBLIOGRAPHY

(1) See for examples, as reviews of progresses :
   a) T.N. Rhodin, see reference 3, pp. 1, 163, 195
   b) A.W. Czanderna, ed., Methods of surface analysis, Elsevier Sci. Pub. Co., Amsterdam (1975)
   c) E. Drauglis and R.I. Jaffee, eds., The Physical Basis for Heterogeneous Catalysis, Plenum Press, New York (1976)
(2) J.W. Mayer, J.M. Poate, and King-Ning Tu, Science, 190 (1975) 228
(3) E.G. Derouane and A.A. Lucas, eds. "Electronic Structure and Reactivity of Metal Surfaces", Plenum Press Corp., New York (1976)
(4) a) O.V. Krylov, Catalysis by Non Metals, Academic Press, New York (1970)
   b) A. Cimino, Chimica e Industria, 56 (1974) 27
   c) F.S. Stone, J. Solid. St. Chem., 12 (1975) 271
(5) R.E. Honig, Thin Solid Films, 31 (1976) 89-122
(6) R.E. Honig, in "Advances in Mass Spectroscopy", vol. 6, A.R. West, ed., Elsevier, 1974, pp. 337-362
(7) J.A. Panitz, C.R.C. Critical Reviews in Solid State Science, (1975), 153-178
(8) R.L. Palmer and J.N. Smith, Jr., Catal. Rev. - Sci. Eng., 12 (1975) 279
(9) R.J. Madix, J. Vac. Sc. Technol., 13 (1976) 253
(10) A.M. Bradshaw, L.S. Cederbaum, and W. Domcke, Structure and Bonding, 24 (1975) 133
(11) P. Cannesson et C. Defossé, J. Micr. Spectr. Electr., (1976) in press
(12) R. Heckingbottom, Physics in Technology, (1975) 47
(13) H.H. Brongersma, F. Meijer and H.W. Werner, Philips Tech. Rev., 34 (1974) 357
(14) M. Barber, see reference 3, pp. 459
(15) a) S. Anderson, see reference 3, pp. 289
   b) B. Holland, Ibid, pp. 267

c) G.E. Rhead, Ibid, pp. 229
(16) D. Bianchi and S.J. Teichner, Bull. Soc. Chim. France (1975) 1463
(17) R.G. Greenler, J. Vac. Sci. Technol., 12 (1975) 1410
(18) K. Horn and J. Pritchard, Surf. Sci., 52 (1975) 437
(19) R.C. O'Handley and D.K. Burge, Surf. Sci., 48 (1975) 214
(20) N.M. Brown and D.G. Nalmsley, Chem. Brit., 12 (1976) 92
(21) R.W. Rice and G.L. Haller, J. Catal., 40 (1975) 249
(22) R. Mason, private communication, to be published
(23) J.H. Lunsford, Catal. Rev., 8 (1973) 135
(24) A.J. Tench, T. Lawson and J.F.J. Kibblewhite, J. Chem. Soc. Faraday Trans. I, 68 (1972) 1169
(25) Y. Ben Taarit, C. Naccache and A.J. Tench, J. Chem. Soc., Faraday Trans. I, 71 (1975) 1402
(26) J.P. Joly, J. Chim. Phys., (1975) 135
(27) B. Halpern and J.E. Germain, C.R. Acad. Sci., Ser. C, 278 (1973) 1287
(28) O. Johnson, Chemica Scripta, 8 (1975) 162
(29) M. Oku, K. Hirokawa, and S. Ikeda, J. Electr. Spectr. Rel. Phenom., 7 (1975) 465
(30) J.P. Contour and G. Mouvier, J. Catal., 40 (1975) 342
(31) N.S. McIntyre and M.G. Cook, Anal. Chem., 47 (1975) 2208
(32) Y. Moro-Oka, Y. Morikawa and A. Ozaki, J. Catal., 7 (1967) 23
(33) J.M. Criado, J. Catal., 37 (1975) 563
(34) J.H. Lunsford, Advan. Catal. Rel. Subj., 22 (1972) 265
(35) E.G. Derouane and J.C. Védrine, Ind. Chim. Belges, 38 (1973) 375
(36) V.A. Shvets, V.M. Vorotsyntsev, and V.B. Kazanskii, Kinet. Katal., 10 (1969) 356
(37) V.A. Shvets and V.B. Kazanskii, J. Catal., 25 (1972) 123
(38) N.B. Wong, Y. Ben Taarit and J.H. Lunsford, J. Chem. Phys., 60 (1974) 2148
(39) R.H. Bartram, C.E. Swenberg, and C.T. Fournier, Phys. Rev., 139 (1965) A941
(40) K. Klier, J. Catal., 8 (1967) 14
(41) M. Boudart and H.S. Hwang, J. Catal., 39 (1975) 44
(42) M. Mahnig and E. Wicke, Z. Naturforsch., 24a (1969) 1258
(43) R.A.B. Devine, J.C.H. Chiu, and M. Poirier, J. Phys. F., 5 (1975) 2362
(44) G. Alquié, G. Sadoe, A. Kreisler and J.P. Burger, 18th Ampere Congress, Nottingham, 1974, p. 99
(45) T. Ito and T. Kadowaki, Physics Lett., 54A (1975) 61
(46) P. Brill and J. Voitlander, Ber. Bunsenges. Physik. Chem., 77 (1973) 1097
(47) T. Ito, T. Kadowaki and T. Toya, Jap. J. Appl. Phys., Suppl.2, Pt. 2, (1974), p. 257
(48) H. Asada, T. Toya, H. Motohashi, M. Sakamoto and Y. Hamaguchi, J. Chem. Phys., 63 (1975) 4078
(49) B.N. Ganguly, Phys. Rev. B., 8 (1973) 1055
(50) T. Ito and T. Kadowaki, Jap. J. Appl. Phys., 14 (1975) 1673
(51) Ph. Yee and W.D. Knight, Phys. Rev. B, 11 (1975) 3261

(52) G.J. Krüger and R. Weiss, 18th Ampere Congress, Nottingham, (1974) 339
(53) H.K. Birnbaum, Scrip. Metallurgica, 7 (1973) 925
(54) T. Kumagai, T. Ito, and T. Kadowaki, to be published
(55) E. Van Meerwall and D.S. Schreiber, Phys. Lett., 27 (1968) 574
(56) Data from "Laboratoire de Spectroscopie Electronique", Fac. Univ. Namur; similar data available in the literature
(57) L.R. Windmiller, J.B. Ketterson, and S. Hörnfeld, Phys. Rev.B, 3 (1971) 4213
(58) J.W. Simons and T.B. Flanagan, Can. J. Chem., 43 (1965) 1665
(59) J. Butterworth, Proc. Phys. Soc. (London), 83 (1964) 71
(60) W.A. Pliskin and R.P. Eishens, Z. Physik. Chem., N.F., 24 (1960) 11
(61) I.D. Gay and J.F. Kriz, J. Phys. Chem., 79 (1975) 2145
(62) J.B. Stothers, Carbon-13 NMR Spectroscopy, Academic Press, New York, 1972
(63) G.C. Levy and G.L. Nelson, Carbon-13 Nuclear Magnetic Resonance for Organic Chemists, Wiley Interscience, New York, 1972
(64) D. Michel, Surf. Sci., 42 (1974) 453
(65) D. Michel, Z. Phyz. Chemie, Leipzig, 252 (1973) 263
(66) D. Michel, W. Meiler, H. Pfeifer, J. Molec. Catal., 1 (1975)85
(67) D. Deininger, D. Geschke and W.D. Hoffman, Z. Phyz. Chemie, Leipzig, 255 (1974) 273
(68) I.D. Gay , J. Phys. Chem., 78 (1974) 38
(69) D. Michel, W. Meiler, and D. Hoppach, Z. Phyz. Chemie, Leipzig, 255 (1974) 509
(70) E.G. Derouane, J.B. Nagy, M. Gigot and A. Gourgue,to be published
(71) S. Sakaki, Theoret. Chim. Acta, 30 (1973) 159
(72) N.D. Chuvylkin, G. M. Zhidomirov, and V.B. Kazanskii, J. Catal., 38 (1975) 214
(73) E.G. Derouane, J. Fraissard, J.J. Fripiat and W.E.E. Stone, Catal. Rev., 7 (1972) 121
(74) A. Ghorbel, C. Hoang-Van, and S.J. Teichner, J.Catal.,33(1974)123
(75) P.A. Jacobs, L.J.Delcerck, L.J.Vandamme,and J.B. Uyterhoeven, J.C.S. Faraday, 71 (1975) 1545
(76) M. Boudart, R.L.Garten and W.N.Delgass, Mem.Soc.Roy.Sci.Liège, 1 (1971) 135
(77) H.M.Gager and M.C.Hobson, Jr., Catal.Rev.-Sci.Eng.,11 (1975)117
(78) W.R.Cares and J.W.Hightower, J. Catal., 39 (1975) 36
(79) M. Boudart, A. Delbouille, J.A. Dumesic, S. Khammouma, and H. Topsøe, J. Catal., 37 (1975) 486
(80) J.A. Dumesic, H. Topsøe, S. Khammouma and M. Boudart, J. Catal., 37 (1975) 503
(81) J.A. Dumesic, H. Topsøe and M. Boudart, J. Catal., 37 (1975)513
(82) R.L. Garten and D.F. Ollis, J. Catal., 35, 232 (1974)
(83) M.A. Vannice and R.L. Garten, J. Molec.Catal., 1(1975/76) 201
(84) C. Kittel, Introduction to Solid State Physics, J. Wiley, New York (1971)
(85) American Institue of Physics Handbook, 3rd Ed., Mc Graw Hill
(86) F. Van der Woude, Phys. Status Solidi, 17 (1966) 417

(87) A.H. Morrish and G.A. Sawatzky, Proc. Intern. Conf. Ferrites, (Japan, 1970), p. 467
(88) T.K. McNab, R.A. Fox, and A.J. Boyle, J. Appl. Phys., $\underline{39}$ (1968) 5703
(89) E.G. Derouane and Ph. Monseur, to be published

THE INVESTIGATION OF SOLID STATE REACTIONS WITH EPR

W. Gunsser and U. Wolfmeier

Institut für Physikalische Chemie der
Universität Hamburg

D-2000 Hamburg 13, Laufgraben 24
Federal Republic of Germany

1. INTRODUCTION

In solid state chemistry a great number of studies has been carried out with metal oxides. Their magnetic properties have been thoroughly investigated: ferrites, and magnetic garnets, for example, play a major role in microwave technology.

The formation of these materials can be studied by using magnetochemical methods because the susceptibility of the reactants differs from that of the product. Ferrimagnetic yttrium-iron garnet (YIG) is formed by solid state reaction between antiferromagnetic α-Fe_2O_3 (hematite) and diamagnetic Y_2O_3.

In diluted magnetic systems, it is more convenient to apply magnetic resonance techniques. The high sensitivity and selectivity of EPR allowed us to obtain notable results on mixed oxide chemistry even if the concentration of magnetic ions could be neglected. First investigations were made by TURKEVICH et al. with sphalerite-wurtzite transformation (1).

In previous studies we examined the Cr_2O_3-Ga_2O_3-system. The EPR spectra of Cr^{3+} indicated the phase transition between hexagonal α-Ga_2O_3 and monoclinic β-Ga_2O_3.

We extended the application of EPR to more complicated reactions. The formation of diamagnetic ternary compounds can be detected after incorporation of very small quantities of tracer ions. Following some work on spinels $MgOCr_2O_3$ (2) we chose $Y_3Ga_5O_{12}$, an end-member of the group of rare earth (RE)garnets, as a model substance. Though garnets have a highly complicated unit cell containing 160 ions there are some advantages which caused us to choose this oxide system. RE-garnets owing the general formula $\{A_3^{3+}\}[B_2^{3+}](C_3^{3+})O_{12}$ have a cubic structure with a nearly close-packed oxygen ion sublattice. The cations are distributed over three sublattices: Trivalent RE-ions are surrounded by 8 O^{2-}-ions forming a distorted cube (dodecahedral c site), the smaller Ga^{3+}-ions prefer tetrahedral and octahedral crystal fields (a site, d site). Since these sublattices are completely filled garnets possess sharp stoichiometry and extremely small electronic conductivity, an important difference to most spinels and perovskites. $Y_3Ga_5O_{12}$ (YGaG) is formed by solid state reaction with yields up to 100 %. Yttrium gallium garnet is thermodynamically stable up to its melting point.

2. EXPERIMENTAL PROCEDURE

We compared several preparation methods. The most reactive starting material was obtained from a thermally decomposed mixture of $Y(OH)_3$. The hydroxides were precipitated from aqueous solutions of the nitrates (including tracer ions) by means of buffered ammonia solution. A second preparation method merely required simple evaporation of an aqueous solution of nitrates followed by thermal decomposition of the residue. A third method made use of mixing powdered Y_2O_3 and Ga_2O_3. Tracer ions had to be incorporated into one of the reactants before firing the material.

EPR measurements of the reacted samples have been carried out on commercial EPR equipment (AEG, BRUKER) using X-band frequencies. All samples were examined after quenching.

3. GARNET FORMATION

The principle way of observing the garnet formation may be demonstrated by figure 1. Trivalent chromium substituting Y^{3+} and Ga^{3+} shows definite EPR spectra in the reactants as well as in the product phase. Thus YGaG formation can be studied by comparing the intensities of increasing garnets lines and vanishing Y_2O_3- or Ga_2O_3-lines.

Cr^{3+} represents merely one of several tracer ions which may be used. Finding a most suitable dopant is generally crucial. The following boundary conditions have to be fulfilled:

1.) The EPR spectra of reactants and product phase must differ from each other.

2.) The paramagnetic ion should prefer o n e site of the lattice (e.g., Cr^{3+} will only be found in octahedral coordination).

3.) The diffusion coefficient of dopant and replaced matrix ions should be approximately the same.

4.) The tracer ion should preferably posess the same valency as the matrix ion. According to SCHWAB et al. (3), different valencies will exert a great influence on reaction rates and possibly on reaction mechanism.

SOLID STATE REACTIONS AND EPR

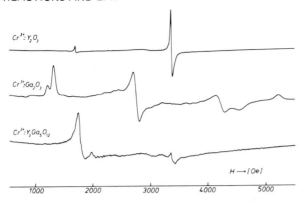

Fig. 1: EPR spectra of Cr^{3+} - containing Y_2O_3, Ga_2O_3, and $Y_3Ga_5O_{12}$ at room temperature.

As mentioned above, Cr^{3+} fulfills these conditions, but there are two disadvantages: Firstly, Cr_2O_3 is volatile, especially at temperatures above 1200°C, and secondly, Cr^{3+} may be oxidized. Due to these effects we found decreasing Cr^{3+} EPR line intensities when firing Cr_2O_3-Ga_2O_3 mixtures at high temperatures.

Manganese ions are not well suited because they are divalent and produce very complicated spectra. More information can be drawn from Fe^{3+} spectra, but Fe^{3+} ions may enter two sites (a, d). Furthermore Fe^{3+} may be reduced at high temperatures.

The best results were found using Gd^{3+} ions as a dopant. Gd^{3+} ions will exclusively occupy dodecahedral sites, and the ionic radii of Y^{3+} and Gd^{3+} differ only slightly. The spectra of Gd^{3+}:YGaG and Gd^{3+}:Y_2O_3 overlap but can be dissolved. We failed to detect Gd^{3+}:Ga_2O_3 powder spectra.

When observing garnet formation by means of the EPR technique we notice great differences between the first two and the third preparation method. On firing, the samples prepared from hydroxides or nitrates seem to convert directly from the amorphous starting material to the garnet phase. We recorded increasing garnet lines from the very beginning of the reaction but no Gd^{3+}:Y_2O_3 spectra. During the reaction between polycrystalline oxides the garnet lines increase, while the Gd^{3+}:Y_2O_3 signals decrease correspondingly (see figure 2).

We tried to derive quantitative results from our experiments. For the first two preparation methods this seems to be very difficult. If decomposed hydroxides react to form a new phase, diffusion paths are relatively short and phase boundary processes will determine the reaction rate. Since many defects are produced caused by the hydroxides loosing water innumerable nuclei of the new phase are formed and the reaction will cease after a short time. The minimum temperatures for garnet formation were found to be about 700°C for the first method and about 1050°C for the second method.

The fraction of reacted powder cannot be measured from peak intensities of the EPR lines. Neither peak-to-peak intensi-

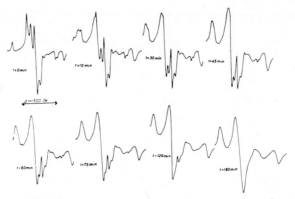

Fig. 2: Development of the central part of Gd^{3+} EPR spectra during YGaG formation

ties nor the integrated areas are directly proportional to the amount of the garnet phase. The problem to gauge the relation between intensities and amount of reaction product, has been solved for the reaction between mixed oxides. We prepared standard samples containing known proportions of reactants and product. Comparison of these samples with our reacted powders enabled us to estimate the fraction of reaction (see figure 3).

In order to get kinetic data from these curves a reaction model has to be postulated which fulfills the boundary conditions. It is difficult to evaluate the velocity constants or the activation energy without knowing the exact reaction mechanism. There are profound reasons for assuming a "Wagner-like" diffusion mechanism: Counterdiffusion of cations will prevail if the electronic conductivity of the oxides in question is extremely small, if the diffusion coefficient of the anions is smallest and if there is no considerable transport via the gaseous phase.

What about the garnet system? As mentioned above, garnets are excellent insulators, and the special structure of the anion lattice - tetrahedrons of O^{2-}, silicate structure - forbids

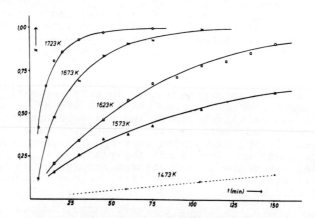

Fig. 3: Fraction of reaction of garnet Formation from Ga_2O_3 and Y_2O_3

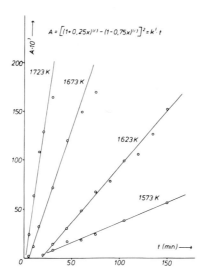

Fig. 4: Plots of KOMATSU equation versus time for our EPR results (x = fraction of reaction)

remarkable diffusion of oxygen. Gas phase transport will only take place at extremely high temperatures in non-oxidizing atmospheres because Ga_2O_3 will be reduced to volatile Ga_2O.

We showed that the equations of KOMATSU (4) can be applied to our problem (fig. 4). They are based on counterdiffusion of cations. The well known models of JANDER (5) and CARTER (6) cannot be used due to principle mistakes and different boundary conditions, respectively.

Although KOMATSU's equations seem to describe our problems sufficiently, we cannot risk to regard this as the sole proof for the Wagner-mechanism. In further experiments using the TUBANDT technique we intend to reveal the exact mechanism.

REFERENCES

1. J. Turkevich, S. Larach, P.N. Yocum Reactivity of Solids ed ed. by G.M. Schwab Amsterdam 1965 p. 115
2. W. Gunsser, S. Elkhouli Physica Status Solids 1976 in press in press.
3. G.M. Schwab, M. Kohler-Ray, S. Ehrenstorfer Reactivity of Solids, ed. by J.H. de Boer Amsterdam 1961 p. 392
4. W. Jander Z. anorg.allg.Chem. 163, 1 (1927)
5. W. Komatsu Z. Phys. Chem. N.F. 72, 59 (1970)

COMPUTER SIMULATION OF CRYSTAL DISSOLUTION

A.I. Michaels[*] and M.B. Ives

Institute for Materials Research, McMaster University

Hamilton, Ontario, Canada, L8S 4M1

ABSTRACT

A computer simulation programme has been developed which models the atomistic processes occurring during free evaporation from a {100} surface of a perfect crystal of copper. The Morse pair-wise potential function has been used to deduce the activation energies for diffusion on, and evaporation from, this surface.

The simulation demonstrates that the activation energies obtained from the Morse potential function produce an overwhelming majority of surface events which do not contribute to a net change in surface morphology. It will be necessary to modify the programme in order to bypass correlated events and speed up the simulation and permit comparisons with real-time observations.

INTRODUCTION

There is a pressing need for a theoretical basis which will allow an understanding of the experimental observations of crystal morphology produced during evaporation and dissolution processes, (1,2). Some progress has been made towards a theoretical understanding of the dissolution kinetics of simple one-dimensional surface structures but an analysis of the multivariant processes occurring during the development of surface morphologies has better

---

[*]Present address: Argonne National Laboratory, Argonne, Illinois 60439, USA.

prospects for success through the application of simulation techniques.

In the last Conference in this series, a simple computer simulation model for evaporation from a simple cubic crystal was described, (3). The study reported here has simulated evaporation from, and diffusion on, the {100} surface of copper.

ACTIVATION ENERGIES FOR DIFFUSION AND EVAPORATION

The pair-wise potential proposed by Morse, (4), gives the potential energy, Q, of an atom as a function of the number and distance of the other atoms in the crystal as,

$$Q = \sum_j N_j^1 D \{\exp[-2\alpha(r_j - r_0)] - 2\exp[-\alpha(r_j - r_0)]\} \quad (1)$$

where N_j^1 is the number of neighbours of the jth. type distance r_j from the atom under study, r_0 is the interatomic spacing and α and D are constants, values of which for copper have been estimated by Girifalco and Weizer, (5).

Activation energies for diffusive motion and evaporation for various types of surface site can be computed by digital computation of the potential energy, Q, of a particular atom as it rolls across the other atoms beneath. Energy maxima are searched for along the path and the difference in the energy at the maximum and at the original lattice site is adopted as the activation energy for motion along that particular path. The initial configuration of surface atoms was taken from the calculations of lattice relaxation by Jackson, (6), but no additional relaxation during atom motion has been permitted in the activation energy calculations reported here. It was found to be necessary to compute the lattice sums of equation (1) to the 7th neighbours. In order to distinguish between the different types of surface site, activation energies were computed directly for the major types of path geometry, and those for other configurations were estimated by inspection.

Consequently, it has been possible to compute activation energies for surface diffusion in each of the four <110> directions in the (100) plane and for evaporation from that plane. Approximately 350 different surface sites have been identified as having a sufficiently different activation energy with most of the proliferation being due to taking account of missing atoms in layers below that in which the particular site is located.
Table I gives some selected values of the activation energies obtained for the more simple and common types of site. The direction of diffusive motion has been identified by the vectors, A, B, C, D as indexed in the Table. Inspection of the values reported

Table I

Activation energies (eV/atom) for diffusion and evaporation of typical surface atom types on a {100} copper surface, using the Morse pair-wise potential.

Type	B ↑ C←X→D ↓ A	DIFFUSION				EVAPORATION
		A	B	C	D	
	Adatoms					
1	⊠	.44	.44	.44	.44	2.3
23	⊠	.28	.49	.40	.40	2.3
	At-ledge					
170		2.6	.69	.48	.48	2.6
220		3.0	.94	.57	.57	3.0
232		3.0	.72	.62	.44	3.0
192		2.8	.82	.65	.38	2.8
	Kinks					
278		3.3	.88	.68	3.3	3.3
291		3.2	1.1	.80	3.2	3.2
296		3.6	.94	.94	3.6	3.6
	In-ledge					
310		3.6	.84	3.6	3.6	3.6
328		3.8	1.0	3.8	3.8	3.8

in Table I reveals that the activation energy for diffusion is always smaller for those directions in which the moving atom increases its numbers of near neighbours. Motion which involves stepping up to a higher plane has been arbitrarily assigned an activation equal to that for the evaporation of that type of site.

The activation energies computed from the Morse potential function are consistent with direct observations of atom motion provided by field emission microscopy, (7). Their significance in controlling the kinetics of crystal dissolution processes will become clear from the initial results of the simulation experiments described below.

COMPUTER SIMULATION EXPERIMENTS

A {100} surface of a face centred cubic crystal was simulated by an array of integers (P) whose position is defined by coordinates (M,N). The value of P denotes the plane in which the atom at that position (M,N) sits. The stacking of atoms in the face centred cubic structure results in atoms in even numbered planes being directly above each other and those in odd numbered planes being similarly stacked in the intervening positions. The "face centred" lattice is outlined in the lower left hand corner of Figure 1(a).

The computer simulation was executed on a CDC 6400 digital computer, with a typical run involving 300 events requiring approximately 16 seconds of computer execution time.

After initialization and reading of the starting surface configuration one surface event is simulated by the sequence of actions:

1. Randomly, determine the type of site to be moved.
2. Randomly, determine which particular atom in that type of site is to be moved.
3. Determine, using another random selection process, which direction that atom is to move.
4. Move the atom one spacing in the selected direction.
5. Update the site types of the moved atom and the affected surrounding atoms.
6. Redefine the instantaneous probabilities for the motion of each mobile atom.

Such a sequence of events is repeated for a predetermined number of events before printing the surface status and/or terminating the simulation run. The programme also provides information and statistics on the surface processes which have occurred in whatever level of detail is desired.

Figure 1 represents aspects of a specific simulation run for copper at 1000°C for a total of 300 surface events. The initial surface consisted of an edge (at n = 19) between {100} and {111} surfaces plus an additional array of kink sites forming a "V" shape within the {100} surface as shown in Figure 1(a). Figure 1(b) indicates the net changes which had occurred after a total of 300 simulation events for the conditions noted. It is clear that the changes in morphology are minimal. A detailed investigation of specific atom motions for the first 50 events is summarized by the arrows in Figure 2(a) demonstrating that the most common sequence of events is for an atom which has moved to a position of lower coordination to immediately move back to its original position.

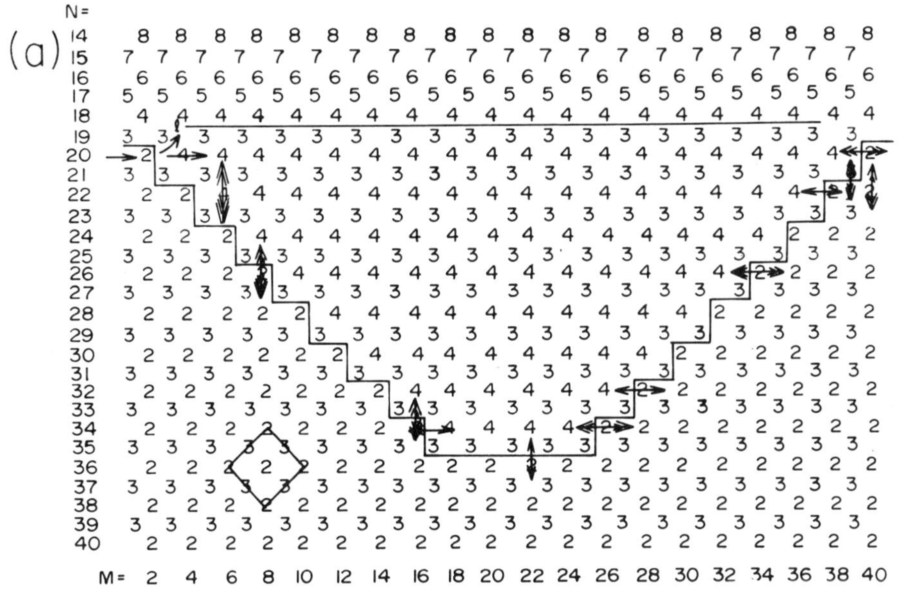

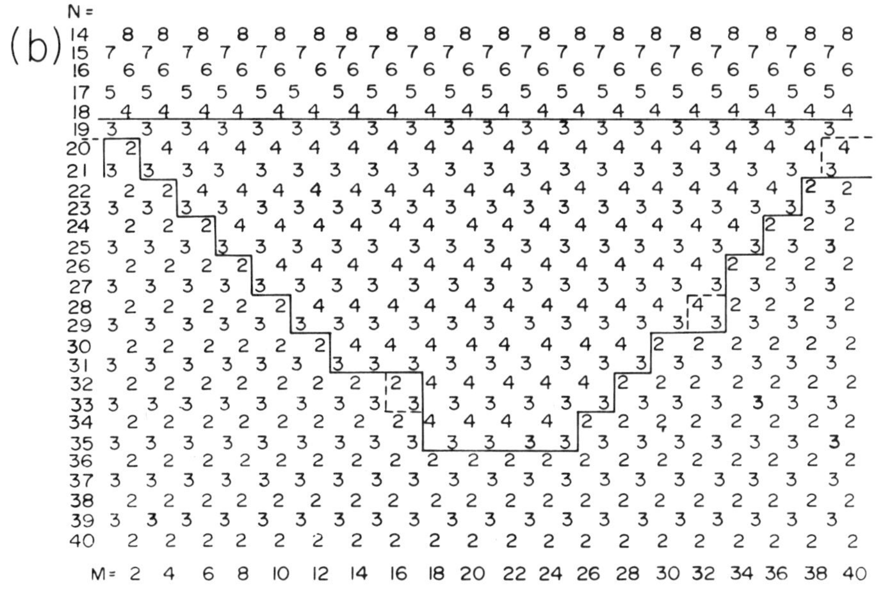

Figure 1

Simulation of evaporation of a {100} copper surface at 1000°C, at (a) the initial (arbitrary) morphology and (b) after 300 events.

These observations are, of course, a direct consequence of the activation energies obtained from the Morse calculation and are to be taken as an appropriate simulation of real surface atom activity for a crystal evaporating into a vacuum. However, in order to simulate the "real-time" events on an evaporating crystal surface it is necessary to speed up the simulation so that those rare combinations of events when, for example, an atom moves away from a kink site to become an adatom and subsequently moves away from the ledge are adequately simulated by the computer run. This requires a modification to the present programme which ignores correlation effects and requires that each atom, once it has moved, subsequently moves to a new position rather than back to the one from whence it came. Such a modification, while simulating less directly the specific atomistic processes, will achieve an acceleration of morphological development which is appropriate to the net morphological changes which occur on real evaporating surfaces.

Further development of the simulation model is planned which will involve the observation of adatom population effects on ledge kinetics and ultimately the effect of adsorbing inhibitors on the dissolution kinetics. These problems are particularly relevant to an understanding of experimental observations in crystal dissolution and evaporation processes, (2).

REFERENCES

1. H. Bethge, Phys. Stat. Solidi $\underline{2}$, 1 (1962).
2. M.B. Ives, p. 78, "Localized Corrosion", NACE, Houston, 1974.
3. C.S. Kohli and M.B. Ives, p. 411, "Reactivity of Solids", Proceedings of the Seventh International Symposium, Chapman and Hall, London, 1972; and J. Crystal Growth $\underline{16}$, 123 (1972).
4. P.M. Morse, Phys. Rev., $\underline{34}$, 57 (1929).
5. L.A. Girifalco and V.G. Weizer, Phys. Rev. $\underline{114}$, 687 (1959).
6. D.P. Jackson, Can. J. Phys. $\underline{49}$, 2093 (1971).
7. D.W. Bassett, Surface Science, $\underline{53}$, 74 (1975).

STUDY OF THE INITIAL STAGE OF THE OXIDATION REACTION OF PURE

TITANIUM AND TA6V4 AT HIGH TEMPERATURE

C. Coddet [*], G. Béranger [*], J. Driole [**] and J. Besson [***]
[*] Laboratoire des Matériaux - Département de Génie Mécanique - Université de Technologie de Compiègne (France)
[**] Laboratoire de Thermodynamique et physico-chimie Métallurgiques
[***] Laboratoire d'Adsorption et de réactions de gaz sur solides
Ecole Nationale Supérieure d'Electrochimie et d'Electrométallurgie de Grenoble (France)

INTRODUCTION

The new highly sensitive techniques of surface analysis (ESCA, SIMS, AES, ...) have opened a large field of research concerning the initial stage of oxidation reactions of metals and alloys.

Although many studies on the early oxidation stage of titanium have been published (1 - 3), we have considered that it would be of interest to apply the new technique of "in situ" photoemission electron microscopy to this problem.

This technique has already been used to study several problems of oxidation (4-5), but it is generally the variation of the work function of the electrons in relation to the oxidation reaction and to the energy level of the incident photons which is measured. So, we shall not deal with this aspect of the problem in this paper, in order to put the emphasis on the structural evolution of the surface.

PHOTOEMISSION ELECTRON MICROSCOPY

The surface of the specimen is excited with a UV beam (2980 $\overset{\circ}{A}$). The photoelectrons emitted are accelerated in a electrostatic field and then ejected into the column of an electron microscope. The resulting magnified electronic image may be viewed on a fluorescent screen or directly recorded under vacuum on photographic plates. The magnifications obtained vary from 200 to 10000.

The distribution of the electronic intensity in photoemission electron microscopy is essentially due to the following phenomena :
- the electron emission differs with phases and substances
- the electron emission varies with crystalline orientation
- the electron emission also varies in space because of the local inclination of the surface to the optical axis of the apparatus.
- the electrical field distortions due to the surface irregularities deflect the electrons and cause a distribution of the intensity of the electronic image, which is different from the real distribution on the surface.

These variations in emission are visualised by contrast effects which occur simultaneously on the image. To eliminate the last two effects with the highest efficiency, the photoemission electron-microscopy is usually limited to the observation of smooth and flat surfaces of unetched samples. Under these conditions, only two effects remain to give the image, and they both depend on the work function of the electrons :
- the contrast due to the materials
- the contrast due to the crystal orientations

SAMPLE PREPARATION AND EXPERIMENTAL PROCEDURE

As the layer thickness which contributes to the emission of electrons is about 100 Å, the evenness and the cleanliness of the observed surface are very important. Moreover, since the structure of titanium and its alloys and the oxidation rates respectively depend on the type of heat treatments undergone by the sample and on its stress state (1) it has been necessary to develop and to use systematically an accurate method for the sample preparation.

First, the specimens were cut out of a cylindrical bar 20 mm in diameter. The disks thus obtained were ground to a thickness of 1,5mm and reduced to 8mm in diameter. They were then polished on abrasive paper, ultrasonically cleaned, annealed under high vacuum at 1100°C for 1 hour, cooled to 600°C and kept at this temperature for 12 hours. The final stage of this preparation was an electrochemical polishing of the surface (6).

The sample was then introduced into the apparatus (Metioscope KE 3 BALZERS) where it was mounted on a small heater capable of raising its temperature to 1200°C. The surface temperature of the sample was measured with a Pt/Pt 10% Rh thermocouple placed in the center of the sample 2 mm under the surface. The gas-solid interaction experiments were performed in the following way : after cleaning the surface by argon ion sputtering, the oxygen was introduced into the system through a leak valve so as to bring the gas pressure to a desired value. Continuous observation of the sample surface during gas admission showed numerous changes occuring on the metal as a result of gas interaction. The experiments were performed with pure titanium and TA6V4 alloy at temperatures between 600 and 900°C and at gas pressures between 1.10^{-7} torr and 8.10^{-5} torr. Changes occuring on the surface were successively recorded on photographic plates.

RESULTS AND DISCUSSION

For both pure titanium and the TA6V alloy, one has to distinguish between two domains : a low temperature domain under about 700°C in which the oxidation reaction is very slow and seems to proceed during the first hours uniformly on each metal grain, and a high temperature domain above 700°C in which the oxidation reaction is more rapid and proceeds according to a nucleation and growth mechanism.

In the low temperature domain ($T < 700°C$), when oxygen is introduced, a strong attenuation of the original crystalline orientation contrast was generally observed probably due to the formation of a microcrystalline oxide film growing at different rates according to the crystalline orientation (Fig. 1). When a vacuum of 10^{-6} torr was again established in the chamber, this primary oxide film remained on the sample even after one hour.

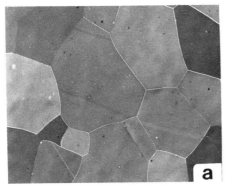 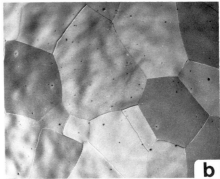

Fig. 1 - Pure titanium observed at 600°C before oxidation (a) and (b) 90 mn after oxygen admission - $P(O)_2 = 4.5 \ 10^{-5}$ torr - X 350

In the high temperature domain ($T > 700°C$), an induction period was observed after the oxygen admission. After this period, the duration of which increased with temperature, nucleii appeared on some grains, their number decreasing with increasing temperature. On the

TA6V alloy, the appearance of these nucleii was preceded by that
of dark areas surrounded with a white fringe (Fig. 2). Then, the
nuclei grew sideways, until the surface was entirely covered.
On some grains, no nucleus appeared which shows the strong influence of the crystalline orientation on the nucleation process ;
but these grains were finally covered by the oxide growing from
the other grains. When the oxygen admission was stopped, the oxide
quickly disappeared, thus showing the fast diffusion of oxygen at
high temperature into the metal from the oxide layer.

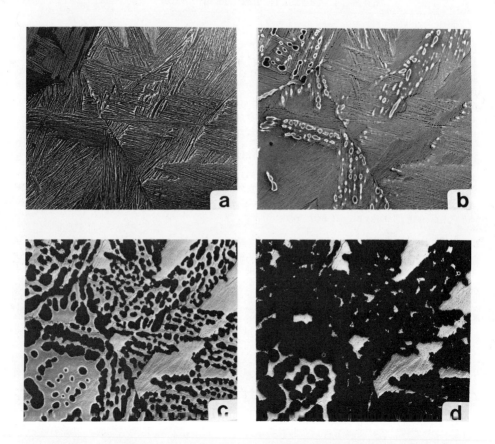

Fig. 2 - TA6V alloy observed at 850°C before oxidation (a)
and after 9mn (b), 34mn (c) and 74mn (d) exposure to an
oxygen pressure of $7 . 10^{-5}$ torr - X 350.

CONCLUSION

Although the shape of the nucleii was not well defined, the
general features seemed to correspond to a nucleation and growth

mechanism, the kinetics of which was strongly influenced by the crystalline orientations. Moreover, the competition between oxygen diffusion into the metal and oxide growth was clearly shown ; the relative importance of the two phenomena varies with temperature, a transition point being observed at about 700°C. Below this temperature, the primary oxide film was quickly formed and remained a long time after keeping the sample under vacuum, while, above this temperature, an induction period was observed and the oxide which formed afterwards disappeared quickly when the sample was kept under a vacuum of 10^{-6} torr, due to the inward oxygen diffusion.

Besides, when cooling the sample, we noticed an important evolution of the surface, probably due to quick growth of the oxide film. Consequently, the appearance of oxide films observed after cooling may not be quite representative of their high temperature structure. Also, "in situ" studies of thin oxide film formation seem to be necessary to decide the correct growth mechanism

Moreover, even for pressures as low as 3.10^{-5} torr, we observed an important reactivity of titanium and its alloy TA6V4 between 600 and 900°C. Therefore, in order to achieve adequate heat treatments of titanium and its alloys, it appears necessary to operate under high vacuum, so as to prevent the surface from any contamination.

The experimental results show that the technique of photoemission electron microscopy is very sensitive to follow continuously an oxidation reaction under low pressure. However, explaining the phenomena only from contrast variations is rather difficult. Therefore, this method should be coupled with a complementary analysis device.

REFERENCES

1. H.M. Flower and P.R. Swann - Acta Metallurgica, 22, 1339 (1974)
2. D.L. Douglass and J. Van Landuyt - Acta Metallurgica, 14, 491 (1966)
3. D.I. :ainer, M.I. Tsypin and A.S. Bai - Kristallografiya 8, (3), 477 (1963)
4. I. Lindau and W.E. Spicer - J. Appl. Phys., 45 (9) 3720 (1974
5. L.F. Wagner and W.E. Spicer - Surf.Sci.,46·(1),301 (1974)
6. L. Rice, C.P. Hinesley and H. Conrad - Metallography,4, 257 (1971)
7. Congres Eurpeen de Metallurgie sous vide - Lille - Octobre 1975

ACKNOWLEDGMENT

The authors wish to thank M. Rouveyre for the preparation of samples and the assistance for using the metioscope.

STUDY OF OXYGEN DIFFUSION IN QUARTZ BY ACTIVATION ANALYSIS

R. Schachtner and H.G. Sockel

Institut für Werkstoffwissenschaften

Lehrstuhl I

Universität Erlangen-Nürnberg

Erlangen, W.-Germany

INTRODUCTION

Oxygen diffusion in crystalline SiO_2 (natural quartz) has been measured by an activation analysis using the reaction $O^{18}(p,n)F^{18}$. Before the activation, the single-crystalline samples with a well-polished surface were annealed in ^{18}O-enriched oxygen gas. The annealed samples were irradiated perpendicular to the polished surface with 2,7 MeV-protons. Part of the O^{18} in the sample is transformed to F^{18} which is a β^+-radiator with a half-life of 110 minutes. The 511 keV annihilation radiation is used as a measure of the ^{18}O-content in the sample.

To establish the relationship between the activity and the diffusion coefficient, the activation probability in dependence on the penetration depth z was determined by computer calculations. The proton-straggling during penetration into the solid was taken into account. Using this calculations the diffusion coefficient could be determined from the activities of annealed and untreated samples with natural content of O^{18}.

Activation Probability as a Function of Penetration Depth

As an incident beam of monoenergetic protons enters the sample, the protons progressively lose their energy through interactions with the solid and at an arbitrary depth z, the initial proton energy is reduced to E(z). However, the interactions leading to this energy loss occur in a random fashion, so that both the energy and the energy spread of the protons are functions of the penetration depth z. In addition, the incident beam is not exactly mono-energetic but has an initial energy spread. Assuming a gaussian distribution for the energy of the protons, the standard deviation (s.d.) of this distribution at a certain depth z can be calculated using Bohr's classical theory of the energy loss of ions in matter. Equations are given by Evans (1). The standard deviation of the depth distribution of protons which have a certain energy E(z) is connected to the s.d. of the energy distribution at depth z by the stopping power of the target material.

The O^{18} (p,n) F^{18} reaction is endoergic with its threshold energy at 2,576 MeV and a resonance peak in the cross-section at 2,643 MeV (2). Because the cross section does not vanish at energies above the threshold, there is a finite probability for protons of interacting with O^{18}, if the protons have energies greater than the threshold energy. Considering the depth distribution of resonant protons, the dn(x) protons which are resonant at depth x will have an energy $E_z(x)$ at depth z and therefore will contribute to the activation at depth z with a cross-section $W_z(X)$. Thus, according to Neild et al. (3), the total activation probability $\Theta(z)$ at depth z is given by

$$\Theta(z) \sim \int W_z(x) \cdot dn(x)/dx \cdot dx$$

Therefore, the total yield of F^{18} in the sample is

$$A \sim \int c(z) \cdot \Theta(z) \cdot dz$$

where c(z) is the O^{18}-concentration versus z. The integrations have to be performed over the overlap of the curves in question.

$\Theta(z)$ versus z curves were determined by computer calculations. Integrations were performed using Simpson's rule. The cross-section versus z, calculated from the cross-section versus energy (2), was used in a tabular form. The stopping power of SiO_2 for protons was calculated using Whaling's additive rule (4) and stopping power data of Williamson et al. (5). The $\Theta(z)$-curves for mean incident proton energies between 2,68 and 2,72 MeV showed a zone of constant value just below the surface due to an approximately constant cross-section at these energies.

DETERMINATION OF THE DIFFUSION COEFFICIENTS

The diffusion profiles were assumed to be of the form

$$c(z,t) - c_o = (c_s - c_o) \cdot \text{erfc}(z/2 \cdot (D_o \cdot t)^{1/2})$$

where c_o is the natural O^{18} concentration and c_s is the surface concentration, assumed to be equal to the O^{18} concentration in the oxygen gas. Therefore, the counting rates (corrected with respect to the radioactive decay) of annealed and untreated samples are given by

$$R = \alpha \int c(z,t) \Theta(z) dz$$

and $\quad R_o = \alpha_o c_o \int \Theta(z) dz$

Assuming a constant counting geometry and equal activation conditions, α and α_o are equal.

According to the $\Theta(z)$ calculations the activation probability shows a constant value $\bar{\Theta}$ in a layer of thickness m below the surface, where m is mainly determined by the stopping power. If the diffusion profile lies in this zone ($(D_o \cdot t)^{1/2} \ll m$), it can be shown that the counting rates and the diffusion coefficient are connected by

$$(D_o t)^{1/2} = b \cdot (R - R_o)/R_o$$

with $b = \Gamma(1.5) \cdot (c_o/(c_s - c_o)) \cdot (\int \Theta(z) dz / \bar{\Theta})$

EXPERIMENTAL

SiO_2 single-crystal wafers with one well-polished surface, oriented parallel or perpendicular to the c axis, were annealed in O^{18}-enriched oxygen gas at 0,21 at in a temperature range from 870 to 1280°C for 25 - 120 h.

The annealed samples, together with some adequately polished samples with natural O^{18}-content, were irradiated perpendicular to the polished surface with 2,7 MeV protons in a Van-de-Graaff accelerator. The 511 KeV-γ-radiation was detected with a NaJ-scintillation-detector equipment.

RESULTS AND DISCUSSION

The obtained data for D_o could be fitted in a plot log D versus 1/T by two parallel lines which are overlapping in the middle temperature range at about 1000°C. In this range the obtained D_o data show a stepwise increase and a strong scattering of data occurs. This effect is probably caused by a slow transformation of β-quartz to β-tridymite, because at the deepest and highest temperatures data-scattering is much smaller. The obtained data for the high and low temperature range can be described by

$$D_o = 1.1 \cdot 10^{-9} \exp(-195 kJmole^{-1}/RT) \, cm^2/s$$

(β-tridymite, 1070 - 1280°C)

$$D_o = 1.1 \cdot 10^{-10} \exp(-195 kJmole^{-1}/RT) \, cm^2/s$$

(β-quartz, 870 - 1180°C)

No relationship between the orientation of the sample-surface relative to c-axis and D_o could be observed. A comparison of the obtained D_o-data with the data of Haul and Dümbgen in polycrystalline quartz (6) shows a quite good agreement.

The condition of constant activation probability leads to the demand $m \gg (D_o \cdot t)^{1/2}$. For annealing times of 25 to 120 h it follows, that the method is applicable for D from 10^{-15} to 10^{-19} cm^2/s if O^{18} is activated in quartz by 2.7 MeV-protons.

Because of the uncertainty of the cross-section, stopping power, and concentration data, due to the necessity of using computed estimates, a possible systematic error of $\Delta D_o/D_o = 0.75$ was estimated. The assumptions of an erfc-diffusion profile and of fast surface reaction compared with the ^{18}O volume diffusion are, of course, possible sources for systematic errors, too.

ACKNOWLEDGMENT

The authors wish to thank the "Deutsche Forschungsgemeinschaft" for financial support.

REFERENCES

(1) D.J. Evans "The Atomic Nucleus" (McGraw-Hill, New York, 1955)

(2) P.M. Beard, P.B. Parks, E.G. Bilpuch, Ann. Phys. (USA), 54, 566 - 97 (1969)

(3) D.J. Neild, P.J. Wise, D.G. Barnes, J. Phys. D: Appl. Phys. 5, 2292 (1972)

(4) W. Whaling; Handbuch der Physik, Vol. 34, p. 13 - 217, Ed.: S. Flügge, Springer, Berlin, 1959

(5) C.F. Williamson, J.P. Boujot, J. Picard, CEA-Report, No. CEA-R-3042

(6) R. Haul, G. Dümbgen Z. Elektrochem. 66, 636 (1962)

THEORETICAL STUDY OF SIMULATED DEHYDRATION OF CRYSTAL SILICA PLANES AND DETERMINATION OF SILICA GEL HYDROXYL DISTRIBUTION BY NMR

J. Demarquay and J. Fraissard

Institut de Recherche sur la Catalyse C.N.R.S.
39, boulevard du 11 Novembre 1918
69 - Villeurbanne - (FRANCE)

Laboratoire de Chimie des Surfaces
Université P. et M. CURIE - Tour 55
75230 Paris Cedex 05 - (FRANCE)

ABSTRACT

Theoretical nuclear dipolar interactions due to OH groups on different silica planes which are more or less dehydrated randomly or homogeneously have been calculated. Comparison with the experimental NMR results reveals the **dehydration** type of the hydroxyl groups for each set of temperatures and the distribution of the remaining OH's on each face of the crystallites.

INTRODUCTION

In this paper it has been shown by NMR spectroscopy that there are no water molecules on the surface of silica gel when the dehydration temperature θ of the sample is higher than 100°C at 10^{-4} torr and that all the hydroxyl groups of this solid are on the surface whenever $\theta \geqslant 350°C$.

Random and homogeneous eliminations of OH groups located on silica planes have been simulated and the corresponding theoretical calculatings of nuclear dipolar interactions have been carried out. Comparison with the experimental results has shown that the distribution of surface hydroxyls is highly inhomogeneous, the inhomogeneity increasing with the level of dehydration. Besides, the dependence of the OH distribution on the different silica planes upon the extent of dehydration has been clarified.

EXPERIMENTAL

Silica gel was prepared by Plank's method (1). Samples were heated for 8 hours under 10^{-4} torr, at a temperature θ between 150 and 800°C. These samples are referred to as Si.θ. In order to determine specifically the surface water properties, hydrogen deuterium exchange has been performed on identical samples, using ND_3 at 25°C. These samples were subsequently reheated under vacuum at the same temperature θ ; they are referred to as Si.θ.D.

The NMR experiments were performed with a VARIAN DP.60 spectrometer operating at 56,4 MHz in the absorption mode.

THEORETICAL DEHYDRATION AND NUCLEAR INTERACTION CALCULATION

Following an idea proposed by Peri, the dehydration of certain planes of β tridymite (110) and (001), (symbol T) and β cristobalite (110), (001) (111) and (100), (symbol C) has been simulated (2). The principle of the method is to assume that, at the beginning, each Si atom which does not bear four siloxane bonds is coordinated with one or more OH groups in order to have its usual coordination number of four. It was also assumed that : a) the dehydration occurs at random, b) all the OH groups are identical, c) the first hydroxyls eliminated are contiguous so that their desorption does not leave any defect on the surface i.e. two adjacent oxygen or silicon atoms.

Let us consider a chart with 30x30 sites, representing a definite crystal plane. Two adjacent sites bearing one OH group each are dehydroxylated so that a bare oxygen and a silicon atoms are formed on the surface. The dehydration would proceed in this manner. Kodratoff and Demarquay showed that it is not necessary to increase the dimensions of the simulated surface on order to obtain reproducible results, but that the average of a large number of random dehydrations must be taken (3). We have therefore chosen a surface with only 30 x 30 sites ; however we have taken the average value of 25 random dehydroxylations. The calculations were made on a IBM 1130 computer, using the program RANDU for IBM 360-75 which has been adapted to the IBM 1130.

For a polycrystalline material, the Van Vleck second moment formula is given by equation 1 (4) :

$$M_2 \cdot G^2 = \frac{6}{5} I(I+1) \gamma^2 \hbar^2 \cdot N^{-1} \sum_{j>k} r_{j,k}^{-6} \quad (\text{Å}) \qquad (1)$$

I and γ are the spin number and the gyromagnetic ratio of the N nuclei at resonance. $r_{j,k}$ is the vector joining the two nuclei.

The theoretical second moment M_2 for several planes and for various degrees of dehydration was determined. In each case we have chosen at random twenty areas having 11 x 11 sites on the previous 30 x 30 surface. For each of them we have calculated the second moment and finally we have averaged the 20 second moments.

RESULTS

It is possible to determine the distribution of hydroxyl groups from the second moment of the signal if the hydrogen atoms are in a rigid lattice. For this reason we recorded the Si 150 and Si 400 signals at 77 K. These are very slightly wider than those recorded at 300 K, the increase of the line width being lower than 10%. Therefore, in view of the difficulties of signal detection at low temperature, we have used the results obtained at 300 K.

When hydrogen-deuterium exchange is performed, we may assume that the non-exchangeable protons Hi of each sample are inside the solid, while the exchangeable ones Hs are on its surface. The comparison of the $Si.\theta$ and $Si.\theta.D$ signals shows that some 28% of the hydroxyl groups are inside the solid Si 150. When θ increases the internal protons migrate to the surface and the relative concentration Hi/Hs decreases. With $\theta \geq 400°C$, all the hydroxyl groups are on the surface. Then, determining the surface area (790 m^2g^{-1} with $\theta \leq 500°C$, 600 $m^2 g^{-1}$ with $\theta=800°C$) and the water concentration, the density $\underline{n}$ of superficial OH groups can be calculated for each sample. When θ increases from 150 to 800°C, this density and the experimental second moment of the corresponding signal decrease from 4.5 to 0.6 OH/100 Å^2 and from 0.21 to 0.02 G^2 respectively. Figure 1 shows the relation between the number $n(OH/100\ \text{Å}^2)$ and the experimental or calculated second moments.

DISCUSSION

For high OH density (about 5 OH/100 Å^2) the experimental M_2 is 0.21 gauss^2, which is greater (by a maximum factor of 2) than that which was calculated for some planes fully hydrated, such as (001)T. With other extensively hydrated planes, such as (110)T or (110)C, on which OH groups are relatively close to one another (approximately 4.2 Å), calculated and experimental M_2 are close. This shows that the latter planes (or others which have similar dipolar interactions) are numerous and highly hydrated when the gel is formed, in comparison with other planes having lower OH densities, therefore low Si densities. In fact, if these latter planes were numerous and covered with OH groups, the experimental M_2 would be lower. This result confirms that the most probable cleavage planes, that is those having a high atomic density constitute the greater part of the surface. We consider separately the plane (100)C which theoretically has the greatest OH density $(8 \text{ OH}/100 \text{ Å}^2)$. Each Si atom can be bonded to two OH's, but these OH's are distributed in pairs. The distance between the two closest OH's on adjacent Si's is shorter than that between the two OH's on the same Si. It is difficult to estimate the value of the angle $Si-O-H$ and the orientation of OH relative to the plane. But the second moments calculated by using reasonable values for these two parameters are always about six times greater than the experimental ones. It follows that the plane (100)C contributes very little to the total OH content, either because it is not present, or because it is easily dehydrated during the first stage of dehydroxylation (θ lower than 150°C) due to its high OH density.

The experimental M_2 value (exp.M_2) decreases from 0.21 to 0.02 G^2 when $n(\text{OH}/100 \text{ Å}^2)$ decreases from 4.5 to 0.6. No M_2 variation relative to the random dehydration of each face is representative of the experimental results for all the OH concentration ranges. This is normal because the crystallite surface is not formed of only one type of plane. In addition, to simplify the calculations only the low index faces were studied although small crystallites would have a great number of the high index faces. First we compare these values with the second moment $(M_2)_m$ calculated on the assumption that the OH distribution is homogeneous. In this case (5):

$$(M_2)_m = 19 \cdot 10^{-4} \, n^3 (\text{OH}/100 \text{ Å}^2) \tag{2}$$

When the OH surface coverage is high, the exp.M_2 value is evidently close to $(M_2)_m$. With this exception however exp.M_2 is always higher than $(M_2)_m$ whatever n. For example the ratio exp.$M_2/(M_2)_m$ increases from about 1-5 with the decreasing of n from 5 to 2.4. This result illustrates clearly that the dehydration of silica gel is very inhomogeneous.

In addition exp.M_2 shows a sharp decrease CD near $n=2 \text{ OH}/100 \text{ Å}^2$, corresponding to a desorption temperature of about 350°C at 10^{-4} torr. This decrease (approximately 50%) caused by a relatively small change in the number of OH groups eliminated ($\Delta n=0.4/\text{H}/100 \text{Å}^2$) can be explained as follows. Exp.M_2 values on the section AC reflect mainly the interaction between the numerous close OH groups; this section AC can be attributed to the random dehydration of the planes having high OH density because of small Si - Si distance [for example (110)T] and which are dehydrated at relatively low temperature. Near 350°C (point C) the preferential desorption of a tiny number of the closest OHs is sufficient to transform a dense OH distribution into another with long distances between hydroxyl groups. For example the hydroxyl groups are located on the plane (110)C at the corner of a 5.09×7.20 Å rectangle with an additional proton at the centre at 4.33 Å from

SIMULATED DEHYDRATION OF CRYSTAL SILICA PLANES

the latters. The desorption of the two closest hydroxyls (4.33 distance) leave the other ones at a relatively long distance. At high treatment temperature exp.M_2 follows the theoretical random dehydration of faces having large intervals between OH groups (variation DG). These intervals are due either to high Si - Si distances on faces, or to the previous dehydration of the closest OH's.

With extensively hydrated surface, the I.R spectrum shows two bands at 3640 and 3400 cm^{-1} associated with weakly and strongly interaction OH groups respectively, pointing out that there are two independent OH distributions (6). The NMR results reveal that the two types of OH are on different planes : the 3640 cm^{-1} band, which disappears completely at 1000°C, corresponds to the OH groups on planes of low Si density. Conversely, the 3400 cm^{-1} band characterises the faces having the closest OH groups, more easily eliminated ($\theta \leqslant 600°C$).

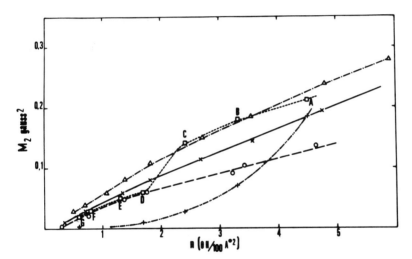

figure 1. Second moment vs the OH concentration. ☐ exp.M_2. A,B...G correspond with θ°C= 150,250,325,400,500,600,800;-..-$(M_2)_m$; simulated random dehydration of planes : ✘(110)C, O(001)T and (111)C, △(110)T.

CONCLUSION

We have measured the nuclear dipolar interactions of the OH groups on silica gel. Using a computer, we have also calculated theoretical values for the different OH distributions on silica planes. Comparison of the results shows that dehydration of silica gel occurs randomly. But it appears that some strongly interacting OH's are eliminated within a narrow range of temperature.

REFERENCES

1. C.I.Plank Catalysis 1, 1954, p 341
2. J.B.Peri J.Phys.Chem.69, 1965, p 290
3. Y.Kodratoff and J.Demarquay C.R.Acad.Sc. Paris 1972, Serie C
    t 274, p 326
4. J.H.Van Vleck Phys.Rev. 1948, 74, 1168
5. J.Fraissard, I.Solomon,R.Caillat, J.Elston and B.Imelik
    J.Chim.Phys. 1963, p 676
6. J.Fraissard and B.Imelik Bull.Soc.Chim.F. 1963, p 1710

THERMAL DECOMPOSITION OF COBALT OXALATE

A. Taskinen, P. Taskinen, and M.H. Tikkanen

Helsinki University of Technology

SF-02150 Espoo, Finland

INTRODUCTION

Cobalt oxalate has two hydrated forms: tetrahydrate $CoC_2O_4 \cdot 4H_2O$ and dihydrate $CoC_2O_4 \cdot 2H_2O$, which dehydrate at about 280 K and 420 K, respectively (1). Cobalt oxalate dihydrate crystallizes during prepicitation as an orthorombic β-phase. However, it recrystallizes in the solution and forms thermodynamically stable monoclinic α-phase (2). Morphology in the crystals is to a great extent due to this transformation.

Co-, Ni-, Pb- and Cu-oxalates have a metallic phase as the primary decomposition product (3-6) as follows

$$MeC_2O_4 \rightarrow Me + 2 CO_2$$

The metal is partly oxidized in secondary reactions by CO_2. It can be shown on the basis of thermodynamics, that CoC_2O_4 produces only 0.1-0.5 mol-% CoO when decomposed at 550-650 K in inert media.

All the previous studies on the decomposition kinetics of CoC_2O_4 have been made in oxidizing atmospheres, the product containing up to 50 mol-% oxide (7-10). The observations show, that the decomposition is a typical nucleation process obeying the Avrami-Erofeev equation.

This study deals with the effects of morphology and particle size on the decomposition kinetics of anhydrous cobalt oxalate in nonoxidizing atmospheres.

EXPERIMENTAL

As raw material two $CoC_2O_4 \cdot 2H_2O$ powders, type 1 and 2, were used. X-ray analysis revealed, that the former contained very small amounts of α-phase, whereas the latter was a well developed α/β-mixture. They were dehydrated at 420 K, under a pressure of 2.7 kPa for 50 h and stored in an exsiccator. Anhydrates were crystalline and their morphology was identical with the corresponding dihydrates, figure 1.

Thermogravimetric measurements were carried out with Mettler H20E semimicrobalance with a sensitivity of ± 0.5 mg. Argon, 10 1/h NPT, was used as the dynamic atmosphere. The gas was purified by passing through silica gel, phosphorus pentoxide and titanium boats at 1200 K. A platinum crucible was used and the amount of the sample was 700 mg.

type 1 (a) type 2 (b)
Figure 1 CoC_2O_4 powder

RESULTS

Decomposition of CoC_2O_4, type 1

Isothermal runs are presented in figure 2 as $\alpha(t)$-plots, where α is the decomposed fraction. The sigmoid shape of the decomposition curve is easily observed. No initial period can be found at the beginning of the main process.

The decomposition obeys the Avrami-Erofeev equation

$$\alpha = 1 - \exp(-kt^n).$$

The best fit with time exponent n = 2 is found at high temperatures, when T > 670 K, over a range α=0.02-0.98 and at low temperatures, when α=0.08-0.98, figure 3.

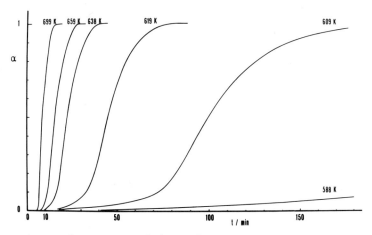

Figure 2   decomposition of cobalt oxalate, type 1

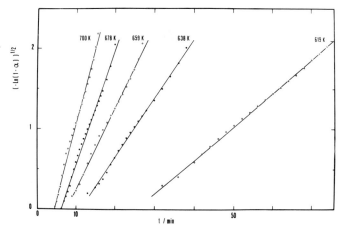

Figure 3   Avrami-Erofeev plot for powder type 1

The activation energy is calculated according to the Arrhenius equation

k   =   A exp(-B/RT),

where the decomposition rate, k, is measured at the fractional decomposition α=0.5; this method was chosen in order to get comparable rate values for both series of decomposition.

Decomposition of CoC_2O_4, type 2

Experimental results are plotted in figure 4. The shape of these curves differs a lot from those of the type 1. A separate initial period is observed and it sets in when the sample has

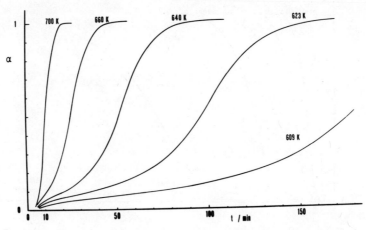

Figure 4 decomposition of cobalt oxalate, type 2

reached the furnace temperature. The reaction is due to surface decomposition obeying the contracting (volume) formula

$$1 - (1 - \alpha/\alpha_o)^{1/3} = kt$$

where a value of $\alpha_o = 0.03$ is estimated. Hereafter the main process is analyzed using corrected fractions of decomposition according to Jacobs (11).

The acceleratory period proceeds immediately after the initial process without separate induction time. This part of the main process can be fitted by the Prout-Tompkins equation

$$\frac{\alpha}{1 - \alpha} = C \exp(kt),$$

when $0.05 < \alpha < 0.2$. However, most of the main reaction obeys the Avrami-Erofeev equation, when the time exponent has the value of n = 3; this fits at high temperatures over a range $0.1 < \alpha < 0.99$ and at low temperatures $0.2 < \alpha < 0.99$. Activation energy for the main process is calculated as previously mentioned.

DISCUSSION

It has been shown, that the limiting step in the thermal decomposition of anhydrous cobalt oxalate in nonoxidizing atmosphere is nucleation. It obeys the Avrami-Erofeev equation, but the mechanism of nucleation and nucleus growth depend on the morphology and particle size.

Coarse grained oxalate decomposes without surface process preceding the main reaction. It is very obvious that nucleus growth takes place in two dimensions if zero nucleation rate - all nuclei initially present- is considered. This mechanism agrees with the results of Kadlec et al.(10) in oxidizing atmospheres.

Fine oxalate powder decomposes in the first stage through a surface reaction. The main process begins after the initial period with chain nucleation, which is presumed to proceed on crystallite surfaces (11). However, the main reaction can be plotted according to the Avrami-Erofeev equation having time exponent $n = 3$. This can be due to three dimensional growth of nuclei -if zero nucleation rate is supposed- or to two dimensional growth with constant nucleation rate. The latter is presumed to be more probable in the view of the mechanism in the acceleratory period.

The activation energy has two values depending on temperature and it is almost identical for different types of oxalates. At low temperatures, $T < 665$ K for type 2 and $T < 625$ K for type 1, we get
$$B(1) = 149.2 \pm 0.3 \text{ kJ/mol}$$
$$B(2) = 120.0 \pm 0.5 \text{ kJ/mol}$$
and at high temperatures respectively
$$B(1) = 56.7 \pm 0.3 \text{ kJ/mol}$$
$$B(2) = 59.0 \pm 0.2 \text{ kJ/mol}.$$
For the acceleratory period in the decomposition of type 2 the activation energy $B'(2) = 130$ kJ/mol is obtained indicating the same limiting factor of the nucleation during both periods. The activation energies obtained here at low temperatures agree fairly well with the literature data (8-10).

This study shows that the decomposition mechanism of CoC_2O_4 is determined by the prior history of the powder.

REFERENCES

1. Avond G.,Pezerat H.,Lagier J-P.,Rev.Chim.Min.,6(1969)1095
2. Deyrieux R.,Berro C.,Peneloux A.,Bull.Soc.Chim.France(1973)1,25
3. Robin J.,Bull.Soc.Chim.France(1957)1078
4. Brown J.,Dollimore D.,Dollimore J.,Heynes A.,in "Progress in Vacuum Microbalance Techniques" Vol 1,Ed. by T.Gast,E.Robens, Heyden&Son,1972,217
5. Dollimore D.,Griffiths D.,Nicholson D.,J.Chem.Soc.(1963)2617
6. Herschkowitsch M.,Z.anorg.allg.Chem.,115(1921)159
7. Kornijenko V.,Ukr.Him.Zh.,23(1957)159
8. Broadbent D.,Dollimore D.,Dollimore J.,J.Chem.Soc.(A)(1966)1491
9. Kadlec O.,Rasmusova J.,Coll.Czech.Chem.Commun.,31(1966)4324
10. Kadlec O.,Danes V.,Coll.Czech.Chem.Commun.,32(1967)1971
11. Jacobs P.,in "Materials Science Research" Vol 4,Ed. by T.Gray, V.Frechette,Plenum Press,1969,37

DISCUSSION

Z.G. SZABO (L. Eötvös University, Budapest, Hungary) In our study of the decomposition of silver oxalate we found the formal kinetics to be the same as the present study and that the rate was influenced by the atmosphere. The acceleration produced by hydrogen or helium was greater than in vacuo, the acceleration decreasing with increasing molecular size for other gases. This can give information on the nature of the nucleaus as was shown in the case of oxygen, carbon dioxide and silver oxalate, where interaction produced a dramatic inhibiting effect.

AUTHORS REPLY: We have also made decompositions in the presence of hydrogen and a N_2-10% H_2-mixture, but practically no accelerating or decelerating effects were observed. The reaction rates were of the same order of magnitude as the results presented above.

J. GUENOT (University of Paris 7, Paris 5, France) In our laboratory we have studied oxalates for several years. We find a number of dihydrated oxalates (Mg,Fe,Co,Ni,Zn,Mn) exhibit the same regular monoclinic structure, the α-oxalates. The β-oxalates, considered as a new phase by some authors, also have the α structure but with some disordered stacking faults (1). For Co-oxalate a new ordered structure has been obtained (2). Structural variations have been found to affect the kinetics of thermal decomposition (3). For anhydrous oxalates, we found that an ex-αoxalate decomposes at a slightly higher rate than an ex- βsample (4). Discrepancies with the results presented here may be due to difficulty in obtaining significant results from kinetic experiments with solid-gas reactions.
1. H. Pezerat, J-P. Lagier, J. Dubernat Bull.Soc.Chim.France 1000,1975
2. G. Avend, H. Pezerat, J-P Lagier, J.Dubernat Rev.Chim.Min. 6,1081,(1975)
3. F. Fievet, J-P. Lagier, H. Pezerat, J. Dubernat C.R.Acad.Sci. 271C,549,(1970)
4. J-P. Lagier. Tesis, Paris 1970. CNRS AO 3492.

AUTHORS' REPLY: We too have studied the precipitation of cobalt oxalate dihydrate and the parameters controlling the morphology, the α/β -ratio and the particle size of the oxalate. In addition we have studied the dehydration of different kinds of oxalates and our results agree very well with the results from your laboratory, if we remember that the correlation between the α/β -ratio and the morphology of the oxalate is not as clear as is assumed in the literature. This paper shows that the β-phase is not always needle-like and the α-phase is not always cube form. According to our results the cube form oxalate anhydrate decomposes with a slightly higher rate than the needle-like one, independently of the α/β -ratio in the original dihydrate.

D. DOLLIMORE (University of Salford, U.K.) Could you give details of the calculation of the relative amounts of Co metal and oxide expected in the residue and indicate how it varies with temperature? I notice a systematic scatter about the drawn line in Fig.3 in the experimental points. We notice similar phenomena in our results I wonder do you have any comment on this observation?

AUTHORS' REPLY: The amount of oxide after decomposition can be calculated as follows: Decomposition of the oxalate constrains a molar ratio of Co:$2CO_2$. Consider the reaction between metallic Co and gaseous CO_2

$$Co + CO_2 \rightleftharpoons CoO + CO \qquad \{1\}$$

The law of mass action gives:

$$K = \frac{P(CO)}{P(CO_2)}$$

which can be calculated as a function of temperature using literature data (1). The stoichiometry of reaction 1 gives: $P(CO) = x$ $P(CO_2) = 2-x$, Hence using the standard Gibbs function,

$$x = \frac{2}{1+\exp(\Delta G^o/RT)}$$

where ΔG^o is the Gibbs Free Energy of reaction $\{1\}$. The amount of the oxide after decomposition of anhydrous cobalt oxalate, x, is plotted as a function of temperature in figure 1.

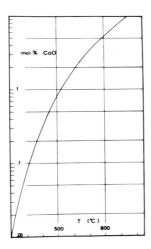

Figure 1. The amount of CoO after decomposition of anhydrous cobalt oxalate.

It is found that x is independent of dilution in inert gases i.e. nitrogen or argon.

From the calculations it can be shown that cobalt oxalate has the metal phase as the primary decomposition product.

To the last question we have no valid explanation.
1. Bugden W. Pratt J. Inst.Min.Met.Trans.(1970) C 221
   Peters H. Möbius H. Z.Phys.Chem. <u>209</u>, (1958) 298.

P. GRANGE (Catholic University of Louvain, Louvain-la-Neuve, Belgium)
It is possible to obtain additional information from your results using MEMPEL theory and the pre-calculated curves developed subsequently (1). For example, comparison of your results presented in Fig. 2 with computed curves shows good agreement for the constant nucleation rate situation and thus confirms your hypothesis. One can also calculate the rate of nucleation and provided that more information on crystallite size were available, it is also possible to get information on the rate of growth of nucleii. This reaction has much practical importance, and a better knowledge of nucleation and growth could help, as it has done in similar reactions, to control the size and texture of the metal powder produced.

1. B. Delmon Introduction a la Cinétique Heterogène. ed. Technip, Paris 1969.

A STUDY OF THE DIFFERENT STAGES DURING NICKEL OXIDE SINGLE

CRYSTAL SULPHURIZATION BY HYDROGEN SULFIDE

A. Steinbrunn, P. Dumas and J.C. Colson

Faculté des Sciences Mirande, Laboratoire de
Recherches sur la Réactivité des Solides associé
au C.N.R.S. 21000 Dijon

The study of metal oxide sulphurization in H_2S is an interesting subject if we consider that a number of metals owe their resistance against corrosion to the protective oxide film which coats them.

The present paper is devoted to the different sulphurization stages of an oxide in H_2S and especially those involved at surface and thick-layer level. We chose the monocrystalline nickel oxide owing to the large amount of data concerning this phase (1). The single crystals were prepared by the Verneuil method and their purity is 99.99 %.

I - $\underline{H_2S\ adsorption\ stages\ on\ oriented\ faces\ \{100\}}$

This investigation was carried out following ultra-vacuum treatment by LEED, RHEED and AES, the three techniques being successively used within the same apparatus. The single crystals, 3x3x15 mm in size, were vacuum cleaved, to produce a (100) face with a device designed in our laboratory. Sulphurization was carried out under pressures of 10^{-6} torr with H_2S, of purity 99.99%, obtained by distillation.

a) Properties of NiO (100) faces

The Auger spectrum obtained immediately following cleavage (figure 1) shows no characteristic impurity feature. Reflexion high energy electron diffraction shows parallel lines (figure 2) confirming a plane and clean surface. The diffractograms obtained from <100> and <110> azimuths allow the calculation of the surface cell parameters of type p(1x1), square cell of 2.93 Å. This value is confirmed in LEED but along with others authors (2)(3)

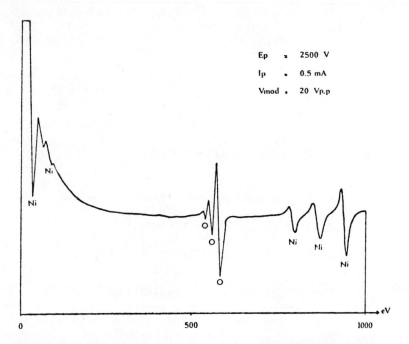

Figure 1 Auger spectra of a **freshly cleaved NiO face (100)**

<100> azimuth, 40kV   <110> azimuth, 40kV

Figure 2 RHEED diagrams of a clean face

we noticed extra, weak diffraction $\frac{1}{2}$, $\frac{1}{2}$ spots, whose intensity is 1 to 3 % of the integer order beam intensities. These are obtained for energies ranging from 30 to 80 eV and are interpreted considering the diffraction by the ordered arrangement of spins of the Ni^{2+} ions, the magnetic unit mesh being a p(2x1). They are removed by slight contamination.

b) Face (100) reactivity towards H_2S

Exposure to H_2S was effected using the following procedure: heating, H_2S exposure, quenching, AES, LEED and RHEED at the ambient.

Structure and composition changes are described referring to figure 3, including values of the ratios of absolute heights of peak S (149 eV) to peak Ni (843 eV), peak O (507 eV) and peak Ni assessed from Auger spectra as a function of temperature fixed. Exposure to H_2S for a given T is 900L (10^{-6} torr - 15 mn).

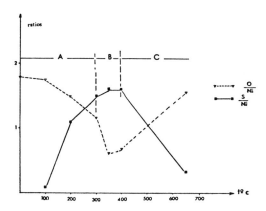

Figure 3 Evolution of ratios

$$\frac{\text{peak height h S (149 eV)}}{\text{peak height h Ni(843 eV)}} \quad \text{and} \quad \frac{\text{h O (507 eV)}}{\text{h Ni(843 eV)}}$$

Over range A, S/Ni increases and O/Ni decreases without any change in surface structure. At 300°C further stains are shown in LEED(figure 4)and trails in RHEED.

At 350°C the superstructure stains disappear in LEED and a change is noticed in RHEED diffractograms.

Over range C, S/Ni decreases until at 650°C the RHEED diffractogram becomes a point diagram.

c) Discussion

The superstructures observed from 300°C probably show the formation of a two-dimensional adsorption layer. The sulphur atoms would be situated at the top of a square cell, 3.5 Å wide, which approximately corresponds to the S^{2-} ion diameter (r = 1.76 Å).

At 650°C the RHEED diagram corresponds to **three-dimensional** blocks which may be constituted by sulphide which developed through NiO after undergoing faceting whose pole would be of type [111] as shown by Allie (4). The latter assumption seems the most likely.

At high temperature, S/Ni ratio decrease should correspond to a sulphur dissolution in the oxide, as shown in Auger analysis of freshly-cleaved surfaces in the neighbourhood of a surface exposed to H_2S.

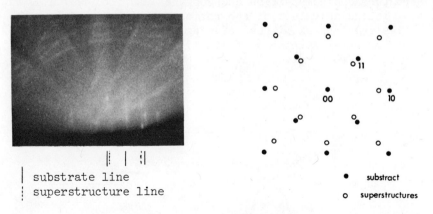

RHEED azimuth <100> 40 kV   LEED schematic representation

Figure 4   Superstructure diffractograms

II - Growth steps of sulphide thick-layer

This part of the work was carried out on small cubes of monocrystalline NiO **cleaved along** (100) or (111) faces. The morphological aspect and phase composition were studied by SEM, EDAX X-diffraction, and reaction rate was measured by thermogravimetry.

a) Composition and morphology of phases formed

In the early stages of growth, the formation of a continuous NiS layer is noticed. The sulphide layer is compact with closed pores (figure 5).

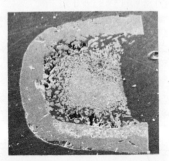

Figure 5   Macroscopic interfacial zone NiO-NiS

At NiS/NiO interface level, the formation of a honeycombed area is evident. Its morphology is the same as that formed during metal sulphurization (5). On reaction completion, a central cavity remains whose volume is less than that of initial NiO.

b) Kinetic aspect of the reaction

Thermogravimetric analysis was effected for 650°C <T<720°C

and 25 < P < 80 torr. The "parabolic" curves, taking account of NiO symmetry, show a diffusion process ($E_{exp} = 18.5 \pm 1$ Kcal.M^{-1}) The rate is pressure dependent following an homographical law.

c) Discussion

The whole set of experimental results allows the general interpretation we put forward for sulphurization of oxides by H_2S (6) to be verified. The sulphide thick layer growth occurs by simultaneous diffusion O^{2-} and Ni^{2+} through NiS. We have proposed how to deal theoretically with this problem by extending Wagner's theory (7).

III - Conclusion

The analogy between the sulphurization mechanisms of metals and oxides tends to be confirmed by these reactions.

At surface level it appears that we may have observed the formation of a two dimension adsorption layer of sulphur on the oxide.

Nickel sulphide growth develops following the same mechanism as that well known in the case of metals. The simultaneous diffusion of Ni^{2+} and O^{2-} towards the external interface accounts for the formation of an internal cavity at the center of compact NiS.

We have not yet characterised with certainty the intermediate stages involved in the passage from the two dimension adsorption layer to the surface continuous layer, the three-dimension blocks obtained not having been identified.

REFERENCES

1) Jarzebski Z.M., 1973, Oxide semiconductors, Oxford, Pergamon press.
2) F.P. Netzer and M. Prutton, J. Phys. c, Solid State Phys. vol. 8, 1975, p. 2401.
3) K. Hayakawa, L. Namikawa, S. Miyake, J. Phys. Soc. Japan, 1971, 31, n° 5, 1403-1417.
4) Allié G. Thèse Paris, 1969.
5) J.C. Colson, M. Lambertin, Oxyde of Metals, 7, n° 3, 1973, p. 165.
6) J.C. Colson, D. Delafosse and P. Barret, Bull. Soc. Chim. Fr. 1, 1968, p. 146.
7) C. Wagner, Z. Phys. Chem. B2, 2542, 1933.

OXYGEN DIFFUSION IN NiO AND ZnO

D. Hallwig, H.G. Sockel and C. Monty,[*]

Institut für Werkstoffwissenschaften,

Lehrstuhl I, Universität Erlangen-Nürnberg

Erlangen, W.-Germany

[*] C.N.R.S. Lab. de Bellevue, 92190 Meudon, France

INTRODUCTION

The oxygen diffusion in the compounds NiO and ZnO has been measured in several previous investigations by other authors and by various methods.

For NiO the ^{18}O exchange technique has been applied by O'Keeffe and Moore (1) at an oxygen pressure of 6.7 kN m^{-2} and in the temperature range from 1100 to 1500°C. The obtained values for single crystals are described by

$$D_o = 6.2 \times 10^{-4} \exp(-240 \text{ kJ mol}^{-1} R^{-1} T^{-1}) \text{ cm}^2 \text{ s}^{-1}.$$

Variation of the oxygen pressure leads to increase of D_o with increasing partial pressure of oxygen.

Three previous publications have concerned the oxygen diffusion in ZnO. Moore and Williams (2) studied the diffusion of ^{18}O at about 100 kN m^{-2} and from 1100 - 1300°C. The diffusion coefficient was estimated to

$$D_o = 6.5 \times 10^{11} \exp(-690 \text{ kJ mol}^{-1} R^{-1} T^{-1}) \text{ cm}^2 \text{ s}^{-1}.$$

D_o was found to increase proportional to $p_{O_2}^{0.5}$.

Hoffmann and Lauder (3) later used the gaseous exchange technique at a pressure 26.5 - 30 kN m^{-2} and in a temperature range 1150 - 1400°C. A diffusion coefficient of

$$D_o = 0.105 \exp(-395 \text{ kJ mol}^{-1} R^{-1} T^{-1}) \text{ cm}^2 \text{ s}^{-1}.$$

is reported. At 1200 and 1300°C a decrease of D_o with increasing p_{O_2} is observed.

An activation analysis was used in an investigation by Robin, Cooper and Heuer (4) in the temperature range 940 - 1140°C and at p_{O_2} of 92 kN m^{-2}. The resulting diffusion coefficient of oxygen is

$$D_o = 1.2 \times 10^{-10} \exp(-124 \text{ kJ mol}^{-1} R^{-1} T^{-1}) \text{ cm}^2 \text{ s}^{-1}.$$

The oxygen diffusion in pure NiO-single crystals were of interest in connection with high temperature deformation studies(5) on the same crystals. In order to test the reliability of data obtained by a special method an investigation of diffusion profiles of ^{18}O by two in principle similar but in experimental application different methods, ion beam microanalysis and SIMS, were planned. In order to exclude influences from different crystal preparation parts from the same single crystal with exactly the same polishing and annealing treatments were used for both the experiments.

The enormous discrepancies of the published data for ZnO with respect to the activation energy, and the absolute value at higher temperatures as well, were reasons for a new investigation of oxygen diffusion in single crystal by SIMS-analysis of ^{18}O-diffusion profiles. Further support of the activation energy is given by the temperature dependence of the high temperature deformation of the same crystals.

EXPERIMENTS

The polished samples were equilibrated at the diffusion temperature in air. The following diffusion annealing was carried out in ^{18}O-enriched gas at 21 kNm^{-2}. The times of equilibrating and annealing were the same. The concentration of ^{18}O was held at about 30% abundance.

Evaporation effects during these procedures were observed by weighing the samples before and after annealing. The investigated temperatures were in the range from 800°C to 1400°C.

At first sight, ion beam sputtering and secondary ion mass spectrometry appears in the field of solid state reactions, especially for the study of very slow processes as an useful experimental technique on account of the high resolution of the penetration depth. For the application of this method the investigated processes must be connected with time dependent concentration profiles in the solid.

A primary ion beam with Ar^+-ions of the kinetic energy of 2 keV up to 10 keV was tilted 60° to the surface normal of the sample. This energy is given to the bulk material near the surface of the target by a collision cascade. Herewith the implantation depth is less than 6 nm for 10 keV Ar^+-ions into ZnO-samples [6].

The present knowledge of the processes involved only allows quantitative analysis of concentrations of isotopic elements. In this case the sputtering yield, ionization yield and other influencing factors are nearly constant. Therefore the concentration of ^{18}O is obtained from the counting rates of ^{18}O and ^{16}O via

$$C(^{18}O) = J(^{18}O) / (J(^{18}O) + J(^{16}O)) .$$

Whereas the SIMS analysis uses a relatively large ion beam (3 mm in diameter), the ion beam microanalyser operates with a small focused ion beam, which scans over the sample surface. The application of the large diameter beam has the great advantage of averaging the ^{18}O concentration over a larger area and of avoiding steps in the concentration at crystal inhomogeneities. In order to eliminate the influence from the boundaries of the sputtered area, the acceptance of the mass spectrometer was reduced to a diameter of 1 mm. In contrast to the ion beam microanalyser the material is removed continuously from the analysed surface by the primary ion beam.

RESULTS

The diffusion profiles and the resulting oxygen diffusion coefficients D_o obtained for NiO by both methods of analysis are compared. Beside small deviations they show good agreement. The measured values of D_o lie two orders of magnitude lower than the data of O'Keefe and Moore. The activation energy is calculated to be 250 kJ mol^{-1}. which is in agreement with the previous investigation.

The measured data for ZnO below 900°C lead to D_O which are in good agreement with values extrapolated from measurements of Hoffmann and Lauder. At higher temperatures the diffusion profiles can be explained by D_O, which agree fairly well with those measured by Robin, Cooper and Heuer. However there is an influence from the rate of evaporation of ZnO in H_2O-containing atmospheres. This rate as a function of temperature has been measured thermogravimetrically. These data in comparison with the D_O of Hoffmann and Lauder, measured in dry atmosphere, allow the conclusion that there is a great influence increasing with temperature on the profil and the diffusion coefficient. Our own data in this temperature region could be corrected using a solution of the differential equation of diffusion which takes into account this evaporation effect. The corrected data show within the limits of error good agreement with the data of Hoffmann and Lauder. Nearly the same activation energy as these corrected data can be evaluated from different sorts of high temperature deformation experiments. It can be assumed that during these processes the necessary mass transport is controlled probably by the mobility of oxygen ions.

ACKNOWLEDGEMENT

The authors wish to thank the "Deutsche Forschungsgemeinschaft" for financial support.

REFERENCES

1. M.O'Keefe and W.J. Moore, J. Phys Chem. 65 1438 (1961)
2. W.J. Moore and E.L. Williams Disc. Far. Soc. 28 86 (1959)
3. J.W. Hoffmann and I. Lauder Trans. Far. Soc. 66 2346 (1970)
4. R.Robin, A.R. Cooper, A.H. Heuer, J. Appl. Phys. Vol.44 No 8, p 3770 - 3777
5. J. Castaing C.N.R.S. Paris peron communication
6. R.G. Wilson, G.R. Brewer: Ion Beams, J. Wiley & Sons, New York 1973

SULFATION MECHANISM OF CoO AND NiO

L.E.K. Holappa and M.H. Tikkanen

Dr. Tech., Research Center, Ovako Group, Imatra, and

Prof., Dept. of Metallurgy, Helsinki University of

Technology, Otaniemi, Finland

INTRODUCTION

Sulfatizing roasting is used when separating small contents of valuable metals from sulfide and composite ores. In sulfatizing roasting the desired constituents are selectively converted to water soluble sulfates, whilst other constituents such as iron remain as insoluble oxides. In Finland the sulfatizing roasting process has been used at the Outokumpu Oy Kokkola Works for the extraction of cobalt and nickel from roasted calcines and pyrite-pyrrhotite concentrates since 1967 (1).

The thermodynamics of the sulfation reactions of cobalt and nickel oxides are well known (3,4,5,6). There is a difference between the sulfation of CoO and NiO since CoO is in equilibrium with the sulfate only in a specific range of temperature and gas composition. At normal sulfatizing temperatures the stable oxide is Co_3O_4 and owing to this an intermediate layer of Co_3O_4 is formed during the sulfation of CoO. In the case of NiO the formation of such an intermediate phase is not possible and $NiSO_4$ is formed directly on the NiO surface.

Both laboratory and industrial sulfation has shown that the sulfation of NiO is much slower than that of CoO and the recovery of Ni also remains low. In this paper the mechanism of sulfation of cobalt and nickel oxides has been discussed based on earlier work on the sulfation kinetics and mechanism (2).

EXPERIMENTAL

Dense cobalt oxide samples were prepared by oxidising 0.10 cm thick Sherritt Gordon cobalt sheet (99.9 pct Co, 0.05 Ni, 0.003 Cu, 0.014 Fe, 0.003 S, 0.008 C) at 1200°C in air for 14 days. The specimens were then cooled, ground and polished. Dense nickel oxide samples were prepared by oxidising 0.02 cm thick Outokumpu cathode nickel at 1250°C in air for 14 days. NiO powder samples were made by oxidising Mond nickel Grade A (0.07-0.10 pct C, 0.15-0.20 O, 0.044-0.01 Fe, residue Ni) in air at 400°C for a day followed by seven days at 1000°C. The powder was then ground and the NiO-compacts were made in a 3 cm^2 steel form and then sintered at 1250°C in air for a week. The density obtained was 4.85 g/cm^3, i.e. a relative density of 65 pct.

The sulfation apparatus and gas system have been described elsewhere (2). The growth mechanisms of $CoSO_4$ and $NiSO_4$ as well as of Co_3O_4 on CoO were studied using gold marker experiments (2).

RESULTS AND DISCUSSION

Sulfation experiments with dense CoO blocks in the temperature range 680-885°C showed parabolic kinetics. Trials with dense NiO blocks showed that the sulfation rate was very low. At 750°C and with a SO_3 partial pressure of 0.255 the parabolic rate constant was of the order of $2-4 \cdot 10^{-3}$ $mg^2/cm^4 \cdot h$, this being about a hundred times lower than that for CoO (0.4 $mg^2/cm^4 \cdot h$). When sintered NiO compacts were sulfatized the simple parabolic form was not valid during the initial period since the surface became covered with a thin sulfate layer. Later when the layer grew dense and uniform, parabolic kinetics were again found.

The temperature and SO_3 partial pressure dependence of the sulfation reactions showed that the sulfation rate is proportional to the sulfating potential gradient across the sulfate layer. The activation energies were calculated using a modified Arrhenius equation the values being 28.1 and 19.3 kcal/mole for $CoSO_4$ and $NiSO_4$, respectively.

The results from the marker experiments are shown in Fig. 1, 2a and 2b. In Fig. 1 the CoO specimen was oxidised to form a Co_3O_4 layer. The gold markers initially situated at the surface of the CoO were located 15-20 pct inside the Co_3O_4 layer. The recorder traces in Figs. 2a and 2b show that the gold markers, which were initially located at the surfaces of the CoO and NiO specimens, can be detected after the sulfation at the $CoSO_4/Co_3O_4$ and $NiSO_4/NiO$ interfaces, respectively.

The sulfation of NiO occurs without any oxidation process

Fig. 1. Position of gold markers in Co_3O_4 layer formed on CoO at 860°C in air after 14 days.

since the higher oxide is not stable in the considered conditions. The SO_3 potential gradient across the sulfate layer causes sulfation to proceed. The marker experiments showed the sulfate growth outwards; the proposed mechanism is the outward diffusion of Ni^{2+} and O^{2-} ions through the sulfate layer. The same mechanism

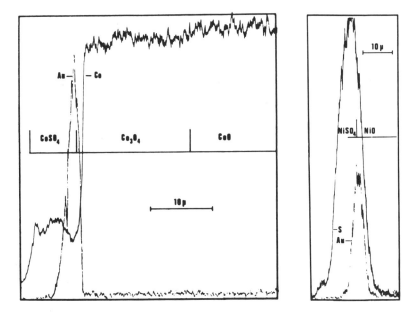

Fig. 2. Micro-analysis showing the position of gold markers in the sulfation experiments a) after CoO sulfation 3 days at 680°C, b) after NiO sulfation 7 days at 680°C.

has been proposed earlier (4) but the inward growth of the sulfate layer has also been suggested (8).

In the case of CoO the intermediate phase Co_3O_4 is formed. The results of the marker experiments in Fig. 1 can be explained by the simultaneous inside and outside growth of Co_3O_4 caused by the outward diffusion of Co-ions. The proposed mechanism is shown in Fig. 3. The following reaction can be presumed at the CoO/Co_3O_4 interface:

$$12CoO = \{3Co^{2+} + 6e^-\} + 6Co^{3+} + 3Co^{2+} + 12O^{2-} + 3\square_{Co^{2+}} \qquad (1)$$

The species shown in the brackets diffuse through the Co_3O_4 scale to form new Co_3O_4 at the oxide/gas interface II as follows:

$$\{3Co^{2+} + 6e^-\} + 2O_2 = Co_3O_4 \qquad (2)$$

The remaining species at the interface I precipitate as Co_3O_4:

$$6Co^{3+} + 3Co^{2+} + 12O^{2-} + 3\square_{Co^{2+}} = 3Co_3O_4 \qquad (3)$$

This is the same reaction as for the precipitation of defects in nonstoichiometric $Co_{1-x}O$. Owing to the reactions (1)..(3), when one Co_3O_4 is formed at the outer surface, three Co_3O_4 is formed at the CoO/Co_3O_4 interface. The calculation is for stoichiometric CoO. An analogous mechanism has been proposed for the growth of Fe_3O_4 (9, 10). On the other hand, inward diffusion of oxygen has been suggested as the growth mechanism for Co_3O_4 (7).

Marker experiments showed $CoSO_4$ to grow outwards. By combining these two mechanisms a model for oxidation-sulfation reactions has been suggested in Fig. 4. Without any sulfating agent the pure oxidation reaction occurs. When small contents of SO_3 are present,

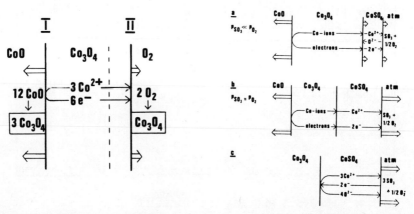

Fig. 3. Proposed mechanism for the growth of Co_3O_4 on CoO.

Fig. 4. Proposed mechanism for the growth of $CoSO_4$.

the growth of $CoSO_4$ starts. When the sulfation potential is low the sulfation rate is slow compared to the oxidation rate and some inward migration of oxygen occurs through the sulfate layer resulting in the Co_3O_4 growth. The growth of $CoSO_4$ occurs by the outward diffusion of Co-ions balanced with electrons (Fig. 4a).

When the SO_3 and oxygen potential are both high, the conditions for sulfation and oxidation are roughly equal and the reaction rates can also be assumed equal (Fig. 4b). Sulfate formation, likewise, consumes Co-ions liberated in the $CoO \rightarrow Co_3O_4$ transformation. Oxidation can be understood as the decrease of cations in the CoO core which thus gradually converts to Co_3O_4.

When pure Co_3O_4 is sulfated, Co^{3+} ions must be reduced and oxygen liberated (Fig. 4c). The sulfation rate, however, is independent of the base oxide and the oxidation reaction (2). According to this and the discussion above it is suggested that the oxygen ion diffusion in the sulfate layer occurs freely, the removal of cobalt ions from the outer surface of Co_3O_4 is relatively easy and the rate-determining factor in sulfation is the diffusion of cobalt ions through the sulfate layer. In the case of NiO the removal of Ni-ions from the $NiO/NiSO_4$ interface seems to be difficult because of the great stability and few defects in the NiO-lattice. This is assumed to be the explanation for the slow sulfation of dense NiO.

SUMMARY

Kinetic measurements and gold marker experiments showed both cobalt sulfate and nickel sulfate grow outwards on the initial oxide surfaces. The much higher sulfation rate of CoO compared to NiO is explained by the formation of an intermediate Co_3O_4 layer and different oxide properties. Oxidation of CoO to Co_3O_4 is proposed to occur by cobalt ion diffusion outwards this causing simultaneous inward and outward growth of the Co_3O_4 layer.

REFERENCES

1 Palperi,M., Aaltonen,O., J.Metals, Febr.(1971), p. 34-38.
2 Holappa,L.E.K., Dissertation, Helsinki Univ. Tech. (1970).
3 Evans,W.H., Wagman,D.D., J.Res.Natn.Bur.Stand. 49(1952), p. 141.
4 Ingraham,T.R., Marier,P., Trans.A.I.M.E. 236(1966), p.1064-1067.
5 Ingraham,T.R., Can.Met.Quart. 3(1964), p. 221-234.
6 Kellogg,H.H., Trans.A.I.M.E. 230(1964), p. 1622-1634.
7 Hocking,M.G., Diss.Univ. London, (1962).
8 Wright,J., Pigott,J.R., Corros.Sci.9(1969),p.121-122,Appendix 2.
9 Davies,M.H. et al, Trans.A.I.M.E. 191(1951), p. 889-896.
10 Hauffe,K., Reaktionen in und an festen Stoffen, Berlin (1955).

HIGH RESOLUTION ELECTRON MICROSCOPY OF REACTING SOLIDS

G. Schiffmacher, H. Dexpert and P. Caro

Laboratoire des Terres Rares du CNRS

92190, Bellevue, (FRANCE)

INTRODUCTION

High resolution electron microscopes like the JEM 100C used in the present study have a resolution close to 3Å between points. This is within the range of interplanar distances for a score of inorganic compounds. It is possible to have images of the periodic potential in the crystal, that is an image of the **atomic distribution**. The images are formed by interference of the direct beam with one, or several, diffracted beams. The images have the same periodicity as the diffracted atomic planes with a resolution which can be as low as 1.4 Å. Very stringent conditions are to be meet, **especially with regard** to the sample thickness and orientation. To obtain a convenient contrast it is necessary to underfocus the objective by some hundreds of Angströms. (1).

Several rare earth compounds exhibit a layered structural character. These are the salts, or oxides, of the rare earth polycations: $(LnO)_n^{n+}$, $(LnOH)_n^{2n+}$, $(LnF)_n^{2n+}$ (2). The cationic layers are separated by anionic layers. A large number of anions are involved: O^{2-}, CO_3^{2-} Cl^-, MoO_4^-, F^-, OH^- etc... . Those materials may yield lattice images showing the sequence of the polycations and anionic layers with various spacings depending on the nature of the anion. As the polycation salts are usually quite thermally stable compounds, it is possible to observe them in the electron microscope because of their relativitely high decomposition temperature. When heat is progressively applied to a **thin flake or thin film sample**, a layered compound involving a thermally unstable component may be decomposed into another, more stable, layered compound, which in turn can be transformed by the same process into another one.

Such conditions are realized on the series of layered hexagonal compounds :

$$B - (LnOH)CO_3 \rightarrow II - (LnO)_2CO_3 \rightarrow A - Ln_2O_3$$

(Ln = Rare Earth, in our case Neodymium).

RESULTS AND DISCUSSION

I. The First Step : Decomposition of the Hydroxycarbonate into the Oxycarbonate.

Fig.1 exhibits a thin flake of B neodymium hydroxycarbonate. The distance between $(LnOH)_n^{2n+}$ layers is 4.95Å. The flake exhibits the corresponding spacing, but roughly round areas in the crystal can be seen which have a larger spacing. Those are "islands" of the neodymium oxycarbonate type II with a 7.82Å distance between the $(LnO)_n^{n+}$ layers (3). Enlargment of area in the photograph shows that the do-

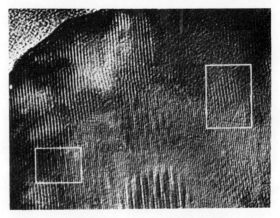

Fig.1 : $B - (NdOH)CO_3 \rightarrow II - (NdO)_2CO_3$ $G=1,2 \times 10^6$

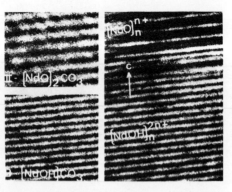

Fig.2 : Unidirectional Syntaxy $G=3,1 \times 10^6$

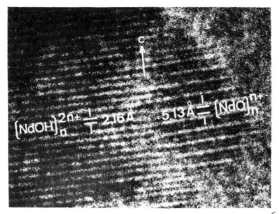

Fig.3 : Bidirectional Syntaxy G=4,3 x 10^6

mains have two different epitaxial relationships, with the initial hydroxycarbonate matrix :

-a syntaxy for which the two $\underline{c}$ axis of the hexagonal structures are common, and the $[1\bar{1}0]$ of $NdOHCO_3$ and $[010]$ of the oxycarbonate parallel (fig.2).

-a syntaxy for which the two $\underline{c}$ axis are distinct but parallel. (fig.3).

In the last case it can be seen that the rare earth atomic planes of the two structures which exhibit an hexagonal symmetry with basically the same neodymium distances (4.11Å in $(NdOH)CO_3$, 3.99Å in $(NdO)_2CO_3$) merge smoothly and continuously from one structure to another. There is an equivalence between 11 $\underline{c}/2$ distances in the hydroxycarbonate with 7 $\underline{c}/2$ distance in the oxycarbonate (54.45Å versus 54.74Å), which makes 12 rare earth planes on the hydroxycarbonate side versus 14 on the oxycarbonate side. Removal of the OH^- group inside the planar $(LnOH)_n^{2n+}$ layers to form the $(LnO)_n^{n+}$ group then yields round domains with roughly 50Å diameter. The process is non destructive for the rare earth arrangement within the (002) planes, as those have only to move coherently perpendicularly to the $\underline{c}$ axis. However the parallelism between the rare earth (002) planes is slightly perturbed in the process (see figure 3) and, as a consequence, the round domains are slightly disoriented with respect to each other, as shown by the mosaic appearance on figure 1. The 50Å domain forms the original nucleus of the oxycarbonate structure when the second type of syntaxy is involved.

II. The Second Step : Decomposition of the Oxycarbonate into the A - Type Oxide.

The decomposition of type II-$(LnO)_2CO_3$ into $A-Ln_2O_3$ is far more

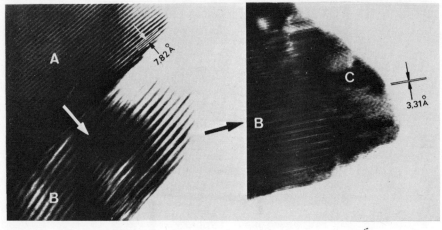

Fig.4 : II-(NdO)$_2$CO$_3$ → A-(NdO)$_2$O   G = 1,3 x 10^6

difficult to observe on the lattice image scale than the preceding one. It occurs at a higher temperature (800°C from thermogravimetric analysis, versus 600°C) (4) and is faster. On fig.4 one can see in zone A the oxycarbonate with the 7.82Å spacing. When heated the crystal takes the appearance on zone B, and at that point very often the crystal tilt itself. However it is sometimes possible to have the image in zone C which corresponds to (100) planes (d = 3.31Å) of the A - type Neodymium oxide. The oxide planes are more or less in continuity with the perturbed sequence of (002) planes of the II-(LnO)$_2$ CO$_3$ in the B zone. (10$\bar{1}$0) diffraction spots of the oxide replace on the diffraction pattern the sequence of (0001) spots of the oxycarbonate. The consequence is that the two c hexagonal axis of the

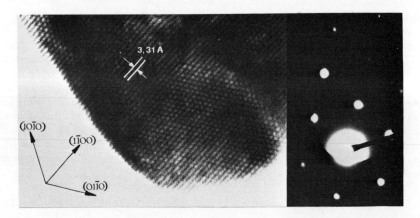

Fig.5 : Lattice imaging of A - (NdO)$_2$O perpendicular to the ternary c axis   G = 4,5 x 10^6

two structures are neither colinear nor parallel,and the epitaxial conditions,if any,cannot be determined.However the A-type oxide has a strong propensity to develop itself as thin crystals elongated perpendicular to the $\underline{c}$ axis (5).Such crystals yield bidimensionnal lattice images on the (1000) planes (d = 3.31Å) in the [001] axis zone (Fig.5).The change in habitus when going from an oxycarbonate crystal with the $\underline{c}$ axis in the plane of the support to an oxide crystal with the $\underline{c}$ axis perpendicular to the plane may explain the frequent tilt observed.Nevertheless the C zone on fig.4 corresponds to an epitaxy of the lamellar structures which is different from the smooth syntaxies observed in the transition from the hydroxy-carbonate to the oxycarbonate.The oxycarbonate to oxide transition involves an intermediate,disturbed,phase (B zone) followed by a cristalline reconstruction.

REFERENCES

1. J.M. Cowley, S. Iijima Zeit. Natur. $\underline{27a}$, 445 1972
2. P.E. Caro N:B:S: 364 5th Materials Research Symp. 1972
3. H. Dexpert, G. Schiffmacher, P. Caro J. Sol. State Chem. $\underline{15}$, 301 1975
4. P. Caro, J.C. Achard, O. De Pous Coll. CNRS $\underline{180}$, 285 1970
5. C. Boulesteix, P. Caro, M. Gasgnier, Ch. Henry la Blanchetais and G. Schiffmacher Acta Cryst. $\underline{A\ 27}$, 552 1971

THE ELECTRONIC STRUCTURE OF AN ADSORBATE LAYER STUDIED BY ANGLE RESOLVED PHOTOEMISSION SPECTROSCOPY

P. Butcher, P.M. Williams, and J.C. Wood*

V. G. Scientific Ltd., East Grinstead, Sussex, England
*presently at Department of Inorganic Chemistry,
Chalmers University of Technology, Goteborg, Sweden

INTRODUCTION

Ultra-violet photoelectron emission spectroscopy (UPS) has developed into an extremely sensitive probe of the valence band electronic structure of solids, and can be used to study the modification of this electronic structure caused by adsorption processes [1,2]. It has been recognised that the additional information to be gained from investigation of the spatial distribution of photoemitted electrons is of extreme importance in fundamental studies of adsorption and gas-solid interactions [3].

Angular resolved investigations so far reported have suffered from two principle disadvantages. Firstly the photoemission has been integrated over a relatively large solid angle because of the type of analyser employed [4]. Secondly, experimental parameters which might be expected to influence the spatially resolved photoemission spectrum have been allowed to vary, in particular the angle of photon incidence [5]. The present paper describes the angle resolved UPS spectrum of a clean and CO-saturated Nickel (111) surface. Particular attention has been paid to eliminating possible instrumental effects, such as those described above.

Considerable controversy has existed in the assignment of photoemission peaks from adsorbed CO because of their near degeneracy and wide variation from the gas phase spectrum [6]. It has previously been pointed out that this controversy could well be resolved by a study of the angular characteristics of the spectrum [3].

Angular resolved spectra have previously been described for this system but not in such a comprehensive form (5).

EXPERIMENTAL

Spectra were recorded using an angular dispersive electron spectrometer model ADES 400 (V. G. Scientific Ltd.). Following bakeout, base pressures 10^{-11} torr were obtained using trapped diffusion and titanium sublimation pumping. Electron kinetic energies were measured with a hemispherical electrostatic analyser having a mean radius of 50mm. This was mounted on a rotating horizontal table, permitting variation in the polar angle of photoemission (θ) from the single crystal sample. Samples were mounted on a universal manipulator, permitting rotation about the surface normal (variation in azimuthal angle ϕ) and about the normal to the plane of the analyser table (photon incidence angle ψ). The photon source was a windowless discharge lamp, operating under HeI (21.21ev) or HeII (40.8ev) conditions. The symmetry of the single crystal surface was confirmed by LEED, and was also apparent in the angular photoemission curves. The additional facility of an AlKα/MgKα twin anode X-ray source was not used. Spatial photoemission resolution was determined by a 1mm diameter aperture to the analyser, 36mm from the sample. This defined an emission cone of semi-angle $3/4^{\circ}$ at the sample.

The sample was cleaned by mild argon ion bombardment, electron bombardment heating to anneal the surface, followed by brief high temeperature flashing. CO was admitted to the sample at saturation coverage >10 Langmuir (torr.sec).

RESULTS

Spectra shewn in Figures 1 and 2 were recorded with a photon incidence angle ψ of 45°. This was found experimentally to give the most intense spectrum of adsorbed CO and was used in all subsequent experiments. Incident photon effects were observed, producing intensity variations in the CO peaks. The similarity between these spectra, recorded using HeI and HeII radiation leads to the expectation that spectral modulation due to final state effects are sensibly absent for the Ni-CO system, and that the spectra of adsorbed CO in Figure 3 are representative of the occupied molecular orbitals of the system.

The observed experimental resolution for this system was 90mev in the spectra recorded close to a 50° polar angle. This can be contrasted with 600 mev observed at normal emission.

ELECTRONIC STRUCTURE OF ADSORBATE LAYER 649

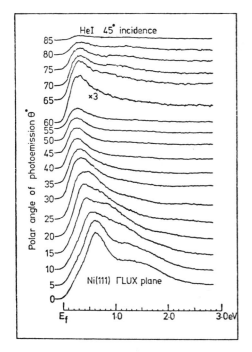

Figure 1 Polar variation of photoemission spectrum. HeI(21.21ev) excitation. Clean nickel (III) surface.

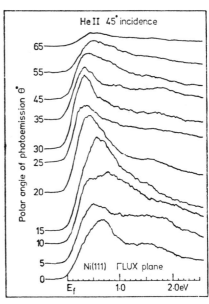

Figure 2 Polar variation of photoemission spectrum. HeII(40.8ev) excitation. Clean nickel (III) surface.

Despite CO being admitted to the surface at saturation dosage, angular effects are observed in the photoemission arising from the adsorbate. This implies that the adsorbed layer is reflecting the symmetry of the underlying crystal and that a condensed multilayer has not been formed. Two principal peaks, generally recognised as characteristic of the presence of adsorbed CO, are observed approximately 8ev and 11ev below the Fermi level. As the polar angle is increased from normal emergence, these two peaks show a similar angular shift and intensity variation. At high polar angles, in excess of $50°$ from normal, a third peak becomes apparent as a shoulder on the low binding energy side of the 8ev peak. The intensity variation of these three peaks as a function of polar angle is shewn in Figure 4.

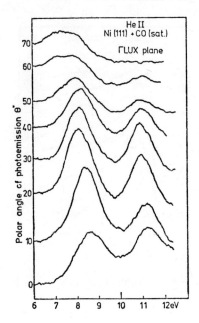
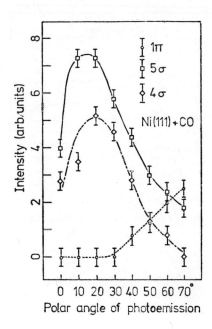

Figure 3 Polar variation of photoemission in the region 6ev-12ev binding energy following adsorption of CO HeII excitation.

Figure 4 Polar variation of intensity of peaks appearing in Figure 3.

DISCUSSION

Angular photoemission results having high spatial resolution have been previously reported for solids, but the present work represents the first comprehensive and systematic study of an adsorption system (7,8). The results permit a revision of the previous orbital assignments for adsorbed CO. From consideration of the gas phase molecule, the three CO molecular orbitals having the lowest binding energy are predicted to be 5σ, principally a lone pair on the carbon atom, 1π, an internuclear bonding orbital, and 4σ, predominantly the oxygen lone pair. Speculation over assignment for adsorbed CO has centred on the appearance of three features at binding energies differing widely from those observed in the gas phase. Various proposals have included overlap of any two of the three orbitals. Most recently Lloyd proposed that the lowest binding energy peak was a combination of 5σ and 1π, with 5σ expected to be the lower BE component, and a unique 4σ orbital (9).

The angular intensity variations, summarised in Figure 4 clearly do not support this assignment. The related intensity variations for the two intense peaks implies that both arise from initial states of the same symmetry and hence orbital momentum components. These peaks must therefore be associated with the two σ molecular orbitals, both having axial symmetry with respect to the C-O bond. The remaining feature of lowest binding energy and unique angular intensity dependence can then be associated with the 1π orbital.

Comparison of binding energy shifts relative to the gaseous molecule are not inconsistent with the above model. Both the 1π and 4σ exhibit similar shifts to lower binding energy as a result of adsorption, the 5σ being only slightly shifted. This results in the reordering of energy levels from the gas phase, and suggests that the 5σ orbital is involved in the chemisorption bond upon adsorption. For CO molecularly adsorbed, with its axis normal to the surface, the angular intensity variation for both types of orbital is in accord with the predictions of Grimley, including the initial increase and subsequent decrease in intensity for the orbitals (3).

A recent SCF-$X\alpha$ cluster calculation for CO adsorbed on a Ni(100) surface, where molecular parameters were preset, also predicted the above order of energies for molecular orbitals (10).

REFERENCES

1) W.E. Spicer, in Vacuum Ultraviolet Radiation Physics, Ed.E.Koch, R. Haensel, C. Kunz, Pergamon/Vieweg, Braunschweig, 1974, p.545.
2) D.E. Eastman, J.K. Cashion, Phys. Rev. Lett. 27 1520 1971
3) T.B. Grimley. Faraday Disc. Chem. Soc. 58, 1, and remarks, 1974
4) D.E. Eastman, J.E. Demuth, Japn. J. Appl. Phys. Suppl. 2 pt 2 827, 1974.
5) W.F. Egelhoff, D.L. Perry. Phys. Rev. Lett. 34 93, 1975
6) C.R. Brundle. J. Elec. Spec. 7 484 1975 and refs. therein.
7) P.M. Williams, D. Latham, J.C. Wood. J. Elec. Spec. 7 281, 1975
8) D.R. Lloyd, C.M. Quinn, N.V. Richardson, P.M. Williams, Comm. Phys. 1, 11. 1976
9) D.R. Lloyd - see Remarks. Faraday Disc. Chem. Soc. 58, 1974.
10) I.P. Batra, P.S. Bagus. Sol. State Comm. 16 1097 1975.

THERMAL DECOMPOSITION OF ALKALINE EARTH CHROMATES, OXALATES

AND CHROMATE-OXALATE MIXTURES. THE Ba, Sr, AND Mg COMPOUNDS

E.G. Derouane, Facultés Universitaires de Namur,
Département de Chimie, 61, rue de Bruxelles,
B-5000-Namur. Belgium

and

R. Hubin and Z. Gabelica, Université de Liège,
Département de Chimie Générale et de Chimie Physique,
B-4000-Sart Tilman par Liège 1. Belgium

INTRODUCTION

Cr(V) and Cr(III), dispersed at the surface of various oxidic materials, play an important role as catalytic species. The present contribution reports a detailed investigation, using a combination of various physical techniques (thermogravimetric analysis, mass spectrometry, e.p.r., ir-spectroscopy, and X-ray diffraction),of some methods of preparing and stabilizing such dispersed ions.

Following an investigation of the thermolysis of the oxalates of Ba, Sr, and Mg (1), we have looked into the progressive decomposition of the chromates and into the reduction of Cr(VI) during the thermal decomposition of chromate-oxalate mixtures of the same alkaline-earth metal.

The advantage and originality of our approach consist mainly in the simultaneous genesis of the support ($M^{II}O$, $M^{II}CO_3$, Cr_2O_3) and that of the active species (Cr(V) and Cr(III)), the reducing agent for Cr(VI) (CO, $M^{II}CO_3$) being formed during the thermolysis. The same approach was already found of particular interest when gaseous species produced during the thermolysis of the pure oxalates, i.e, CO and CO_2, were retained at the support surface as molecular or ionic ad-species (1).

All the intermediate species have been identified in the temperature range from 20 to 800 °C,which enable the proposition of the decomposition mechanisms.

EXPERIMENTAL

The preparation of the starting materials and the experimental conditions and procedures have been described previously (1-3).

RESULTS AND DISCUSSION

I. Thermal decomposition of the oxalates of Ba and Sr.

Thermogravimetric analysis shows that the oxalates of Ba and Sr decompose according to the following scheme:

$$MC_2O_4 \rightarrow MCO_3 + CO(g) \quad (1)$$
$$MCO_3 \rightarrow MO + CO_2(g) \quad (2)$$

Mass spectrometry analysis of the evolved gases (i.e., the determination of the ratio $n(CO)/n(CO_2)$), as a function of temperature or percentage of decomposition of the starting material, indicates in addition the occurence of the following intermediate reactions :

$$2\ CO(g) \rightarrow C + CO_2(g) \quad (3)$$
$$MCO_3 + C \rightarrow MO + 2\ CO(g) \quad (4)$$

In agreement with the e.p.r. results, this shows that CO is first adsorbed and dismuted on the carbonate, the latter being further reduced **by the carbon so-formed.** The CO^+ adsorbed species which has been observed by e.p.r. is only present when C is present, thereby showing the acidic character of the C-covered surface. Fig.1 illustrates the results obtained in the case of $BaC_2O_4 \cdot \frac{1}{2}H_2O$.

II. Thermal decomposition of the Mg oxalate.

The mass spectrometric determination of the $n(CO)/n(CO_2)$ ratio (see fig. 2) has enabled for the first time to show that MgC_2O_4 decomposes probably with the intermediate formation of $MgCO_3$:

$$MgC_2O_4 \rightarrow MgCO_3 + CO(g) \quad (5)$$
$$MgCO_3 \rightarrow MgO + CO_2(g) \quad (6)$$

E.p.r. shows the presence of two different adsorbed species:
- CO_2^- radicals formed on the basic MgO surface and
- CO neutral or slightly positive species bound to surface electrodonor sites by σ bonding and π backbonding.

The initial adsorption of CO and its further desorption at higher temperature explain quantitatively the variation of the $n(CO)/n(CO_2)$ ratio as a function of temperature.

III. Thermal decomposition of the Ba, Sr, and Mg chromates.

While the Ba and Sr chromates are stable up to 850°C, $MgCrO_4$ is autodecomposed according to the scheme :

$$2\ MgCrO_4 \rightarrow MgO + MgO + Cr_2O_3 + \frac{3}{2}O_2(g) \quad (7)$$
$$\searrow \swarrow$$
$$MgCr_2O_4$$

E.p.r. enables in addition the identification of intermediate Cr(V) species which probably correspond to distorted CrO_6^{7-} octahedrons.

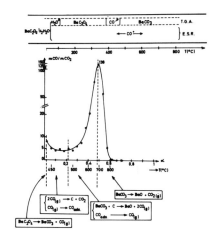
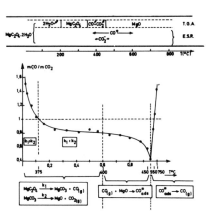

Fig. 1. Thermolysis data for Ba oxalate

Fig. 2. Thermolysis data for Mg oxalate

IV. Thermal decomposition of the chromate-oxalate mixtures of Ba and Sr.

Thermogravimetric and e.p.r. data, with additional information provided by mass spectrometry, ir-spectroscopy, and X-ray diffraction lead to the decomposition scheme proposed in table 1 (2)

It appears that :
- Cr(III) dispersed species are formed and stabilized when the thermolysis is carried out between 450 and 550°C,
- Cr(V) dispersed species are not generated during the decomposition,
- the end product of the thermolysis, at 850°C, is the Cr(V) chromate $M_3(CrO_4)_2$ (M = Ba or Sr).

V. Thermal decomposition of the chromate-oxalate mixtures of Mg.

The proposed mechanism based on the various experimental data is schematized in table 2 (3). It has been possible to show, in addition, that Cr(V) dispersed species as CrO_6^{7-} octahedrons are formed and stabilized when the thermal decomposition is realized at about 350-400°C. Similarly Cr(III) dispersed species are present for thermolysis temperatures in the range 350-550°C. Above the latter temperature, Cr(III) enters the MgO lattice and forms a solid solution.

In contrast to the reduction selective role of CO and MCO_3 which has been observed in the thermolysis of the Ba and Sr oxalate-chromate mixtures (see table 1), the reduction of the Cr(VI) species, for the Mg oxalate-chromate mixtures is spontaneous and leads to Cr(V) and Cr(III) (see table 2), even in the absence of the oxalate. However, CO which is evolved during the oxalate decomposition can reduce Cr(V) to Cr(III). Both reduction mechanisms are compared in table 3.

TABLE 1

Thermolysis of oxalate-chromate mixtures of Ba and Sr

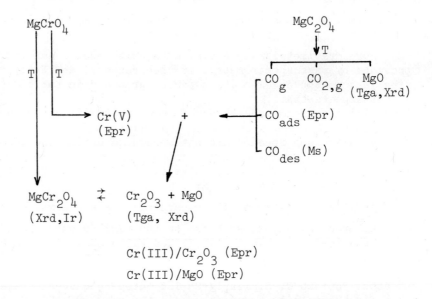

TABLE 2

Thermolysis of oxalate-chromate mixtures of Mg

Note : Thermogravimetric analysis, X-ray diffraction, and Ir-spectroscopy are used to determine the overall reaction steps. Mass spectrometry and Electron paramagnetic resonance enable the identification of intermediate species.

TABLE 3

Compounds of Ba and Sr

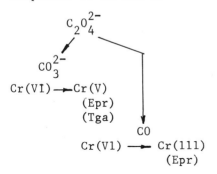

Compounds of Mg

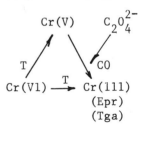

CONCLUSIONS

This contribution shows how various physical techniques can be applied to elucidate complex thermal decomposition mechanisms.

For the thermolysis of the oxalates and chromates, and their mixtures, of Ba, Sr, and Mg, there is experimental evidence for the possibility of preparing well characterised Cr(lll) and Cr(V) species.

This investigation also shows that e.p.r. could be a quite powerful technique to study thermal decomposition mechanism when a paramagnetic species is present, as well as to identify gaseous species which are retained at the solid surface during thermolysis. The latter are likely to influence the reactive properties of the accessible surface.

BIBLIOGRAPHY

1. E.G. Derouane, Z. Gabelica, R. Hubin, and M.J. Hubin-Franskin, Thermochim Acta, 11 (1975), 287.
2. E.G. Derouane, Z. Gabelica, and R. Hubin, Thermochim Acta 14 (1976), 315
   E.G. Derouane, Z. Gabelica, and R. Hubin, Thermochim Acta 14 (1976), 327

APPENDIX

The following abbreviations have been used in the figures and tables

Thermogravimetric analysis : T.G.A. or Tga
Electron paramagnetic resonance : E.S.R. or Epr
Mass spectrometry: Ms   X-ray diffraction: Xrd
Ir-spectroscopy: Ir   Gaseous species: sub"g"
Adsorbed species: sub"ads"   Desorbed species: sub"des"

HIGH-RESOLUTION ELECTRON MICROSCOPY STUDIES OF FLUORITE-RELATED CERIUM OXIDES

O. Toft Sørensen

Danish Atomic Energy Commission, Risø, Denmark

INTRODUCTION

Previously, non-stoichiometric ceria, CeO_{2-x}, were considered to be a single grossly non-stoichiometric phase extending over a considerable composition range at higher temperatures. Recently, however, thermogravimetric measurements in atmospheres of controlled oxygen pressures have shown that the non-stoichiometric phase range can be divided into several subregions, some of which consist of an apparent single non-stoichiometric phase with a characteristic defect structure, whereas other regions apparently consist of a whole series of ordered intermediate phases. In order to confirm the existence of such ordered phases in the non-stoichiometric phase range, high-temperature X-ray studies as well as high-resolution electron microscopy studies on reduced single crystals of ceria, have been started; the results obtained so far will be discussed in this paper. As a basis for the discussion, the results of the thermogravimetric experiments will also be received.

RESULTS AND DISCUSSION

Thermogravimetric Experiments

According to the classical defect theories, the relation between the composition of a non-stoichiometric oxide - x in CeO_{2-x} - and the partial pressure of oxygen in an atmosphere in equilibrium with the oxide can be expressed as: $x \propto p_{O_2}^{-1/n}$, which means that

$$\Delta \bar{G}_{O_2} = RT \ln p_{O_2} \propto -n RT \ln x,$$

where n depends on the type of defect formed in the oxide — e.g. if double-charged oxygen vacancies are formed, $V_O^{\cdot\cdot}$, n = 6 whereas n = 4 or 2 when respectively single-charged or neutral oxygen vacancies are formed. If this treatment is valid, a straight line should thus be observed when isothermal $\Delta \bar{G}_{O_2}$ values are plotted against ln x provided that n is constant, i.e. only one type of defect is present in the oxide. From experimental data obtained by thermogravimetric measurements of the composition of CeO_{2-x} as a function of partial pressure of oxygen at temperatures in the range 900-1450°C, it has been observed that linear relationships really exist between $\Delta \bar{G}_{O_2}$ and ln x but from the data it is also evident that all the data cannot be described by the same straight line. Apparently the non-stoichiometric range covered in these experiments can be divided into subregions each with a characteristic slope of the linear $\Delta \bar{G}_{O_2}$ - ln x relationships corresponding to the fact that a characteristic defect is predominantly formed within each subregion. Finally, a closer examination of the finer details of the experimental points within the subregions with large n-values - e.g. n ≃ 15 or 18, which cannot be described by the classical defect theories - also shows that the steep curves can be considered as composed of a step curve, where the vertical lines indicate the presence of an ordered intermediate phase, whereas the horizontal lines indicate the presence of two-phase regions between the intermediate phases. An interesting feature is that the composition of the intermediate phases seems to follow the homologous series M_nO_{2n-2}, which describes the ordered phases formed at lower temperatures in this oxide system.

High-Temperature X-Ray Examinations

High-temperature X-ray examinations were carried out in vacuum in a Philips high-temperature diffractometry attachment. In order to correctly determine the sample temperatures, ThO_2

was used as an internal calibrant. At sample temperatures up to about 750°C the diffraction patterns closely corresponded to the fluorite structure, but at about 800°C and above splitting of the fluorite peaks, as well as extra peaks between the fluorite peaks, were observed. The splitting and the rather large number of extra peaks observed indicate that a new structure of low symmetry is formed at higher temperatures and after having examined several possible structures (rhombohedral, triclinic) it appeared to be possible to account for all the peaks by assuming that the new structure was of monoclinic symmetry similar to the structure previously reported for Pr_6O_{11} (monoclinic cell with the space group $P2_1/n$, C_{2h}^5). A least squares calculation gave the following values for the lattice parameters of this cell for CeO_{2-x}:

$a_o = 6.781 \pm 0.006$ Å, $b_o = 11.893 \pm 0.009$ Å,

$c_o = 15.823 \pm 0.015$ Å and $\beta = 125.04 \pm 0.04°$,

which closely correspond to the lattice parameters reported for Pr_6O_{11} that is a typical superstructure cell of a rather large size.

High-Resolution Electron Microscopy Studies

High-resolution electron microscopy studies were carried out on a JEM 100C microscope equipped with a double-tilting side-entry goniometer (±45° both on x and y tilt). Single crystals of ceria, which were grown in a PbF_2 melt, were reduced prior to the examination either by heating the particles in the electron beam in the microscope itself or in a separate vacuum furnace at about 1450°C - the conditions used in this heat treatment were fixed so that a composition of $\sim CeO_{1.80}$ could be obtained. The observations made on the two types of reduced crystals, which will be thoroughly discussed in the paper, can be summarized as follows:

1. Structures observed on beam-heated particles,

    (a) superstructures corresponding to the monoclinic superstructure observed in the high-temperature X-ray examination

(b) lamellae structure which can be tentatively interpreted as a shear structure.

2. Structures observed on heat-treated particles,

(a) twins.

The electron microscopy studies, as well as the high-temperature X-ray measurements, show that there is much more order in the non-stoichiometric cerium oxides than previously envisaged. What is clearly needed is a systematic examination of the whole non-stoichiometric range and there is no doubt that further work using, for instance, high-resolution electron microscopy will be particularly fertile for future research.

THE DEVELOPMENT OF SIMULTANEOUS TA-MS FOR THE STUDY OF COMPLEX THERMAL DECOMPOSITION REACTIONS

P.A. Barnes

Department of Chemistry, Leeds Polytechnic
Leeds LS1 3HE, England

INTRODUCTION

As thermal decomposition is clearly a temperature dependant process thermal methods of analysis are of considerable value in both kinetic and mechanistic studies. Accordingly, differential thermal analysis (DTA) and thermogravimetry (TG) have been widely used but the results, although yielding a wealth of information, are difficult to interpret unambiguously. Hence simultaneous techniques have been developed in which additional experimental methods are used to identify the phenomena revealed by DTA and TG (1). One such method, evolved gas analysis (EGA), has been used frequently to identify gas-loss processes and to characterise the materials evolved in thermal analysis. This paper describes the development of sophisticated EGA techniques using simultaneous DTA-MS and TG-MS to obtain qualitative and quantitative information on the gaseous products of thermal decomposition reactions.

The use of a mass spectrometer to monitor evolved gases is a relatively recent development and much work has been done by monitoring a particular m/e ratio during the course of a decomposition reaction (2) (3). The equipment described here holds several advantages over more conventional apparatus. The ability to scan the mass range rapidly during the lifetime of a DTA peak permits the analysis of gas phase components. Once these have been identified magnetic scanning can be used to obtain quantitative information over a selected area of the mass spectrum. Finally, the use of a peak selector enables the rate of evolution of up to four different gases to be monitored simultaneously with the DTA or TG curve. This is a major advantage because, as thermal analysis is a dynamic method, exact reproduction of experimental conditions is difficult and therefore it is not easy to correlate precisely the results of separate experiments. The equipment, although designed for temperature programmed work is equally suited for isothermal studies.

EXPERIMENTAL

The TA equipment used was a Stanton Redcroft thermobalance, type TG-750. The M.S. inlet from the TG-750 is via a heated gasline and is of the capillary and by-pass type which gives low mass discrimination at the expense of some decrease in sensitivity.

The mass spectrometer, a V.G. Micromass MM601, is of modular construction and this particular instrument was tailored to the requirements of simultaneous TA-MS. Modification of the source slit has increased the resolution to approximately 1000 (10% valley) and the addition of an electromagnet has improved the mass range and accuracy attainable. An externally adjustable collector slit was used in conjunction with a four channel peak selector and output hold device to give continuous scanning of four pre-selected flat-topped peaks.

The maximum speed available for voltage-scanning is 0.3S per 1000 mass units which is sufficiently rapid for many analyses to be carried out during the lifetime of a DTA peak which can extend to several minutes. The precision of ion current detection, using magnetic scanning is $\pm$ 1%. The sensitivity of the instrument is such that components present at an inlet pressure of 10^{-6} torr can be sensed at levels of 0.01 ppm and measured at 1 ppm with reasonable accuracy. Thus it is possible to follow the evolution of trace quantities of products formed during side reactions, in addition to the main components.

APPLICATIONS OF TA-MS

In order to demonstrate the potential of simultaneous TA-MS two systems were chosen which had been the subject of previous study and which were suspected of being more complex than earlier workers had realised.

The thermal decomposition of silver carbonate is a simple process to judge by the generally accepted reaction scheme

$Ag_2CO_3 \longrightarrow Ag_2O + CO_2$ - reaction temperature 150 - 230 C

$Ag_2O \longrightarrow 2Ag + \frac{1}{2}O_2$ - reaction temperature 380 - 410 C

This two-stage mechanism was based on the interpretation of early DTA curves and the analysis for carbon dioxide and oxygen during isothermal experiments (4). Considerable kinetic interpretation of both isothermal and temperature programmed work has been undertaken

COMPLEX THERMAL DECOMPOSITION REACTIONS

based on this scheme (5), (6), (7), (8),

A more detailed analysis of the thermal decomposition of silver carbonate by the author revealed that the apparently simple mechanism outline above was not followed in practice and significant differences were found in the behaviour of samples, depending on the method of preparation (9).

In the present work similar samples of silver carbonate were subjected to TA-MS. The equipment was set up to give simultaneous readings of sample weight, sample temperature, evolved carbon dioxide and evolved oxygen. Not only were the various samples of silver carbonate clearly distinguished, (fig 1) but the reactions responsible for the differences were immediately identified. The results show the decomposition of silver carbonate to be a complex process. In even the simplest case, (curve D, fig.1) there is a small amount of oxide breakdown during the major loss of carbon dioxide and, further, two minor carbonate decompositions are observed at higher temperatures. Hence any kinetic interpretation of the rate of thermal decomposition by conventional weight loss or pressure increase methods would be invalid. However the rate of reaction of silver carbonate and silver oxide can be followed from the EGA curves for each individual stage.

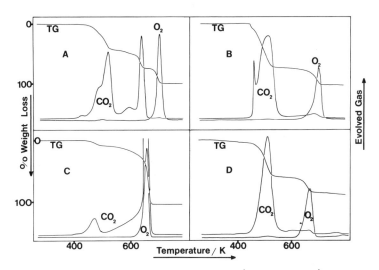

Fig 1. Simultaneous TG and evolved gas (CO_2 and O_2) curves for silver carbonate prepared from (A) Na_2CO_3 (1M) + $AgNO_3$ (1M), (B) Na_2CO_3 (0.003M) + $AgNO_3$ (0.003M), (C) ex BDH, (D) Na_2CO_3 (1M) + $AgNO_3$ (1M) doped with 3 mole % Gd. Heating rate 20 K min^{-1}, purge gas N_2 at 50 cm^3 min^{-1}, 4 mg FSD, sample weights 10 mg.

Another example of the use of simultaneous TA-MS is found in the decomposition of nickel oxalate dihydrate. From the work of Ugai this appeared to be a simple two-stage process (10). More recent investigations, using DTA equipment capable of high resolution, suggested that, after the initial dehydration, the final process was not a simple reaction as the corresponding endothermic peak exhibited a definite shoulder (11). This is in agreement with the comprehensive studies of Dollimore et al which show the decomposition products to be dependent on the reaction atmosphere (12).

The thermal decomposition of nickel oxalate dihydrate was investigated, therefore, by using the mass spectrometer in the peak select mode simultaneously with derivative thermogravimetry. The thermobalance was purged initially with helium until the mass spectrometer indicated a low level of oxygen. The results (fig. 2) clearly show the major processes observed in DTG to be dehydration and decarboxylation. However, evolution of carbon monoxide was observed also during the loss of carbon dioxide. Thus the elementary mechanism of direct metal formation via decarboxylation would seem to be an over simplification (12).

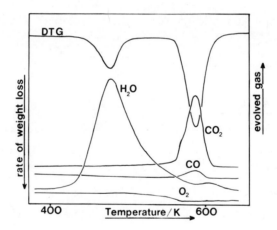

Fig 2. Simultaneous DTG and evolved gas curves (H_2O, CO, O_2, CO_2) for nickel oxalate (ex B.D.H.). Heating rate 20 K min^{-1}, purge gas helium at 50 cm^3 min^{-1}, sample weight 6 mg, relative gain, H_2O and O_2-3x; CO and CO_2-1x.

It is interesting to note that the level of oxygen present in the purge gas decreased further during the second weight loss process suggesting the occurrence of an oxidation reaction. This was supported by XRD of the solid remaining which indicated the presence of nickel oxide. Finally, the evolved water-vapour trace shows a second dehydration step, probably of moisture adsorbed on the product surface.

CONCLUSIONS

Thermoanalytical methods are useful techniques for investigating thermal decomposition reactions, frequently uncovering detail not revealed in isothermal studies. The combination of DTA and TG with mass spectrometry provides a powerful tool for the investigation of this additional information. The use of the sophisticated equipment described enables such work to be performed simultaneously, thus eliminating the experimental uncertanties associated with dynamic methods. Finally the selectivity and high precision of the equipment enable quantitative information to be obtained so that the kinetics of individual reaction steps can be followed.

ACKNOWLEDGMENT

The author wishes to thank the SRC for providing finance for the mass spectrometer.

REFERENCES

1. P.D. Garn, in Thermoanalytical Methods of Investigation, pp. 350-352 Academic Press 1965.
2. B. Kushlefsky, I. Simmons and A. Ross, Inorg. Chem.2 (1963) 187
3. J.G. Brown, J. Dollimore, C.M. Freedman and B.H. Harrison, Thermochim. Acta, 1 (1970) 499.
4. S. Tobisawa, Bull. Chem. Soc.(Japan), 32 (1959) 1173.
5. W.D. Spencer and B. Topley, J. Chem. Soc.,131 (1929) 2633.
6. T. Wydeven and M. Leban, Anal. Chem., 40 (1968) 363.
7. Yu. A. Zakharov, V.V. Boldyrev and A.A. Alekseenko, Kinetica i Kataliz, 2 (1961) 365.
8. T. Wydeven, J. Catal., 16 (1970) 82-89.
9. P.A. Barnes and F.S. Stone, Thermochim. Acta, 4 (1972) 105-115.
10. Ya, A. Ugai, Zh. obshch. Khim., 24 (1954) 1315-1321.
11. P.A. Barnes and A. Wormald, unpublished results
12. D. Dollimore in Differential Thermal Analysis Vol. 1 (ed. R.C. MacKenzie) Academic Press (1970) pp. 398, 406.

DISCUSSION

M.E. BROWN (Rhodes University, Grahamstown, South Africa) Can the small CO peak in Fig. 2 be assigned unambiguously to a primary decomposition product or could it be part of the fragmentation pattern of CO_2 in the mass spectrometer?

AUTHOR'S REPLY: I agree the analysis of CO by mass spectrometry is complicated by the presence of CO_2. However, the problem can be reduced by choosing a suitable ionising energy. In the present case, while there is undoubtedly a contribution to the CO peak from the fragmentation of CO_2, this does not appear to account for the total amount of CO produced.

THE STATUS OF UNDERSTANDING DIFFUSION CONTROLLED

SOLID STATE SINTERING, HOT PRESSING AND CREEP

R. L. Coble

Massachusetts Institute of Technology

Cambridge, Massachusetts   02139

INTRODUCTION AND SUMMARY

Since Kuczynski's pioneering work on the development of models to predict neck growth rates in sintering by diffusion, and his experimental efforts to test their applicability, there have been numerous authors who have contributed a plethora of models covering the different stages of the process with varying assumptions regarding the geometries, the diffusion fields, and the different atomic transport paths which contribute to the overall behavior (1-5). Ashby's recent paper is a useful source to the past modelling work on single phase materials (6). The selection of the models cited there was based on simplicity and reasonable accuracy; both the initial and later stages of the process are covered there, although the importance of grain growth and atmosphere effects are not considered. (7,8). Experimental data on different materials, initial particle sizes, dopants, and temperature schedules have provided a relatively convincing view that for most of the range of variables, the sintering process is diffusion controlled. However, in spite of the extensive work to date, a lack of precise quantitative agreement between predicted behavior and the observed time dependences of shrinkage, size effects, temperature dependences, and dopant effects has lead to the question of what is needed: improved models; more thorough characterization of the morphology changes which occur; or should this interpretation be rejected in favor of other mechanisms for the process? Plastic flow is an alternative, but is not well modelled for any stage of the process. Another alternative is that the process may be controlled by interface kinetics, but modelling has not been available for use in analysis of sintering kinetics. However, Cannon has concluded that the creep of fine grained Al_2O_3 is interface controlled (9). The author has argued that improved characterization is one of the most important needs; it is recognized that the morphology changes which occur during sintering are complex and that the models only represent first order approximations of the behavior (10). Because of variable particle sizes and shapes, it should not be surprising that the time dependences for initial stage shrinkage in powder compacts do not follow the prediction of models based on spheres of constant size, or that densification does not follow that predicted by intermediate stage models either, because the actual geometry is different than that assumed.

In this paper, other sources of difficulty in understanding the size effects and temperature dependences are emphasized; they principally result from the idea that rather than a single controlling mechanism a mixture of mechanisms is, in general, operative and that transformations among rate controlling ions and paths are expected. For ceramics, interpretation of dopant effects in sintering could only be done properly if the influences on the surface and boundary and lattice diffusion coefficients of both cations and anions were documented or predictable. It is emphasized below that no basis now exists to model or predict dopant effects on surface or boundary diffusion. Thus, an improved data base is still needed, on diffusion coefficients and dopant effects, as well as on the microstructure changes during sintering. The latter could provide the guidance needed for fomulation of statistical models for the process. Understanding grain growth is also obviously important, but is similarly hampered by the lack of boundary diffusion data in addition to other problems (11).

The creep behavior of numerous materials at low stresses has been found to be diffusion controlled, i.e., quasi-steady state models have been found to apply at stresses below the yield stress as will be discussed below (9, 12, 13). It is now assumed that any difficulties which arise in interpretation of the

DIFFUSION CONTROLLED SOLID STATE SINTERING

low stress behavior also originates from our incomplete documentation or theoretical base to interpret dopant and impurity effects on the lattice and boundary diffusivities of both species, including their temperature dependences. Also, interface kinetics control may become important at very small particle sizes (9).

Grain boundary sliding is to be recognized as a necessary adjunct to lattice or boundary diffusion controlled grain shape changes but is not emphasized here except to note that it may be of greater importance in early stage hot pressing than in sintering (9). Creep in dense polycrystalline materials is a simpler situation than is sintering or hot pressing. Diffusion controlled creep has been quantitatively demonstrated at low stresses and justifies the assumption above because there are no other mechanisms to which recourse could be taken. Therefore, understanding diffusional creep can be taken as a basis to conclude that diffusion also controls sintering and hot pressing for most of the range of variables (particle size, temperature, pressure, etc.) under which these operations are conducted. The basic assumption is that when the stresses in a powder compact (due to surface curvatures) are below the yield stresses the material can deform only by diffusive transport, and the rate limiting processes are either the surface defect reactions or the defect transport at surfaces, grain boundaries or in the lattice. The geometrical change occurring in sintering or hot pressing is more complicated than in creep and has precluded a precise quantitative description of the kinetics, particle size effects, temperature dependences, and diffusivity comparisons based on simplistic models. Impurity effects, initial particle shape and packing effects, space charge effects at grain boundaries, grain growth simultaneously with densification, and the pore shape evolution are other recognized factors contributing to our inability to predict the kinetics from start to finish of the sintering and hot pressing processes.

The intermediate and final stages of sintering are thought to be simpler than the initial stage, because problems associated with inadequate descriptions of broad distributions of initial particle (and therefore pore) sizes, as well as variable agglomerate densities are reduced significantly by a few percent shrinkage. Very different initial structures in compacts evolve to structures with considerable similarity at 80 to 90% of theoretical density.

SINTERING MODELS

Numerous newcomers to this field have presumed the need to contribute new, improved models. The set now available shows inconsequential differences among the time dependences predicted for shrinkage, neck growth or densification, particle size effects, and the numerical factors associated with the respective models for any assumed mechanism of transport. Multiple modelling has been done for each of the diffusion paths, from which one may conclude that no further simplistic modelling is needed. The first order models are summarized in Table 1. Further consideration should be given, however, to the predictions and implications of the results of combined models, i.e., those in which simultaneous transport by surface diffusion, lattice diffusion and grain boundary diffusion is assumed, as Johnson has done by computer synthesis for the initial stage of sintering under constant rates of heating (14). He showed that the predicted temperature dependence for shrinkage can be larger than that for the atomic mechanism that causes shrinkage and the time dependence is also altered from that predicted by the single models taken alone. These results provide insight as to why quantitative understanding of the observed kinetics has been so elusive. Combinations of competing mechanisms of transport can give rise to a continuously varying time dependence for shrinkage; i.e., no simple constant integer as the time exponent results as is predicted by the single mechanism models. In Table 1, the models for neck growth (X/R) and shrinkage ($\Delta L/L_o$), or densification ($\Delta \rho$) are indicated to be functions of the lattice diffusion coefficient (D_L), the surface energy (γ), the atomic volume (Ω), time (t), particle size (R), Boltzman's constant (k) and temperature (T). The coefficients n, N, and m depend on the assumed mechanism of transport; for lattice diffusion $n = 4$, $m = 3$ while for boundary or surface diffusion $n = 6$, $m = 4$.

Table 1 then indicates the important experimental variables which are used to attempt to deduce the operative atomic transport mechanism. The most important single observation is whether shrinkage (or densification) occurs: that **can result from lattice diffusion or boundary diffusion, or viscous or plastic flow** (when the effective stress exceeds the yield stress).

TABLE 1

SINTERING MODELS AND EXPERIMENTAL VARIABLES

Neck growth, or shrinkage,
or densification predicted

$$\frac{X}{R}^n \; ; \; \frac{\Delta L}{Lo}^{n/2} \; ; \; \Delta \text{ density} = \text{Fn's} \; \frac{ND\gamma\Omega t}{R^m kT}$$

Experimental variables:

Occurrence of shrinkage?

Yes: D_L, D_b, η, $\tau > \tau_y$

No: D_s, p^o

Kinetics:

Neck Growth
Shrinkage Rate
Particle/Grain Size Effects
Additive Effects
Atmosphere Effects
Temperature Dependence
Grain Growth Rates

With models and data, calculate D_{sint}, ΔQ_{sint}

$D_{sint} \approx D^*$, and $\Delta Q_{sint} \approx \Delta Q^*$ for Cu, Al_2O_3 (15)

Needs:

Complete descriptions of microstructure evolution
Pore shapes and size distribution
Grain size distributions
Enlarged data base on diffusivities

Neck growth without shrinkage may be manifested by substantial increases in the strength of powder compacts; that can occur either from vapor transport or surface diffusion of material to inter-particle contacts. Neither of these latter mechanisms give rise to shrinkage. Quantitative measurements of the other variables listed are then used with the models to calculate the diffusivities to compare with independently measured tracer values. A common result is given at the bottom of Table 1; calculated D's sometimes agree with the tracer diffusivities (D*) such as for alumina or copper, but frequently the calculated D's have been larger, and the observed activation energies also have been larger than that for the presumed diffusion mechanism (15). The latter might arise from heating rate effects and competetive mechanisms, as Johnson has illustrated (14).

Because of the disagreement between measured and calculated diffusivities, we have considered the possible need for transient solutions to the diffusion equation instead of the quasi-steady state solutions which are almost universally employed for solid state reactions. Because the pore and grain boundary surfaces move during the process, and the defect concentrations at all sites in the diffusion field change with time, transient solutions should formally cover the total behavior with greater precision. By computer simulation, transient solutions have been generated for a final stage model in which a spherical pore was assumed to be located within a concentric spherical shell (16). The outer boundary simulates the grain boundaries, it was assumed to be fixed or moving at velocities typical of those observed during grain growth. The results showed that the quasi-steady state solution is approached rapidly during early time intervals that were orders of magnitude smaller than the time required for pore closure. In general, the results showed for observed values of pore and grain boundary surface velocities that the quasi-steady state solutions are adequate. Therefore, most of the behavior that is measurable in sintering experiments should be interpretable based on earlier modelling, as summarized in Table 2. Thus, the discrepancy between D's cannot be ascribed to a failure to consider the transients for the process.

TABLE 2

SINTERING/HOT PRESSING MODELS

Transient solutions: $dC/dt = D\nabla^2 C$

  Blendell, Onorato, and Coble: not necessary

Quasi-steady state: $\nabla^2 C = 0$

  D_s Kuczynski, Cabrera, Nichols,- - - -
  D_L Kuczynski, Kingery, Coble, Johnson,- - - -
  D_b Coble, Johnson,- - - -
  p^o Kingery - - - -
  η Frenkel
  τ_y Mackenzie, Shuttleworth

Various diffusion models, with different assumptions show <u>quantitative</u> <u>agreement</u>!

PROBLEMS

A summary of some of the problems in understanding sintering is collected in Table 3, where suggestions for resolving them are advanced. For the various cases in which the discrepancy in D's exist there are several possibilities. For cases in which the process is interpreted to be lattice diffusion controlled, it may be governed by boundary diffusion, particularly when small particle sizes are being used. The influences of impurities on the defect concentrations and transport in compound crystals is presumed to be understood, although recently, increased importance has been attached to association of defects, and to space charge effects at interfaces that have only recently been combined with impurity drag models to predict grain growth rates (11,17). However, the general lack of data on grain boundary diffusion, and the influence of impurities on boundary transport will require much more work before this problem area can be discounted or quantitatively resolved. It should be noted that smaller contents of impurities could produce more substantial effects at grain boundaries than would be required to cause extrinsic behavior in the matrix crystals.

Typical lattice diffusion behavior in a metal oxide with Schottky disorder is shown in Fig. 1; an impurity cation with higher valence than the host is presumed to be present. In materials like MgO, cation diffusivities are generally found to be greater than the anion diffusivities. The expected temperature dependences for both ions are illustrated for the intrinsic and extrinsic regions. More complex behavior is expected in materials with mixed Schottky-Frenkel disorder as may be expected for Al_2O_3 or in more complex compounds such as the spinels, for which the applicability of the modeling for alkali halides is unlikely to be correct. However, for lattice diffusion it is presumed that improved characterization of the defects and complexes present (by ESR, NMR, etc.) might demonstrate the applicability of existing defect models. In contrast, we have no adequate structural models for grain boundaries or for the nature of the defects in them to interpret intrinsic or extrinsic grain boundary behavior. Further, there are insufficient experimental data on boundary diffusion to provide empirical correlations for analysis or predictions of

TABLE 3

PROBLEMS IN SINTERING, HOT PRESSING

$D_{sint} \neq D^*$, and $\Delta Q_{sint} \neq \Delta Q^*$, frequently

Geometrical complexity sintering
    Documentation of microstructure evolution

Small particle sizes used 100Å - 10µ
    Thermodynamic approach breaks down
    Statistical mechanics approach?

Grain growth
    Space charge effects on boundary "drag"
    Boundary diffusion documentation needed
      Structures: ionic/covalent/metallic
      Chemistry effects on D_b, "core", and space charge regions

Sintering covalent materials: γ_{sv}/γ_{gb}
    D's, P° boundary diff, liquids?

Kinetics not understood
    Not expected for simple models
    Microstructure evolution needed

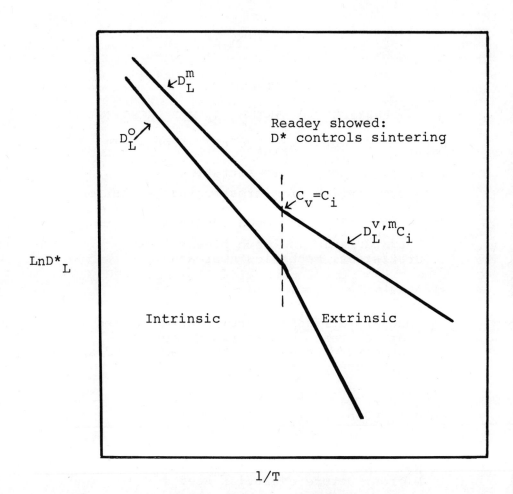

Figure 1: Aliovalent impurity effects on D_L's in a compound with Schottky disorder.

impurity or dopant effects on the boundary transport of both species as functions of temperature.

Another important conclusion from earlier theoretical work is noted in Fig. 1. Readey analyzed sintering by lattice diffusion (only) using a statistical mechanics approach, and considered the coupled fluxes of the different charge species (ambipolar diffusion) (18). He showed that the ion that controls the process is the slower species as would be observed in a tracer diffusion experiment (in samples with identical impurity contents). For sintering, hot pressing or creep, the slower species is enhanced by a small amount ($\sim 2x$) due to ambipolar coupling.

When transport can take place by boundary diffusion in addition to lattice diffusion, the process may then take place by transport of the cations in the lattice while the anions move along grain boundaries. In this instance, Readey's conclusion that the phenomenological diffusivity governs the process must be extended to apply to all paths and ions simultaneously. Hence, the uncertainties in boundary transport in the extrinsic range become very important in governing our interpretation of the behavior, because the shrinkage may take place by any combination of paths for both species that will provide the fastest net flow. For example: lattice diffusion of both ions, or boundary diffusion of both ions, or lattice diffusion of the cation with boundary transport of the anion, or vice versa: the fastest combined path(s) operates with the slower ion in whichever path controlling the rate.

The point to be emphasized is that the sintering process is quite complicated apart from the geometrical problems. If we consider surface diffusion as well, the above schema also apply to neck growth behavior, but with lattice (or boundary) and surface diffusion paths interacting cooperatively. Since both neck growth and shrinkage phenomena must be understood, with all three diffusion paths contributing, there should be little wonder that attempts to understand the kinetics, and the temperature dependences of observed behavior have rarely been conclusive. Consider that for intrinsic and extrinsic transport of two species along three paths that there are 12 possible activation energies to be considered. For each process (neck growth or shrinkage) for different species and paths may control the progress as the size scale as changed; eight could conceivably control each process is impurities/dopants and temperature are also changed (i.e., if D_L^o could become greater than D_L^m, etc.). In the latter case, sintering <u>could</u> be controlled by any of the twelve different <u>ions</u>, transport paths, and regimes; transitions among them are expected as neck growth takes place, or as changes in particle size, temperature, dopant concentrations, or oxygen pressures are manipulated as experimental variables.

It has been assumed that the initial stage models are applicable for shrinkage less the ~5%. However, it has not been recognized that the pressures predicted from the models exceed reasonable bounds when applied to samples made from powders of small particle size. For sintering between spheres by lattice/boundary diffusion, the minimum radius at the neck (ρ) may be equated to one-half the overlap (y) between the spheres (19). The pressure diffusion across the neck surface (γ/ρ) may then be set equal to γ/y, and since $y/R = \Delta L/L_o$, $P = \gamma/R(\Delta L/L_o)$. The pressures calculated for two ranges of particle size (R = 10^{-6}, and 10^{-4} cm, with $\gamma = 10^3$ ergs/cm^2) are plotted as functions of shrinkage in Fig. 2. The size ranges selected are those typical of sol-gel powders, and numerous dielectric bodies, respectively. The results show that at small shrinkage values the pressures exceed the yield stresses and plastic deformation may therefore be expected. In Al_2O_3, the range of non-basal yield stresses from ~1200°C to 2000°C is indicated. Therefore, for shrinkage greater that 1%, the diffusion models are expected to apply. However, that covers most of shrinkage of interest. The curves are shown to turn upwards to represent, schematically, the pressure increases at later stages in the process when pore radii decrease (6).

DIFFUSIONAL CREEP

The highlights of the results obtained on creep in oxides are intended only to justify the conclusion above: since creep is diffusion controlled at low stresses, we conclude that most of the range of sintering is also diffusion controlled. The combined lattice and boundary diffusion models as derived for an

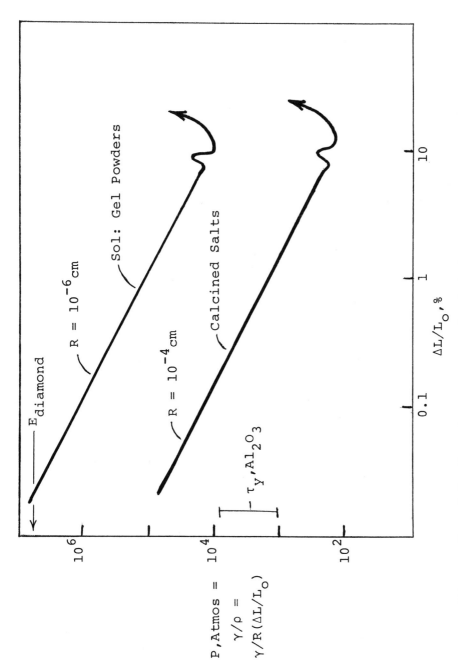

Figure 2: Calculated pressures in initial stage sintering

TABLE 4

DIFFUSIONAL CREEP

IN A DENSE POLYCRYSTALLINE:

Metal: Lattice and grain boundary contributions additive:

$$\overset{\circ}{\varepsilon} = \frac{14\sigma\Omega D_L}{kT\ G^2} + \frac{14\pi\sigma\Omega D_b W}{kT\ G^3} = \frac{14\sigma\Omega}{G^2 kT}(D_{eff})$$

Effective D for combined paths

$$D_e^i = D_L^i + \pi D_b^i\ W/G$$

Ceramic: Effective D in compound $M_\alpha O_\beta$

$$D_{eff} = \frac{(\alpha+\beta)\ D_e^m D_e^o}{\beta D_e^m + \alpha D_e^o}$$

elemental metal are presented in the first equation in Table 4. The second equation represents the uncommon terms after factoring common terms and is to be evaluated independently for anions and cations. Those values inserted in the third equation give D_{eff} for use in the expression on the right hand side of the first equation; this then covers the combined paths with ambipolar coupling of the respective ion fluxes. Note that the strain rate is linear in stress (σ), which distinguishes diffusional creep from dislocation motion controlled processed, for which higher stress dependences are expected.

The stress dependence is the first experimental variable listed in Table 5 among those used to test the applicability of the models. The second important variable is the grain size (G) dependence; the explicit effects for lattice and boundary diffusion control are given in Table 4, and their effect on the creep rate for a compound is shown in Fig. 3. It is assumed that the cations move more rapidly through the lattice than do the anions ($D_L^m > D_L^o$); either could control the rate at values fixed by the curves with slopes equal to -2. Independently, boundary transport gives rates shown by the curves with slopes: equal to -3, but the assumed order is reversed, i.e.: the oxygen diffusivity in the boundary (D_b^o) is assumed to be greater than that for the cation (D_b^m). The fastest transport of both species using the combined paths, is indicated for the slower species by the dash-dot trace. As the grain size increases, evolution in control of creep is predicted from: boundary diffusion of the cation at small grain sizes; to lattice diffusion of the cation; to oxygen control in the boundary; and finally to oxygen control in the lattice at large grain sizes. The range which has been observed most frequently is for control of the process by lattice diffusion of the cations (with anions transported (faster) at the grain boundaries). Reasonable agreement has been found between the diffusivities calculated from the creep rates, and tracer values in MgO and Al_2O_3 (9, 12, 13). At intermediate grain sizes, Cannon concludes that creep in Al_2O_3 is controlled by transport of Al in the lattice with oxygen moving along grain boundaries (9). At small grain sizes (G < 10µ) and at low temperatures, Al_2O_3 creep is controlled by Al transport at the grain boundaries. Similarly, in Fe doped MgO, Gordon et al conclude that the process changes from control by Mg transport in the lattice (at low P_{O_2}) toward control by

TABLE 5

VARIABLES USED TO DEDUCE CREEP MECHANISMS

1. Stress: $\sigma^1 \to \sigma^4$
    High stress Dislocation motion
                 Pore nucleation
                 Fracture

2. Grain Size
    Diffusional creep: Explicit effects:
                                  G^{-2} or G^{-3}
    Dislocation climb/glide: Cell structure
                                  documentation needed

3. Temperature
    Increase T, Intrinsic: D_L/D_b increases
                 Extrinsic: D_L/D_b increase/decrease?
    Decrease T → Extrinsic range, and initial
                 transient dominant for overall
                 creep strain

4. Change P_{O_2} (Kroger:Vink) changes stoich, and valence of impurities

5. Increase dopant → Solid solution effects: D_L's, D_b's
                    2nd phase: solid, liquid

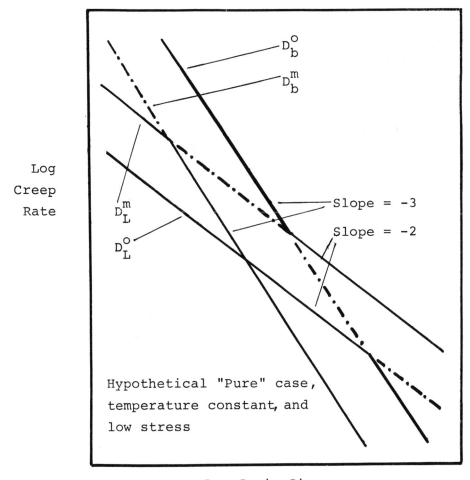

Figure 3: Diffusional creep in a dense, polycrystalline compound as a function of grain size.

oxygen transport at the boundaries in samples with median grain sizes (13). Intermediate exponents for the grain size effects are interpreted as being indicative of mixed control and thus substantiate the assumption of additive contributions from mixed paths for the respective ions as a valid hypothesis. The main point is that transitions from one set of controlling paths to another have been demonstrated for creep as a function of one of the variables (grain size) making quantitative interpretation quite complex.

Other variables investigated for creep include dopant effects and oxygen pressure dependencies which further support the conclusion that creep is diffusion controlled in-so-far as most of the effects can be interpreted in terms of expected changes in defect concentrations, atom mobilities, and hence creep rates (9,12,13).

In summary, creep studies with variable stresses, grain sizes, dopants, and changes of oxygen pressure with and without dopants, and the observed temperature dependences have provided a broad range of conditions under which diffusional creep is observed and for which good agreement is obtained between calculated diffusivities and tracer values for the cation lattice diffusivities. Cannons' re-analysis of the published data for Al_2O_3 yielded a spread of only $\sim 2x$ for the lattice diffusivities for intermediate grain sized material in contrast to the wide spread disagreement among the independent papers (9).

It is noteworthy that diffusional creep in dense polycrystalline material is, of itself, quite complicated. Although the sintering or hot pressing behavior is recognizably more complicated, these densification processes should also be diffusion controlled in the ranges of stress, grain size, and temperature for which the creep behavior has been documented and shown to be diffusion controlled. That range covers most of the evolution in structure during sintering of Al_2O_3, and it is presumed for many other less-well-documented materials as well. We conclude that for the initial stages of sintering of sub-micron particles, boundary diffusion is probably dominant. Lattice diffusion should take over during the intermediate and late stages in the process as a result of neck growth or grain growth. The evolution in control of the process among paths and ions for sintering should follow the creep analogy shown in Fig. 3.

STATUS, OR CONCLUSIONS

1. Transient solutions are not needed for modelling diffusion controlled sintering.
2. Improved characterization of the microstructure evolution in sintering and hot pressing is still needed to guide realistic geometrical or statistical remodelling.
3. Numerous quasi-steady state models show quantitative agreement, and when applied to sintering data, and yield diffusivities which sometimes agree with independently measured diffusivities.
4. Steady state creep at low stresses is diffusion controlled.
5. For most of the range of variables in sintering and hot pressing at moderate pressures, diffusion must also be controlling.
6. The contributions from the independent diffusion paths are additive, and transitions in operative paths are expected with changes in neck size, grain growth, dopants, oxygen pressures, and temperature.
7. Transitions among paths have been demonstrated for creep.
8. An expanded data base on diffusivities in crystal lattices and at grain boundaries, and dopant effects upon them will be required before quantitative interpretations of the sintering behavior can be improved significantly.
9. Space charge effects at grain boundaries influence grain growth.
10. Grain boundary sliding is a necessary adjunct process to diffusional grain shape changes and may be important as an alternate to interface control by defect equilibration under presently unspecifiable conditions.

The author regrets that the space limitations preclude an exhaustive review for this broad subject and is apologetic to numerous other significant contributors to the field not cited herein.

REFERENCES

1. G. C. Kuczynski, Trans A.I.M.E., **185** (1949), 169.
2. T. L. Wilson and P. G. Shewmon, Trans. A.I.M.E., **236** (1966), 48.
3. F. Thummler and W. Thoma, Met. Rev., **115** (1969), 69.
4. R. L. Coble, J. App. Phys., **32** (1961), 787.
5. D. L. Johnson, J. Am. Ceram. Soc., **53** (1970), 574.
6. M. F. Ashby, Acta. Met., **22** (1974), 275.
7. J. E. Burke, J. Am. Ceram. Soc., **40** (1957), 80.
8. R. L. Coble, J. Am. Ceram. Soc., **45** (1962), 123.
9. R. M. Cannon and R. L. Coble in "Deformation of Ceramic Materials", Plenum Press (1975), 61.
10. R. L. Coble, *Review of Understanding Sintering and Related Phenomena*, Vol. 6, Plenum Press, (1973).
11. M. Yan et al.,"Deformation of Ceramic Materials", Plenum Press (1975), 549.
12. G. Hollenberg and R. S. Gordon, J. Am. Ceram. Soc., **56**, (1973), 140.
13. P. Lessing and R. S. Gordon, "Deformation of Ceramic Materials", Plenum Press (1975), 271.
14. D. L. Johnson in "Ultrafine-Grain Ceramics", Syracuse University Press (1970), 173.
15. R. L. Coble and T. K. Gupta, "Sintering and Related Phenomena" Gordon & Breach, N. Y. (1967), 123.
16. J. E. Blendell, P. K. Onorato and R. L. Coble unpublished research.
17. W. D. Kingery, J. Am. Ceram. Soc., **57**, (1974), 1.
18. D. W. Readey, J. Am. Ceram. Soc., **49**, (1966), 366.
19. W. D. Kingery and M. Berg, J. Appl. Phys., **26** (1955) 1205.

ACKNOWLEDGEMENTS

The author gratefully acknowledges helpful review of the manuscript by H. Kent Bowen, R. M. Cannon, Jr., J. E. Blendell, E. Skaar, and D. Kramer, and for financial support under ERDA Contract No. E(11-1)2390.

KINETIC STUDY OF THE REDUCTION OF Ni^{2+} IONS IN AN X TYPE ZEOLITE

M. Kermarec, M.F. Guilleux, D. Delafosse, M. Briend-Faure and J.F. Tempère

E.R. 133 "Cinétique des Réactions Superficielles"

Tour 55, 4 Place Jussieu, 75230 Paris Cedex 05

INTRODUCTION

There is a theoretical and practical interest in studying small metal particle formation. The electronic structure, the nature of metal-metal bonding and the crystallographic arrangement of the small particles can be modified as compared to these properties observed in bulk solids. In this field, a limited number of metallic elements have been investigated (1) (2) (3). Zeolites, which allow the introduction of transition cations in to well defined crystallographic sites constitute very good supports to obtain highly divided metallic particles. Recent work suggests that in certain cases the dispersion of these particles involve particular catalytic properties (3).

Few kinetic studies have been undertaken on the formation of small particles in these supports and we now report some results on the reduction by hydrogen of Ni^{2+} ions introduced into an X type zeolite (4). These ions can be located in sites S_1, S_1', and S_{11} (5). In this work, we study particularly the influence of the cation location and the modifications caused in the reduction process by the presence of an easily reducible cation such as Pd^{2+}.

EXPERIMENTAL

The samples are obtained by successive exchanges from NaX zeolite. Their composition is shown in Table 1. Before being reduced, the samples are pretreated at $500°C$ under a residual

Na X	Na_{86} $SiO_{2\,106}$ $AlO_{2\,86}$
Ni_8 X	Na_{64} Ni_8 H_6 $SiO_{2\,106}$ $AlO_{2\,86}$
Ni_{31} X	Na_{24} Ni_{31} $SiO_{2\,106}$ $AlO_{2\,86}$
Pd Ni X	$Na_{12,4}$ $Ni_{24,4}$ $Pd_{0,3}$ $SiO_{2\,106}$ $AlO_{2\,86}$

Table I

pressure of 10^{-3} torr. In these conditions, most of Ni^{2+} ions in Ni_8X are located in S_I and $S_{I'}$, while in $Ni_{31}X$ the Ni^{2+} ions occupy all three types of site (6).

The reduction rate is followed by the volumetric measurement of hydrogen uptake under a constant pressure. An infra-red study was also made using a Perkins-Elmer 521 spectrometer and EPR measurements were made with a Varian E 9 X-band spectrometer.

RESULTS

NiX Samples

In the case of Ni_8X and $Ni_{31}X$ the reduction was investigated in a temperature range from 200 to 400°C and a pressure range from 10 to 100 Torr. When the pressure is increased above 50 Torr, the reduction rate no longer varies. α, which expresses the extent of the reduction at t time, is defined by $Ni^°_{(t)}/Ni^{2+}$ total per gram of anhydrous zeolite. The curves α= f(t) show that the rate is continuously decreasing. As a first approximation, these curves are superimposable after a {α(t) ⟶ kα(t)} transformation (where k is a constant) (Fig. 1). It can therefore be considered that in these temperature and pressure ranges, there is only one rate limiting process. The activation energy value is 28 ± 2 Kcal mole^{-1} for the two sets of samples. However, for Ni_8X, when α is below 0.06 the activation energy decreases to a limiting value of 14 Kcal mole^{-1} at α = 0,03, which implies a more complicated mechanism in this range of α.

Fig. 1 shows that the lower the degree of the exchange, the higher is the rate of the reduction. When the outgassing treatment is below 400°C, a large decrease in the reduction rate is observed. Furthermore, it is shown that if a partially reduced sample is then outgassed for 15 hours, the rate of a subsequent reduction made in similar conditions is higher than that observed before outgassing. The higher the temperature of outgassing, the higher is the reduction rate.

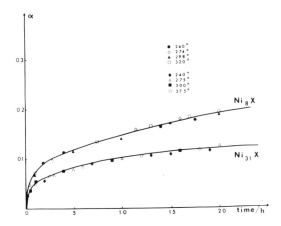

Fig. 1 Super imposition of the α(t) reduction curves at 50 Torr after a {α(t) → k(α)t} transformation for Ni_8X and $Ni_{31}X$. Reference curves are taken at 275°C.

The infrared spectra of a partially reduced sample show the presence of a band centered about 3595 cm^{-1} (Fig. 2). The relative intensity of this band assigned to OH or H_2O molecules bonded to Ni^{2+} ions (7) increases when the reduction temperature increases.

The E.P.R. spectra show no evidence for Ni^+ state. A large ferromagnetic Ni^0 band only is observed.

Ni Pd X Sample

The presence of reduced Pd atoms in the neighbouring of Ni^{2+} ions leads to an important increase in the extent of the reaction, as shown by Fig. 3. The lower the reduction temperature, the higher is this increase. Above 350°C the α(t) curves are the same as those observed without Pd^0 atoms : in this temperature range, Pd^0 atoms migrate quickly towards the external surface of the zeolite and no longer activate the reduction process.

DISCUSSION

The experimental results suggest a diffusion mechanism in which OH or H_2O groups are involved.

For the two sets of NiX samples the α(t) curves can be transformed into straight lines of the form $F_D = kt$ where:

$$F_D = 3/(1-\alpha)^{1/3} + \log(1-\alpha)$$

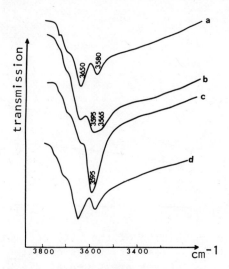

Fig. 2 Infrared spectra of $Ni_{31}X$ sample.
a) $Ni_{31}X$ evacuated at 500°C
b) after a subsequent reduction at 275°C during 3 h.
c) after a subsequent reduction at 300°C during 15 h.
d) evacuated at 500°C after c)

From the slopes of these lines it is possible to find the same value of the experimental activation energy : E. The kinetic analysis expressed by this equation is obtained from a defined model for a gas-solid reaction in a microporous medium (8). This model can be applied to zeolite structures, assuming that the sample is constituted of spherical particles and the repartition of Ni^{2+} ions is homogeneous. The steady state approximation is made and the diffusion coefficient varies as $D = D_o (1 - \alpha)$.

From an energetic point of view, it does not appear that the initial cation location affects the reduction process. Thus, the same value of E was observed for the two samples over a large range of α. A lower value of E should be found for $Ni_{91}X$, in which most part of the ions being located in S_{11} sites are very accessible. It can be assumed that the presence of OH or H_2O groups occuring during the reduction process, enables the migration of Ni^{2+} ions from S_1 and S_1' towards S_{11} sites (9). Therefore the sites appear to be energetically homogeneous. In order to explain the linear decrease of D, we suggest that a part of OH or H_2O groups are fixed in the neighbouring of Ni^{2+} ions as $Ni(OH)^+$ or $(Ni_n H_2O)^{2+}$. This leads to a decrease of the accessibility of the Ni^{2+} ions which will be reduced. The higher reduction rate observed for Ni_8X can be explained taking in to account the greater dispersion of Ni^{2+} ions in the zeolite cages. Therefore the probability of hindrance of these ions by OH or H_2O groups must be weaker.

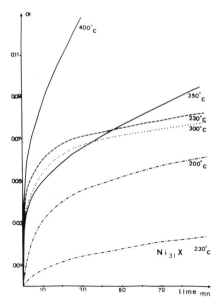

Fig. 3 α(t) curves of the reduction at 50 Torr of PdNiX and $Ni_{31}X$ samples.

For PdNiX sample, an increase of the reaction rate was observed in the presence of Pd^o particles. This effect can be understood from a spill-over mechanism (10). The atomic hydrogen formed on Pd^o particles is more reactive than molecular hydrogen which allows a faster reduction process.

In conclusion, it is possible to accelerate the reduction process by adding cations which act as co-catalysts such as Pd^{2+}. Another way may consist in introducing a cation whose affinity towards OH or H_2O groups is greater than that of Ni^{2+} ions. The first experiments made with Ce^{3+} seem to confirm this suggestion. Furthermore, the extent of the reduction rate of Ni^{2+} ions appear to depend on the cation-lattice bonding which varies with the type of the zeolite. Therefore the environment of Ni^{2+} ion has to be taken into account when studying the formation of highly dispersed Ni^o particles in three dimensional matrices.

ACKNOWLEDGEMENT

The authors thank M. Patel for carrying out kinetic measurements.

REFERENCES

1) P. Ratnasamy, A.J. Leonard, L. Rodrique and J.J. Fripiat, J. Catal., 1973, **29**, 374.
2) R.A. Dalla Betta and M. Boudart, Int. Congress of Catalysis, Miami, 1972, paper 96.
3) P. Gallezot, A. Alarcon-Diaz, J.A. Dalmon, A.J. Renouprez and B. Imelik, J. Catal., 1975, <u>39</u>, 334.
4) A.C. Herd and C.G. Pope, J.C.S. Faraday I, 1973, 69, 833.
5) G.R. Eulenberger, D.P. Schoemaker and J.G. Keil, J. Phys. Chem. 1967, 71, 1812.
6) P. Gallezot and B. Imelik, J.Phys. Chem., 1973, 77, 652.
7) M.F. Guilleux and J.F. Tempère, C.R. Acad. Sci., 1971, 272, 2105.
8) A. Boisselier, F. Caralp and M. Destriau, Bull. Soc. Chim. France, 1974, 9-10, 1735.
9) P. Gallezot, Y. Ben-Taarit and B. Imelik, J. Phys. Chem., 1973, 77 (21), 2556.
10) R.B. Levy and M. Boudart, J. Catal., 1974, 32, 304.

DEVELOPMENT OF CONDUCTIVE CHAINS IN RuO_2-GLASS THICK FILM RESISTORS

R. W. Vest

Professor, School of Materials Engineering

Purdue University, West Lafayette, Indiana   (U.S.A.)

INTRODUCTION

Thick film resistors can satisfy certain design functions (e.g. high sheet resistance, power dissipation, and high voltage requirements) that are difficult or impossible to achieve with silicon monolithic integrated circuits. They are fabricated by: 1) preparing a formulation consisting of a glass frit and an electrically conducting oxide powder dispersed in an organic screening agent; 2) printing the formulation on to a ceramic substrate; and 3) firing in a tunnel kiln with the proper time-temperature profile to remove the organic constituents, fuse the glass frit and bond the resistor to the substrate. The system considered in this study was a lead-borosilicate (63% PbO-25% B_2O_3-12% SiO_2) glass frit, ruthenium dioxide (RuO_2) conductive, ethyl cellulose in butyl carbitol screening agent, and 96% Al_2O_3 substrate.

BLENDING CURVE

The glass is an electrical insulator with a room temperature resistivity in excess of 10^{14} ohm-cm, and the RuO_2 is a metallic conducting oxide with a room temperature resistivity of 3.5×10^{-5} ohm-cm. One of the most desirable features of thick film resistors is the ability to vary the sheet resistance (resistivity/25 μm thickness) by many orders of magnitude by varying the ratio of conducting powder to glass powder. (See the experimental curve in Figure 1). A typical firing schedule includes eight minutes at temperatures above the softening point of the glass (∿450°C) and ensures that a continuous matrix of the glass is formed. The system of conducting particles dispersed in an insulating medium was first considered by Maxwell and an adaptation applicable to the volume

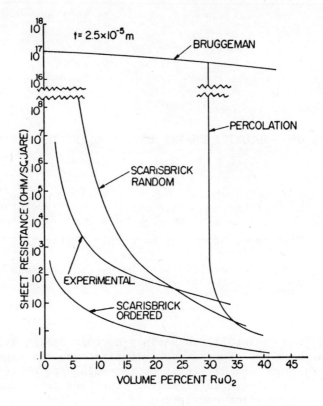

Figure 1. Experimental and Theoretical Blending Curve

fraction range of interest was developed by Bruggeman [1]. His equation is plotted in Figure 1, and predicts a decrease from the sheet resistance of the glass by a factor of less than 10, whereas a decrease by a factor of 10^{16} is observed experimentally. Percolation theory [2] is a more recent approach to describing the resistivity of a system consisting of a conductive phase dispersed in an insulating phase. Depending upon the statistical model chosen, a decrease from a resistivity near that of the insulating phase to one near that of the conducting phase is predicted over a very narrow volume fraction range at a percolation threshold within the volume fraction range of interest, and a typical result is plotted in Figure 1. An alternate statistical approach was developed by Scarisbrick [2], who assumed that the conducting particles were touching, and considered two limiting cases: an ordered model and a random model. These two limiting cases are plotted in Figure 1. Since Scarisbrick's model based upon the formation of conducting chains seems to describe the experimental blending curve better than any other model, this study was undertaken to determine the driving forces for formation of the chains in terms of physical properties of the ingredient materials.

MATERIALS PREPARATION AND PROPERTIES

The ruthenium dioxide was prepared by drying the hydrous oxide to give a powder with an average particle size of 60Å. The glass was fritted in distilled water and ground to pass through a -325 mesh sieve. The viscosity and the surface tension of the glass were measured [4] and the equilibrium solubility of RuO_2 in the glass was also determined [5]. The wettability of the glass to RuO_2 was studied by contact angle measurements at 700, 800, 900 and 1000°C. It was found that the equilibrium contact angle was 0 (complete wetting) at all temperatures, but the spreading rate varied strongly with temperature (e.g. complete wetting was achieved in five minutes at 900°C whereas 30 minutes at 800°C was required). Initial stage sintering kinetics for the glass were determined [6] and it was found that the neck growth between glass spheres followed the behavior predicted for Newtonian viscous flow. Ostwald ripening of RuO_2 particles in the glass was studied [6], and it was concluded that growth of RuO_2 particles in the glass occurs by a phase boundary reaction controlled solution-precipitation process. From these studies it was also concluded that the initial stage sintering of RuO_2 particles in the presence of the glass occurs with the same limiting step.

MICROSTRUCTURE

Since the glass is optically transparent, resistor microstructure can be studied using transmitted light, and Figure 2 shows that the macrochains are composed of a fine structure of microchains. While the glass is optically transparent, it is opaque to an electron beam unless the lead is leached out of the surface. An example of

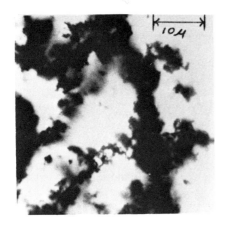

Figure 2. Transmitted Light Microstructure

Figure 3. SEM Microstructure

the microstructure observed after leaching is shown in Figure 3, and a structure of interconnected loops of RuO_2 particles can be seen. Observations on the hot stage microscope during resistor firing revealed that the networks, such as those shown in Figure 2, were continually moving and forming during the firing; even at 800°C (the maximum firing temperature) parts of the network were observed moving distances greater than 10μm in less than 1 minute. Observations on the hot stage microscope also revealed that the resistor was in a dynamic state of agitation at all temperatures above 600°C due to escaping gas bubbles.

DISCUSSION

In order for the microstructure development to proceed the glass must spread and wet the RuO_2 particles. A theoretical equation to describe the kinetics of spreading which most closely approximates the situation in thick film resistors is that proposed by Newman [7] for the rate of penetration of liquids into microirregularities when capillary pressure is the driving force. This model was utilized in the present study to calculate the rate of penetration of the glass between two adjacent RuO_2 particles as a function of surface tension and viscosity of the glass, the particle size of the RuO_2 and the time dependent contact angle. This model predicts that a glass bridge will be formed between two 60Å RuO_2 particles separated by 30Å in one minute over the temperature range 450 to 500°C. The force acting between two particles separated by a liquid bridge has been calculated by Huppmann and Reigger [8] as a function of surface tension of the glass, radius of the particles and contact angle. The force was calculated from this model and substituted in Stoke's equation to determine that the velocity with which the particles would be pulled together at 450°C was greater than 0.1 μm per second. The rate of glass sintering in this temperature range (450 - 500°C) was calculated from earlier results [6] for the particle size range present in the glass frit (1 - 10 μm), and it was found that the ratio of neck radius to particle radius will reach approximately 0.3. Therefore, RuO_2 particles which are adjacent to glass particles are pulled together on the surface of the glass particles which remain near to their original size, and loop structures having diameters of 1 to 10 μm would be expected.

Intermediate stage sintering of the glass occurs above 500°C, and when RuO microchains come in close proximity due to the motion generated by the glass sintering, they will rapidly be pulled together and held by surface tension forces to produce macrochains. As the intermediate stage of glass sintering proceeds, the glass-vapor interfacial area decreases very rapidly and with it the driving force for the rearrangement process which was responsible for formation of the microchains and macrochains. Therefore, the structure of the microchains and macrochains will not change appreciably above 550°C.

The final stage of glass sintering is driven by pore collapse and bubble release. Pore collapse is a result of the negative pressure inside the pores due to the surface tension which leads to an effective hydrostatic compressive force on the outside of the system [9]. If the pore contains an insoluble gas, equilibrium wil be reached when the capillary pressure is equal to the gas pressure inside of the pore. In this case a force will be exerted on the bubble due to its buoyancy, and its rate of motion can be calculate from Stoke's law. Carrying out this calculation it was found that pore with an initial radius of 1 μm containing an insoluble gas wil move at the rate of 0.3 μm per minute at 800°C. This velocity is consistent with the observed rate of bubble release. When two macr chains of RuO_2 particles are brought into close proximity by the stirring action caused by the bubble release, they will remain in contact due to the net compressive force, and sintering by a soluti disolution process will begin. The initial stages of sintering between 60Å RuO_2 particles as calculated from earlier results [6] will be completed in less than 4 minutes at 800°C.

SUMMARY

The development of conducting chains in RuO_2-glass thick film resistors can be described in terms of five physical processes: glass sintering; glass spreading; microrearrangement; glass densification; and conductive sintering. The materials properties which influence the kinetics of these processes include: surface tension viscosity, and particle size of the glass; particle size of the RuO_2; solubility of RuO_2 in the glass; and contact angle of the glass on RuO_2. These physical properties determine the kinetics of the formation of the chain structure, and this microstructure is directly responsible for the observed blending curve with the only additional property needed being the resistivity of RuO_2.

REFERENCES

1. D.A.G. Bruggeman, Ann. Phys. (Leipzig) 24, 636 (1935).
2. I. Webman and J. Jortner, Phys. Rev., B11, 2885 (1975).
3. R.M. Scarisbrick, J. Phys. D:Appl. Phys., 6, 2098 (1973).
4. A. Prabhu, G.L. Fuller, R.L. Reed and R.W. Vest, J. Amer. Ceran Soc., 58, 144 (1975).
5. A. Prabhu, G.L. Fuller and R.W. Vest, J. Amer. Ceram. Soc., 57, 408 (1974).
6. A. Prabhu and R.W. Vest, Materials Science Research, 10, 399 (1975).
7. S. Newman, J. Colloid. Interfac. Sci., 26, 209 (1968).
8. W.J. Huppmann and H. Riegger, Acta Met., 23, 965 (1975).
9. J.K. Mackenzie and R. Shuttleworth, Proc. Phys. Soc. (London), B62, 833 (1949).

EVALUATION AND EFFECTS OF DISPERSION AND MIXING METHOD OF REACTANT PARTICLES ON THE KINETICS AND MECHANISM OF SOLID STATE REACTIONS

T. Yamaguchi, S. H. Cho, H. Nagai and H. Kuno

Department of Engineering, Keio University

Yokohama, Japan

ABSTRACT

Effects of starting materials and mixing conditions have been studied in the solid state reaction between $BaCO_3$ and TiO_2, and kinetics and mechanism of the reactions involving intermediate phases are discussed with special emphasis on the effect of mixing performance.

INTRODUCTION

Effects of reactant particle size on the solid state reactions have been studied extensively and kinetic equations describing the obtained isotherms have been proposed (1). It is felt however, that reported activation energies and kinetic equations describing the isotherms suffers from discrepancies even in the same system(2, 3). Too fine reactant particles, for instance, gives even poor reactivity in some cases(4). The present work was undertaken to understand better the effect of dispersion of reactant particles on the solid state reactions.

EXPERIMENTAL

Two kinds of TiO_2(rutile) powders, tentatively disignated as "fine TiO_2" and "coarse TiO_2" were used: both with similar primary particle size (0.2-0.5μ) but the latter contained secondary particles of which size range up to 100μ. Rod-shaped $BaCO_3$ particles were prepared by pouring 0.15 mol/l $(NH_4)_2CO_3$ solution into 0.10

mol/l $Ba(OH)_2$ solution at 30°C while stirring rapidly. Equimolar mixtures of $BaCO_3$ and TiO_2 were prepared by dry ball-mill mixing and precipitation mixing: the latter was prepared by precipitating $BaCO_3$ in TiO_2-$Ba(OH)_2$ solution suspensions under the same conditions as those for preparing $BaCO_3$ starting materials. Reaction isotherms were obtained by isothermal gravimetry in air using loose powder mixtures. The degree of reaction was conveniently calculated from the weight loss of the mixtures, since no free BaO was observed throughout the reaction. All the recorded data are for equimolar mixtures unless otherwise specified.

RESULTS AND DISCUSSION

Coarse TiO_2 on precipitation mixing gave poor reactivity. Fig. 1 shows the results of X-ray diffraction analysis during the course of reaction for mixtures obtained by precipitation mixing using coarse TiO_2. Initial interaction of reactants yields $BaTiO_3$ and the subsequent decrease in $BaTiO_3$ is accompanied by the formation of Ba_2TiO_4 at about 5 min. After $BaCO_3$ is consumed, $BaTiO_3$ increases gradually. Apparently, this mixture would take prolonged heating to reach an equilibrium phase constitution. It must be added that a small amount of $BaTi_4O_9$ was detected also. For mixtures using fine TiO_2, on the other hand, $BaTiO_3$ was the only product throughout the course of the reaction and the amount of $BaTiO_3$ increased in accordance with the reaction isotherms indicating the consumption of $BaCO_3$. The above results indicate the effect of particle size ratio of reactant powder particles on the reaction scheme as well as on the reactivity, and obviously, the dispersion of reactant particles. X-ray diffraction analysis of the coarse TiO_2-mixtures after 1h-heating at 1000°C showed that prolonged ball-mill mixing resulted in the similar reaction behavior to that for fine TiO_2-mixtures. Electron microscopy of the mixtures showed that the disintegration of $BaCO_3$ particles occurred up to 1h-ball-mill mixing, resulting in a few-micron sphere-like particles, without any significant dimensional change on further milling. Thus, the effect of ball-mill mixing time on the reaction behavior should be understood by the improved dispersion of TiO_2 particles caused by the disintegration of TiO_2 secondary particles during the milling performance.

Fig. 2 shows the effect of ball-milling time on the reaction yields after 3 min-heating for fine and coarse TiO_2 powders. In view of the similar primary particle size for both of them, a sharp increase in reaction yield observed after the short-term ball-milling of fine TiO_2 reflects easy-to-break nature of fine TiO_2. On the other hand, coarse TiO_2 was found to have strong interparticle forces as demonstrated by the gradual increase in reaction yield up to 5h-ball-milling. In Fig. 2 is also shown the effect of

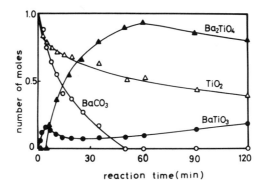

Fig. 1 Effect of reaction time on the relative amounts of reactants and products, prepared by precipitation mixing using coarse TiO_2, 1000°C in air.

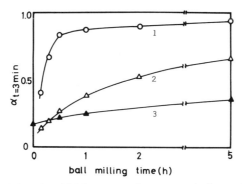

Fig. 2 Effect of ball-milling on the reactivity at 1000°C in air, ball-mill mixing with, 1; fine and 2; coarse TiO_2, 3; wet ball-milling of coarse TiO_2 prior to precipitation mixing.

preliminary wet ball-milling of coarse TiO_2 prior to precipitation mixing. It should be noticed that the ball-milling mixing is more effective in improving reactivity. These findings lead us to the remark that $BaCO_3$ to TiO_2 size ratio in powder mixture is closely related to the formation of intermediate phases. Kato reported that increasing TiO_2 particles resulted in the formation of Ba_2TiO_4 in equimolar mixtures(5).

Results of X-ray diffraction analysis (Fig. 1) suggest that for coarse TiO_2 the reaction proceeds in the following three steps:

$$BaCO_3 + TiO_2 = BaTiO_3 + CO_2 \tag{1}$$

$$BaTiO_3 + BaCO_3 = Ba_2TiO_4 + CO_2 \quad (2)$$

$$Ba_2TiO_4 + TiO_2 = 2BaTiO_3 \quad (3)$$

In fact, X-ray diffraction analysis of the $BaCO_3$-fine TiO_2 mixtures containing a large excess of $BaCO_3$ showed that the formation of $BaTiO_3$ preceded that of Ba_2TiO_4. Hence, it is reasonable to assume the above scheme for equimolar mixtures even with fine TiO_2 and in spite of the essential absence of Ba_2TiO_4 during the reaction.

Hancock and Sharp proposed a method for predicting a kinetic equation by means of the slope m of the straight line in ln t vs. $\ln[-\ln(1-\alpha)]$ plot, where α is the fraction reacted(6). This method is valid if α is less than 0.5. A break observed at $\alpha=0.38$ in the plot for $BaCO_3$-coarse TiO_2 mixtures prepared by precipitation mixing suggests that the process occurring before the break could correspond to the reaction (1). One hour-ball-mill mixing of $BaCO_3$-coarse TiO_2 shifted the break to higher α and for fine TiO_2 such a break was no longer observed.

Although they are nothing but the overall effect averaged over the numerous reaction sites, the observed reaction isotherms reflect the relative contributions of reactions (1) and (2). Unfortunately the gravimetry does not give any information on how the reaction (3) proceeds, yet X-ray diffraction analysis could be a useful means. The dependence of the exponent m on the mixing and also on the starting TiO_2 powders given in Table 1 is possible evidence supporting the different contributions of reactions (1) and (2) to the observed isotherms. Combination with Fig. 2 suggests the usefulness of the exponent m as a criterion on the dispersion or reactivity of powder mixtures. Another indication in favor of this point of view is two-step reaction isotherms for $2BaCO_3-SiO_2$ mixtures(7).

It must be mentioned however, that such an approach also applies for reactions involving intermediate steps towards an equilibrium phase constitution. Although this is the case for the systems

Table 1 Effect of ball-milling time on the exponent m, 1000°C.

ball-milling time (h)		0	1/2	1	2	5	10
ball-mill mixing	fine TiO_2	—	1.12	1.14	1.20	1.30	—
	coarse TiO_2	—	0.80	0.84	0.90	0.95	1.0
precipitation mixing	coarse TiO_2	0.69	0.75	0.78	—	0.82	—

Table 2 Kinetic parameters in the solid state reaction in $BaCO_3$-TiO_2 equimolar mixtures.

mixing method	fine TiO_2		coarse TiO_2	
	pptn.mixg	ball-mill	pptn.mixg	ball-mill
m	1.04	1.14	0.69	0.84
α (t=3min)	0.59	0.85	0.17	0.38
Arrhenius E(kcal/mol)	50.5	67.9	49.4	47.8
parameters log A	5.8	8.8	5.4	5.6

$BaCO_3$-TiO_2 and $BaCO_3$-SiO_2, no clear-cut dependence of m on the mixing conditions has been observed in the reaction of equimolar mixtures of $BaCO_3$ and ZrO_2, in which the exponent m remained unchanged regardless of the mixing method and conditions. No phase other than $BaZrO_3$ was detected. Thus, the dependence of m on starting materials and also on the mixing method is characteristic of the reactions involving intermediate phases.

As a conclusive summary, in Table 2 are given some characteristic parameters in the reaction between $BaCO_3$ and TiO_2 for different mixtures. A possible choice of kinetic equations enabled us to estimate apparent activation energies. Accepting the complicated nature of powder reactions, the authors take the position that the choice of kinetic equation is merely a matter of convenience for obtaining temperature dependence of reactivity, and furthermore, that no model is appropriate on which kinetic equations are based.

ACKNOWLEDGMENT

We thank the Asahi Glass Science Foundation and Nippon Denki Co. for their financial support of this work.

REFERENCES

1. B. Serin and R. T. Ellickson, J. Chem. Phys., **9**, 742 (1941).
2. K. Kubo and Y. Jinriki, Kogyo Kagakuzasshi, 55, 49 (1952).
3. H. Kuno, A. Suzuki and M. Yokoyama, J. Powder Met. Soc. Japan, **13**, 47 (1966).
4. S. L. Blum and P. C. Li, J. Amer. Cer. Soc., **44**, 611 (1961).
5. Y. Suyama and A. Kato, Ceramurgia International, **1**, 5 (1975).
6. J. D. Hancock and J. H. Sharp, J. Amer. Cer. Soc., **55**, 74 (1972).
7. T. Yamaguchi, H. Fujii and H. Kuno, J. inorg. nucl. Chem., **34**, 2739 (1972).

CARBOTHERMAL REDUCTION OF SILICA

J. G. Lee, P. D. Miller and I. B. Cutler

Department of Materials Science and Engineering

University of Utah, Salt Lake City, Utah 84112 (U.S.A.)

ABSTRACT

Silicon carbide, silicon nitride or silicon monoxide may be produced by carbothermal reduction of silica depending on conditions. The mechanism for the reduction of silica involves a chain reaction, gas-solid scheme. Both silica and carbon surfaces are involved. An explanation for the role of iron as a catalyst is given.

INTRODUCTION

The reaction between silica and carbon has both scientific and practical importance. Despite numerous studies, however, the reaction mechanism is not fully understood. The complexity of the reduction process makes it difficult to designate the various steps that are involved.

Kinetic studies of the carbothermal reduction of silica showed a rather fast and complete reaction (1,2). This cannot be explained by the direct solid-solid contact between carbon and silica. A liquid-solid scheme is eliminated in the absence of a liquid below about 1700°C. Gas-solid reaction mechanisms were proposed by several workers introducing SiO as an intermediate gas phase (1,2,3).

We propose that silica can be reduced by carbon via gas phase as follows:

$$SiO_2 + CO \longrightarrow SiO + CO_2 \qquad (1)$$

$$C + CO_2 \longrightarrow 2CO \qquad (2)$$

$$SiO + 2C \longrightarrow SiC + CO \qquad (3) \text{ or}$$

$$SiO + C + 2/3\, N_2 \longrightarrow 1/3\, Si_3N_4 + CO \qquad (4)$$

By this mechanism the surface areas of the reacting solids are both involved in the rate of reaction. The initial CO gas for Reaction (1) can be obtained from reaction at the solid-solid contacts between carbon and silica particles. Once the reaction is initiated, carbon will regenerate CO through Reaction (2). This mechanism is believed to take place with the carbothermal reduction of iron oxides (4,5). Using a thermogravimetric technique, the controlling steps in the reaction were investigated by varying surface areas of reactants, catalysts, etc.

RESULTS AND DISCUSSION

Figure 1 shows the rates of three reactions taking place in the SiO_2-C-N_2 system. As predicted by thermodynamic calculations, there was a temperature boundary between the formation of SiC and Si_3N_4; SiC was formed at temperatures above about 1400°C and Si_3N_4 was formed at temperatures below about 1400°C.

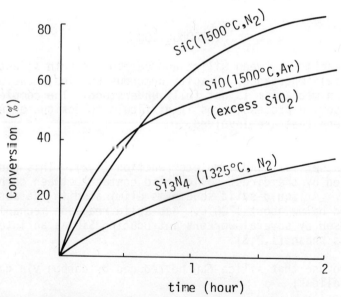

Figure 1. The rates of carbothermal reduction of silica.

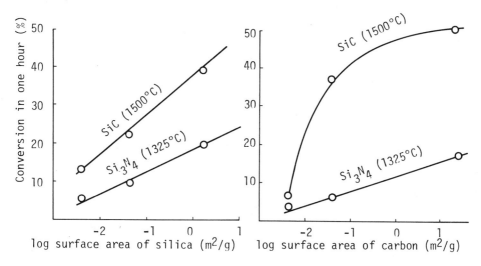

Figure 2. Effect of surface areas of reactants on the formation of SiC and Si_3N_4

The formation of SiC and Si_3N_4 was confirmed by X-ray diffraction. Silicon monoxide escaped from the reaction zone when excess silica was present. The two segments in the curve suggest that SiC forms first followed by a reduction of SiO_2 by SiC in the second segment; i.e., Reactions (1) (2) (3) followed by

$$2SiO_2 + SiC \longrightarrow 3SiO + CO \qquad (5)$$

According to the proposed reaction scheme, the carbothermal reduction of silica is composed of two major gas-solid reactions, (1) and (2). If correct, the overall rate will depend on the surface areas of both silica and carbon. A set of samples was prepared to check this assumption. Figure 2 shows the effect of surface areas of reactants on the overall rate. The surface areas of both silica and carbon have a marked effect on both SiC and Si_3N_4 formation. This result clearly shows that both Reactions (1) and (2) are important for the overall reaction.

Reactions (1) and (2) occur through a chain mechanism; CO_2 produced from Reaction (1) becomes a reactant in Reaction (2) while CO gas produced by Reaction (2) serves as a reactant in Reaction (1). This explains why intimate mixing is important to the rate of reaction. The local partial pressure of CO_2 must be low. When the carbon and silica were more than a mean free path apart, the reaction rate was small and difficult to measure.

Under these conditions the local partial pressure of CO_2 would rise appreciably.

Iron is known as a catalyst for the formation of SiC by the carbothermal reduction of silica (2,3,6). Iron is also a catalyst in the formation of Si_3N_4 as shown in Figure 3. It is a general rule that a catalyst can be effective only when it catalyzes the controlling step(s). This means iron is acting on Reaction (1) and/or (2). Mn, Mo and Co are also known as catalysts for the reaction between carbon and CO_2 (7). Experiments showed that they also catalyzed the overall reaction of silica with carbon. It was also found that iron acted as catalyst for the carbothermal nitridization of alumina. These results clearly indicate that iron enhances the overall reaction rate between silica and carbon by catalyzing Reaction (2). The catalytic effect of iron on Reaction (2) has been well recognized although the detailed mechanism of this effect is still not clear (8).

Using this information it has been demonstrated that these reactions can be used to form stoichiometric amounts of either SiC or Si_3N_4 from carbon and silica of high surface area in a few hours. Sialon, a solid solution of Si_3N_4 with Al_3O_3N, can be produced from clay by a similar reaction.

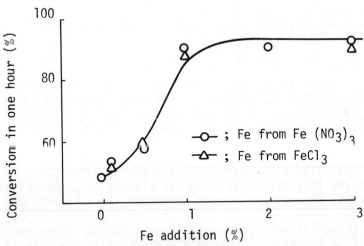

Figure 3. Effect of Fe addition on the formation of Si_3N_4 from SiO_2-C-N_2 system at 1350°C.

CONCLUSIONS

1. The carbothermal reduction of silica proceeds through the gas-solid mechanism composed of Reaction (1),(2),(3) and (4).

2. The controlling steps are Reactions (1) and (2).

3. Iron effectively enhances the overall reaction of silica with carbon by catalyzing Reaction (2).

ACKNOWLEDGEMENT

The authors gratefully acknowledge the financial support of the National Science Foundation through Grant No. DMR 75-02893.

REFERENCES

1. N. Klinger, E. L. Strauss and K. L. Komarek, "Reaction Between Silica and Graphite," J. Amer. Ceram. Soc., 49 (7) 369-75 (1966).

2. J. G. Lee and I. B. Cutler, "Formation of Silicon Carbide from Rice Hulls," Ceram. Bull., 54 (2) 195-8 (1975).

3. J. D. Chidley and J. D. Seader, "Effect of Additives on Ablation of Phenolic-Silica Composites," J. Spacecraft and Rockets, 10 (1) 7-14 (1973).

4. K. I. Otsuka and D. Kunii, "Reduction of Powdery Ferric Oxide Mixed with Graphite Particles," J. Chem. Eng. Japan 2 (1) 46-50 (1969).

5. Y. K. Rao, "The Kinetics of Reduction of Hematite by Carbon," Met. Trans. 2 (5) 1439-47 (1971).

6. I. B. Cutler; unpublished report of work conducted at the University of Utah, April 1956.

7. P. L. Walker, M. Shelef and R. A. Anderson, Chemistry and Physics of Carbon, Marcel Dekker Inc., New York, 1968, 287-383.

8. J. F. Rakszawski, F. Rusinko and P. L. Walker, Proceedings of the Fifth Conference on Carbon, Penn. State, 1961, Vol. 2, Pergamon Press, New York, 1962, 243-250.

ELECTROCHEMICAL CELLS WITH SULPHATE-BASED SOLID ELECTROLYTES

B. Heed, A. Lundén and K. Schroeder

Department of Physics, Chalmers University of Technology, S-402 20 Gothenburg 5, Sweden

ABSTRACT

Both mono- and divalent cations have a high mobility in a number of sulphate-based systems. Conductivities exceeding 10^{-3} (ohmcm)$^{-1}$ can be obtained at room temperature. A number of cell tests have been made. A power density of 400 W/kg was obtained at 745° C for the cell $Mg/Li_{1.76}Mg_{0.12}SO_4/MnO_2$. At ambient temperatures Ag/electrolyte/I_2 cells can be run in air without any precautions to remove moisture.

INTRODUCTION

While solid ionic compounds usually have a very low electrical conductivity compared to their melts, there are in some systems solid phases characterized by an appreciable ionic conductivity due to either some cation or anion species having a high mobility. Provided that the electronic conductivity of these phases remains low, they become of interest for applications in electrochemical devices. The high ionic mobility is closely related to the structure of the compound. Thus for ceramics consisting of non-stoichiometric mixtures of oxides of different valencies, there may be an excess of vacancies either in the cation or the anion lattice. In the first case cations of the "right" size are highly mobile, such as Na and Ag ions in beta-alumina, consisting of Al_2O_3 with a certain amount of Na_2O. This ceramic is used as the electrolyte in sodium-sulphur batteries. In the second case we have the mobile oxygen ions of calcia-stabilized zirconia and similar ceramics used as electrolytes for fuel cells, oxygen-sensing devices etc.

A third type of solid electrolyte consists of salts where the anions form a rigid lattice, while the cations are disordered. Thus, in recent years a high mobility of silver ions has been reported for a number of double salts with AgI as one of the components, and many applications of these electrolytes are reported in the literature, including patents.

The three mentioned types of solid electrolyte are those that have been given most interest so far when it comes to technical applications. Solid electrolyte behaviour has also been reported for many other systems, including fluorides, silicates and Cu(I) double salts. The alkali sulphate based systems studied by us differ in one important aspect from three main groups mentioned above, since in the sulphates the high ionic mobility is not limited to one or two species, but instead many mono- and divalent cations have high mobilities. This is a clear advantage, since it means that a number of metals can be used as the anode of a power source.

PHYSICAL PROPERTIES

It was reported already in 1921 that pure lithium sulphate has a very high electrical conductivity at high temperatures, but little attention was given to this observation. Førland and Krogh-Moe found for the high-temperature phase that the sulphate ions from a face centered cubic or pseudo-cubic lattice in which there is a large excess of possible cation positions (1). This explains the high electrical conductivity of this phase, ranging from 0.9 to 3 $(ohmcm)^{-1}$. Both the volume change and the latent heat are larger at the phase transition (573° C) than at the melting point (860° C).

Medium sized cations such as Na^+, Ag^+, Mg^{2+} and Zn^{2+} show a high solubility in the fcc phase, called I in the phase diagrams given in Fig. 1 - 4, while the solubility is small for K^+ and larger cations (3). In addition, other high-conducting phases are known such as body centered cubic $LiNaSO_4$ (VI in Fig. 1) and $LiAgSO_4$ (II in Fig. 2) and orthorhombic $Li_4Zn(SO_4)_3$ (IV in Fig. 4). An example from a ternary system is $Li_{0.22}Na_{1.33}Zn_{0.22}$ which exists as a single phase from the eutectoid at 325° C up to 660 - 690° C, where it melts incongruently. Its electrical conductivity is 0.1 $(ohmcm)^{-1}$ at 325° C. Diffusion studies have been performed in both one-phase and two-phase regions (6). E.g. for fcc Li_2SO_4 at 750° C $D(Li) = 4.2 \times 10^{-5}$, $D(Na) = 3.7 \times 10^{-5}$ and $D(Zn) =$

SULPHATE-BASED SOLID ELECTROLYTES

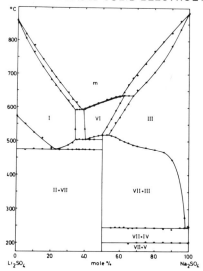

Fig. 1 Phase diagram of the system Li_2SO_4-Na_2SO_4. Phases I and VI have a high conductivity (2,3).

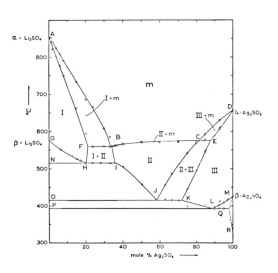

Fig. 2 Phase diagram of the system Li_2SO_4-Ag_2SO_4. Phases I and II have a high conductivity (4).

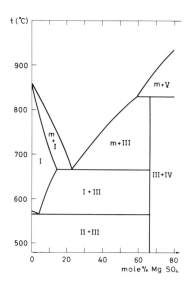

Fig. 3 Phase diagram of the system Li_2SO_4-$MgSO_4$. Phase I has a high conductivity (5).

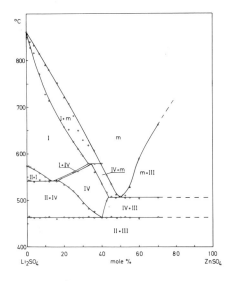

Fig. 4 Phase diagram of the system Li_2SO_4-$ZnSO_4$. Phases I and IV have a high conductivity (3).

= 0.22 x 10^{-5} cm^2/s. Electromigration studies confirm that several cations have a high mobility in sulphate systems.

Rheology studies show that fcc Li$_2$SO$_4$ deforms fairly easily under a load, and its behaviour is similar to a non-Newtonian liquid (7.8). Rather small additions can change the apparent viscosity of the salt drastically, and this is of interest for some applications.

It is possible to partly replace sulphate ions by iodide, and in the system (Li,Ag)$_2$(SO$_4$,I$_2$) a single phase with a bcc structure and a high conductivity exists over an appreciable concentration and temperature region (5). However the solid miscibility of Li$_2$SO$_4$ with LiCl or LiBr is small, and the high conductivity found above 484° C is due to mixed solid-melt phases (5). Both these mixtures and some sulphate-nitrate and sulphate-nitrite ones have been used in electrochemical cells (9,10).

ELECTROCHEMICAL CELLS

Cells with a metal anode and a cathode of MnO$_2$ or I$_2$ have been made by pressing layers of powder in a steel die (11). The electrode layers contained also electrolyte powder and often graphite to give both ionic and electronic conductivity. Flat spirals of silver wire were often used as current leads, see Fig. 5. The whole cell weighed a couple of grams. The

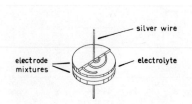

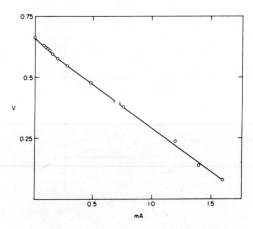

Fig. 5 Test cell with incorporated current collectors of silver wire. Diameter 20 mm.

Fig. 6 Current-voltage characteristics for a Ag/electrolyte/I$_2$ cell at 20° C. Diameter 20 mm; electrode distance 4 mm.

Table 1. Experimental results for some cells (diameter 20 mm)

Cell no	Anode material	Cathode material	Electrolyte composition	Electrolyte thickness mm	Temperature °C	EMF V	Internal resistance ohm
1	Mg	MnO_2	$Li_{1.76}Mg_{0.12}SO_4$	1	745	2.3	1.4
2	Ca	MnO_2	$Li_{1.85}Ca_{0.075}SO_4$	2	650	2.6	1
3	Zn	MnO_2	$Li_{1.28}Zn_{0.36}SO_4$	3.5	500	1.3	53[x]
4	Mg	MnO_2	$Na_{1.6}Mg_{0.2}SO_4$	4	360	2.4	1000
5	Zn	MnO_2	$Na_{1.6}Zn_{0.2}SO_4$	1.5	380	1.2	50
6	Mg	MnO_2	$Na_{0.9}Li_{0.9}Mg_{0.1}SO_4$	1	540	2.3	5
7	Zn	MnO_2	$Li_{0.22}Na_{1.33}Zn_{0.22}SO_4$	1	385	1.2	10
8	Ag	I_2	$LiAg_{0.8}Mg_{0.1}I_{0.4}(SO_4)_{0.8}$	4	22	0.67	350

[x]This cell had a diameter of 15 mm.

cells were tested in air without any encapsulation. The internal resistance was independent of the load, see Fig. 6, and it was proportional to the thickness of the electrolyte layer. Results for some cells are shown in Table 1. The open-circuit voltage is in general in good agreement with thermodynamic data. For cell no 1 the theoretical energy density is 1170 Wh/kg, and in our test run a power density of 400 W/kg was obtained.

Comparison With Other Alternatives

Concerning power and energy densities cells with sulphate-based electrolytes are comparable with other high-temperature cells with either beta-alumina or a molten salt as electrolyte. For ambient temperatures the sulphate-based electrolytes studied so far have a lower conductivity than the AgI-double salts, but the former have the advantage that they are not affected by moisture. Thus for both temperature regions it seems worth while to consider sulphate-based electrolytes as a possible alternative, at least for some applications.

<u>Acknowledgements.</u> This investigation is supported by the Swedish Board of Technical Development.

REFERENCES

1. T. Førland and J Krogh-Moe, Acta chem. scand. 11, 565 (1957).
2. L.-I. Staffansson, Acta chem.scand. 26, 2150 (1972).
3. K. Schroeder, Thesis, Gothenburg, 1975.
4. H.A. Øye, Acta chem. scand. 18, 361 (1964).

5. H. Ljungmark, Thesis, Gothenburg, 1974.
6. A. Bengtzelius, Thesis, Gothenburg, 1973.
7. B. Augustsson and A. Lundén, Z. Naturforsch. 21a, 1860 (1966).
8. B. Jansson and C.-A. Sjöblom, unpublished.
9. E.S. Buzzelli, US Pat 3,506,490 and 3,506,491 (1970);
   E.S. Buzzelli and R.A. Rightmire, US Pat. 3.506,492 (1970).
10. G.W. Mellors, US Pat. 3.726,718 (1973).
11. B. Heed, Thesis, Gothenburg, 1975.

REVERSIBLE TOPOTACTIC REDOX REACTIONS OF LAYERED CHALCOGENIDES

R. Schöllhorn

Anorganisch-chemisches Institut der Universität Münster

Gievenbecker Weg 9, D-4400 Münster / Germany

INTRODUCTION

A systematic investigation has been undertaken by us on room temperature reactions which proceed by simultaneous uptake and loss of electrons and ions by electronically conducting solids able to act as a host matrix. The term "reversible topotactic redox reactions" which has been proposed by us earlier, seems to be an appropriate general description for this type of solid state process which receives increasing attention especially in connection with the search for improved electrode materials of primary and secondary batteries (1-5).

LAYERED CHALCOGENIDES

We found that dichalcogenides MX_2 (M=metal, X=S,Se) of transition elements of group IVb to VIb are particularly suitable solids for model studies of the type of reaction considered here. Galvanostatic and cyclovoltammetric techniques combined with X-ray and neutron diffraction procedures are suitable methods to study reaction kinetics and composition and structure of the reaction products.

The dichalcogenides crystallize in closely related layer lattices built up by two dimensional X-M-S sandwich units with strong forces between the constituent atoms. These units are held together in the lattice by weak van der Waals forces only. On chemical or electrochemical reduction in electrolyte solutions, electrons are taken up by the dichalcogenide layers which thus become negatively charged and cations enter the interlayer space

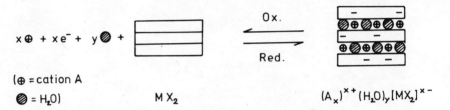

Fig. 1 Scheme of reversible topotactic redox reactions of layered dichalcogenides

with or without solvent molecules depending on polarity of the solvent phase (Fig. 1).

Cathodic reduction e.g. in aqueous alkali or alkaline earth salt solutions results in the formation of hydrated phases $A_x^+(H_2O)_y[MX_2]^{x-}$; x may vary from 0.1 to 0.6 with H_2O as solvent. During this process the basal spacing (i.e. the distance between the layers in direction of the hexagonal c-axis) increases from ca. 6 Å for the binary MX_2 phases to 8-12 Å for the hydrated compounds. In Fig. 2 basal spacings are listed for hydrated phases $A_{0.3}^+(H_2O)_y[TaS_2]^{0.3-}$ (A^+=alkali or alkaline earth ion) obtained by reduction of TaS_2. It turns out that the basal spacings are strongly dependent on cation radius and charge i.e. on hydration energies. Small or bivalent ions are able to stabilize bilayers of solvate molecules (11.4-11.9 Å; $e_o/r > 1$), whereas for large monovalent cations only monomolecular water layers are observed (8.8-9.3 Å; $e_o/r < 1$). High mobilities of the cations in the interlayer space are characteristic for these compounds and allow rapid cation exchange with ambient electrolyte solutions i.e. the hydrated phases behave as polyelectrolytes. Depending on the type of cation introduced, basal spacings far above 12 Å may be obtained e.g. 58 Å for A^+= n-octadecylammonium $(C_{18}H_{37}NH_3)^+$. In the same way the water molecules may be replaced by other polar solvents e.g. NH_3, alcohols, ethers, acid amides. Cation and solvent exchange result in strong variations of the physical properties of these phases as has been shown by our studies on the superconductivity of compounds with $[TaS_2]^{x-}$ and $[NbS_2]^{x-}$ layer units (6,7).

On anodic oxidation or reaction with oxidants such as O_2 the hydrated ternary phases yield the corresponding binary compounds MX_2 according to the reaction scheme shown in Fig.1. The high degree of order retained in these redox reactions is accentuated by the fact that different stacking modifications revert to the original lattice e.g. $1T-TaS_2 \xrightarrow{red} A_{0.3}^+(H_2O)_y[TaS_2]^{0.3-} \xrightarrow{ox.} 1T-TaS_2$.

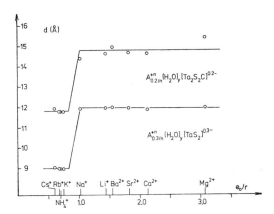

Fig. 2 Dependence of basal spacings d on charge/radius ratio for hydrated phases obtained on reduction of TaS_2 and Ta_2S_2C.

Similar reactions are observed with the carbide sulfide Ta_2S_2C which is built up by S-Ta-C-Ta-S sandwich type layers. Reduction proceeds reversibly e.g. in aqueous electrolyte solutions according to eq. 1. In H_2O as solvent x may reach values up to 0.5.

$$Ta_2S_2C + x\ e^- + x\ A^+ + y\ solv \rightleftharpoons A_x^+(solv)_y [Ta_2S_2C]^{x-} \quad (1)$$

In Figure 1 basal spacings are given for reduced phases with alkali and alkali earth ions which demonstrate a hydration behaviour in close analogy to the dichalcogenides. The difference in the absolute values of the basal spacings is due to the difference in the layer thickness of MX_2 phases and Ta_2S_2C.

Chromium is the only element among the group IVb to VIb metals which does not form binary chalcogenides of the stoichiometry CrX_2. Ternary phases $ACrX_2$ (A=Li,Na,K; X=S,Se) are known however; they are built up by X-Cr-X layers with the A atoms between the chalcogenide sheets. We found that on partial oxidation of these chalcogenides in aqueous electrolytes hydrated thiochromates $A_x^+(H_2O)_y[CrX_2]^{x-}$ are obtained. These phases react in a similar way to the layered MX_2 compounds described above with respect to hydration properties, cation and solvent exchange. For the range of $0.1 < x < 0.5$ reduction and oxidation of these hydrated chalcogenides was shown to be reversible. Further oxidation beyond x=0.1 results in the formation of the paramagnetic metastable dichalcogenides CrS_2 and $CrSe_2$ which show considerable lattice disorder.

Among the group VIII elements Co, Ni, Pd and Pt are known to form ternary chalcogenides with alkali metals which crystallize in layer structures. We found that these compounds may undergo topotactic redox reactions similar to those described for $NaCrS_2$. A more detailed investigation was performed by us on the sulfides $K_2Pt_4S_6$ and $K_2Ni_3S_4$. In both phases the K atoms are interleaved between transition metal chalcogenide sheets, the latter being, however, structurally more complex as compared to the CdI_2 type layers of group IVa to VIa elements (8). $K_2Pt_4S_6$ undergoes spontaneous hydration in aqueous suspension and can be oxidized to a binary phase Pt_4S_6 (basal spacing = 4.48 Å) which is supposedly metastable. The latter may reversibly be reduced in electrolyte solutions according to eq. 2 up to x=1. $K_2Ni_3S_4$ yields Ni_3S_4 on

$$Pt_4S_6 + x\ e^- + x\ A^+ + y\ H_2O \rightleftharpoons A_x^+(H_2O)_y[Pt_4S_6]^{x-} \qquad (2)$$

oxidation which may still be reduced but hydrolysis and irreversible oxidation are competing processes in aqueous medium. The chalcogenides of the main group element tin, SnS_2 and $SnSe_2$, are known to crystallize in a simple CdI_2 type lattice. Although the metal does not provide empty d-orbitals, these chalcogenides may be reduced e.g. by aqueous sodium dithionite to give hydrated phases $Na_x^+(H_2O)_y[SnX_2]^{x-}$ which on oxidation revert to the binary chalcogenide.

LAYERED OXIDES

Transition metal oxides which crystallize in layer structures were found to undergo topotactic redox reactions similar to those described above for layered sulfides, selenides and carbide sulfides. Kinetic aspects are, however, of enhanced importance in the case of oxide lattices. MoO_3, a typical layer oxide, may be reversibly reduced e.g. in neutral aqueous electrolytes containing small cations such as Li^+ or Na^+ according to eq. 3. Between

$$MoO_3 + x\ e^- + x\ A^+ + y\ H_2O \rightleftharpoons A_x^+(H_2O)_y[MoO_3]^{x-} \qquad (3)$$

$0.4 < x < 1$ this reaction is easily reversible, but reoxidation to MoO_3 proceeds only slowly. The uptake of large cations such as Cs^+ is kinetically hindered and protons penetrate the lattice under these conditions. Hydrogen bronzes H_xMoO_3 are formed as reaction products. Depending on pH and on current density, x may reach values up to 1.7. Closely related reactions were observed for a series of oxides with quite different structures e.g. $Li_xV_3O_8$ and $MoO_3 \cdot 2H_2O$.

CONCLUSIONS

All solid state reactions considered above proceed at room tem perature and do have in common that electronically conducting twodimensional layer units are being retained during the redox conversions as topotactic matrix elements. High ionic mobilities are found for the guest cations. A large number of new solid compounds with considerable phase ranges may be obtained by these reactions which are of preparative, theoretical, electrochemical and technological interest.

REFERENCES

1. R. Schöllhorn, E. Sick and A. Lerf, Mat. Res. Bull. 10, 1005 (1975)
2. R. Schöllhorn and W. Schmucker, Z. Naturforsch. 30b, 975 (1975)
3. R. Schöllhorn, R. Kuhlmann and J.O. Besenhard, Mat. Res. Bull. 11, 83, (1976)
4. M.B. Armand in "Fast Ion Transport in Solids", W. van Gool, editor North Holland Publ. Co, Amsterdam 1973
5. M.S. Wittingham, J. Electrochem. Soc. 122, 713 (1975)
6. R. Schöllhorn, A. Lerf and F. Sernetz, Z. Naturforsch. 29b, 810 (1974)
7. F. Sernetz, A. Lerf and R. Schöllhorn, Mat.Res.Bull. 9, 1597 (1974)
8. W. Rüdorff, A. Stössel and V. Schmidt, Z. anorg. allg. Chem. 357, 264 (1968)

PREPARATION AND PROPERTIES OF FeOCl-PYRIDINE DERIVATIVE COMPLEXES
AND THEIR REACTIVITIES WITH METHYL ALCOHOL

S. Kikkawa, F. Kanamaru, and M. Koizumi

The Institute of Scientific and Industrial Research

Osaka University, Osaka (Japan)

INTRODUCTION

Reaction mechanisms between layer-type silicates and organic molecules have been extensively studied on clay-organic complexes from the viewpoints of crystal chemistry and earth science. In recent years, it was found that the superconducting transition temperatures of some transition metal dichalcogenides (MX_2) were strongly influenced by absorbing pyridine molecules between the successive MX_2 layers. Since this discovery, the reaction mechanism of formation and properties of intercalated inorganic compounds have extensively studied from the viewpoints of crystal chemistry and material science. In this report, preparation and partial characterization of FeOCl-pyridine-derivative complexes and the reactivity of the complexes with methyl alcohol are presented.

EXPERIMENTAL

Preparation

FeOCl was prepared by heating the mixture of α-Fe_2O_3 and $FeCl_3$ with the mole ratio of one to four-thirds in a sealed pyrex glass tube at 370°C for two days. The product was washed with water and dried. Reddish violet and thin blade-like FeOCl crystals were obtained.

The reactions with pyridine and its derivatives [4-aminopyridine (AP), 2,6-dimethylpyridine (DMP) and 2,4,6,-trimethylpyridine (TMP)] were conducted at 60°C for two weeks in the closed system, in which FeOCl was soaked into each pyridine derivative in aceton solution. After this reaction, black crystals which were intercalated compounds of pyridine derivatives and FeOCl were obtained.

These sorption complexes were again soaked with methanol in a sealed pyrex glass tybe. The duration of reactions was ten to twenty days at 100°C. Finally, brown crystals were produced.

Measurement

X-ray analysis was conducted with a Rigaku-Denki diffractometer using Cu-Kα and Co-Kα radiations. Determination of C, H, N and Cl contents of this complex was made with elementary analysis. Infrared absorption spectra were obtained using a Japan Spectroscopic Co. Ltd. DS-402G spectrometer by usual KBr pellet technique and nujol method. For differential thermal analysis, a Rigaku-Denki apparatus fitted with a platinum-platinum-rhodium thermocouple was used: Al_2O_3 was used as a reference. The rate of temperature increase was 10°C/min. The electrical resistivity of FeOCl and its complexes were measured in the temperature range between 200 and 373K with using a silver electrode.

RESULTS AND DISCUSSION

FeOCl belongs to the orthorhombic space group Pmnm with a=3.780, b=7.917. c=3.302Å, and Z=2. The crystal structure consists of a stack of double layer sheets of oxygen octahedra linked together with shared edges. The outermost atoms on each side of the layers are Cl^- ions. The interlayer forces between adjacent layers are weak van der Waal's type. When FeOCl was heated with pyridine or with each solution of pyridine derivative, pyridine, AP, DMP and TMP molecules could penetrate into the van der Waals gap, resulting in the remarkable expansion of the FeOCl lattice along the b axis to give the basal spacing as indicated in Table 1. The chemical compositions of the intercalated compounds were determined to be $FeOCl(py)_{1/4}$, $FeOCl(AP)_{1/4}$, $FeOCl(DMP)_{1/4}$ and $FeOCl(TMP)_{1/6}$ by both thermogravimetry and the chemical analysis of the compounds.

One dimensional electron density projections on the b axis were synthesized using (0k0) reflections of the complexes. As shown in Fig. 1, the intercalated organic molecules are placed between the FeOCl layers so that the plane of the pyridine ring is perpendicular to the layers and the nitrogen atom faces the layer.

Table 1. Basal spacings and electrical resistivities of FeOCl-pyridine derivative complexes

	pKa	b(Å)	$\rho(\Omega cm)$	Ea(ev)
FeOCl		7.92	10^6	0.6
FeOCl(Py)$_{1/4}$	5.2	13.45	10	0.2
FeOCl(DMP)$_{1/4}$	6.8	14.98	10^2	0.3
FeOCl(AP)$_{1/4}$	9.2	13.57	10^3	0.2
FeOCl(TMP)$_{1/6}$	9.6	11.79	10^3	0.2

The results of thermogravimetry of the complexes showed that deintercalation temperatures rise in the order of FeOCl(py)$_{1/4}$, FeOCl(DMP)$_{1/4}$, FeOCl(AP)$_{1/4}$ and FeOCl(TMP)$_{1/6}$, indicating that, in pyridine derivative complexes, interaction between Cl ions on the surface of the FeOCl layer and NH_2 or CH_3 of the organics played an important role to stabilize the complex. This assumption is supported by the result of IR absorption of FeOCl(AP)$_{1/4}$, in which a remarkable shift in νNH band towards lower wave numbers was observed.

FeOCl is a semiconductor with resistivity of $10^7 \Omega cm$ at room temperature, but its pyridine and pyridine derivative complexes exhibit good electrical conductivities along the c-axis. Electrical conductivities of the complexes were 10^{-3} to $1 \Omega^{-1} cm^{-1}$ at room temperature, the value of which were larger by factors of 10^4-10^7 than that of FeOCl. This marked change in electrical conductivity may be explained by considering interactions between the intercalated organic molecules and FeOCl layer, i.e. a kind of charge transfer from the lone pair electrons on the nitrogen of pyridine ring to FeOCl layer. The thermoelectric power of the complexes was negative, so that carriers for electrical conductivity might be electrons.

The above mentioned charge transfer type complexes were soaked with methyl alcohol in a sealed pyrex glass tube at 80°C

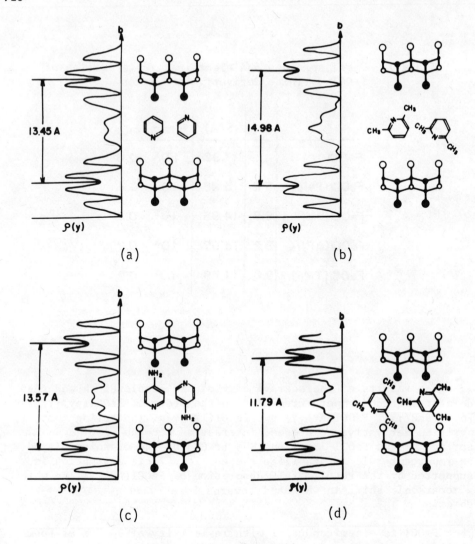

Fig. 1. One dimensional electron density maps projected on the b-axis of the complexes: a) $FeOCl(p)_{1/4}$, b) $FeOCl(DMP)_{1/4}$, c) $FeOCl(AP)_{1/4}$ and d) $FeOCl(TMP)_{1/6}$

for ten to twenty days. During the treatment, methanol molecules were easily intercalated in the interlayer spaces of FeOCl, and finally, brown crystals were produced. The lattice constants of the last phase were a=3.83, b=9.97 and c=3.99Å. From chemical analysis of C.H.N. and Cl, the chemical formula of the compound

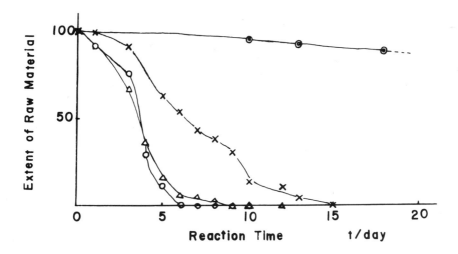

Fig. 2. Conversion rate from FeOCl-pyridine complexes to FeOOCH$_3$ at 100°C; x : FeOCl(py)$_{1/4}$, △: FeOCl(DMP)$_{1/4}$, ○: FeOCl(AP)$_{1/4}$, ◉: FeOCl(TMP)$_{1/6}$

is evaluated to be FeOOCH$_3$. The infrared spectrum of FeOOCH$_3$ gave the absorption band at 1050 cm^{-1} assigned to the stretching vibration of C-O bond, but none to O-H stretching vibration, indicating that methanol molecule absorbed in the interlayer region of FeOCl changed to a methoxide ion, CH$_3$O$^-$, which substituted a chloride ion the outermost layers of the FeOCl. The crystal structure of FeOOCH$_3$ is closely related to that of γ-FeOOH, even though hydrogen ions of the latter are substituted for CH$_3$ in the former. The detail of structural analysis of FeOOCH$_3$ will be presented elsewhere. On the other hand, no reaction was observed between FeOCl and methanol. This fact indicates that the intercalated organic molecules make it easy to penetrate methanol molecules into the interlayer region of FeOCl by expanding the interlayer space. Furthermore, the conversion rates from the charge transfer type complexes to FeOOCH$_3$ were measured by X-ray quantitative analysis using (010) reflections of the complexes and FeOOCH$_3$. As seen in Fig. 2, the reaction rate in the case of the pyridine complex was more sluggish than those of the pyridine derivative complexes with the exception of that of FeOCl(TMP)$_{1/6}$, whose interlayer distance is shorter than those of other complexes. This fact indicates that the large pK value and the interaction between side groups of pyridine ring and the Cl$^-$ ions on the surface of FeOCl layers enhance the replacement of Cl$^-$ ions with OCH$_3^-$ ions by reducing the inter atomic force between Fe and Cl ions.

FORMATION AND PROPERTIES OF SOME SUBHALIDES OF TELLURIUM

A. Rabenau

Max-Planck-Institut für Festkörperforschung

Stuttgart (Germany)

INTRODUCTION

Subhalides of tellurium with the general formula Te_yX (X = Cl, Br, I; $y \geq 1$) show unusual properties. From the six examples of this new group of compounds:

$Te_3Cl_2^*$, Te_2Cl, Te_2Br^*, Te_2I, ß-TeI, α-TeI*

only those with an asterisk are thermodynamically stable crystalline phases with respect to the appropriate phase diagrams Te-TeX$_4$, which form pseudobinary peritectic systems. In spite of the thermodynamic facts, all compounds can be obtained as crystalline solids. The "unsaturation" with respect to tellurium is reflected in the crystal structures of the subhalides, whose main characteristic is bonding in various geometries between tellurium atoms (1). The structural unit of elemental tellurium - a screw with 3_1 or 3_2 symmetry - is stepwise modified to a series of other infinite arrangements and finally to a molecular type structure (α-TeI) Fig. 1. Even in the latter case, the Te_4I_4 molecules form chains parallel to the c-axes. The strong relationship of the molecular type to the chain structures is demonstrated by the topotactical transformation of a small single crystal of ß-TeI into a crystal of the stable form α-TeI.

FORMATION OF SUBHALIDES

The phase diagram Te-TeI$_4$ (Fig. 2) is based on DTA (circles) and X-ray powder patterns on samples annealed below the peritectic

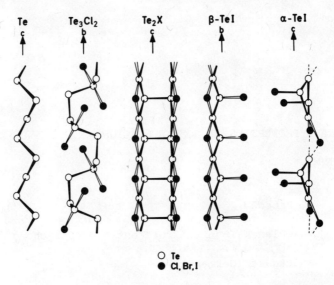

Fig. 1 Structural Units of Tellurium Subhalides

and eutectic line resp. With the exception of high tellurium contents the liquidus had to be derived from vapour pressure measurements (crosses), the latter also confirming the presence of only one stable intermediate phase (2). All three subiodides can be obtained, however, under hydrothermal conditions depending on the thermal treatment above the temperature of the peritectic reaction isotherm. The stable phase forms under isothermal conditions over a wide range of temperatures. The metastable phases require a temperature gradient during treatment (265/280°C and 192/198°C for Te_2I and ß-TeI resp.) and cooling, the length of the needles depending on the extent of the temperature gradient. It could be shown, however, that crystals do not form at these high temperatures. Pure melts behave differently when cooled down. High tellurium contents (o-32 at.-%I) favour metastable ß-TeI, whereas after an intermediate region (32-43 at.-%I) α-TeI is predominant and is the only intermediate phase above 5o at.-%I. Te_2I is not observed. The latter forms, however, stable solid solutions $Te_2Br_{1-x}I_x$ (o≤x≤o.75); melts within this range solidify in a glassy state which recrystallizes when annealed below the melting point. Big crystals of stable α-TeI have been grown by the Bridgman technique from the stoichiometric melt, the nucleus formation of elemental tellurium being completely suppressed in spite of the low cooling rate of 9 mm per day. Occasionally even pure metastable ß-TeI is formed under these conditions. The formation of the subhalides demonstrates the important role the state of the liquid plays in the formation of solids and may be an excellent model for studying these mechanisms.

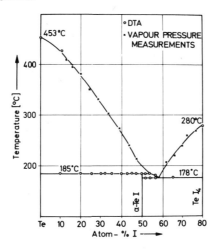

Fig. 2 Phase diagram Te-TeI$_4$

IONIC CONDUCTIVITY OF α-TeI

The electric conductivity decreases in the sequence of Fig.1 over many decades, the last member, α-TeI, being an insulator with a room temperature resistivity comparable to that of elemental iodine in agreement with its molecular structure (3). At higher temperatures under the influence of the current (dc), an irreversible increase in conductivity by five decades is observed, caused by partial decomposition. As this may be due to ionic conductivity, solid state galvanic cells were measured with AgI as auxiliary electrolyte. The chemical potential of α-TeI was fixed by adding Te and TeI$_4$ resp. (see Fig. 2).

Pt/Ag/AgI/TeI,Te/C/Pt E_1

Pt/Ag/AgI/TeI,TeI$_4$/C/Pt E_2

With the emf (E) the free energies of the cell reactions Ag + TeI = AgI + Te (E_1) and TeI$_4$ + 3Ag = TeI + 3AgI (E_2) resp. have been obtained ($\Delta G = nFE$) (4). These measurements were proved with the arrangement (difference cell):

Pt/C/TeI,Te/AgI/TeI,TeI$_4$/C/Pt E_3.

Experimental results are shown in Fig. 3 together with the calculated curve $|E_2 - E_1|$. Crosses represent measurements of the cell

Pt/Ag/TeI,Te/TeI,TeI$_4$/C/Pt E_4

without auxiliary electrolyte. The agreement with $|E_2 - E_1|$ proves
α-TeI to be a pure ionic conductor. (The relatively high iodine
pressures of the electrode / TeI,TeI$_4$ / restricts these measurements to the low temperature region). Estimates of the heat and
entropy of reaction of α-TeI from the values of Fig. 3, using

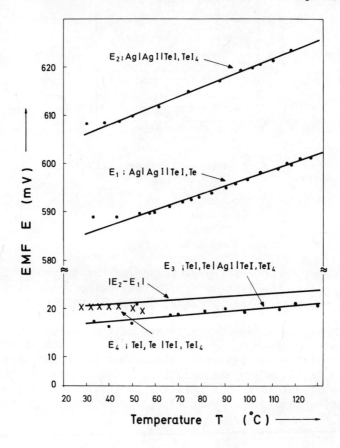

Fig. 3 Galvanic cells with α-TeI: EMF versus temperature

tabulated data for AgI, are in good agreement with data derived
from vapour pressure measurements (2). The unexpected ionic conductivity of α-TeI must be related to its crystal structure which
exhibits layers of terminal iodine parallel to the a-axes (1).

ELECTRONIC BANDGAP OF Te$_3$Cl$_2$ and α-TeI

Representative of optical measurements, Fig. 4 shows the
absorption spectrum of Te$_3$Cl$_2$ and α-TeI single crystals. The spectral dependence of the absorption constant may be interpreted as

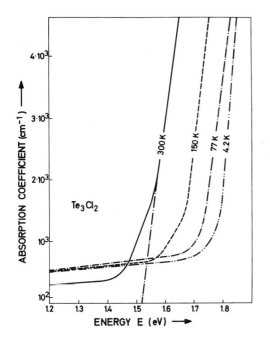
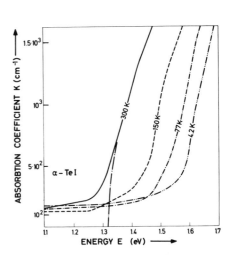

Fig. 4 Optical absorption spectrum of Te_3Cl_2 and α-TeI

an allowed transition. These results show the gap energy at 300 K to be E_{gap} = 1.52 eV and 1.32 eV resp. and the temperature dependence to be $\Delta E/\Delta T$ = -9.31 . 10^{-4} eV/K and -8.78 . 10^{-4} eV/K. The dielectric constants of the subhalides show a characteristic increase from Te_3Cl_2 (ε_∞ = 6) to α-TeI with (ε_∞ = 12) and to Te ($\varepsilon_{\infty\parallel}$ = 23; $\varepsilon_{\infty\perp}$ = 36) reflecting the increase of the polarizability.

ACKNOWLEDGMENTS

Those who participated in the work included above all U. von Alpen, R. Kniep, D. Mootz, H. Rickert, H. Rau, E. Schönherr, W. Stetter.

REFERENCES

1. R. Kniep, D. Mootz and A. Rabenau, Z. Anorg. Allg.Chem., 422, 17, (1976) R. Kniep, Thesis Braunschweig (1974)
2. R. Kniep, A. Rabenau and H. Rau, J. Less-Common Metals, 35, 325 (1974)
3. U. von Alpen and R. Kniep, Solid State Comm. 14, 1o33 (1974)
4. U. von Alpen, J. Haag, and A. Rabenau, Mat.Res. Bull. 11, 7 (1976)

THERMAL STUDIES ON THE BORON-MOLYBDENUM TRIOXIDE PYROTECHNIC DELAY SYSTEM

E. L. Charsley and M. R. Ottaway

Consultancy Service, Stanton Redcroft
London, SW17, U.K.

INTRODUCTION

Pyrotechnic delay compositions are designed to undergo self-sustained combustion in the absence of air and are used as timing devices. They consist typically of an intimate mixture of a metal powder with an inorganic oxidant. This work reports a preliminary study on the boron-molybdenum oxide delay system using thermal analysis techniques, supplemented by X-ray diffraction analysis.

The reaction between boron and molybdenum trioxide does not appear to have been studied previously, although the reaction of MoO_3 with a number of transition metal powders has been recently investigated by Kirshenbaum and Beardell [1]. They concluded that in each case a multistage reaction was given, in which MoO_3 was reduced first to the dioxide and then to elemental molybdenum, the transition metals being oxidised to different levels during these reactions.

For most pyrotechnic compositions two types of reaction can be distinguished depending on the DTA conditions used. The first is a non-ignition reaction, normally obtained with slow heating rates and small sample sizes, giving one or more exothermic peaks. The second is an ignition reaction, favoured by faster heating rates and larger sample sizes, where a single large exothermic reaction is given, often involving a measured rise in sample temperature of over 100°C. This ignition reaction normally embraces all the exothermic reactions obtained under controlled conditions. Comparison of the results of DTA runs under controlled conditions and temperature profile measurements on burning compositions for the tungsten-potassium dichromate system [2], has shown that similar reaction stages are given and that the much slower heating rate used in DTA allows greater resolution.

EXPERIMENTAL

Amorphous boron of 90-92% purity and mean particle size 0.8μm [Fisher Sub-sieve sizer] (Trona brand, American Potash and Chemical Corporation) was used for preparing the mixes. This brand is used in pyrotechnics since high purity boron is not readily available and crystalline boron is less reactive. A limited number of experiments have been carried out using 99% purity boron and the results did not differ significantly from those reported here. Analytical reagent grade molybdenum trioxide and 99.9% purity molybdenum dioxide (Research Organic/Inorganic Chemicals Corp.) were used. Mo_4O_{11} was prepared by heating a mixture of MoO_3 and MoO_2 in the required proportions in a sealed evacuated quartz tube for two hours at 600°C. DTA studies were carried out on a Stanton Redcroft 673 DTA unit. All runs on compositions were carried out in dimped quartz crucibles, 6mm diameter, 30mm long, using an atmosphere of flowing argon to avoid atmospheric oxidation of the boron. The reactions were investigated under non-ignition conditions using a heating rate of 10°C/minute and a sample weight of 50 ± 2mg. Calcined aluminium oxide was used as the reference material. Hot stage microscopy was carried out under ignition conditions on a Stanton Redcroft HSM-5 Unit, using the techniques described previously [3]. Residues for X-ray diffraction analysis were obtained by arresting runs at the desired temperatures and cooling in an argon atmosphere at approximately 50°C/minute. Photographs of powder patterns were obtained using a Debye-Scherrer camera (ENRAF-Nonius Ltd.). CuK_α and CrK_α radiation were used.

RESULTS

Boron-molybdenum trioxide mixes containing from 0.5% to 50% by weight of Trona boron have been studied by DTA under non-ignition conditions from ambient to 900°C. All compositions were found to give a complex but reproducible exothermic reaction consisting of a number of unresolved peaks starting in the region of 430°C and extending to above 800°C, thus indicating several overlapping reactions, these are designated exotherms 1 - 5 as shown in Fig. 1 (c). In addition endothermic peaks were observed above 700°C for samples containing 4% or less boron. The 0.5% and 1% boron mixes gave an endotherm in the region of 780°C, which is close to the reported melting temperature of the $MoO_3 - Mo_9O_{26}$ eutectic. The endotherm at 820°C for the 1.25% and 2% boron mixes was found to correspond to melting of Mo_4O_{11}. Runs on the individual components, boron and molybdenum trioxide showed only an endotherm corresponding to fusion of the latter in the region of 795°C. Typical curves are shown in Fig. 1 and it can be seen that for the lower boron compositions the main reaction takes the form of a double exotherm at 600-650°C (exos 3 and 4). The higher

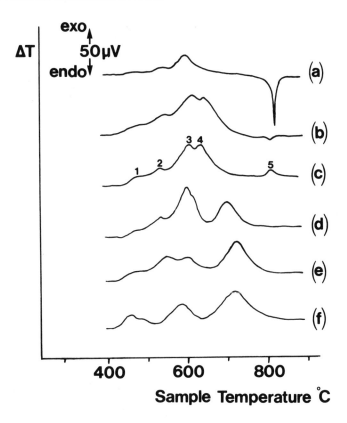

Fig. 1 DTA curves of B–MoO$_3$ compositions heated under non-ignition conditions: (a) 2% boron, (b) 4% boron, (c) 7% boron, (d) 13% boron, (e) 25% boron, and (f) 50% boron

temperature peak reduces in magnitude with increasing boron content, until it merges into the main peak for compositions with 13% and above of boron. Compositions containing in excess of 7% boron also show an additional exothermic reaction (exo 5) which increases in magnitude with increasing boron content reaching a maximum in the composition containing 25% boron.

X-ray diffraction analysis of the reaction products carried out at selected temperatures indicated that exotherms 1 and 2 were associated with a direct but slow reduction of MoO$_3$ to MoO$_2$, exotherm 3 with formation of the intermediate oxide Mo$_4$O$_{11}$ and exotherm 4 with the reduction of this oxide to MoO$_2$. At a higher temperature, exotherm 5, MoO$_2$ was reduced to metallic molybdenum.

In order to confirm the roles of the lower oxides in the

overall reaction, mixes of $B-MoO_2$ and $B-Mo_4O_{11}$ were prepared, containing from 3 - 50% boron and 3 - 10% boron respectively. DTA runs on the $B-MoO_2$ compositions gave a single broad exothermic reaction with a peak temperature in the region 840°C to 863°C, X-ray analysis of the product showing the presence of molybdenum. The results therefore confirm that exotherm 5 is associated with the reduction of MoO_2 to molybdenum metal, the broader exotherm given by the $B-MoO_2$ mix indicating that the MoO_2 is less reactive than the MoO_2 formed in situ during the $B-MoO_3$ reaction. With $B-Mo_4O_{11}$ mixes, DTA runs showed that exotherm 3, thought to be associated with the formation of Mo_4O_{11} from MoO_3, was absent and that small peaks corresponding to exotherms 1 and 2 and a major peak corresponding to exotherm 4, were given. For the 7% and 10% boron mixes, exotherm 5 was also observed.

Samples of $B-MoO_3$ mixes (4 - 50% Trona boron) were heated on the hot stage unit at 100°C/min. All samples ignited and the ignition temperature decreased from 600°C to 565°C, on increasing the boron content from 4% to 25% and rose again above this level to 580°C for the 50% boron mix. A plot of the measured rise in temperature versus percentage boron shows two maxima at 11% and 25% boron respectively and suggests a change in mechanism of ignition above 20% boron. Residues from ignition runs were characterised by X-ray diffraction and the results show that above 13% boron, molybdenum borides were formed on ignition (MoB or MoB_2 depending on the boron content). Below this value the products were the same as those obtained under non-ignition conditions. DTA studies on B-Mo mixes have shown that an exothermic reaction takes place in the range 900-1300°C, forming Mo_2B_5 and it is possible that boride formation is being observed on the hot stage film as the glowing reaction following ignition.

DISCUSSION

From the results to date, the following reaction scheme may be postulated for runs under non-ignition conditions:

Exotherm	Reaction	% Boron for complete reaction
1 and 2	$2B + 3MoO_3 \rightarrow B_2O_3 + 3MoO_2$	4.77
3	$2B + 12MoO_3 \rightarrow B_2O_3 + 3Mo_4O_{11}$	1.24
4	$2B + Mo_4O_{11} \rightarrow B_2O_3 + 4MoO_2$	3.72
5	$4B + 3MoO_2 \rightarrow 2B_2O_3 + 3Mo$	10.14

Although incorporating the formation and reduction of Mo_4O_{11} the reaction scheme resulting from this investigation is essentially similar to that advanced for reduction of MoO_3 by Ti, Zr, V and Fe(1). The temperature at which reduction commences has been found to be independent of boron content and to occur at 430–440°C. For compositions above 13% boron, the exotherms 3 and 4 merge and a DTA curve similar to that observed for the Ti – MoO_3 mixes is seen. It is considered that the Mo_4O_{11} is still a reaction intermediate, but that in the presence of excess boron it reacts as soon as it is formed. The role of Mo_4O_{11} in the reaction scheme is supported by the DTA curves for low boron content mixes, since runs on 0.5% – 1.25% boron mixes show only exotherms 1, 2 and 3. The absence of exotherm 4 being explained by the fact that there is insufficient boron present to reduce the Mo_4O_{11}, formed in reaction 3.

Exotherms 1 and 2 show a gradual increase in peak height with boron content and have not reached a maximum at the 50% boron level. Since the theoretical amount of boron needed for complete reaction is 4.8%, this implies a slow reaction with the rate dependent on the boron content or more specifically on the free surface area of the boron. These reactions are being investigated using temperature programmed X-ray diffraction with a view to determining if molybdenum oxides with compositions between MoO_3 and Mo_4O_{11} are formed as unstable reaction intermediates. The reactions under ignition conditions form molybdenum borides in the presence of excess boron, thereby increasing the overall exothermicity of reaction.

ACKNOWLEDGEMENT

This work was carried out with the support of the Procurement Executive, Ministry of Defence.

REFERENCES

1. A. D. Kirshenbaum and A. J. Beardell, Thermochim. Acta, 4 (1972) 239.

2. T. Boddington, P. G. Laye, H. Morris, C. A. Rosser, E. L. Charsley, M. C. Ford, and Diane E. Tolhurst, Combustion and Flame, 24 (1975) 137.

3. E. L. Charsley and Diane E. Tolhurst, Microscope, 23 (1975) 227.

THE KINETICS OF THE REACTION BETWEEN MAGNESIUM OXIDE
AND IRON(III) OXIDE: THE EFFECT OF SIZES OF PARTICLES

J. Beretka, T. Brown and M.J. Ridge

Division of Building Research, CSIRO, Graham Road

Highett, Victoria, Australia, 3190

INTRODUCTION

Many authors have accounted for the rates of reactions similar to those studied here in terms of various equations based on physical models[1-5], but not always with great success. One study carried out with very fine powders (<1 µm across) yielded an empirical linear relationship between the amount reacted and the log of time[6].

Most of the work published on the kinetics of reactions between solids has been carried out on powders composed of particles of a range of sizes. In such systems the fine powders usually react before the larger particles. The few results available on systems of sized particles of metal oxides[7] do seem to be described adequately by Jander's[1] or similar equations although not always to complete reaction. To attempt to clarify this situation, we have studied the comparatively simple system of magnesium oxide-iron(III) oxide not only using finely ground powders with particles 0.1 to 1 µm across, but also with three lots of carefully sized, "large" particles roughly 50, 100 and 200 µm across.

EXPERIMENTAL

Materials

The finely divided magnesium oxide (MgO, 90.5; CO_3, 0.95;

R_2O_3, 0.12; Ca, 0.01; Na, 0.13; Cl, 0.01; SO_4, <0.01, and H_2O, 8.00%) and the similar iron(III) oxide (Fe_2O_3, 99.1; FeO, 0.20; SiO_2, <0.05; Na_2O, 0.02; H_2O, 0.11; halides, <0.01; R_2O_3, <0.01; and CO_2, 0.03%) were of reagent grade.

Large particles of magnesium oxide were made by sieving fused magnesium oxide (composition: MgO, 97.9; CO_3, 0.07; Al_2O_3, 0.20; Fe_2O_3, 0.35; Ca, 0.61; Na, <0.01; Cl, <0.01; SO_4, 0.03; H_2O, 0.24%; Norton Company, Worchester, Mass.) in water, and collecting three fractions, viz. 45-54, 90-105 and 180-210 μm in diameter. Similarly, iron(III) oxide was obtained by sieving hematite in water (composition: Fe_2O_3, 99.4; FeO, 0.10; SiO_2, <0.05; Al_2O_3, 0.14; TiO_2, 0.15; CaO, 0.045; MgO, <0.01; S, 0.015, and H_2O, 0.02%), from Koolan Island (Western Australia) and collecting the same fractions as for magnesium oxide. X-ray diffraction revealed that the samples of magnesium oxide and iron(III) oxide were in the forms of periclase and hematite, respectively.

Procedures

The systems of small particles were made by mixing the finely divided oxides in the stoichiometric ratio of 1:1, and grinding them to about 1 μm. The systems with large particles were carefully (to avoid breakage) mixed until homogeneous. Optical microscopy showed that no breakage of the particles had occurred.

Lots of about 3 g of the mixtures were heated in porcelain crucibles lined with platinum foil at selected temperatures for up to 168 h, the temperature of the furnace being controlled to within ±1.5°. Crucibles were removed after selected periods (about 1,2,4,6,8,24,48 h etc.), quenched in air, weighed, ground and sieved. The extent of reaction, α, defined as the ratio $(MgFe_2O_4)/(MgO + Fe_2O_3)$, was determined from the residual amount of Fe_2O_3, measured by quantitative X-ray diffraction. The intensity of the reflection for $d = 2.67$ Å was determined with Co Kα radiation, using a Siemens X-ray diffractometer fitted with a scintillation counter and digital printer. Calibration curves were prepared using finely ground mixtures of MgO, Fe_2O_3, and $MgFe_2O_4$ in which the ratio 1:2 for Mg:Fe was maintained over the calibration range.

Mixtures of small particles, before and after calcination, were inspected by transmission electron microscopy, and those consisting of large particles by optical microscopy.

EFFECTS OF PARTICLE SIZES

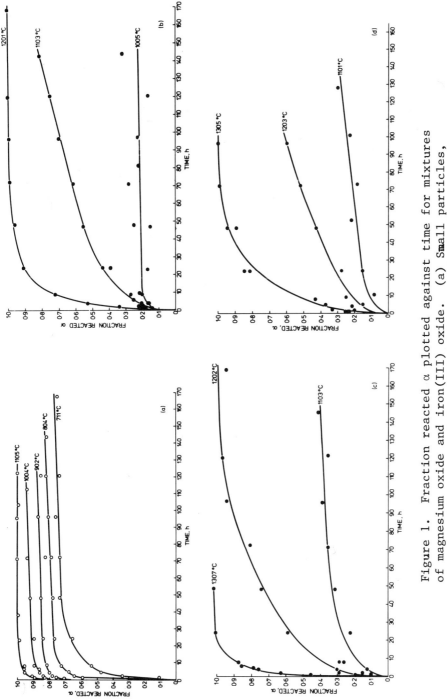

Figure 1. Fraction reacted α plotted against time for mixtures of magnesium oxide and iron(III) oxide. (a) Small particles, <1 μm, and large particles, (b) 45–54 μm, (c) 90–105 μm, and (d) 180–210 μm.

RESULTS

Fig. 1 (a-d) shows the fraction α of the product formed plotted against time for the four initial sizes reacting at a series of temperatures. There is scatter in the experimental results particularly at the beginning of the runs, but it is clear that the reaction is not autocatalytic and that the rate progressively decreases with time. Reaction is slower the larger the particles and higher temperatures are required to produce complete reaction.

Both optical and electron microscopy failed to provide any positive information about the mechanism of the reaction. With mixtures of small particles it was impossible to distinguish the particles of iron oxide and magnesium oxide by electron microscopy. It was possible only to observe that overall sintering had occurred and that finally, the material became agglomerated and well sintered without idiomorphic crystals developing.

With the systems of large particles, it was easy to distinguish by optical microscopy the transparent magnesium oxide and the dark particles of iron oxide. When heated, the magnesium oxide gradually became opaque and then each particle sintered.

DISCUSSION

With systems of small particles it was found that the reaction at each temperature could be best described by a purely empirical relationship, viz. α ∝ log$\underline{t}$, where $\underline{t}$ is the time elapsed[6,8]. The equation of Avrami[9] appears to fit the data to some extent, but the system under investigation does not seem to accord with the model upon which this type of relationship is based. None of the well known equations based on diffusion described the kinetics at all well. These findings agree with those already obtained on the system barium carbonate-iron oxide[6].

With systems of large particles it was found that the reaction can be reasonably described by Jander's equation[1] for diffusion-controlled reactions. Neither the logarithmic dependence on time[6,8] nor the Avrami equation[9] applied in these cases. These findings agree with those of Fresh et al[7] who used sized particles for studying the kinetics of reactions like the one investigated here. However, our findings are in disagreement with those of researchers who used fine powders with random distribution of sizes in the range 1-10 μm and fitted their kinetic data to Jander's equation.

Figure 2 shows a plot at α against $t/t_{0.5}$, ($t_{0.5}$ being the time at which α = 0.5) for data from Figure 1 with systems where

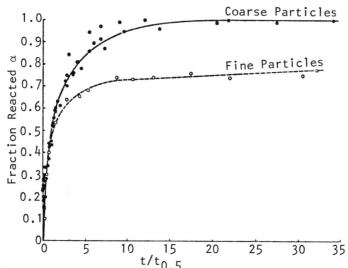

Figure 2. Fraction reacted, α, plotted against reduced time $t/t_{0.5}$ for some of the results shown in Figure 1. The solid line is a calculated curve for Jander's equation and it is seen that the experimental data for systems of coarse particles are distributed about it. The dotted line is drawn through the data for small particles.

α did reach 0.5. It is seen that the points for large particles fall around Sharp, Brindley and Achar's[5] theoretical curve for Jander's equation and are quite separate from those for the small particles. There are slight differences in composition between the coarse and fine systems (See Experimental). The latter contains more impurities than the former so there will be a difference in reactivity. However, the plots of α against $t/t_{0.5}$ indicate a difference in mechanism. It seems probable that this is due to greater structural damage and more extensive surface with systems of small particles as compared with coarse ones.

REFERENCES

1. W. Jander, Z.anorg.Chem., 1927, 163, 1.

2. R.E. Carter, J.Chem.Phys., 1961, 34, 2010; 1961, 31, 1137.

3. G. Valensi, Compt.Rend., 1936, 202, 309.

4. A.M. Ginstling and B.I. Brounshtein, J.Appl.Chem.USSR, (English Translation), 1950, 23, 1327.

5. For a review see e.g. J.H. Sharp, G.W. Brindley and B.N. Narachari Achar, J.Amer.Ceram.Soc., 1966, 49, 379.

6. J. Beretka, M.J. Ridge and T. Brown, Trans.Faraday Soc., 1971, 67, 1453.

7. D.L. Fresh and J.S. Dooling, J.Phys.Chem., 1966, 70, 3198.

8. J.S. Anderson and K.J. Gallagher, Proc. 4th Int.Symp. Reactivity of Solids, ed. T.H. de Boer, (Elsevier, London 1961, p.222.

9. M. Avrami, J.Chem.Phys., 1939, 7, 1103; 1930, 8, 212.

CO-PRECIPITATION OF METAL IONS DURING FERROUS HYDROXIDE GEL FORMATION

Shoichi Okamoto and Shoko I. Okamoto

The Institute of Physical and Chemical Research

Wako-shi, Saitama, 351 Japan

INTRODUCTION

It has been found that metal ions in solutions are coprecipitated with $Fe(OH)_2$ gel and are removed quite satisfactorily from the solutions. When Fe^{2+} solutions containing divalent metal ions such as Mg^{2+}, Mn^{2+}, Co^{2+}, Ni^{2+}, Zn^{2+} and Cd^{2+} are brought to pH≥10, these metal ions are incorporated into ferrous hydroxide precipitates. Rapid oxidation of the precipitates, for example, by a H_2O_2 solution, leads to the formation of cation-substituted δ-FeOOH which is ferrimagnetic at room temperature (1) and stable against elution of the metal ions by washing. Thus, the metal ions incorporated in the ferrimagnetic solids are easily removed from the solutions with the aid of high gradient magnetic field. Residual concentrations of the metal ions in the solutions were of the order of 0.1 to 0.01 ppm depending on the coprecipitation and oxidation conditions.

It has been found further that $Fe(OH)_2$ gel is very efficient for removing Hg^{2+} and Cu^{2+} from solutions. The coprecipitation occurs neither by formation of solid solution nor by adsorption, but by redox reaction with the formation of mixtures of magnetite and metallic mercury or Cu_2O (or metallic copper). The mixtures could easily be removed from the solutions by magnetic separation. It seems that this type of coprecipitation has not yet been analyzed.

Utilization of $Fe(OH)_2$ gel as a magnetic scavenger (2) with the aid of the recently developed magntic separation techniques (3) offers a novel technique for the recovery of heavy metal ions from dilute solution systems.

EXPERIMENTAL

One hundred milliliters of a solution containing $FeSO_4$ and $HgCl_2$ were brought to pH 10 by slow addition of 1N NaOH while stirring. Concentrations of Fe^{2+} and Hg^{2+} in the mixed solutions were varied from 0.2 to 1.8 mMol and from 0.01 to 0.1 mMol, respectively. After stirring for another 15 minutes the suspensions thus formed were filtered. The filtrates were analysed to determine residual concentration of mercury by flameless atomic absorption spectroscopy. The amounts of mercury coprecipitated were obtained by subtracting the residual concentration of mercury from the initial one.

The whole procedures were carried out at room temperature in a pure N_2 atmosphere for suppressing oxidation of the system by oxygen in air. Triply distilled water was used for the sample preparation. Prior to experiments the distilled water was boiled again and then cooled to room temperature to avoid the influence of oxygen dissolved in the water.

RESULTS

The residual concentration and percent coprecipitation of mercury are shown in Fig.1. When the concentration of Fe^{2+} was

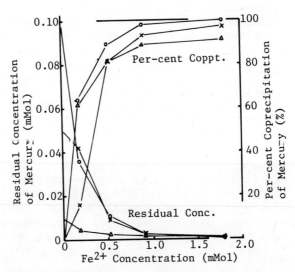

Fig. 1   Influence of Fe^{2+} concentration on the coprecipitation of Hg^{2+} with Fe^{2+}. Initial concentration of Hg^{2+}, Δ ; 0.01 mMol, × ; 0.05 mMol and ○ ; 0.10 mMol.

very low, for instance <0.9 mMol, coprecipitation of mercury was incomplete. At the Fe^{2+} concentration of 1.8 mMol, mercury was eliminated quite satisfactorily from the solution by coprecipitation and the residual concentrations of mercury ranged from the minimum value of 0.0003 to 0.0015 mMol. Even though the concentration of Fe^{2+} was higher than 1.8 mMol, the residual concentration was not reduced to less than 0.0003 mMol. This value corresponds to the solubility of metallic mercury (0.0003 mMol or 61 ppb) at pH of about 10 (4).

In Fig.2, the residual concentrations of mercury were plotted against molar mixing ratios of Fe^{2+} and Hg^{2+} in the solutions. The concentrations of Fe^{2+} were kept constant at 1.8 mMol and those of Hg^{2+} were varied from 0.05 to 1.0 mMol. A marked change in the residual concentrations of mercury is seen in the figure at the mixing ratio of Fe^{2+}/Hg^{2+} of about 3.

The precipitates formed from the solutions containing Fe^{2+} and Hg^{2+} were dark green or black. They were strongly magnetic and filtered off easily and almost completely by applying a magnetic field of about 5 kOe provided by an electromagnet. The Debye patterns obtained at room temperature revealed that the precipitates from the solutions with the mixing ratio Fe^{2+}/Hg^{2+} of more than 3 were mainly composed of magnetite together with ferrous hydroxide, and that the diffraction intensities of ferrous hydroxide increased with increasing amount of Fe^{2+} added. Careful inspection of the Debye patterns detected no crystalline mercuric or mercurous compound, but strong blackening of the X-ray films at low diffraction angles was observed suggesting the presence of metallic mercury in the precipitates. Therefore, the Debye patterns of dried sample were taken at liquid nitrogen temperature

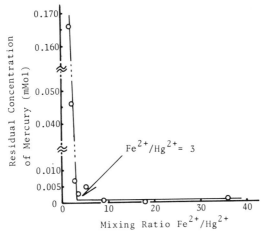

Fig. 2   Removal of mercury vs. mixing ratio Fe^{2+}/Hg^{2+} (Fe^{2+} concentration = 1.8 mMol).

using a low temperature Weissenberg camera. The presence of metallic mercury together with magnetite was verified.

It has been found that ferrous hydroxide gel is easily oxidized by mercuric ions at room temperature. Suspensions of white ferrous hydroxide precipitates were prepared by first adding 1N NaOH to 1.8 mMol $FeSO_4$ solutions to bring the pH to 10, and then appropriate quantities of 10 mMol $HgCl_2$ solutions were mixed into the suspensions so as to make the amounts of mercury ranging from 0.01 to 0.15 mMol. Within a few seconds after mixing, the white precipitates changed gradually into strongly magnetic dark green or black precipitates. The existence of metallic mercury and magnetite in the magnetic precipitates was confirmed by means of X-ray. More than 95 % of mercury added was removed by magnetic separation together with magnetite.

DISCUSSIONS

A reaction between Fe^{2+} and Hg^{2+} hardly occurs at all in acidic solution at room temperature. In alkaline solutions, however, the coprecipitation resulted in the formation of metallic mercury and magnetite, which involved the redox reaction between Fe^{2+} and Hg^{2+}. This may be due to the strong reduction potential of $Fe(OH)_2$. Although texts referred to only the half reaction (5)

$$Fe(OH)_3 + e^- = Fe(OH)_2 + OH^- \qquad E^o = -0.56 \text{ V},$$

we offer the following reaction

$$Fe_3O_4 + 4H_2O + 2e^- = 3Fe(OH)_2 + 2OH^- \qquad E^o = -1.025 \text{ V}.$$

From the reference,

$$HgO + H_2O + 2e^- = Hg^o + 2OH^- \qquad E^o = +0.098 \text{ V}.$$

Thus, we can obtain +51.8 kcal/mole for the free energy change in the reaction

$$Hg^o + Fe_3O_4 + 3H_2O = 3Fe(OH)_2 + HgO.$$

It may be considered that the oxidation products of $Fe(OH)_2$ crystals suspended in alkaline solution differ depending upon "topos" where the redox reaction occurs. At high oxidant concentrations, it occurs at the surface layer of the crystals. Oxidant may arrive at the surface of the $Fe(OH)_2$ crystals and is reduced by accepting proton and electron released from the crystal. During the oxidation, the oxygen framework of the crystal is kept unchanged, but cations may migrate to the neighboring sites. Thus, ferrous hydroxide is oxidized topotactically to δ-FeOOH. In contrast, at a low oxidant concentrations the redox reaction occurs in the solution layer in the crystal-solution interfacial region where the oxidant is reduced by ferrous ions dissolved in the solution. Ferric ions thus formed may associate with ferrous ions to form magnetite, if sufficient ferrous ions are supplied from the $Fe(OH)_2$ crystals through the interfacial region. If the rate of diffusion of ferrous ions is reduced, for example, by lowering temperature, association between ferric ions would result in the formation of α-FeOOH. Such reactions are characterized by the dissolution and recrystallization type reactions.

From the above discussions together with the observation that ferrous hydroxide gel was oxidized by mercuric ions to magnetite, the coprecipitation of Hg^{2+} with Fe^{2+} in alkaline solutions can be considered to take place in three stages. The first stage is the formation of ferrous hydroxide crystal. Then, the redox reaction between Hg^{2+} and Fe^{2+} takes place in the crystal-solution interfacial region. The last stage is the formation of magnetite by association of ferrous and ferric ions. The reaction takes place almost quantitatively between Hg^{2+} and Fe^{2+}, as is shown in Fig.2.

A similar reaction has been observed when solutions containing Cu^{2+} and Fe^{2+} are added to NaOH solution. The precipitates comprised Cu_2O and Fe_3O_4. The residual concentration of copper ions was as low as about 0.07 ppm at room temperature under the wide varieties of coprecipitation conditions. On the other hand, when the solutions containing Cu^{2+} and Fe^{2+} are added to ammonia solution, the precipitates comprised metallic copper and magnetite with considerable amounts of copper ions unremoved in solution. Further, it has been found that ferrous hydroxide gel in ammonia solution is oxidized topotactically to δ-FeOOH when an excess amount of Cu^{2+} is added using a solution of high Cu^{2+} concentration. In this case, however, almost all of copper ions remained in solution. These are interpreted on the basis of the discussion of described above.

REFERENCES

(1) S.Okamoto, J.Chem.Soc.Japan,Industr.Chem.Sect.,67(1964)1855.
(2) S.Okamoto, IEEE Trans.Magnetics,MAG-10(1974)923.
(3) For example, J.Oberteuffer, IEEE Trans.Magnetics,MAG-9(1973) 303.
(4) E.Onat, J.inorg.nucl.Chem.,36(1974)2029.
(5) For example, W.M.Latimer, "The Oxidation States of the Elements and Their Potentials in Aqueous Solutions", Prentice-Hall, Inc. (1952).

DEHYDRATION-REDUCTION COUPLING EFFECTS IN THE TRANSFORMATION OF

GOETHITE TO MAGNETITE

M.L.GARCIA-GONZALEZ, P. GRANGE, B. DELMON

Université Catholique de Louvain

Groupe de Physico-Chimie Minérale et de Catalyse

Place Croix du Sud 1, 1348 Louvain-la-Neuve, Belgium

INTRODUCTION

The quality of $\gamma\text{-Fe}_2O_3$ (maghemite) used in magnetic tapes strongly depends on the various stages of preparation. The present communication concerns the first stages, namely dehydration and reduction of goethite (γ-FeOOH) to magnetite (Fe_3O_4). We have focused our study on the effect of water pressure on these two stages and on the coupling effect between dehydration and reduction. Particular attention was paid to consequences of this coupling effect on the texture of the products.

EXPERIMENTAL DETAILS

The goethite used was an industrial goethite "Bayer 910", prepared by precipitating ferric sulfate in an acidic medium. This goethite is constituted of acicular crystallites (0.5 µ length 0.05 µ diameter). The specific surface area is $22 \pm 3 m^2.g^{-1}$.

Kinetic measurements have been made using a microbalance "SETARAM MTB 10-8." A separate connected system permits a constant water vapour pressure to be maintained inside the reactor in a flux of inert gas (argon or hydrogen). The inert gas flows through a saturator. For all the "wet" experiments reported in the present communication, the saturator was maintained at 55 °C. The corresponding water vapor pressure was 112 torr.

RESULTS

Figure 1 illustrates experimental results in which dehydration and reduction reactions of goethite were done simultaneously either with "dry" (curve 1) or "wet" hydrogen (curve 2). In these experiments, hydrogen was introduced in to the balance at ambient temperature after outgassing the sample. The sample was then brought very rapidly to the temperature of reaction by introducing the reaction tube into a preheated oven. The kinetics of transformation into magnetite are not much affected by the presence of water. In particular, the degree of transformation is the same. However, in accordance with the thermodynamic expectations, the reaction goes on to the formation of metallic iron in "dry" hydrogen, whereas it is stopped in the "wet" conditions.

In the second experiment, we have voluntarily separated the dehydration and the reduction steps.

In a first stage, isothermal dehydration of goethite at 340°C is done either under a vacuum of 10^{-4} torr (fig.2 curve 1) or a par-

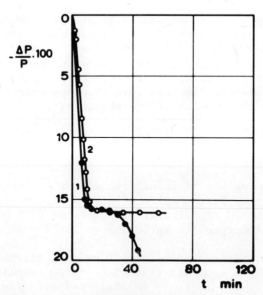

fig 1. Simultaneous dehydration and reduction of goethite

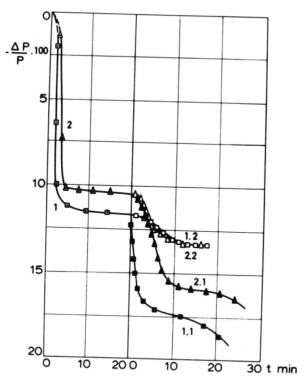

fig.2. Dehydration of goethite (curves 1 and 2) followed by "dry" reduction (curves 1.1. and 2.1.) or "wet" reduction (curves 1.2 and 2.2.)

tial water vapor pressure of 112 torr in argon (fig.2 curve 2). As in the previous experiments, the sample is instantaneously brought to the reaction temperature (340°C) by introducing it in to a preheated oven. When dehydration takes place in "wet" conditions the rate of reaction and weight loss are less than in the case of dehydration under vacuum. In addition, X-ray diffraction patterns show noticeable differences. Only hematite is detected in the samples obtained by dehydration under vacuum, whereas, in the other case, it is possible to detect traces of goethite together with hematite. Another important difference concerns the texture. Specific surface area of vacuum dehydrated samples is 75 $m^2 g^{-1}$; that of samples dehydrated in water vapor is 37 $m^2 g^{-1}$. The volumes measured by the method of M.M.Dubinin (1) are, respectively, 1.78 10^{-2} and 1.82 10^{-3} $cm^3 g^{-1}$. We have noted the absence of any desorption hysterisis in the sample dehydrated under partial water pressure. This indicated that micropores are absent in the latter sample. The method described by R.Sh.Mikhail, S.Brunnauer and E.E.Bodor (2) shows that the sample dehydrated under vacuum probably contains slit shaped pores, mainly of 6 to 10 Å width, and for a smaller proportion, of 20 Å width. Computing the cumulative area according to this model gives a value of 71 m^2g^{-1}, which substantiates our slit-pore hypothesis. The electron microscopic examination shows big differences between the two samples. The one dehydrated under vacuum shows small white dots, which are not observed in the other sample.

The right hand part of Fig.2 corresponds to the second step. Curves 1.1 and 2.1 illustrate the reduction of hematite by "wet" hydrogen, while the curves 1.2 and 2.2 show the reduction by "dry" hydrogen. It can be observed that the rate and amount of reduction are markedly different for these two reduction conditions. In the reduction by "dry" hydrogen, an intermediary plateau is observed; it corresponds to the composition $Fe_{3.3}O_4$ (if the dehydration is carried out under "wet" conditions) or $Fe_{3.5}O_4$ (if the dehydration is carried out under "dry" conditions). In both cases, the reduction continues further to metallic iron. In "wet" reduction (curves 1.2, 2.2), the phenomenon is characterized by a simple sigmoidal curve. The gross composition of the reduced product is the same in both cases, although the starting composition is different. X-ray patterns of the samples reduced under "wet" conditions, but dehydrated differently, also show important differences. The sample prepared by "wet" dehydration and "wet" reduction contains traces of goethite in addition to magnetite, whereas the sample prepared by "dry" dehydration and "wet" reduction contains only magnetite.

DISCUSSION

Some of our observations correspond to those of J.J.Jurinak (3) and A.Claveau and C.E.Beaulieu (4), who reported an increase in surface area caused by the dehydration phenomenon. A.Claveau and C.E.Beaulieu carried out the dehydration of goethite under nitrogen at 275°C and proved the formation of pores of 7 Å diameter. It seems that the dehydration of goethite under nitrogen or vacuum proceeds along the same reaction path to microporous hematite. By dehydration under water it is possible to produce a less porous hematite. It is possible that the presence of water vapor modifies the processes of water diffusion and the rearrangements of superficial OH.

The kinetics of "dry" reduction are strongly influenced by water vapor pressure during the preliminary dehydration stage, if the two reactions occur consecutively. The intermediary products obtained by "dry" reduction have compositions corresponding to $Fe_{3.3}O_4$ and $Fe_{3.5}O_4$. The difference is approximately equal to the composition difference after the initial dehydration stage (corresponding to hematite and traces of goethite). After reduction under "wet" hydrogen, the gravimetric loss is the same, but X-ray analysis gives different results. Two explanations can be given. It could be postulated that residual goethite remains amorphous in the samples reduced under "dry" hydrogen. Another possibility would be that residual OH have a different repartition between goethite and magnetite in samples reduced in "dry" and "wet" conditions (an OH containing spinel analogous to defective magnetite is mentioned by various authors)(5-6).

It is striking that so strong coupling effects between dehydration and reduction are observed.

Comparing fig.1 and fig.2, one notes that the rates of dehydration in comparable conditions (i.e. with, or without water being present) are almost identical, whereas differences appear in the amount of dehydration attained when hydrogen is not present (plateau of curves 1 and 2, fig.2). It is surprising how hydrogen can influence dehydration. We have no explanation for this phenomenon.

On the other hand, there is a symmetric type of effect, namely the influence of preliminary dehydration on the course of reduction. This influence may be ascribed either to the different textures of the samples depending on dehydration conditions or to a different degree of absorbance of water on the surface of the dehydrated samples.

ACKNOWLEDGEMENTS

We thank Usines E.HENRICOT for having supported this work. Discussions with Dr.R. HOUREZ are gratefully acknowledged.

REFERENCES

1. M.M. DUBININ, J.Colloid and Interface Sci., 23, 1967, 487.
2. R.Sh.MIKHAIL, S.BRUNAUER, E.E.BODOR., J.Colloid.and Interface Sci., 26, 1968, 45.
3. J.J.JURINAK, J.Colloid Science, 19, 1964, 477.
4. A.CLAVEAU and C.E.BEAULIEU, Mem.Sci.Rec.Metal.,60, 1973, 173.
5. G.FAGHERAZZI, Chim.Ind.(Milan), 47, 1965, 75.
6. V.J.ARKHAROV, V.N.BOGOSLUVSKY, F.S.Met.Metallov., 3, 1956, 254.

STEPS IN LOW TEMPERATURE DEHYDROXYLATION OF CLAY

M. Gábor, J. Wajand, L. Pöppl and Z.G. Szabó

L. Eötvös University, Budapest

INTRODUCTION

Thermal changes in the kaolinite structure belonging to the group of layer silicates are well known. On heating dehydroxylation of silicates begins, resulting in the Metakaolinite structure. However, the mechanism of the solid phase transformation cannot be taken as understood, not even on the basis of the latest literary data (1,2).

Hitherto the dehydroxylations were investigated generally at about $500^\circ C$, although the complex process begins already below $400^\circ C$, (3). Consequently, the aim of our research was to approach the transformation of the kaolinite structure both below $400^\circ C$ and between $400^\circ - 500^\circ C$. These processes were followed by the simultaneous application of TG, IR, and MS and X-ray methods. Our main interest has been directed - with regard to our previous investigations on the preservation of the kaolinite structure, (4) - to determine the thermal changes of the different bonded and situated OH^- groups.

EXPERIMENTAL

The clay mineral used was a Georgian Kaolin (GC), of kaolinite $-M_d$ type (4). TG investigations were carried out by METTLER vacuum thermoanalyser, connected with quadrupole mass spectrometer permitting weight changes of < 0.01 mg to be detected. Therefore the process with very low loss of weight can easily be followed even below $400^\circ C$. The samples were heated in an inert gas atmosphere with a flow rate 5 l/h. The heating rate was $4^\circ C$/min and thereafter at temperatures under isothermal conditions for 60 minutes.

RESULTS AND DISCUSSION

From IR investigations on kaolins and the group-theoretical considerations Si-O, Al-O and OH bands were assigned, Table 1 (2). These are also characteristic for GC.

Table 1
Interpretation of IR Spectra of GC in 3700-400 cm^{-1} Range

Kaolinite	Metakaolinite	Qualitative characterization of vibrations	
cm^{-1}			
3695	--	OH	Situated on the surface of di-octahedral layer
3660	--	H-bonded OH	
3650	--	H-bonded OH between the layers	
3620	--	H-bonded OH	Situated on the tetrahedral layer
1110		ν_{as} /Si-O-Si/	
1040	1100	ν /Si-O/	
1010			
940		δ/Al.... O-H/	H-bonded OH
920		δ/Al.... O-H/	
800	820		
755		ν_s/Si-O-Si/	
700			
540		Al-O-Si	
472	470	δ Si-O	
430			

Our results show that on heating between 150-400°C the intensity of all bands decreases in parallel and nearly proportionally to the increase in temperature. The slight change in these band intensities is unambiguously confirmed by the quantitative TG data. (Table 2).

The MS data prove that when heating GC up to 400°C, water and hydrogen are evolved from the sample according to the reactions shown below:

$$M-OH + M'-OH \rightarrow H_2O + M-O-M'$$
$$OH^- + OH^- \rightarrow O^- \ldots O^- + H_2$$

M and M' = different metal ions

Among the OH bands the intensity of the 3660 cm^{-1} band decreased more than others. Contrary to the literary data it did not reappear on cooling the sample. The TG curves also show a sudden decrease of weight between 350-400°C. According to X-ray investigations the disorder of the samples increased up to 400°C along the direction of the b axis.

The changes in IR spectra between 400-500°C are summarised on Table 3. With the aid of J. G. Miller's method semi-quantitative conclusions can be deduced from the IR spectra concerning the changes of the hydroxyl- and amorphous content of the clay occuring as a result of heating. It can be clearly seen (Fig. 1) that with the decrease of the n values the extinction (E) values of the bands are decreasing as well. The fact that the E values decrease more abruptly between n=1.94-1.50 than n<1.5 is also shown on Fig.1. The line of the 920 cm^{-1} band intersects the x-axis at n=0.4 (T>450°C) but the E value of the 3695 cm^{-1} band decreases to zero only at n=0.15 (T>500°C). These data prove that OH$^-$ group can be found in

Table 2
TG Data of GC

Heated at °C	Weight loss %	n*
250	0.519	1.86
300	0.590	1.85
350	1.34	1.74
400	2.23	1.62
425	4.24	1.34
450	9.51	0.58
475	12.37	0.17
500	12.41	0.17

*The value of n in $Al_2O_3 \cdot SiO_2 \cdot nH_2O$ = 1.94 from loss on ignition

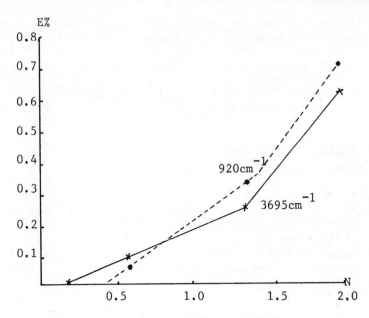

Figure 1 E = F(N) of heat treated G.C.

the kaolin even after collapse of the octahedral structure. Moreover the extinction ratio of the 800 and 700 cm^{-1} bands indicates the amorphous content of the kaolin. The silicon-oxygen bonds are weakening between 400-425°C. At 425 and 450°C the 3695 and 3650 cm^{-1} bands can still be assigned though their intensity decreases. The loss in weight is remarkably high during this period. The Si-O bands, separated until now also overlap at 450°C. It is worth noting that the amount of the amorphous material is already quite considerable at 425°C. This is indicated by band broadening with the increase of temperature. At 475°C the metakaolinite structure can already be identified.

Relying on the MS results M=2 (H_2), 18 (H_2O), 28 (CO), 44 (CO_2) products were identified between 100-500°C. Water can be detected between 150-475°C (maximum at 390°C) and this is simultaneously accompanied by hydrogen over this temperature range although the amount is very small compared to water (approx 1%). The hydrogen formation can most probably be attributed to the thermal dissociation of OH⁻ groups. A similar phenomenon was observed by F. Freund (6), in the case of dehydroxylation of $Mg(OH)_2$.

The CO_2 and CO detected above 500°C can originate a) from carbon dioxide from between the kaolin layers which can only escape from the solid material after the complete collapse of the kaolinite structure, or b) the reaction product of organic materials oxidised by water above 500°C.

Table 3
The Changes in IR Spectra of Preheated GC

cm^{-1}	400°C	425°C	450°C	475°C	500°C
3695	s	s	w	-	-
3660	-	-	-	-	-
3650	s	s	w	-	-
3620	s	s	w	-	-
1110	w	b	-		
1040	w	w	b	b	b
1010	w	w	w		
940	s	s	w	-	-
920	s	s	w	-	-
800	s	s	b		
755	s	s	w	b	b
700	s	s	w		
540	s	s	w	-	-
472	s	s	w	b	b
430	s	s	w	b	b

s = strong   w = weak   b = broadened

Thus on the basis of IR spectra it can be concluded that in low temperature dehydroxylation, the concentration of different OH$^-$ groups decreases with the increase of temperature. Nevertheless, the loss of weight at each temperature cannot be ordered to a certain OH$^-$ group. The separation of dehydroxylation into single steps and a more correct interpretation of this is going to be approached by computerised evaluation of TG data.

REFERENCES

1. P.D. Garn, J. of Am.Cer.Soc. 57, 132, (1974)
2. F. Freund, Ber.Deut.Ker.Ges. 44, 392, (1967)
3. F. Freund, Proc. of the Intern. Clay Conf. Tokyo 1969 Vol.I. 121
4. Z.G. Szabo et al., Proc. of the Intern. Conf. on Thermal Analysis Budapest 1975, Vol 2, 569
5. J.G. Miller, J. Phys. Chem. 65, 800, (1961)
6. F. Freund, J.Phys.Chem. 78, 758, (1974)

SEPARATION OF PHASES IN SOME CARBIDE SYSTEMS

V. S. Stubican and M. Brun

Department of Material Sciences, The Pennsylvania

State University, University Park, PA 16802 U.S.A.

ABSTRACT

The solubility of WC in ZrC and TiC was redetermined. Tungsten carbide precipitated from (W, Zr)C and (W, Ti)C solid solutions as well alligned ribbon-shaped particles. The crystallographic orientation of the hexagonal precipitate and cubic matrix established by electron diffraction techniques is $[100]_h \,||\, [110]_c$ and $[010]_h \,||\, [001]_c$. The interfacial planes were (001) for WC and (111) for the cubic matrix. The rate of precipitation of WC from solid solutions follows a square root time-dependence. An apparent orthorhombic symmetry for the precipitated WC from (W, Ti)C solid solutions was observed. An increase in the microhardness to ∼23% was observed in the phase separated materials.

INTRODUCTION

Strengthening techniques based on the precipitation of a second phase have not yet been used in carbide systems. The strengthening effects of the interfaces present and the resulting coherency strains in multiphase materials may be applied to enhance mechanical properties of carbides. Many data are available for precipitation phenomena in metals and some for oxides, but the same phenomena have not been exploited in carbide systems, (1)(2). Both nucleation and spinodal types of phase separation could be expected in these systems when the appropriate solid solutions are heat treated.

EXPERIMENTAL

The materials used were (i) a commercially available WC with a grain size of 1-5 μm, containing 6.6% total C, 0.06% free C, and 700 ppm O, (ii) a ZrC synthesized from Zr-metal powder containing 2190 ppm O, and 700 ppm N, (iii) TiC powder containing 19.5% total C, 0.24% free C, 1200 ppm O, and 700 ppm N. Pressed pellets were reacted in an induction furnace equipped with a quenching arrangement. A protective atmosphere of purified Ar or He was used in all runs. The precision of the temperature measurements was better than ±10°C. The limits of solid solubility were found by the lattice-parameter method using a diffractometer. The lattice parameters were accurate to ±0.002 Å.

Samples for transmission electron microscopy were thinned in the ion-beam thinner using argon accelerated by a 7-kV field. The samples for microhardness determination were arc-welded to obtain sufficiently large grain-size and then annealed at 1900°C. Hardness was measured on polished surfaces using a Knoop diamond with 100 g. load.

RESULTS AND DISCUSSION

The solubility curves of WC in ZrC and TiC are shown in Fig. 1A and B. The higher solubility of WC in TiC results from the fact that the atomic radius of W(1.41Å) is nearer that of Ti(1.46Å) than that of Zr(1.67Å).

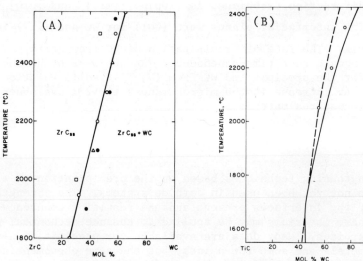

Fig. 1 (A) Solubility limit of WC in ZrC. Open circles and triangles ref, 3, squares ref. 4, filled circles ref. 5. (B) Solubility limit of WC in TiC. Open circles ref. 6, solid line ref. 7, dashed line ref. 8.

The compositions chosen for the precipitation studies were (a) 55 mol % WC and 45 mol % ZrC, (b) 75 mol % WC and 25 mol % TiC. Solid solutions were formed at 2450°C and annealed at 1900°C for 15 min. to 75 hrs. The rate of precipitation of WC from the supersaturated (W, Ti)C solid solutions was slower than in the system WC-ZrC and followed a square-root time dependence. The precipitated WC as well aligned ribbon-shaped particles is shown in Figure 2A. By comparing the electron diffraction patterns (Fig. 2B) and images of the areas from which they were obtained, it was concluded that the indices of the interfacial planes and the orientation of the two phases are $(001)_h = (111)_c$ and $[100]_h || [110]_c$ ($\underline{h}$ refers to the hexagonal WC and $\underline{c}$ to the cubic matrix). The close similarity between the atomic arrangement in the (001) plane of WC and (111) plane of ZrC or TiC is evident in Figure 3. Examination of the precise x-ray peak profiles of precipitated WC from (W, Ti)C solid solutions showed that diffraction peaks are split in all but 001-type reflections. The fact that the [001] direction is unaffected shows that deformation occurs within the basal plane of WC. One possible type of deformation that would account for this behavior is a slight shortening of one of the a-axes and formation of an orthorhombic cell. Using d-values corresponding to the separated peaks and unchanged values of the $\underline{a}$ and c-axis (2.9076 and 2.8403Å, respectively) the length of the b-axis was calculated as 5.0230Å. The experimental and calculated $\underline{d}$ values agree then to 0.0006Å. The unit cell of the deformed WC is orthorhombic c-centered and belongs to the space group C 2/m 2/m 2/m as shown in Fig. 4.

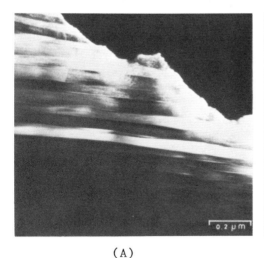   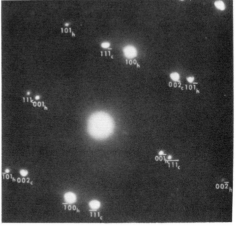

(A)   (B)

Fig. 2 The solid solution 75 mol % WC-25 mol % TiC annealed 20 hrs at 1900°C. (A) Dark field transmission electron micrograph, (B) selected-area electron diffraction pattern

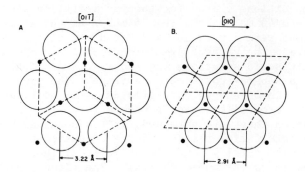

Fig. 3 Arrangement of metal and carbon atoms in (A) (111) plane of ZrC and (B) (001) plane of WC. Large circles are Zr and W atoms in the plane; small solid circles are C atoms 1.61Å below the (111) plane of ZrC and 1.42Å below the (001) plane of WC.

X-ray peak splitting was not observed for WC precipitated from (W, Zr)C solid solutions, probably because the specific volume of these solid solutions is larger than that of WC. Also WC precipitated in the regular hexagonal form from (W, Ti)C solid solutions to which 10% Co was added. In this case WC precipitated from the liquid phase in the form of platelets with no mutual orientation in an essentially stress-free condition.

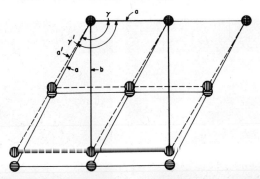

Fig. 4 Projection along c-axis of precipitated WC. Light solid lines show hexagonal unit cell, broken lines deformed quasi-hexagonal unit cell and heavy solid lines resulting orthorhombic unit cell. Circles represent W atoms, horizontally hatched if they are in hexagonal unit cell and vertically hatched if they are in orthorhombic cell.

The microhardness of the solid solutions annealed at 1900°C increased initially and then decreased as is typical of precipitation hardened materials. In some cases the hardness increase was 23% as shown in Fig. 5.

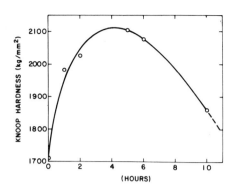

Fig. 5 Knoop microhardness of the solid solution 75 mol % WC - 25 mol % TiC as a function of time of annealing at 1900°C

ACKNOWLEDGEMENTS

This work was supported by the National Science Foundation under grant No. GP-11808 and by the Pennsylvania Engineering and Science Foundation.

REFERENCES

1. J. W. Christian, Theory of Transformations in Metals and Alloys, Pergamon Press, Oxford, 1965.

2. M. E. Fine, in Advances in Materials Research, Vol. 4 edited by H. Herman, Interscience Publishers Inc., New York, 1974.

3. M. Brun and V. S. Stubican, J. Amer. Ceram. Soc. 57(1974) 117.

4. P. Schwarzkopf and R. Kieffer, Refractory Hard Metals, The MacMillan Co., New York, 1953.

5. F. Trombrel, in Plansee Proc., Pap. Plansee Senia. De Re Matal. 2nd, 1955, 1956, 205.

6. M. Brun and V. S. Stubican, J. Amer. Ceram. Soc. 58(1975) 392.

7. A. G. Metcalfe, J. Inst. Met. 73(1947) 591.

8. O. T. Shulishova and I. A. Shcherbak, Poroshk. Metall. 6(1967) 16.

MASS TRANSPORT BY SELF-DIFFUSION AND EVAPORATION-CONDENSATION IN HIGH TEMPERATURE KINETIC PROCESSES IN $UO_2.PuO_2$ NUCLEAR FUEL

Hj. Matzke

European Institute for Transuranium Elements, EURATOM

D-75 Karlsruhe, Postfach 2266, Germany

INTRODUCTION

Uranium dioxide, UO_2, is the conventional fuel of thermal nuclear reactors, and the mixed oxide $(U,Pu)O_2$ will be the fuel for the first generation of fast breeder reactors. New results on metal self-diffusion and a critical review of kinetic measurements on UO_2 show that many data need re-interpretation. This has also consequences for the mixed oxide. Most laboratory data were obtained on coprecipitated solid solutions of $(U,Pu)O_2$ whereas in technological fuel fabrication frequently mechanically blended UO_2 and PuO_2 powders are sintered. Homogenization of these blends is important to avoid local overheating during reactor operation since fission and thus heat generation occur almost entirely with Pu only.

Extensive data on metal self-diffusion as function of temperature and composition (oxygen to metal ratio, O/M, or Pu-valence) were obtained both in laboratory experiments and in a nuclear reactor. Chemical interdiffusion was also studied. The data show that diffusion and evaporation-condensation processes are competing for rate-determination in the high temperature phenomena of interest, i.e. sintering, grain growth, creep, formation of solid solution, surface relaxation etc, depending on atmosphere and temperature.

EXPERIMENTAL

Sinters of $(U_{0.85}Pu_{0.15})O_{2\pm x}$ fabricated in the Transuranium Institute or single crystals of $(U_{0.8}Pu_{0.2})O_{2\pm x}$ obtained from AL-KEM were used. The diffusion experiments were performed with Pu-238 as tracer applying high resolution α-spectroscopy to determine diffusion profiles (1-3). This method is non-destructive and enables the time dependence of diffusion to be determined. This proved to be of extreme importance in determining reliable diffusion coefficients, D.

Annealing was done in controlled atmospheres of CO/CO_2 in ratios between 10000:1 and 1:100, in H_2/H_2O with different H_2O content or in high vacuum at temperatures between 1400 and 2000°C.

RESULTS ON METAL DIFFUSION

The metal atoms are much less mobile in all actinide dioxides of the fluorite structure (such as UO_2, ThO_2, $(U,Pu)O_2$) than oxygen. Their transport is therefore rate determining for high temperature kinetic processes. Unfortunately, most published data have to be rejected either because of the use of out-of-date techniques, poor specimen quality or because they represent "artifacts". Such artifacts may be tracer evaporation when the "surface decrease method" is applied, they may be due to using thick tracer layers (e.g. with U-235 as tracer), or simply there might be disturbing effects of short circuit paths (grain boundaries or dislocations). In normal sinters of UO_2, ThO_2 or $(U,Pu)O_2$ with 5 to 20 µm grain size and at the (comparatively low) temperatures normally used (1300 to 1700°C, i.e. ~0.5 to 0.6 T_m, T_m=melting point in K), 50 to 80% of the tracer penetrates the sinter via short circuit paths.

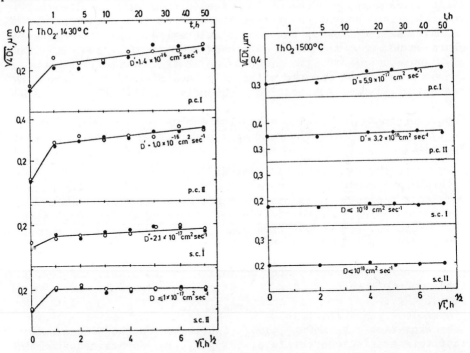

Fig.1: Experiments on the diffusion of U-233 in ThO_2 sinters (pc) and single crystals (sc). At 1430°C, some initial tracer penetration might simulate diffusion, as do grain boundaries. At subsequent anneals at 1500°C, the absence of measurable diffusion becomes obvious (I and II are for the two parts of each diffusion couple).

The most serious error source consists in an effect that might be called "surface smoothing" or "accommodation of the tracer layer to the surface". It consists in an initial tracer penetration of 0.05 to 0.2 μm and depends on the state of the surface. It is most likely due to the accommodation of the tracer layer to the thermal equilibrium shape of the surface (5,6). The time dependence of the evolution of the tracer profile must be carefully followed to allow for this effect (see Fig.1). Additional errors originate from tracer transfer by evaporation-condensation from one specimen to the other of the diffusion sandwich. Therefore, not only the t, but also the x-y dependence of diffusion should be followed. This has never been done before this study.

If all these errors are avoided, the resultant diffusion coefficients D are much lower than previously reported ones, and the activation enthalpies ΔH are much higher, e.g. 150 and 130 kcal/mole for ThO_2 and UO_2, than the previously reported ~ 100 kcal/mole. The mixed oxide $(U,Pu)O_2$ is of basic particular interest since it can accommodate both substantial amounts of oxygen interstitials in $(U,Pu)O_{2+x}$ by oxidizing U^{4+} to U^{5+} or U^{6+} as well as substantial amounts of oxygen vacancies in $(U,Pu)O_{2-x}$ by reducing Pu^{4+} to Pu^{3+}. Metal diffusion in $(U,Pu)O_{2\pm x}$ depends strongly on x, as indicated by the arrow at 1600 °C in Fig. 2 where O stands for oxidizing conditions (e.g. $(U,Pu)O_{2.15}$), R stands for strongly reducing conditions (e.g. $(U,Pu)O_{1.90}$, and M indicates a minimum in mobility inbetween (e.g. for $(U,Pu)O_{1.96}$).

Fig.3 shows this minimum for 1500°C. Similar minima are also observed for vapour pressures, grain growth and creep.

Further experiments were performed to study the interdiffusion between UO_2 and PuO_2 as well as the effect of irradiation on diffusion by separate experiments in a nuclear reactor. Space limits do not allow to present these results here in detail. Finally, rates of free evaporation were measured (Fig. 4). A coefficient for efficiency of condensation α of 0.1 is seen to fit both experiments and rates calculated from vapor pressure data (16). Note also, that the ΔH's for diffusion and evaporation are very similar.

As a consequence of the new diffusion data, any previous suggestion of rate-controlling processes based on ΔH's needs revision. This is true for much of the literature on sintering, plastic deformation, grain growth, surface relaxation etc. Also, a dependence on x as observed for diffusion is not necessarily a proof for volume diffusion being rate-controlling. For instance, the grain growth data of Fig.3 are better explained by an evaporation-condensation mechanism of pore mobility than by volume diffusion. The same can be shown for many of the interdiffusion results. Also, much of the sintering work needs reconsideration.

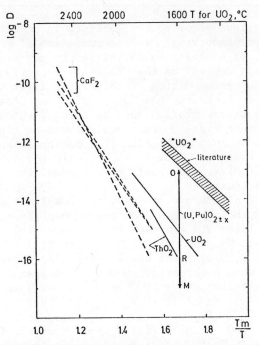

Fig.2: Normalised Arrhenius diagram for metal atom diffusion in UO_2, ThO_2, $(U,Pu)O_2$ and, for comparison, in CaF_2 (7-9).

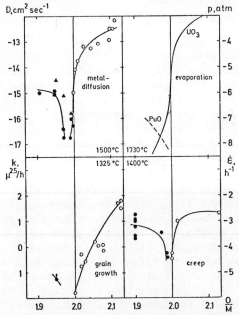

Fig.3: Metal diffusion (triangles from (10)), evaporation (11), grain growth (12,13) and creep (14,15) in UO_{2+x} (open symbols) or $(U,Pu)O_{2-x}$ (full symbols)

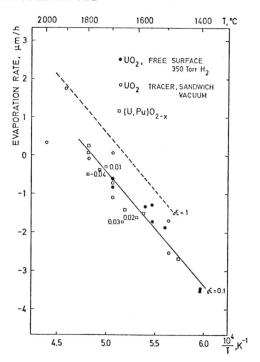

Fig.4: Evaporation rate from UO_2 and $(U,Pu)O_{2-x}$. The dashed line was calculated for $\alpha=1$ from evaporation data (16). $\alpha=0,1$ fits most of the results. The numbers on the data points of $(U,Pu)O_{2-x}$ are x-values.

The minimum in metal diffusion rates indicated in Figs. 3 and 4 at an O/M of about 1.96 to 1.98 can be explained by vacancy diffusion on the right side and interstitial diffusion on the left side (17). Alternatively, formation of defect clusters (e.g.10) or effects of valence on mobilities can be invoked. The existence of a minimum is important for technological application of blended UO_2-PuO_2 powders since in the early stages of homogenization and formation of solid solutions the composition of the minimum can build up in the outer zones of the grains thus forming diffusion barriers, slowing down the desired processes and changing the rate-controlling mechanism. It can be predicted that suitable choice of oxygen potentials as well as the effect of irradiation can help to overcome these difficulties.

REFERENCES

/1/ F.Schmitz and R.Lindner, J.Nucl.Mater. 17(1965)259
/2/ A.Höh und Hj.Matzke, Nucl.Instr.Methods 114(1974)459
/3/ V.Nitzki and Hj.Matzke, Phys.Rev. B8(1973)1894
/4/ R.A.Lambert, Thesis submitted to University of Surrey (1976)

/5/ Hj.Matzke and C.Ronchi, Phil.Mag. 26(1972)1395
/6/ H.Bay and Hj.Matzke, to be published
/7/ Hj.Matzke and R.Lindner, Z.Naturforschg. 19a(1964)1178
/8/ M.F. Berand, J.Amer.Ceram.Soc.54(1971)144
/9/ A.D.King and J.Moerman, Phys.stat.sol.(a)22(1974)455
/10/ F.Schmitz and A.Marajowski, in Thermodynamics of Nuclear Materials 1974, IAEA Vienna I(1975)467
/11/ M.H.Rand and T.L.Markin, in Thermodynamics of Nuclear Materials 1967, IAEA Vienna (1968)637
/12/ J.R.McEwan and J.Hyashi, Proc.Brit.Ceram.Soc.7(1967)245
/13/ C.Sari, private communication
/14/ B.Burton, in Thermodynamics of Nuclear Materials 1974, IAEA Vienna I(1975)415
/15/ J.L.Routbort, N.A.Javed and J.C.Voglewede, J.Nucl.Mater. 44(1972)247
/16/ M.Tetenbaum and P.D.Hunt, J.Nucl.Mater.34(1970)86
/17/ Hj.Matzke, J.Physique (France) 34(1973)C9-317

THE STRUCTURAL DEGRADATION OF CARBON FIBRES IN NICKEL AT ELEVATED TEMPERATURE

R. Warren and J. Wood

Department of Engineering Metals and Department of Inorganic Chemistry, Chalmers University of Technology

Gothenburg, Sweden

ABSTRACT

The interactions of various types of carbon fibre with coatings of nickel have been studied in a temperature range of 800 - 1300°C. For all fibre types, the presence of nickel catalyses the graphitization of the fibre structure. The graphitization is accompanied by the penetration of nickel to the centre of the fibre and by a drastic reduction in the fibre strength. It occurs rapidly above a characteristic temperature that is dependent on the fibre type and structure. The results are discussed with respect to possible graphitization mechanisms.

INTRODUCTION

The possible use of carbon fibres as a reinforcement for high-temperature, nickel-base composites has led to an interest in the interactions of carbon fibres and nickel. Though little chemical reaction is to be expected, earlier studies have shown that, in the presence of nickel, carbon fibres are structurally degraded at temperatures between about 800° and 1250°C (1-5). The process is a nickel-catalysed graphitization and is accompanied by an almost complete loss of fibre strength. Apparent disagreement exists in the earlier work with regard to the exact mechanisms of the process and the factors affecting it. For example, the effect of impurities in the system or the effect of the fibre structure have not been systematically investigated.

There exists a range of fibre types with a wide variety of structures and properties dependent both on the polymer precursor from which the fibre is prepared and on the conditions of preparation, the most significant being the final heat-treatment temperature of the fibres, which can lie between $1000°$ and $2700°C$ (6,7). Briefly, the fibre structure can be described as a distorted array of graphite basal planes. With increasing HTT the basal planes become increasingly ordered and increasingly oriented, both along the fibre axis and parallel to the fibre surface (7). With increased order, the chemical stability of the structure increases (8.9). In practice, the fibre structure can be characterized by measurement with X-ray diffraction of, for example, an effective crystallite size and an average basal plane spacing.

The intention of the present work was to study the interaction of a variety of fibre types with nickel coatings, deposited by a variety of techniques.

EXPERIMENTAL

The fibre types studied, together with details of their structure are included in Table 1. Nickel coatings, having thicknesses between about 0.5 and 4 microns, were deposited on the fibres by electrolyte deposition, decomposition of nickel carbonyl, vacuum evaporation and electroless deposition. Electroless deposition, as described in (10), gave consistently the best-quality coatings and this was chosen for the majority of the work.

Coated fibres were heat-treated between 800 and $1300°C$ in an alumina tube furnace under high-purity argon. Such treatments did not affect uncoated fibres significantly. Before and after treatment the samples were examined metallographically and by X-ray diffractometry. The latter technique was used to determine the c-direction crystallite size, the average (0002) planar spacing, and the degree of structural degradation (graphitization). The latter was estimated from the relative areas under the (0002) peak, characteristic of the fibre, and that of the crystalline graphite that develops during graphitization (1). Both the strength and Young' modulus of the treated fibres were measured using a technique described earlier (11).

RESULTS

All the PAN-based fibres graphitized between $800°$ and $1250°C$. This could be observed metallographically (Figures 1, 2) as well as by X-ray diffraction (Figure 3). The process involved penetration of the fibre by nickel and was accompanied by an almost complete loss of strength and Young's modulus. For a given heat-treatment time (viz. 1 h), the fibres remained relatively unaffected up to a characteristic temperature above which the graphitization occurred very rapidly. As can be seen from Table 1, the stability of the fibres (i.e. the temperature of rapid graphitization) was strongly dependent on the fibre structure, increasing with structural order. The rayon-base, Thornel 50 fibres, were more stable than PAN fibres of comparable structure; for a 1 h treatment, they did not graphitize below the nickel-carbon eutectic temperature. Sarian reported that these fibres graphitized after 200 h at 1250°C (4).

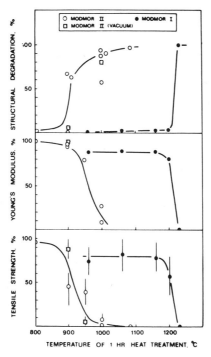

Fig. 1. Degradation of nickel-coated carbon fibres.

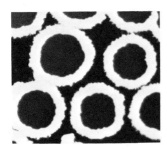

Fig. 2. Electroless Ni-coated fibres, untreated (1000 x).

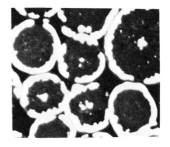

Fig. 3. Type II fibres after 1 h at 1000°C (1000 x).

Table 1. The structure of various carbon fibres and their graphitization temperatures in nickel.

Fibre type and (precursor)	(0002) planar spacing (Å)	crystallite size L_c (Å)	Orientation of (0002) planes*	Graphitization temperature °C
Modmor II (PAN-base)	3.51	16	23.5°	850-950
Courtaulds HTS (PAN-base)	3.49	16	22.7°	920-1080
Courtaulds HMS (PAN-base)	3.43	60	13.8°	1050-1120
Modmor I (PAN-base)	3.41	72.5	7.8°	1200-1230
Thornel 50 (Rayon-base)	3.41	52.5	8.1°	> 1270

* Average angular deviation from fibre axis, taken from ref. (6).

The type I fibres (Modmor I and Thornel 50) exhibited a small amount of graphitization (up to about 10%) at temperatures below that of rapid graphitization (Figure 3). This will be discussed below, but it may be noted these fractions of graphitization corresponded closely to the equilibrium solubility of graphite in nickel and could therefore be explained by precipitation of dissolved carbon during cooling.

Experimental studies of the isothermal kinetics of the graphitization are difficult because the process occurs either very slowly (below the graphitization temperature) or very rapidly. Analysis of such studies on the basis of a kinetic model are hampered by the inhomogeneity of the fibre structure. In the present work, limited studies of the isothermal graphitization of Modmor II fibres were made in the region of the graphitization temperature. These showed that the graphitization rates were consistent with known rates of material transport, e.g. with the diffusion of nickel into graphite or of carbon in nickel (12,13). Analysis of the kinetics for type I fibres is even more difficult but the observed behaviour suggests that some form of induction process is involved in the graphitization.

The nature and thickness of the nickel coatings appeared to have little effect on the graphitization temperature. With electroless nickel, Modmor II fibres graphitized at somewhat higher temperatures (~ 900°C) than with the other nickel types (~ 850°C).

DISCUSSION

A general review of catalytic graphitisation emphasises the fact that no two systems, even closely related fibre types, necessarily behave in the same way. Four general mechanisms have been proposed (14,15,16) of which two would appear to be particularly relevant to the graphitisation of carbon fibres. The first, dissolution of carbon in metal followed by reprecipitation as graphite, is generally accepted as the mechanism responsible for the degradation. This suggested mechanism followed observations that the reaction had many similarities with activated sintering processes for tungsten (1). Some micrographs of degraded fibres support this model, as carbon can be seen transported to the exterior surface of the nickel coated fibre. A general description would be the occurrence of localised reactions in which carbon is dissolved in excess nickel as it moves through the fibre. Calculations employing nickel/carbon diffusion rates and solution concentrations are in good agreement with such a model. The observed dependence of fibre stability on structure can be explained by differences in the internal energy of the different fibre types leading to corresponding differences in solubility. The apparent induction behaviour of type I fibres can be explained by the existance of a very stable surface layer.

The second mechanism is via the catalytic removal of crystal imperfections such as interstitial atoms, crystal imperfections and carbon atoms involved in cross-linkages between crystallites (16).

The preparation of the carbonised fibre is via a free radical process. ESR and oxygen effects confirm the nature of the carbonisation process, which appears to be analogous to the degradation of polyvinylidene chloride (17). Recent work using calcium and increasingly sub-divided nickel (15,16) has focussed attention on the possible free-radical nature of catalytic graphitisation. Thus examining the inverse of the first mechanism, nickel transported through excess carbon, it is found that the effective diffusion coefficient is high and that it occurs along grain-boundaries and defects by a free-radical type of chemical combination/decomposition mechanism (12).

Applying these ideas to the expected type I and II fibre structures, an alternative model for fibre degradation can be proposed. Nickel dissolves and rapidly diffuses through the structure via a series of chemically bonded traps (12) along the many defects. This can be regarded as an inverse of the solution-precipitation mechanism occurring in the alternative concentration range. Both carbon-carbon bonds and free valencies are attacked, subsequently reforming into more ordered structures and leading to progressive graphitisation of the fibre. The driving force is again the beneficial free energy change (14). In type I fibres, single crystallites will promote strained chemical bonds and defects. Simultaneous rupture of several bonds will be required before reorientation can occur, although only a small reorientation is required to give a graphitic structure. This occurs at a critical temperature; the simultaneous removal of cross-linking chemical bonds replaced by van-der-Waals bonds explains the loss in fibre strength. In type II fibres, the structure retains more of the characteristics of the polymer precursur. Graphitisation in this case requires both significant structural rearrangement and crystallite coalescence. Thus the critical temperature and loss of strength do not occur over a so sharply defined temperature interval.

Optical microscope studies in the present work indicate that both types of mechanism occur. The factors that decide which mechanism occurs are still unclear and further studies to resolve this are in progress.

REFERENCES

1. P.W. Jackson and J.R. Marjoram, J. Matls. Sci., 5 (1970) 9.
2. R.B. Barclay and W. Bonfield, J. Matls. Sci., 6 (1971) 1076.
3. S.V. Barnett, S.J. Harris and J.F. Weaver, Faraday Special Discussions of the Chem. Soc., No. 2 (1972) 144.
4. S. Sarian, J. Matls. Sci., 8 (1973) 251.
5. I. Shiota and O. Watanabe, J. Jap. Inst. Met., 38 (1974) 794.
6. G.A. Cooper and R.M. Mayer, J. Matls. Sci., 6 (1971) 60.
7. R.J. Defendorf and E.W. Tokarsky, Techn. Report AFML-TR-72-133 (1973).
8. F.S. Galasso and J. Pinto, Fibre Sci. and Techn. 2 (1970) 303.
9. D.J. Thorne and A.J. Price, Fibre Sci. and Techn. 4 (1971) 9.
10. J.W. Dini and P.R. Coronado, Plating, 54 (1967) 385.
11. R. Warren and M. Carlsson, Proc. Vth Internat. Conf. Chemical Vapour Deposition (1975), Electrochemical Society, Princeton.
12. J.R. Wolfe, D.R. McKenzie and R.J. Borg, J. Appl. Phys., 36 (1965) 1906.
13. J.J. Lander, H.E. Kern and A.L. Beach, J. Appl. Chem., 20 1952) 1305.
14. H. Marsh and A.P. Warburton, J. Appl. Chem., 20 (1970) 133.
15. S. Otani, A. Oya and J. Akagami, Carbon, 13 (1975) 353.
16. A. Oya and S. Otani, Carbon, 13 (1975) 450.
17. D.H. Davies, D.H. Everett and D.J. Taylor, Trans. Far. Soc. 67 (1971) 382.

SOLID STATE REACTION STUDIES ON $La_2O_3-NH_4F$ and $La(OH)_3-NH_4F$

G. Adachi, B. Francis, K. Rajeshwar, E. A. Secco and J. C-S. Wong

Chemistry Department, St. Francis Xavier University
Antigonish, Nova Scotia, Canada

INTRODUCTION

The fluorination of the lanthanide oxide (1,2,3) represents an important step in the preparation of the lanthanide metal by the reduction process.

The formation of an intermediate fluoride ammine $LaF_3 \cdot 3/2NH_3$ was reported (4,5) in the course of the solid state reaction between La_2O_3 and NH_4F leading to the final product of the fluoride Unsuccessful attempts to isolate the intermediate ammine in the $La_2O_3-NH_4F$ reaction were followed by a study of the $La(OH)_3-NH_4F$ solid state reaction. This paper, therefore, reports the results of our investigation on these fluorination reactions using the methods of thermal analyses, infrared spectroscopy and X-ray diffraction, with the goal of identifying the various steps.

EXPERIMENTAL

La_2O_3 and $La(OH)_3$ used in this work were of >99.99% purity supplied by Alfa Inorganics Ventron, Beverly, Mass., U. S. A., NH_4F was Fisher Certified grade. The oxide was ignited to ~800°C before using; no thermal effect was observed to 800°C after ignition. Reaction mixtures in various molar ratios were prepared by grinding in a mortar.

Thermal analyses by DTA and TG were performed with the instrumentation and platinum ware described elsewhere (7,8).

The intermediate and final products of the reaction were characterized and identified by X-ray diffraction patterns and infrared spectra recorded as previously described (6,9).

RESULTS AND DISCUSSION

La_2O_3-NH_4F

DTA and TG heating curves assignments and the numerical results for the various reactant ratios are given in Table 1.

The strong endotherm at 170°C is attributed to NH_4F disproportionation (10), viz.

[1] $\quad 2NH_4F_{(s)} \longrightarrow NH_4HF_{2(\ell)} + NH_{3(g)}$

The liquid phase NH_4HF_2 was confirmed by i) observation on the melting point apparatus and ii) cooling the mixture after partial disproportionation to reveal the freezing exotherm - the melting endotherm appeared on second heating.

The shallow second endotherm, 225°-250°C, is assigned to partial fluorination in presence of excess oxide, i.e.

[2a] $\quad 1/3 La_2O_{3(s)} + NH_4HF_{2(\ell)} \longrightarrow 2/3 LaF_{3(s)} + NH_{3(g)} + H_2O_{(g)}$;

lanthanum fluoride appeared amorphous to X-ray diffraction.

The formation of tetragonal LaOF, confirmed by its X-ray

TABLE 1a

DTA Results For The Various Reactant Ratios

Mole Ratio	Peak Number	Peak Temperature °C	Reaction
1:2	1	170 (S)[†]	[1]
	2	240 (W)	[2a]
	3*	330 (VS)	[2b]
1:4	1	170 (S)	[1]
	2	225 (M)	[2a]
	3	250 (M)	[3]
	4*	330 (VS)	[2b]
1:6	1	170 (S)	[1]
	2	225 (S)	[2a]
	3	245 (M)	[3]
	4	330 (S)	[2b]
	5*	350 (W)	Crystallization LaF_3
1:8	1	170 (S)	[1]
	2	225 (S)	[3a]
	3	250 (M)	[3b,c]
	4*	370 (M)	Crystallization LaF_3

* Exotherm Peak
† VS-very strong; S-strong; M-medium; W-weak

TABLE 1b
TGA Results For The Various Reactant Ratios

Mole Ratio	Weight Loss to Solid Intermediate			Weight Loss to Final Residue		
	Temp.Range °C	Calc.†%	Obs. %	Temp.Range °C	Calc.%	Obs.%
1:2	--	--	--	90 - 200	13.0*	12.6
1:4	90 - 170	21.6	22.3	170 - 220	24.5	25.2
1:6	90 - 170	28.4	28.6	170 - 225	31.8	32.4
1:8	90 - 180	33.0	34.4	180 - 240	37.0	38.8

† Intermediate assumed $NH_4La_3F_{10}$
* LaOF

diffraction pattern (11) and infrared spectrum (12), is assigned to the intense exotherm at 330°C, i.e.

$$[2b] \quad 2/3 La_2O_{3(s)} + 2/3 LaF_{3(s)} \longrightarrow 2\ LaOF_{(s)}.$$

The heating trace of 1:1 molar ratio of La_2O_3-LaF_3 mixture displayed a similar exotherm at 480°C confirming LaOF formation. The lower temperature LaOF formation is attributed to the higher reactivity state of LaF_3 from [2a] as evidenced by its amorphous character and by a parallel effect reported earlier (13).

Thermogravimetric traces for higher ratios reveal an intermediate compound corresponding to the stoichiometric composition $NH_4La_3F_{10}$. X-ray diffraction pattern of the intermediate consists of LaF_3 lines plus additional lines. The infrared spectrum of the intermediate showed intense absorption bands at 3140 $\overline{cm}^1$ and 1400 $\overline{cm}^1$ characteristic of ν_3 and ν_4 of NH_4^+ respectively. Both bands were triplets; the resolved ν_4 triplet occurred at 1484, 1465, and 1400 $\overline{cm}^1$. The splitting of the degenerate ν_4 mode strongly suggests the removal of the NH_4 group degeneracy. The spectrum of the partially deuterated intermediate analogue displayed the triplet shift to lower frequencies confirming the presence of NH_4 group.

Isothermal pretreatment of La_2O_3-NH_4F stoichiometric mixtures at 160°C revealed the initial intermediate to be $NH_4La_2F_7$ according to the equation,

$$[3a] \quad La_2O_{3(s)} + 4NH_4HF_{2(\ell)} \longrightarrow NH_4La_2F_{7(s)} + 3NH_{3(g)} + 3H_2O_{(g)} + HF_{(g)}$$

followed by the decomposition steps confirmed by DTA and TG traces,

$$[3b] \quad NH_4La_2F_{7(s)} \longrightarrow NH_4LaF_{4(s)} + LaF_{3(s)},$$

$$[3c] \quad NH_4LaF_{4(s)} \longrightarrow LaF_{3(s)} + NH_{3(g)} + HF_{(g)}$$

The initial intermediate $NH_4La_3F_{10}$ observed in thermogravimetry was a fortuitous 1:1 mixture of $NH_4La_2F_7$ and LaF_3 as evidenced by its X-ray diffraction pattern.

The disparity between our fluorolanthanate intermediate and the intermediate fluoride ammine reported by Markovskii et al (5) is obvious. The formation of ammonium fluorolanthanate in the oxide-bifluoride reaction consistent with our study, was reported by Mikhailov et al (14).

$La(OH)_3 - NH_4F$

The decomposition of $La(OH)_3$ occurs via the intermediate LaOOH, viz.

[5a] $2La(OH)_{3(s)} \longrightarrow 2H_2O_{(g)} + 2LaOOH_{(s)}$,

[5b] $2LaOOH_{(s)} \longrightarrow H_2O_{(g)} + La_2O_{3(s)}$.

The DTA traces and the data are consistent with the following equations for the specified molar ratios:

1:1

[6a] $La(OH)_{3(s)} + NH_4F_{(s)} \longrightarrow 1/3 La(OH)_{3(s)} + 1/6 La_2[NH_3]_3F_6 + 1/2 NH_{3(g)} + H_2O_{(g)}$

[6b] $1/6 La_2[NH_3]_3F_6 \longrightarrow 1/3 LaF_{3(s)} + 1/2 NH_{3(g)}$

[6c] $1/3 LaF_{3(s)} + 2/3 La(OH)_3 \longrightarrow LaOF_{(s)} + H_2O_{(g)}$

1:3

[7a] $La(OH)_{3(s)} + 3 NH_4F_{(s)} \longrightarrow 1/2 La_2[NH_3]_3F_{6(s)} + 3/2 NH_{3(g)} + 3H_2O_{(g)}$

[7b] $1/2 La_2[NH_3]_3F_{6(s)} \longrightarrow LaF_{3(s)} + 3/2 NH_{3(g)}$

The $La(OH)_3-NH_4F$ results reveal a striking resemblance to the findings of Markovskii et al (5), viz. the formation of the fluoride ammine intermediate and the endothermic effects in DTA traces. In contrast to the $La_2O_3-NH_4F$ results we note i) the direct action of NH_4F on $La(OH)_3$ and the absence of NH_4HF_2 as the active fluoridating agent, ii) the action of LaF_3 on $La(OH)_3$ to yield LaOF with an endothermic effect, and iii) the absence of any exothermic effect.

ACKNOWLEDGMENT

Grateful acknowledgment is made to the National Research Council of Canada and the University Council for Research for financial support of this research program.

REFERENCES

1. A. Zalkin and D. H. Templeton, J. Am. Chem. Soc. $\underline{75}$, 2453 (1953)
2. O. N. Carlson and F.A. Schmidt, The Rare Earths, Edited by F.H. Spedding and A.H. Doane, John Wiley and Sons, New York, 1961. Chpt. 6, p. 77.
3. R. E. Thoma, G. M. Herbert, H. Insley, and C. E. Weaver. Inorg. Chem. $\underline{2}$, 1005 (1963).
4. L. Ya. Markovskii, E. Ya. Pesina, L. M. Loev, and Yu. A. Omelichenko Russ. J. Inorg. Chem. $\underline{15}$, 2 (1970).
5. L. Ya. Markovskii, E. Ya. Pesina, and Yu. A. Omelichenko. Russ. J. Inorg. Chem. $\underline{16}$, 172 (1971).
6. K. C. Patil and E. A. Secco. Can. J. Chem. $\underline{50}$, 567 (1972).
7. O. K. Srivastava and E. A. Secco. Can. J. Chem. $\underline{45}$, 597, 1375 (1967).
8. P. Ramamurthy and E. A. Secco. Can. J. Chem. $\underline{47}$, 2185 (1969).
9. E. A. Secco. Can. J. Chem. 42, 2143 (1963).
10. H. Remy. Treatise on Inorganic Chemistry. Elsevier, London 1956. Vol. 1, p. 795.
11. W. H. Zachariasen. Acta Cryst. 4, 231 (1951).
12. L. R. Batsanova and G. N. Kustova. Russ. J. Inorg. Chem. 9, 330 (1964).
13. P. Ramamurthy and E. A. Secco. Can. J. Chem. $\underline{48}$, 1619 (1970).
14. M. A. Mikhailov, D. G. Epov, V. I. Sergienko, E. G. Rakov, and G. P. Shchetinina. Russ. J. Inorg. Chem. $\underline{18}$, 794 (1973).

THE CORROSION RESISTANCE OF Fe-Cr ALLOYS AS STUDIED BY ESCA

I. Olefjord and B.-O. Elfström

Department of Engineering Metals, Chalmers University

of Technology, Fack, 402 20 Göteborg

ABSTRACT

The chemistry of passive films formed on Fe-Cr alloys in acid, oxygenated neutral water and dry oxygen for different times at room temperature has been studied by ESCA. The composition of the film is dependent on the environments. In acid at the passivation potentials 0.1, 0.5 and 0.9 V vs. SCE an homogeneous film with about 60 pct. Cr is formed. A stationary state is reached after a few minutes. In water the initial composition of the passive layer corresponds to that of the alloy but is changed by dissolution and by enrichment of Cr in the outermost layer during prolonged exposure. After a few tens of hours a stationary state is reached and the film consists of a top layer of $Cr(OH)_3$ with Fe-Cr oxides beneath. Exposure in dry oxygen gives an oxide with the same average composition as the metal and due to low diffusivity at room temperature the composition does not change with time.

INTRODUCTION

The main property of stainless steel is that it attains a passive state by forming a thin layer of reaction products on the surface. The chemistry of these products is of great importance for the chemical resistance of the alloy. In spite of this very little is known about the composition of the passive layers, the reason being that they are only a few tens of Ångström thick.

During recent years new surface sensitive electron spectroscopic methods have been available. The most powerful is ESCA due to its ability to detect chemical shifts.

EXPERIMENTAL

The main chemical composition of the specimens and the experimental conditions are given in table 1.

Table 1: The chemistry of the specimens and the environments.

Spec.no	Cr	Fe	Plane	Environments
1,2	18.0	balance	(100),(110)	0.5 M H_2SO_4; 0.1, 0.5, 0.9 V (SCE)
3	7.8	"	polycrystal	neutral oxygenated H_2O
4	15.7	"	"	"
5	5.8	"	"	room temp. 0.2 atm. O_2
6	11.0	"	"	"

In all studies, it was attempted to obtain clean metal surfaces free from preoxides before exposure to the liquid or gas. Precautions were taken to avoid changes of the composition of the film due to oxygen in air. In the potentiostat-controlled work (specimen 1 and 2) this was achieved by applying to the specimens the corrosion potential at which the preoxide is dissolved for half an hour before the test potential was switched on. The latter was held constant for times ranging from 2 min. to 24 h. Then the specimens were washed in acetone and moved into the vacuum system under a stream of pure nitrogen. (The cell is attached to the vacuum chamber.)
In the neutral water experiment (1), specimens were polished on emery paper no. 600 in oxygenated water and then exposed for times ranging from 2 min. to 300 h. The specimens were moved from the reaction cup to the ESCA instrument under a stream of nitrogen. The oxidation studies were performed in the vacuum system. The clean surfaces were obtained by ion bombardment. The surfaces were oxidized for 2 min., 1 h and 50 h.

Chemical gradients in the films were measured by successive ion bombardments and ESCA analyses. The composition and the thickness of the passive films were established by semiquantitative evaluation of the ESCA spectra (2).

RESULTS

Acid environments

Figure 1 illustrates the ESCA spectra recorded. The example is the potentiostat-treated single crystal (sample no. 1). Row 1 shows the spectra from the sample pretreated at the corrosion potential ($\approx$ -0.5 vs SCE). The peaks of Fe and Cr at low binding energies are the metallic, and those at high binding energies are the oxidic states of the elements.

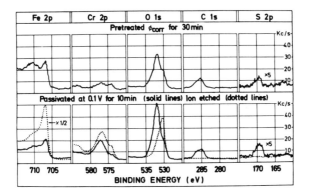

Fig. 1. ESCA-spectra of spec. 1 (100) before (row 1) and after (row 2) passivation for 10 min. at 0.1 V in H_2SO_4. The dotted peaks are recorded after ion etching.

The ratio between Cr and Fe is the same in the metallic and the oxidic states. The thickness is estimated as 10 Ångström, which is 2 to 3 times less than an air-formed oxide. It is therefore supposed that the oxide is formed after washing in the vacuum chamber and that the surface was clean and in its active state before the potential was switched to the passivation potential.

Row 2 shows the spectra after the specimen had been passivated at +0.1 V for 10 min. in the cell. The ratio between the oxidic and the metallic state has increased due to surface coverage of a thicker oxide layer. The estimated thickness for the passivated specimen is 15 Ångström. The dotted peaks show the spectra recorded after a short time of ion bombardment where the contaminants have been removed and only oxygen in the oxidic state is present.

Figure 2a shows the Cr content in the outermost layer of the film vs. passivation times ranging from 2 min. to 24 h for the potentials 0.1, 0.5 and 0.9 V (SCE). The figure shows that Cr is enriched from 18 pct. in the metal to about 65 pct. in the surface layer. The large increase in Cr content occurs during the first minutes. A slight potential dependence is observed. A "saturation" value is reached after half an hour. With prolonged exposure the Cr content increases very slightly. The ringed signs represent measurements on the (100) plane while the unringed represent the (110) plane. The squares represent investigations on electropolished specimens (0.1 mA/cm^2 for 30 min.). These demonstrate that the chemical composition of the film is independent of the pretreatment and the planes.

The thicknesses of the passive films (2) increase slightly with time (fig. 2b). At the lowest potential (0.1 V) the layer is thinnest. Figure 3 shows that the Cr content through the layer is approximately constant, only in the outermost layer a slightly higher Cr content is noticed.

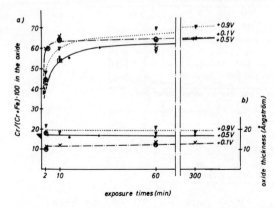

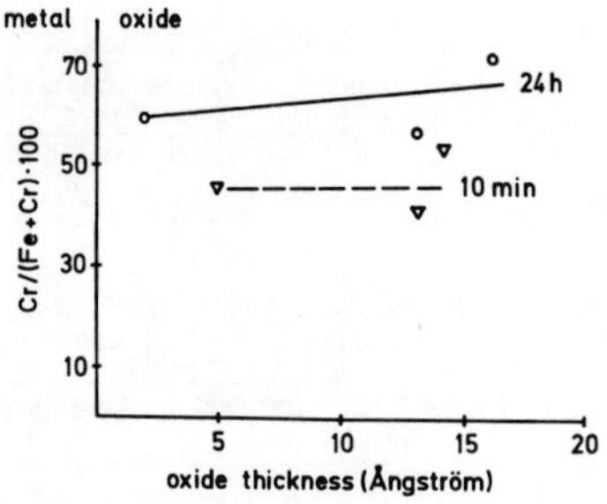

Fig. 2. a) Integral Cr content in the outermost layers vs. passivation time for the potentials (SCE): x-- + 0.1 V o--- + 0.5 V, ▽······ 0.9 V. Signs arrounded by o are (100) - spec. 1 - and unringed signs are (110) - spec. 2.
b) The thickness of the passive film.

Fig. 3. Integral Cr content at various depths in the oxide film.

The current was plotted on a x-t recorder. Figure 4 shows the i-t diagram. An inflection point occurs after 2 min. It is proposed that the layer is built up during this period. By approximating the two line segments in the diagram, using Faraday's law and assuming only one anodic reaction, the time dependence of dissolution

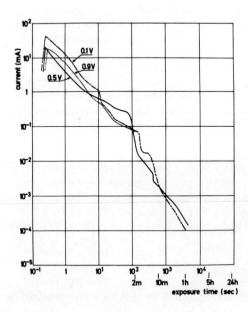

Fig. 4. Passivation current at 0.1 V, 0.5 V, 0.9 V (SCE) vs. time.

($\Delta \ell$) in Ångström units are:

$$\Delta \ell = 40 \ln \frac{t}{0.3} \qquad 0.3 < t < 100 \text{ sec.}$$

$$\Delta \ell = 2800 \left(\frac{1}{100} - \frac{1}{t}\right) \qquad t > 100 \text{ sec.}$$

The maximum dissolved metal during the two periods is 230 and 28 Ångström respectively. From the investigation of the thickness of the layer it appears that only a fraction of the dissolved species forms oxides. The low increase in thickness indicates that a quasi-stationary state is reached where the passive film is dissolved at nearly the same rate as it is built up.

Neutral environments

Figure 5a shows the Cr content through the film after passivation of specimens 3 and 4 in oxygenated water (pH ≈ 7) for different times (1). For short exposure the composition corresponds to that of the alloy (solid lines), but for prolonged exposure the Cr content increases with time. A chemical gradient is set up with the highest Cr content in the outermost atomic layers. A steady state is reached after a few tens of hours. After this little or no changes of the composition occur.

A correlation is obtained between the Cr content in the outermost region of the film and the shape of the oxygen peak. The latter consists of a dual signal corresponding to oxygen in hydroxide and oxide. The former dominates when the Cr content is high indicating that the outermost layer consists of a thin layer of $Cr(OH)_3$ and underneath is an Fe-Cr oxide.

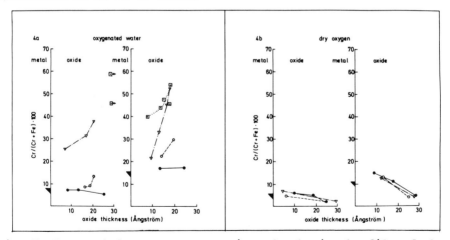

Fig. 5. Integral Cr content at various depths in the film of the specimens: a) no 3 and 4 passivated in neutral oxygenated H_2O; b) no 5 and 6 oxidized in 0.2 atm O_2. Exposure times: ● - ● 2 min., o---o 1 h, ▽ -·- 50 h. ▫·········300 h.

Figure 5b shows the composition profiles of the oxides after oxidation at room temperature in oxygen at 0.2 atm. (samples 5 and 6) for oxidation times ranging from 2 min. to 46 h (2). The average composition of the oxide corresponds to that of the metal. The change with exposure time is within the accuracy of the method. In this case a small enrichment of Cr at the metal/oxide boundary and a depletion at the outermost boundary is noticed in contrast to the behaviour in oxygenated water.

DISCUSSION

From the thermodynamic point of view one could expect preferential oxidation of Cr. The study confirms that the reaction products formed at room temperature are controlled by kinetic effects dependent on the environments. During oxidation in dry oxygen an oxide film is formed. Due to the low mobility of the atoms at room temperature, the reaction is controlled by the availability rather than the reactivity of the elements. The layer is built up very quickly and only small changes of composition occur. In neutral water, the initial reaction products are again controlled by the availability of the elements in the outermost atomic layers of the metal. However, in this case the outermost layers of the inital iron-rich film is dissolved due to presence of water and replaced by the more stable Cr-product ($Cr(OH)_3$).

When a clean surface is passivated in an artificial way by applying a potential to the specimen, the high initial current density dissolves the outermost layer of the metal and only the most stable compound is formed on the surface initially. The composition of the film is homogeneous and does not change drastically with prolonged exposure as it does in the neutral environments.

The current-time relation shows that the formation of the high Cr content film is an excellent obstacle for the reaction. Only a few monolayers of the metal are dissolved during prolonged exposure. The overall logarithmic law during the first stage indicates that the rate controlling step would be tunneling of electrons through the film (Cabrera-Mott (3)). However, this model is worked out for oxidation in dry atmosphere. In an accurate interpretation one has to consider the current contribution from non-precipitated species.

ACKNOWLEDGEMENT

Financial support from the Swedish Board for Technical Development (STU) is gratefully acknowledged.

REFERENCES

1. I. Olefjord, H. Fischmeister, Corrosion Science 15 (1975) 697.
2. I. Olefjord, Corrosion Science 15 (1975) 287.
3. N. Cabrera, H.F. Mott, Rep. Progr. Phys. 12 (1948-49) 163.

THE INFLUENCE OF OXYGEN ON THE CURRENT POTENTIAL CHARACTERISTICS OF Zn/- AND ZnO/ELECTROLYTE CONTACTS

O.Fruhwirth, J.Friedmann, G.W.Herzog

Institute of Inorg.-chem.Technology
and Anal.Chemistry, TU Graz (Austria)

INTRODUCTION

With regard to corrosion phenomena of Zn in solutions containing 0,1 M borax (pH 9,2), potentiodynamic current potential characteristics and open circuit potentials of rotating disk electrodes made of Zn and ZnO have been investigated. Measurements have been carried out at a constant temperature of $25°C$ under different conditions of Zn^{++} concentration, O_2 partial pressure and rotation speed. It is observed, that below Zn^{++} saturation and without dissolved oxygen corroding Zn electrodes behave like metal electrodes in equilibrium with ZnO, and with Zn^{++} saturation like pure n-conducting ZnO electrodes. Experimental details were described in other papers (1),(2),(3),(4).

RESULTS

At Zn electrodes a remarkable potential increase is caused by dissolved oxygen below Zn^{++} saturation (c_s is about 10^{-4} mol l^{-1}) amounting to nearly 1 V. This increase is much less at $c_{Zn^{++}} = c_s$ as well as with ZnO pellets (fig.1). In the first case the increase depends on the rotation speed (fig.2), thus pointing to a non equilibrium potential. But the higher the speed the more negative the potential becomes, although one would expect a more positive potential.

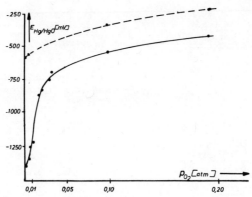

Fig.1: Dependence of open circuit potential of a corroded Zn electrode and of a ZnO electrode on O_2 concentration; $c_{Zn^{++}} < c_s$, —— Zn, --- ZnO pellet

Current potential curves of a Zn electrode at $c_{Zn^{++}} = c_s$ and at $c_{Zn^{++}} < c_s$ are shown in fig.3. Two steps for O_2 reactions are noted, corresponding to limiting currents caused by the following diffusion controlled electrode reactions:

$$O_2 + 2H_2O + 2e^- \rightarrow H_2O_2 + 2OH^- , \qquad (1)$$

$$O_2 + 2H_2O + 4e^- \rightarrow 4OH^- . \qquad (2)$$

Both limiting currents are proportional to the O_2 concentration (fig.4) and proportional to the square root of the rotation speed, thus proving diffusion control. At a concentration of about 10^{-4} mol O_2 l^{-1} there is a knee in the current/O_2 concentration curve due to different slopes. Below the knee the slope corresponds approximately to a 1 and a 3 electron transfer and above the knee to a 2 and a 4 electron transfer reaction. Based on the theory of rotating disk electrodes the following wellknown equation holds for limiting currents:

$$j = 0{,}62 F D^{2/3} \nu^{-1/6} \omega^{1/2} zfc_{O_2} . \qquad (3)$$

j is the current density in $mAcm^{-2}$, c_{O_2} is the O_2 concentration in $mol\ l^{-1}$, z is the electron transfer number and f is a roughness factor. With the diffusion coefficient for the dissolved O_2 molecules $D = 1 \cdot 10^{-5}$ $cm^2 s^{-1}$, the kinematic viscosity $\nu = 1 \cdot 10^{-2}$ $cm^2 s^{-1}$ and the rotation speed $\omega = 130\ s^{-1}$ equ.3 yields for the slope 1082 zf.

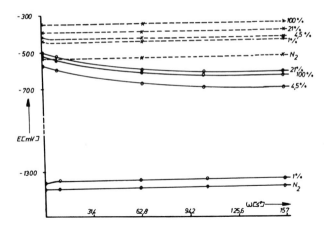

Fig.2: Dependence of open circuit potential of a corroded Zn electrode at different O_2 concentrations on rotation speed; —— $c_{Zn^{++}} < c_s$, --- $c_{Zn^{++}} = c_s$,

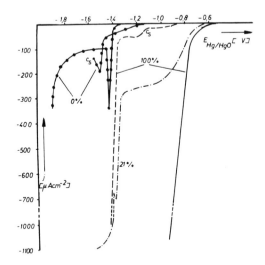

Fig.3: Current potential curves of a Zn electrode at $c_{Zn^{++}} < c_s$ (except where indicated) and $c_{Zn^{++}} = c_s$ with different O_2 concentrations; rotation speed 130 s^{-1}, scan rate 0,25 mV s^{-1};

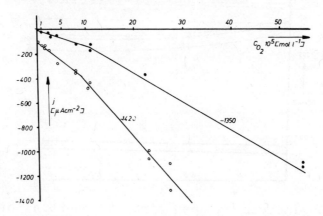

Fig.4: Dependence of O_2 limiting currents at -1350 mV and -1420 mV on O_2 concentration; rotation speed 130 s^{-1} scan rate 0,25 mVs^{-1}, $c_{Zn^{++}} < c_s$

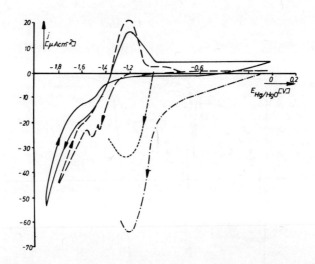

Fig.5: Current potential curves of a ZnO pellet and of a ZnO single crystal; rotation speed 130 s^{-1}, scan rate 0,25 mV s^{-1}, without oxygen: --- $c_{Zn^{++}} < c_s$, ―― $c_{Zn^{++}} = c_s$; with 100 % O_2: -.- ZnO pellet, ... ZnO single crystal

From the current potential curves in fig.3 it is noted that with Zn^{++} saturation the first reaction step is depressed to a small current peak and that the second step is left unchanged. The small current peak, however, is observed only with much more sensitive recording and is probably caused by an adsorption of a still unknown O_2 species or of O_2 itself.

The current potential curves carried out with ZnO pellets and the ZnO single crystal show no appreciable difference between Zn^{++} saturation and nonsaturation. The current density is much smaller and diffusion control does not occur(fig.5), thus pointing to current limitation by the resistance of the pellet and the crystal.

DISCUSSION

The dependence of the open circuit potential on O_2 concentration originally lead to the conclusion, that both adsorption and mixed potential could be potential controlling. The latter assumption is now found to be true for concentrations below Zn^{++} saturation, because the total current potential curves are composed additively by partial curves. This is noted by the fact, that at small O_2 concentrations the characteristic anodic peak is pushed to the cathodic side. At O_2 concentrations of more than 10^{-4} mol l^{-1} the cathodic O_2 partial curve is mainly composed of a diffusion controlled 2 electron reaction (equ.1) and a 4 electron reaction (equ.2). The first one obviously requires a high electron density at the surface, because it is degenerated to a small current peak, which is believed to be caused by adsorption under Zn^{++} saturated electrolyte conditions.

With regard to technical corrosion phenomena on Zink($25°C$, stirred weak alkaline electrolyte) following conclusions can be drawn: In the presence of dissolved O_2 the corrosion current density is five times greater than without O_2 and the corrosion potential is a mixed potential of

$$ZnO + H_2O + 2e^- \longrightarrow Zn + 2OH^- \quad (4)$$

and some O_2 reaction. At highly doped oxide states the $O_2/H_2O_2^-$ reaction is going on, at sites without Zn or with pure Zn density as well as with Zn^{++} saturation some O_2 adsorption is assumed to control the potential. Best prohibition in the presence of O_2 is therefore also gained only with Zn^{++} saturation.

LITERATURE

(1) G.W.Herzog: Electrodep.Surface Treat., <u>1</u>(1972/73)175
(2) O.Fruwirth: Dissertation TU Graz,1976
(3) J.Friedmann, O.Fruhwirth, G.W.Herzog: Electrodep. Surface Treat., <u>3</u>(1975)343
(4) O.Fruhwirth, J.Friedmann, G.W.Herzog: Surface Technology, to be published 1976

ACKNOWLEDGEMENT

The authors are indebted to Prof.Dr.H.Grubitsch and to the Fonds zur Förderung der wissenschaftlichen Forschung in Österreich for support of this work.

AUTHOR INDEX

Adachi, G.	785
Alario-Franco, M.A.	519
Anderson, J.S.	261
De Angelis, B.A.	77
Arean, C.O.	69
Arndt, G.	487
Aubry, J.	391
Baird, T.	143,337
Barnes, P.A.	663
Barret, P.	125
Bartha, L.	449
Behrens, E.	481
Benavent, W.E.M.	305
Béranger, G.	599
Beretka, J.	743
Berthod, L.	311
Besson, J.	171,599
Biagini, E.	119
Bilen, C.S.	475,499
Billy, M.	107
van den Boom, H.	529
Bracconi, P.	311
Brandt, W.W.	547
Briend-Faure, M.	689
Brown, A.M.	83
Brown, M.E.	221
Brown, T.	743
Brun, M.	767
Buljan, S.T.	421
Burggraaf, A.J.	297
Burlamacchi, L.	49
Butcher, P.	647
Caillet, M.	171
Cammenga, H.K.	481
Carlsson, R.	541
Caro, P.	641
Carpy, A.	439
Carrión, J.	267
Charsley, E.L.	737
Chianelli	89
Cho, S.H.	701
Coble, R.L.	669
Coddet, C.	599
Colson, J.C.	625
Cordischi, B.	379
Criado, J.M.	267

Cutler, I.B.	707
Dalgleish, B.J.	373
Dance, J.M.	289
Delafosse, D.	689
Delmon, B.	221,755
Demarquay, J.	611
Derouane, E.G.	559,653
Desimoni, E.	137
Desmaison, J.G.	107
Dexpert, H.	641
Dollimore, D.	373
Domokos, L.	209
Driole, J.	599
Drofenik, M.	113
Dufour, L.C.	311
Dumas, P.	625
Elfström, B.-O.	791
Eveno, P.Y.	433
Falb, K.	191
Fayard, M.	439
Felix, F.W.	305
Fievet-Vincent, F.	415
Figlarz, M.	415
Fraissard, J.	611
Francis, B.	785
Friedmann, J.	797
Frischat, G.H.	427
Fruwirth, O.	797
Gaballah, I.	391
Gabelica, Z.	653
Gabor, M.	761
Gál, S.	209
Galerie, A.	171
Galwey, A.K.	221
Galy, J.	439
Garcia-Clavel, M.E.	343
Garcia-Gonzalez, M.L.	755
Gayoso, M.	519
Gillot, B.	125
Gleitzer, C.	391
Götz, J.	525
Grabke, H.J.	55
Graf, H.-J.	487
Grange, P.	755

Guarini, G.G.T.	349	Lalauze, R.	155
Guenot, J.	415	Laqua, W.	367
Guilhot, B.	155	Lee, J.G.	707
Guilleux, M.F.	689	Lengyel, J.	449
Gunsser, W.	587	Leute, V.	63
Günthard, Hs.H.	285	Levy, P.W.	355
		Lieser, K.H.	215
Haber, J.	331,409	Liljestrand, G.	385
Hadjicostantis, D.C.	161	Loehman, R.E.	203
Hadjoudis, E.	493	Longo, J.M.	43
Hagen, A.P.	69	Lundén, A.	713
Hagenmuller, P.	289		
Hallwig, D.	631	Magnéli, A.	1
Haul, R.	101	Mari, C.M.	161
Hayishi, K.	397	Marinkovič, V.	113
Heed, B.	713	Martinez-Esparza, A.	343
Hennicke, H.W.	427	Martini, G.	49
Hennings, D.	273	Matzke, Hj.	773
Herley, P.J.	355	Maurin, M.	535
Hermansson, L.	541	McGinn, M.J.	221
Herrera, E.J.	267	McKinstry, H.A.	421
Herzog, G.W.	797	Michaels, A.1.	593
Hieltjes, A.H.M.	297	Michalak, S.	443
Hoebbel, D.	525	Miller, P.D.	707
Holappa, L.E.K.	635	Montel, G.	403
Horowitz, H.S.	43	Monty, C.	631
Hrabě, Z.	325	Morán, E.	519
Hubin, R.	653	Morantz, D.J.	475,499
Hübner, K.	101	Morawetz, H.	457
Hutchison, J.L.	261	Mrowec, S.	177
		Mutin, J.C.	131
Iguchi, E.	77		
Ives, M.B.	593	Nagai, H.	701
		Nakahira, M.	397
Jesenák, V.	325	Nakano-Onoda, M.	397
Jóvér, B.	191	Nemeth, T.	449
Juhász, J.	191	Niepce, J.C.	131
		Nishino, T.	149
Kanamuru, F.	725	Nishiyama, S.	149
Kermarec, M.	689	Nowell, D.V.	373
Kikkawa, S.	725		
Kircher, O.	101	Okada, M.	197
Ko, H.W.	547	Okamoto, S.	749
Kofstad, P.	15	Okamoto, S.I.	749
Koizumi, M.	725	Olefjord, I.	791
Kolar, D.	113	Ottaway, M.R.	737
Kosłowska, A.	331	Paniccia, F.	137
Kotera, Y.	227,233	Paris, J.	125
Kriegesmann, J.	427	Passerone, A.	119
Kuno, H.	701	Paulitschke, W.	55
Küter, B.	367	Paulus, M.P.	433

AUTHOR INDEX

Pepe, F.	183	Stone, F.S.	69
Perugini, G.	167	Stratman, W.	63
Philippot, E.	535	Strnad, Z.	553
Piekarska-Sadowska, H.	409	Stubican, V.S.	421,767
Pittermann, U.	95	Surman, P.L.	83
Pizzini, S.	161	Szabó, Z.G.	191,761
Pokol, Gy.	209	Szalay, T.	449
Pöppl, L.	761	Sztatisz, J.	209
Porta, P.	183		
Portier, J.	289	Taskinen, A.	617
Portier, R.	439	Taskinen, P.	617
Przybylski, K.	177	Tejedor-Tejedor, M.I.	343
Pungor, E.	209	Tempère, J.F.	689
		Tesi, B.	49
Rabenau, A.	731	Thomas, G.	155
Rajeshwar, K.	785	Thompson, R.C.	499
Rao, C.N.R.	261	Tikkanen, M.H.	617,635
Reuter, B.	367	Tilley, R.J.D.	77
Ribes, M.	535	Tofield, B.C.	253
Ridge, M.J.	743	Torres, C.	267
Rieke, E.M.	361	Tressaud, A.	289
De Rossi, S.	77	Trombe, J.C.	403
Rousset, A.	125		
		Valbusa, G.	119
Sakurai, T.	149	Valigi, M.	379
Schachtner, R.	605	Verweij, H.	529
Schiavello, M.	77,183	Vest, R.W.	695
Schmalzried, H.	237	Vogel, W.	505
Schöllhorn, R.	719	de Vries, K.J.	297
Schroeder, K.	713		
Schweiger, A.	285	Wajand, J.	761
Secco, E.A.	785	Walther, G.C.	203
Sestak, J.	553	Warren, R.	779
Shibata, K.	397	Watelle, G.	131
Schiffmacher, G.	641	Wegner, G.	487
Simonetti, S.	49	Weil, K.G.	95
Słoczynski, J.	331	Wernicke, R.	279
Smeltzer, W.W.	107	Whittingham, M.S.	89
Smith, A.G.C.	337	Wieker, W.	525
Snell, D.S.	337	Williams, P.M.	647
Sockel, H.G.	605,631	Wojtczak, L.	443
Soria, J.	519	Wolf, E.	481
Sørensen, O.T.	659	Wolfmeier, U.	587
Soustelle, M.	155	Wong, J.C-S.	785
Spath, H.T.	319	Wood, J.C.	647,779
Spinicci, R.	349		
Srinivasan, S.R.	55	Yamaguchi, T.	197,701
Steinbach, M.	487	Yonemura, M.	227,233
Steinbrunn, A.	625		
		Zambonin, P.G.	137

SUBJECT INDEX

Activation analysis, 605,632.
    energy, 108,128,210,218,
        229,235,331,431,483,553,
        564,593,619,690.
Adsorption,496,576,647.
    isotherm,374,450,520.
Aggregation,222,375.
Alkalai halides,305.
Alloys, 21,169,791.
Aluminium,
    oxide, 19,285,670,686,713
    ions, 125.
    sulphide, 37.
Ammonium,
    perchlorate, 355.
Angular resolved spectra, 647.
Antiferromagnetism, 254,291,313.
Apatites, 403.
Arrhenius equation, 103,144,369,
    547,557,619,636.
Auger electron spectroscopy, 55,
    560,625.
Autocatalysis, 110,138,331.
Avrami-Erofeyev, 358,619,621,746.

Barium, 254.
    azide, 319.
    borosilicate, 508.
    chlorate, 351.
    chromate, 654.
    oxalate, 654.
    titanate, 273.
BET method, 50,70,75,269,311,332,
    376.
Bismuth,
    bismuthate, 261.
    oxide, 3,81.
Bromination, 493.

Cadmium, 49.
    oxide, 131,343.
    telluride, 63.
Calcium, 43.
    carbonate, 209.
    oxide, 233.
Carbon, 143.
    carbides, 16,197,199,767.
    carburisation, 26,55.
    dioxide, 49.
    disulphide, 171.
    fibres, 779.
    monoxide, 647.
Catalysis, 49,76,137,183,267,331,
    565.
Cerium,
    oxide, 659.
Chemical potential, 18,249.
Chemisorption, 16,49,55,562.
Chromatography, 79,482,525.
Chromium, 125,168.
    chromite, 38.
    chromate, 38,149,653.
    hydroxide, 791.
    oxide, 19,22,149,238,314,
        411,653.
Cisazobenzene, 481.
Clustering, 226,240,334,656.
Cobalt, 168,177,331,417,617.
    alloy, 22,24,201.
    ions, 69,73,285.
    oxide, 19,22,69,177,242,312,
        367,617,635.
Colloidal solutions, 339.
Computer simulation, 593.
Copper, 95,197.
    sulphate, 343.
    sulphide, 172.
Corrosion, 31,205,791,797.
Creep, 669,682,684,773.

Debye function, 430.
Debye - Hückel theory, 241.
Defects, 6,52,76,160,237,240,244,
    267,270,273,305,312,361,367,
    391,431,439,458,483,660.
Desorption, 547,614.
Dielectric, 444.
D.S.C. 349,484.
D.T.A. 48,127,343,373,394,521,530,
    541,553,663,737,785.
D.T.G. 343,521,530.

Diffusion, 18,21,64,101,127,200,
    207,246,263,280,367,433,
    436,591,607,631,638,671,
    683,676,679,773
    coefficient, 19,21,63,103,
    127,179,280,297,607,631,
    670,674,773.
Dislocation,238,240,247,262,265,
    267,356,365,391,432.
Doping, 49,499,537,588.

E.G.A. 663.
Electrochemical cell, 90,101,713,
    733.
    potential, 18.
Electron,
    density, 443.
    diffraction, 10,257,767.
    micrograph, 340.
    microprobe, 63,210,299,
    325,411.
    microscopy, 6,78,221,253,
    261,337,385,399,416,439,
    505,520,641,659,746,768,
    781,
    tunnelling, 18.
Electronic bandgap, 735.
Ellingham diagrams, 17.
ENDOR, 285.
Epitaxy, 645.
EPR/ESR, 49,285,379,465,478,499,
    519,559,562,587,653,678,
    690.
ESCA, 77,791.
Expansion coefficient, 111,120,
    505,541.

Ferrites, 433,434,577,587.
Field Emission Microscopy, 595.
FMR, 559,577,581.
Formates, 191,221,267.
Free radicals, 138,783.

Gas chromatography, 50.
Gas-solid reaction, 15,209.
Gel, 519,749.
Germanium, 227,233.
    germanates, 227.

Gibbs',
    adsorption isotherm, 58.
Glass, 119,227,505,514,526,529,
    541,547,553,695.
Grain boundary, 58.
    diffusion, 18,281,671,676,
    681.
Grinding, 46,267,484.

Heat of adsorption, 566.
Heterogeneous catalysis, 49,77,183.
    ion-exchange, 449.
    reactions, 155,171.
Hot pressing, 669.
Hydrogen, 331,361,562,567.
    deuterium exchange, 613.
    sulphide, 625.
Hysteresis, 249.

IETS, 560.
Induction period, 476,483,601.
Infra-red spectroscopy, 285,293,
    337,343,403,535,560,615,
    653,690,761,785.
Inhibitor, 158,224.
Intercalation compounds, 726.
Intrinsic defects, 63.
Ion beam microanalysis, 632.
Iron, 83,361.
    alloys, 24.
    -gas reactions, 24,55.
    hydroxide, 749.
    (III) ions, 53,126,289,455.
    oxides, 83,125,176,391,743.
    single crystal, 55,85.
    -sulphur reaction, 55,172,
    174.
Irradiation, 355,475,487,503,605,
IRS, 560.
Isomerisation, 463,481,484,494.
Isotopic exchange, 101,215,631.
ISS, 560.

Jander's equation, 591,743.

Kaolinite, 761.
Kinetics,18,43,109,125,137,171,
    173,177,210,216,221,227,
    233,279,319,331,369,409,
    636,689,707,743.

SUBJECT INDEX

dissolution, 598.
of solid-gas reactions, 18,210,331.
of unmixing, 508.
Knight shift, 567.
Komatsu equation, 591.

Lanthanum, 273,279,297.
    oxide, 273,279,297,785.
Laser raman spectroscopy, 529.
Lattice image, 6,399,644.
Layer structure, 134,400.
Lead, 297,319,343,346.
LEED, 55,625,648.
Lithium, 89,716.
Luminescence, 500,537.

Magnesium, 3,337,654.
    oxide, 69,427,577,683, 743.
Magnetic analysis, 70,254,291,311.
Magnetite, 84,125,242,577,753,755.
Manganese, 43,51,379,381,434.
Mass spectrometry, 548,633,653, 663,761.
Mempel theory, 624.
Microhardness, 197,767.
Microglass phases, 510,514.
Microscopy, 203,325,697.
Molybdenum, 385,525,563.
    oxide, 38,49,81,92,385, 737.
Mössbauer spectroscopy, 245,291, 505,535,577.

Neck growth, 674,675,676.
Néel temperature, 254,291,313.
Neutron diffraction, 253.
Nickel, 29,113,143,145,168,221, 267,436,647,666.
    alloys, 21,24,31.
    oxides, 17,19,20,29,35, 115,184,238,344,625,631, 635.
Nitridation, 26,55,107.
NMR, 464,559,567,611.
Nucleation, 221,556,601,620.

Opal glasses, 514.
Organic reactions in solids, 457.
Oxalic acid, 341.
    oxalates, 131,343,627,653.
Oxidation, 15,18,22,24,49,77,83, 107,122,149,177,571,599, 625.
Oxygen, 16,49,74,101.
    diffusion, 101,280,605,631.
    exchange, 101,631.

Particle size, 48,106,416,554, 617,743.
Passivation, 790.
Perovskite structure, 44,253,261, 280,289,297,439.
Peroxides, 137.
Phase transition, 227,421,541, 608.
Phosphorescence, 475,477,499, 537.
Photoemission, 599,647.
Plutonium, 773.
PMMA, 475,478,499.
Polymerisation, 457,487,489,490, 525.
Powder, 167,207,267.
Pyrotechnic delay, 737.

Quartz, 325,410,605,608.
    spring balance, 398,494.

Radiotracers, 369,452.
Raman spectroscopy, 488,505,529, 535,
Redox processes, 379,456,719.
Reduction, 13,44,311,331,385, 689,707.
Reflectance spectroscopy, 70,78, 560.
Rhodium, 523.
    oxide, 519.
Rotating disc electrode, 797.

SEM, 83,301,311,393,697.
Silica, 325,337,511,547,553,605, 611,705,709.

silicate, 150,525.
Silver, 95,319,664.
SIMS, 560,632.
Sintering, 55,168,434,669,672, 677,697,773.
Sodium, 33,215,233,319,507.
Solid-gas reactions, 15,209,215, 391,415,493,600,647.
   -state reactions,368,379, 391,409,457,535,559,587.
Spinel, 22,69,125,183,191,249, 311.
Strontium, 101,233,253,254,273, 280,654.
Sulphur, 27,55,57.
   sulphide, 26,60,89,173,635.
Surface, 55,58,106,216,223,373, 447,565,593,627.
   area, 69,106,548,613.
   reaction, 15,49,69,103,218, 559,621.
Symmetry, 187,403.

Tellurium, 116,731.
   tellurides, 113.
Thermal decomposition, 46,191,209, 355,373,519,582,617,641,653, 663,737,761,779.
Thermodynamics, 15,99,172,484, 506.
Thermoluminescence, 475,478, 499.
Thin films, 95,559,603,641.
TGA, 44,114,143,193,209,267,275, 311,333,343,373,393,520, 530,618,628,644,659,663, 761,785.
Titanium, 90,119,167,199,203 599.
   dioxide, 78,122,149,701.
Topotaxy, 89,238,415,422,457, 493,719,752.
Tungsten, 197.
   oxide, 6,13,77,289,449.

Ultra violet,
   irradiation, 475,482,489, 500,563.
   photoelectron spectroscopy, 647.
   reflectance spectra, 70,78,
Uranium, 773.

Vacancies, 128,257,273,280.
Vitreous solids, 505.
Voids, 19,305.

Wadsley defects, 6,247.
Wagner theory, 19,179,247,370,590, 629.
Weiss constant, 71,75,291.

XPS, 78,562,569.
Xray diffraction, 541.
   powder, 2,10,44,70,78,234, 257,267,273,290,311,332, 343,653,702,739,744,757, 761,780,785.
   single crystal, 3,132,659.

Zeolite, 69,571,689.
Zinc,
   oxide, 553,631,797.
Zirconium, 149,297.
   nitride, 107.
   zirconia, 421,705.